Understanding Numbers
in Elementary School Mathematics

Understanding Numbers in Elementary School Mathematics

Hung-Hsi Wu

American Mathematical Society
Providence, Rhode Island

2010 *Mathematics Subject Classification.* Primary 97–01, 00–01, 97F30, 97F40, 97F80.

For additional information and updates on this book, visit
www.ams.org/bookpages/mbk-79

Library of Congress Cataloging-in-Publication Data
Wu, Hongxi, 1940–
  Understanding numbers in elementary school mathematics / Hung-Hsi Wu.
    p. cm.
  Includes bibliographical references and index.
  ISBN 978-0-8218-5260-6 (alk. paper)
  1. Number concept—Study and teaching (Elementary) 2. Mathematics—Study and teaching (Elementary) I. Title.

QA141.15.W82 2011
372.7′2—dc22
                                                                    2010053021

Copying and reprinting. Individual readers of this publication, and nonprofit libraries acting for them, are permitted to make fair use of the material, such as to copy a chapter for use in teaching or research. Permission is granted to quote brief passages from this publication in reviews, provided the customary acknowledgment of the source is given.

Republication, systematic copying, or multiple reproduction of any material in this publication is permitted only under license from the American Mathematical Society. Requests for such permission should be addressed to the Acquisitions Department, American Mathematical Society, 201 Charles Street, Providence, Rhode Island 02904-2294 USA. Requests can also be made by e-mail to reprint-permission@ams.org.

Copyright © 2011 by the author.
Printed in the United States of America.

∞ The paper used in this book is acid-free and falls within the guidelines
established to ensure permanence and durability.
Visit the AMS home page at http://www.ams.org/
     10 9 8 7 6 5 4 3 2 1   16 15 14 13 12 11

To N^2 and W^4

Contents

Preface	xv
Acknowledgments	xxi
To the Reader	xxv
Some Conventions in this Book	xxxi

Part 1. Whole Numbers

Part Preview	3
Chapter 1. Place Value	5
1.1. How to Count	6
1.2. Place Value	19
1.3. The Use of Symbolic Notation	21
1.4. The Number Line	22
1.5. Comparing Numbers (Beginning)	25
1.6. Multiplication and the Expanded Form of a Number	27
1.7. All about Zero	32
1.8. The Hindu-Arabic Numeral System	33
Exercises	35
Chapter 2. The Basic Laws of Operations	39
2.1. The Equal Sign	39
2.2. The Associative and Commutative Laws of +	40
2.3. The Associative and Commutative Laws of ×	43

2.4.	The Distributive Law	44
2.5.	Comparing Numbers (Conclusion)	48
2.6.	An Application of the Associative and Commutative Laws of Addition	51
	Exercises	53

Chapter 3. The Standard Algorithms — 57

Chapter 4. The Addition Algorithm — 61
4.1.	The Basic Idea of the Algorithm	61
4.2.	The Addition Algorithm and Its Explanation	62
4.3.	Essential Remarks on the Addition Algorithm	65
	Exercises	69

Chapter 5. The Subtraction Algorithm — 71
5.1.	Definition of Subtraction	71
5.2.	The Subtraction Algorithm	73
5.3.	Explanation of the Algorithm	75
5.4.	How to Use the Number Line	79
5.5.	A Special Algorithm	81
5.6.	A Property of Subtraction	82
	Exercises	83

Chapter 6. The Multiplication Algorithm — 85
6.1.	The Algorithm	85
6.2.	The Explanation	87
	Exercises	92

Chapter 7. The Long Division Algorithm — 95
7.1.	Multiplication as Division	97
7.2.	Division-with-Remainder	103
7.3.	The Algorithm	107
7.4.	A Mathematical Explanation (Preliminary)	110
7.5.	A Mathematical Explanation (Final)	117
7.6.	Essential Remarks on the Long Division Algorithm	119
	Exercises	123

Chapter 8. The Number Line and the Four Operations Revisited — 127
8.1.	The Number Line Redux, and Addition and Subtraction	127
8.2.	Importance of the Unit	129

8.3.	Multiplication	131
8.4.	Division	132
8.5.	A Short History of the Concept of Multiplication	134

Chapter 9.	What Is a Number?	135

Chapter 10.	Some Comments on Estimation	139
10.1.	Rounding	140
10.2.	Absolute and Relative Errors	144
10.3.	Why Make Estimates?	147
10.4.	A Short History of the Meter	150
Exercises		152

Chapter 11.	Numbers in Base b	155
11.1.	Basic Definitions	155
11.2.	The Representation Theorem	158
11.3.	Arithmetic in Base 7	161
11.4.	Binary Arithmetic	164
Exercises		168

Part 2. Fractions

Part Preview		173

Chapter 12.	Definitions of Fraction and Decimal	177
12.1.	Prologue	179
12.2.	The Basic Definitions	183
12.3.	Decimals	187
12.4.	Importance of the Unit	188
12.5.	The Area Model	190
12.6.	Locating Fractions on the Number Line	195
12.7.	Issues to Consider	196
Exercises		199

Chapter 13.	Equivalent Fractions and FFFP	203
13.1.	Theorem on Equivalent Fractions (Cancellation Law)	204
13.2.	Applications to Decimals	207
13.3.	Proof of Theorem 13.1	209
13.4.	FFFP	212
13.5.	The Cross-Multiplication Algorithm	214
13.6.	Why FFFP?	216

Exercises	218
Chapter 14. Addition of Fractions and Decimals	**221**
14.1. Definition of Addition and Immediate Consequences	222
14.2. Addition of Decimals	224
14.3. Mixed Numbers	225
14.4. Refinements of the Addition Formula	227
14.5. Comments on the Use of Calculators	229
14.6. A Noteworthy Example of Adding Fractions	230
Exercises	232
Chapter 15. Equivalent Fractions: Further Applications	**235**
15.1. A Different View of a Fraction	236
15.2. A New Look at Whole Number Divisions	238
15.3. Comparing Fractions	239
15.4. The Concept of $\frac{m}{n}$ of $\frac{k}{\ell}$	245
Exercises	250
Chapter 16. Subtraction of Fractions and Decimals	**253**
16.1. Subtraction of Fractions and Decimals	253
16.2. Inequalities	256
Exercises	258
Chapter 17. Multiplication of Fractions and Decimals	**261**
17.1. The Definition and the Product Formula	263
17.2. Immediate Applications of the Product Formula	269
17.3. A Second Interpretation of Fraction Multiplication	272
17.4. Inequalities	278
17.5. Linguistic vs. Mathematical Issues	278
Exercises	280
Chapter 18. Division of Fractions	**283**
18.1. Informal Overview	283
18.2. The Definition and Invert-and-Multiply	286
18.3. Applications	291
18.4. Comments on the Division of Decimals	296
18.5. Inequalities	302
18.6. False Doctrines	303
Exercises	306

Chapter 19.	Complex Fractions	309
19.1.	The Basic Skills	310
19.2.	Why Are Complex Fractions Important?	315
Exercises		317
Chapter 20.	Percent	319
20.1.	Percent	319
20.2.	Relative Error	325
Exercises		328
Chapter 21.	Fundamental Assumption of School Mathematics (FASM)	331
Chapter 22.	Ratio and Rate	335
22.1.	Ratio	336
22.2.	Why Ratio?	344
22.3.	Rate	344
22.4.	Units	348
22.5.	Cooperative Work	349
Exercises		355
Chapter 23.	Some Interesting Word Problems	357
Exercises		364
Chapter 24.	On the Teaching of Fractions in Elementary School	367

Part 3. Rational Numbers

Chapter 25.	The (Two-Sided) Number Line	375
Chapter 26.	A Different View of Rational Numbers	379
Chapter 27.	Adding and Subtracting Rational Numbers	381
27.1.	Definition of Vectors	382
27.2.	Vector Addition for Special Vectors	383
27.3.	Addition of Rational Numbers	386
27.4.	Explicit Computations	387
27.5.	Subtraction as Addition	390
Exercises		394
Chapter 28.	Adding and Subtracting Rational Numbers Redux	395
28.1.	The Assumptions on Addition	396
28.2.	The Basic Facts	397
28.3.	Explicit Computations	399

28.4. Basic Assumptions and Facts, Revisited	400
Exercises	402
Chapter 29. Multiplying Rational Numbers	**403**
29.1. The Assumptions on Multiplication	404
29.2. The Equality $(-m)(-n) = mn$ for Whole Numbers	405
29.3. Explicit Computations	408
29.4. Some Observations	410
Exercises	412
Chapter 30. Dividing Rational Numbers	**413**
30.1. Definition of Division and Consequences	413
30.2. Rational Quotients	416
Exercises	420
Chapter 31. Ordering Rational Numbers	**421**
31.1. Basic Inequalities	421
31.2. Powers of Rational Numbers	425
31.3. Absolute Value	426
Exercises	429

Part 4. Number Theory

Part Preview	433
Chapter 32. Divisibility Rules	**435**
32.1. Review of Division-with-Remainder	435
32.2. Generalities about Divisibility	436
32.3. Divisibility Rules	439
Exercises	444
Chapter 33. Primes and Divisors	**445**
33.1. Definitions of Primes and Divisors	445
33.2. The Sieve of Eratosthenes	448
33.3. Some Theorems and Conjectures about Primes	450
Exercises	454
Chapter 34. The Fundamental Theorem of Arithmetic (FTA)	**455**
Exercises	461
Chapter 35. The Euclidean Algorithm	**463**
35.1. Common Divisors and Gcd	463

35.2.	Gcd as an Integral Linear Combination	466
	Exercises	472

Chapter 36. Applications — 473
  36.1. Gcd and Lcm — 473
  36.2. Fractions and Decimals — 478
  36.3. Irrational Numbers — 480
  36.4. Infinity of Primes — 483
  Exercises — 485

Chapter 37. Pythagorean Triples — 487
  Exercises — 489

Part 5. More on Decimals

Part Preview — 493

Chapter 38. Why Finite Decimals Are Important — 495

Chapter 39. Review of Finite Decimals — 497
  Exercises — 504

Chapter 40. Scientific Notation — 505
  40.1. Comparing Finite Decimals — 505
  40.2. Scientific Notation — 507
  Exercises — 510

Chapter 41. Decimals — 511
  41.1. Review of Division-with-Remainder — 511
  41.2. Decimals and Infinite Decimals — 513
  41.3. Repeating Decimals — 518
  Exercises — 523

Chapter 42. Decimal Expansions of Fractions — 525
  42.1. The Theorem — 525
  42.2. Proof of the Finite Case — 528
  42.3. Proof of the Repeating Case — 529
  Exercises — 539

Bibliography — 541

Index — 545

Preface

> *...no music is vulgar,*
> *unless it is played in a way that makes it so.*
> Herbert von Karajan
> ([**Mat82**, page 226])

This is a textbook for the mathematics curriculum of grades K–6, and is written for elementary school teachers. Specifically, it addresses the most substantial part of the curriculum, namely, numbers and operations.

How does this book differ from textbooks written for students in K–6? The most obvious difference is that, because adults have a longer attention span and a higher level of sophistication, the exposition of this book is more concise; it also offers coherent logical arguments instead of sound bites. Moreover, it does not hesitate to make use of symbolic notation to enhance the clarity of mathematical explanations whenever appropriate, although it must be said in the same breath that symbols are introduced very carefully and very gradually. Because the present consensus is that math teachers should know the mathematics beyond the level they are assigned to teach ([**NMP08a**, Recommendation 19, page xix]), this book also discusses topics that may be more appropriate for grades 7 and 8, such as rational numbers (positive and negative fractions), the Euclidean Algorithm, the *uniqueness* of the prime decomposition of a positive integer, and the conversion of a fraction to a (possibly infinite) decimal.[1] Because teachers also have to

---

[1] In the context of the current requirement that all elementary teachers be generalists, one may question whether it is realistic to expect all elementary teachers to possess this much mathematical knowledge. For this reason, the idea of having *mathematics teachers* in elementary school is being debated and examined (see, for example, [**NMP08a**, Recommendation 20, page xx], and [**Wu09b**]).

answer questions from students, some of which can be quite profound, their knowledge of what they teach must go beyond the minimal level. Ideally, they should know mathematics in the sense that mathematicians use the word "know": *knowing* a concept means knowing its precise definition, its intuitive content, why it is needed, and in what contexts it plays a role, and *knowing* a skill means knowing precisely what it does, when it is appropriate to apply it, how to prove that it is correct, the motivation for its creation, and, of course, the ability to use it correctly in diverse situations. For this reason, this book tries to provide such needed information so that teachers can carry out their duties in the classroom.

The most noticeable difference between this book and student texts is, however, its comprehensive and systematic *mathematical* development of the numbers that are the bread and butter of the K–12 curriculum: whole numbers, fractions, and rational numbers. Such a development acquires significance in light of the recent emphasis on *mathematical coherence* in educational discussions. Coherence in mathematics is not something ineffable like Mona Lisa's smile. It is a quality integral to mathematics with concrete manifestations affecting every facet of mathematics. If we want a coherent curriculum and a coherent progression of mathematics learning, we must have at least one default model of a logical, coherent presentation of school mathematics which respects students' learning trajectory. It is unfortunately the case that, for a long time, such a presentation has not been readily available. The mathematics community has been derelict in meeting this particular social obligation.

The result of this neglect is there for all to see: infelicities abound in mathematics textbooks and in the school mathematics curriculum. For example, it is common to see the use of fraction multiplication to "explain" why $\frac{m}{n} = \frac{cm}{cn}$ for any fraction $\frac{m}{n}$ and any nonzero whole number c (see pages 206 and 270–271 for a discussion of this line of reasoning). Another example is the teaching of the arithmetic operations on fractions as if "fractions are a different kind of number" and as if these arithmetic operations have little or nothing to do with those on whole numbers. Yet another example is the teaching of decimals *parallel* to, and distinct from, the teaching of fractions in the upper elementary grades instead of correctly presenting the mathematics of decimals as part of the mathematics of fractions.

This book does not call attention to coherence per se, but tries instead to demonstrate coherence by example. Its systematic mathematical development makes it possible to point out the careful logical sequencing of the concepts and the multiple interconnections, large and small, among the concepts and skills.[2] Thus, it points out the fact that the usual algorithm for converting a fraction to a decimal by long division, if done correctly, is in

---

[2]One should not infer from this statement that the systematic development presented in this book is the only one possible. This book follows the most common school model of going

fact a consequence of the product formula for fractions, $\frac{m}{n} \times \frac{k}{\ell} = \frac{mk}{n\ell}$. It also points out the overwhelming importance of the theorem on equivalent fractions (i.e., $\frac{m}{n} = \frac{cm}{cn}$) for the understanding of every aspect of fractions. On a larger scale, one sees in this systematic development the *continuity* in the evolution of the concepts of addition, subtraction, multiplication, and division from whole numbers to fractions, to rational numbers, and finally—in the context of school mathematics—to real numbers.[3] Although each arithmetic operation may look superficially different in different contexts, this book explains why it is fundamentally the same concept throughout. Thus with a systematic development in place, one can step back to take a global view of the entire subject of numbers and gain some perspective on how the various pieces fit together to form a whole fabric. In short, such a development is what gives substance to any discussion of coherence.

The universities and the education establishment[4] have been teaching teachers *mathematics-without-coherence* for quite some time now. In fact, the importance of content knowledge in the training of teachers has been (more or less) accepted as part of the education dogma only in the last few years. Thus far, we have not been serving our teachers very well in terms of providing them the minimal mathematical knowledge they need to carry out their teaching duties (see, for example, the discussion in [**Wu11**]). Although there seems to be an increasing awareness of this problem (to cite but one example, [**GW08**]), it is nevertheless worthwhile to note the danger of "teaching content" without also being alert to the multitude of flaws in school mathematics. It does not matter whether this is school mathematics handed down to us by tradition or recent reform, the flaws are there. The presence of these flaws is the inevitable consequence of the long separation between the mathematics and the education communities. For example, something as absurd as "$27 \div 6 = 4$ R 3" would have been caught decades ago by any competent mathematician had the two communities been in constant communication (see page 106). Since many such errors are mentioned in the text proper, there will be no need to overdo a good thing by repeating them here. How did this separation come about? I cannot speak for the education community. What I can say with some confidence is that mathematicians generally avoid getting involved in education for two reasons: they believe that education is a bottomless pit in which infinite hard work can lead nowhere, and that the mathematics is trivial. About the former I have nothing to say. The latter is wrong, however. School mathematics may be elementary, but trivial it is not, unless it is written

---

from whole numbers to fractions and then to rational numbers, but it would be equally valid, for example, to go from whole numbers to integers and then to rational numbers.

[3]See Chapter 21 on the Fundamental Assumption of School Mathematics (FASM).

[4]They are responsible for teachers' pre-service and in-service professional development, respectively.

in a way that makes it so.⁵ The fact that school mathematics has been trivialized in innumerable books and articles should actually be a rallying point for some truly competent mathematicians to step into the education arena and stop the bleeding. It is time for both communities to learn to minimize the damage that has resulted from this separation.

This book is one mathematician's attempt at a systematic presentation of the mathematics of K–6. Subsequent volumes written for middle school and high school teachers will round out the curriculum of the remaining grades. My fervent hope is that others will carry this effort further so that we can achieve an overhaul of the *mathematical education of teachers* as we know it today. Our teachers deserve better, and our children deserve no less.

I also hope that this book, together with its companion volumes for the higher grades, will serve two other purposes. One is to give mathematics educators a more solid starting point for their research. Doing education research on the basis of faulty mathematics is no different from trying to formulate a theory in physics on the basis of faulty experimental data. Regrettably, faulty mathematics is mostly what educators have to work with thus far,⁶ and the time for change is now. A second purpose is to provide a resource for textbook publishers. The quality of our textbooks has to improve,⁷ but publishers have a valid excuse that there is no literature to help them do better. Perhaps this and other volumes to follow will begin to give them the help they need.

This book is the product of over ten years of experimentation in my effort to teach mathematics to elementary and middle school teachers. The starting point was the workshop on fractions that I conducted in March of 1998 ([**Wu98**]). Part 2 of this book, and arguably its most important part, is nothing more than an expansion of [**Wu98**].⁸ Over the years, I have taught from different parts of the book to in-service teachers, but in terms of a regular college course, there is too much material in the book for one semester. A suggested syllabus of such a course on numbers is the following:

$$\text{Chapters 1–9, 12–22, 32.}$$

A second course on numbers could be based on

$$\text{Chapters 10–11, 25–42.}$$

Elementary teachers also need to know some geometry; I plan to post a file on geometry on www.ams.org/bookpages/mbk-79, which may be downloaded for use as a supplementary textbook.

---

⁵See von Karajan's remark at the beginning of the Preface.

⁶See, for example, the discussion on pages 33–38 of [**Wu08**].

⁷To get a very rough idea of the quality of school mathematics textbooks in the most scientifically advanced nation of the world, see Appendix B in [**NMP08b**].

⁸The reader may be startled by how little I have deviated from the original blueprint.

Solutions to the "Activities" scattered throughout this book as well as solutions to the exercises will be posted on the same website www.ams.org/bookpages/mbk-79. However, the latter will be accessible only to course instructors, and an instructor can receive a link to the solutions by sending a link to the department webpage that lists him or her as being on the teaching staff.

Hung-Hsi Wu
Berkeley, California
December 1, 2010

Acknowledgments

Many books have acknowledgments that span several pages. Each time I cast my eyes on one of those, my heart would sink and I told myself that the author was abusing the printed page and most, if not all, of what followed had to be nothing but vulgar insincerity. Now that I am going to write about my immense debt to the many people who have helped me when help was needed, I know better.

For a research mathematician to teach teachers, there is unavoidably a long learning curve. But I made it even longer by committing more rookie mistakes than others. The first time I did serious professional development for teachers was the year 2000, and it was three weeks long, eight hours a day, and content based. Neither the length nor the emphasis was what Californian teachers were accustomed to, and sparks began to fly almost from day one. I would not have survived that experience without the extraordinary understanding and support of "The A Team",[9] my three assistants at the time: Bev Braxton, Ada Wada, and Mary Burmester. I was making perhaps unreasonable demands on the teachers, and I did so like a tyrant. Somehow, Bev, Ada, and Mary understood that, all the problems notwithstanding, what I tried to do had some merit and would ultimately be good for the teachers. So they did all they could to protect me and interceded between me and the teachers. In retrospect, they helped launch my career in mathematics education. Words fail me when I try to express how much I owe each of them.

I have been conducting three-week mathematics professional development institutes (MPDI) every summer since 2000, and such institutes are expensive. I was funded for the first four years because Elizabeth Stage

---

[9]Terminology of Bev Braxton.

of the University of California Office of the President appointed me to the Design Committee to formulate policy for funding mathematics professional development in California. The policy that was eventually promulgated by the Committee insisted on a strong content component. On that basis, I got funded for my MPDIs. So thank you, Elizabeth.

The high-tech bust of 2002 brought an end to California's funding of professional development. After that, I could not get federal or state funding for my work. The fact that I managed to continue with the mathematical education of teachers, and indirectly the writing of this book, was mainly the result of decisions by three far-sighted individuals. In 2004, Henry Mothner and Tim Murphy invited me to the Los Angeles County Office of Education (LACOE) to teach their math teachers. That I did, for the next three years, 2004–2006. I was aware from the beginning that the invitation was a courageous move, because "content" was not a popular word in education and certainly not in state or federal educational agencies. In 2006, Mr. Stephen D. Bechtel, Jr. personally funded a three-week institute for math teachers in the Bay Area, and the institute took place at the Mathematical Sciences Research Institute (MSRI). This was an uncommon act of generosity. Subsequently, beginning with 2007, I have received funding from the S. D. Bechtel, Jr. Foundation to give a three-week institute every summer at MSRI. To all three, Henry, Tim, and Mr. Bechtel, and to the Bechtel Foundation, I want to express my heartfelt thanks. I also wish to thank MSRI for its unfailing hospitality and, of course, for its magnificent learning environment.

Through the years, I have also received invitations from other sources to try out my ideas on professional development. I would like to thank the following people and institutions for their kindness and generosity:

> 2004: Garth Gaudry, Tony Guttmann, and Jan Thomas of the Australian Mathematical Sciences Institute.
> 2006–2007: Joan Ferrini-Mundy and Bill Schmidt of the Michigan State University PROM/SE Project.
> 2007: Judy Carlton of the Hacienda La Puente Unified School District in California.
> 2009: Ruth Anne Hodges and the Michigan Council of Teachers of Mathematics.
> 2010: Allison Coates of the Middle School Mathematics Institute in St. Paul, Minnesota.

None of the institutes I conducted would have been possible without expert assistance. For ten years, I had the good fortune of being surrounded by a group of dedicated teachers, colleagues, and administrators, and it was they who helped to make my vision of professional development a reality. My gratitude goes to them all, including (in addition to Bev Braxton, Ada Wada, and Mary Burmester): Hana Huang, Kay Kirman, Jaine Kopp, Devi Mattai,

Bruce Simon, Paul Toft, and the three who are still working with me, Winnie Gilbert, Stefanie Hassan, and Sunil Koswatta. For the three years 2004–2006 at LACOE, I benefited from Yvonne Koga's phenomenal managerial skills, and Jim Sotiros of MSRI has supported the Bechtel institutes in every conceivable way in recent years. I want to thank them both most warmly for their help and their friendship.

This book has gone through more drafts than I care to remember. Each draft was used as lecture notes for at least one institute, and the next draft would bear the imprint of what the teachers had taught me in the previous institute. I would like to thank my teachers collectively for their critical contributions. A special thanks goes to Larry Francis, who, more than once, gave me line-by-line corrections of the notes, and to him and Betty Lewis for their help with the thankless task of proofreading at the eleventh hour.

Other people have made comments on the early drafts and, among them, I should single out Ralph Raimi and Ken Ross. Beyond making very detailed comments, both had a significant impact on the final outcome of this book. Ralph took me to task for the lack of clarity in my language; I am afraid to show him my final draft now, but I *hope* it meets with his approval! Ken unceremoniously told me that, instead of treating decimals in a separate chapter (the forerunner of the present Part 5), I should simply blend as much of the decimal material as possible into Part 2. I have since done that. These are decisive contributions, and I think the readers have even more reason to thank them than do I.

Many of my personal friends also helped in the writing of this book in different ways. I am grateful to Richard Askey and Stevens Carey for the interest they showed in my work, and to Madge Goldman who volunteered to give me linguistic corrections. I owe Scott Baldridge a tremendous debt for his critical comments that come from his own experience in professional development; they helped me avoid some embarrassing errors. Andrew Fire, a former student and now a friend, assured me that what I was doing was worthwhile. When a research mathematician ventures into the somewhat uncharted territory of education, there is unavoidably doubt and anxiety. Such words of encouragement from someone outside mathematics, therefore, meant a lot to me.

Finally, my special thanks goes to David Collins, who is the editor of this book. In the final days of preparation of the manuscript, his contribution went way beyond the editorial. He did all the needed reorganization of the text and watched over literally every word. Such dedication can only come from a shared vision and the same missionary zeal. David: thank you very much.

To the Reader

"A major sin of modern education, in all branches of the curriculum, has been the unwillingness to demand serious effort." So said the literary historian Douglas Bush [**Bus59**, page 115].

Let it be said at the outset that reading this book requires serious effort. Having demanded serious effort, I do want to assure you that serious effort is essentially all it takes to master everything in this book. In a literal sense, you don't need to know any mathematics to start reading. After all, how much can be required for something that begins by explaining why 273 stands for two *hundreds* and seven *tens* and three *ones*, and then proceeds to give a careful explanation of the "hows" and "whys" of adding two whole numbers? In practice, however, this book may be too much of a challenge if you are unfamiliar with the procedural aspect of elementary school mathematics. While all the standard procedural skills will be carefully reviewed, it is assumed that you are familiar with routine computations.

For those who think they are math-phobic, I have a special message. Your math-phobia may well be the result of having been repeatedly asked to do things without benefit of sufficient preparation or explanation. You were told to just follow orders and then left to fend for yourself. It would be a wonder if you were not math-phobic, but let me assure you that this book will not indulge in such irrational behavior. *It will not ask you to do anything without first giving you all the information you need, as well as explaining to you why what you must do is correct.* So why not give this book a try and make a fresh start?[10]

What can you expect to learn from this book? Let me give you a preview.

---

[10]Just because your first movie was a monster movie that scared you silly, it does not follow that all the movies in this world are out to get you.

You will learn about the *precision* that is inherent in mathematics. In its avoidance of explicit language and its penchant for the evocative, mathematics education literature tends to discuss mathematics in almost poetic terms. In contrast, this book eschews poetry, choosing to present mathematics using unglamorous, precise prose so that it can be understood without ambiguity. For example, an article in mathematics education, attempting to help middle school teachers understand fraction multiplication, can unabashedly describe the multiplication of fractions as "finding multiplicative relationships between multiplicative structures". The seductive cadence of such writing puts to shame anything this book has to offer, but for the purpose of communicating mathematics, it is far more productive to first give a *precise* meaning to $\frac{a}{b} \times \frac{c}{d}$, and then demonstrate how this precise meaning together with logical reasoning can explain common procedures about multiplication and be helpful in the solving of problems. Indeed, a main focus of this book is on using *precise* language and *precise* reasoning to make logical deductions and solve problems.[11]

Your learning experience up to this point may not have prepared you for this kind of precision. Standard school texts and professional development materials routinely teach skills or concepts by engaging in drawn-out and vague discussions of said skills or concepts in as many contexts as possible. The wishful thinking behind this strategy is that, just as children learn to speak by uttering complete sentences without comprehending the individual words used, people can learn mathematics in like manner. There is of course no validity to this strategy for any but the most elementary portion of mathematics, and the present crisis in school mathematics education bears eloquent witness to that. But because it is likely that you have become inured to such imprecision, the demand for precise thinking and precise articulation that this book imposes on you may first impress you more as a burden than an aid to learning. However, you will soon come to see why precision is both necessary and beneficial in the learning and teaching of mathematics.

Concomitant with the demand for precision, this book insists on the *primacy of precise definitions*. The present aversion to definitions in school mathematics is the inevitable consequence of the slighting of definitions during the mathematical development of most school textbooks and K–12 professional development materials. For example, a definition of a rational number $\frac{a}{b}$ could be presented *purely formally* as a solution of the equation $bx = a$,[12] and yet the ensuing development of rational numbers ignores this

---

[11]Or as mathematicians prefer to say, *to prove theorems*. There is no logical distinction between problem-solving and theorem-proving.

[12]Such a formal definition, one that relies solely on algebraic reasoning, is a bad choice of a definition for rational numbers in the context of school mathematics because the *mathematical* discussion based on such a definition requires more abstraction than is appropriate for the K–12 classroom. Of course, this kind of definition is routinely employed in advanced mathematics.

definition completely, even in the discussion of equivalent fractions for which the above definition would be tailor made. It therefore comes to pass that, in a school setting, a definition becomes nothing more than one more thing to memorize for taking standardized tests.

This book puts the definitions of concepts into their proper place in mathematics, namely, front and center as the *foundation for all reasoning and discussions*. For example, a fraction is considered by some mathematics educators to be a concept so complex that its meaning should not be given succinctly, but instead gradually expanded upon in the course of subsequent discussions, so that its different "personalities"[13] can emerge unhurriedly, one by one. Let it be said in no uncertain terms that such an approach to fractions does not constitute acceptable mathematics at the fifth or sixth grade level, much less beyond it. It is bad pedagogy, because it leaves students in a state of perpetual confusion about what they are dealing with. How can they be expected to come to grips with the concept of a fraction if they know that whatever they learn will be revised a few days or a few weeks later?

By contrast, this book defines a fraction at the beginning of the discussion: it is one of a precisely prescribed collection of points on the number line, with each subsequent assertion about fractions carefully explained by use of this definition. In other words, if there is anything about fractions that cannot be explained using only the fact that a fraction is such a point on the number line, then it will not be found in this book. The same holds true for all the concepts to be introduced, e.g., division-with-remainder among whole numbers, decimal, rational number, etc. In each case, the definition dictates completely what can be said about that concept. The far-reaching impact that such a mathematical environment exerts on mathematics learning and teaching is that, at any point of a discussion, the learner can never be in doubt as to where she stands. Whatever she needs to know about a concept is already contained in its definition, and any fear that she may not be able to learn it because she has been denied access to some secret information has been laid to rest once and for all. In this way, everybody is given a level playing field to learn mathematics: the information needed for learning is always on the table, and there are no surprises or hidden agendas.

It has been said, aptly, that mathematics is not a spectator sport. Those who want to learn it must get their hands dirty by doing it. In this light, how does knowing the importance of definitions impact the learning of mathematics in terms of concrete actions? It mandates that, once you have accepted the definition of a given concept, you methodically *rearrange* all the facts related to this concept in a logical order until you can trace these facts back to the definition as your starting point. *You also learn the definition*

---

[13]This is not a piece of ad hoc terminology that I made up, but rather a standard one used in the education research literature.

by heart, to the point of instant recall. None of this is easy, but this book will do its best to smooth the transition from your old knowledge to the new one. This transition will not take place, however, without the effort of memorizing the definition and reorienting your mental compass so that you can embrace the way the old facts now flow naturally and logically from this definition. While this is hard work, it is hard work well worth the trouble.

If this approach to learning about numbers seems too stark and unforgiving, and therefore too restrictive to be useful, fear not: *Every* standard fact about numbers will be found in this book. Moreover, with the availability of precise definitions, rigorous reasoning now becomes possible.

This then leads us to the next principal feature of this book, its unremitting emphasis on *reasoning*. A reason is given for every assertion: Why is the invert-and-multiply rule correct for fraction division? Why is the product of two negative numbers a positive number? Why is a fraction, defined as parts-of-a-whole, also a division? Why is every fraction equal to one and only one reduced fraction? Etc.

At the beginning, you may find it oppressive to be bombarded by so many "whys" at every turn,[14] but with some serious effort, you will get used to it and begin to appreciate what an asset it is to make reasoning part of your basic repertoire. This is easy to explain. Teachers of mathematics must constantly maneuver to win their students' trust, because mathematics is ultimately about abstract and intangible concepts. Students need someone they trust to lead them out of the jungle of abstractions. When they see a teacher so sure of her footing that she can provide a reason every step of the way, they will readily put their trust in her. Then, mathematics learning becomes possible.

Reasoning disarms students' perception that mathematics is a game played with rules they know nothing about.

A final feature of this book that is worthy of mention is its emphasis on the *structure of mathematics*. Because structure may be a new concept to you, let me begin with a linguistic analogy. Suppose a teacher of ESL students is approached by one of them for the meaning of the word "huge", and he replies with "enormous". Do you think he has done his job? You would probably say *no*. Common sense tells you that if an ESL kid is coming across a simple word like "huge" for the first time, the chances of his having learned the meaning of the longer word, "enormous", are slim. The teacher's explanation most likely doesn't make sense to this student. Note that in part your judgment here is based on your overall understanding of

---

[14]A teacher in the Los Angeles area, Winnie Gilbert, who has worked with me for a number of years, recently told me about her students' comment on her style of teaching: "We have never been asked so many whys in our lives!"

the normal trajectory of learning a language, progressing from the simple to the complex.

Now consider an analogous situation in mathematics. A popular explanation of *the fundamental fact* of equivalent fractions, namely, $\frac{m}{n} = \frac{km}{kn}$ for all whole numbers k, m, n ($k, n \neq 0$), goes as follows:

$$\frac{m}{n} = 1 \times \frac{m}{n} = \frac{k}{k} \times \frac{m}{n} = \frac{km}{kn}.$$

As a teacher, do you think it is a good explanation? Because it is **a mathematical explanation**, you can no longer just use common sense, but **must** now draw on your content knowledge to imagine *the normal trajectory of learning about fractions*. You would reason as follows:

The fundamental fact must be taken up at the beginning of any discussion of fractions because equivalent fractions are basic. On the other hand, the concept of fraction multiplication requires an elaborate discussion for it to make sense, and such a discussion cannot take place near the beginning.[15] Such being the case, the use of something mysterious and not yet defined (multiplying fractions) to explain something else that is more elementary (the fundamental fact) is logically unsound. Your conclusion, therefore, should be that this explanation will not work.

The preceding reasoning is an illustration of the *structure of mathematics*. Mathematical statements are part of a *hierarchical* pattern, not thrown together at random. Once a particular development for a given subject has been chosen, there will be a certain rigidity in the unfolding of mathematical ideas: some concepts and skills must precede others because logical reasoning demands it. For example, we have just seen that, since the normal development of fractions would have equivalent fractions precede fraction multiplication, use of fraction multiplication is no more valid in explaining equivalent fractions than is the use of "enormous" in explaining "huge". This particular aspect of the structure of mathematics impacts the teaching of fractions.

More generally, the usual approach to numbers in the school curriculum begins with whole numbers, then fractions, and then rational numbers.[16] It makes no sense to try to teach students fractions if they do not already have a firm grasp of whole numbers. In a similar vein, one defines addition of fractions before subtraction, multiplication before division, division before percent, and so on. For this reason, one cannot successfully teach division of fractions before multiplication or teach percent before division. A teacher must always be aware of this hierarchical structure so that, when she gives an

---

[15] The reason may be paraphrased by a question: "How to multiply two pieces of pizza?" (I have stolen this phrase from the British educator Kathleen Hart [**Har00**].)

[16] These topics can be sequenced differently, of course; for example, whole numbers, integers, fractions, and rational numbers.

explanation, she does not do so by making indiscriminate use of everything under the sun. Rather, the mathematical explanation must respect this structure and only make use of facts that are compatible with this structure.

I hope you are beginning to see that this book demands serious effort from you solely to ensure that you learn correct mathematics and, subsequently, teach correct mathematics. Correct mathematics is much easier to teach than incorrect mathematics for the same reason that a well-written article is much easier to read than a badly written one. Your efforts in learning correct mathematics will ultimately help you become a better teacher. And *that*, after all, is what this book is all about.

Some Conventions in this Book

- Each chapter is divided into sections. Titles of the sections are given at the beginning of each chapter.
- When a new concept is first defined, it appears in **this typeface** but is not often accorded a separate paragraph all by itself. You will have to look for many definitions in the text proper.
- Equations are labeled with numbers inside parentheses, for example:

$$(2.3) \qquad m + n = n + m.$$

  The text often uses just an equation's label inside parentheses to refer to the equation; for example, "The commutative law of addition is given in (2.3)."
- The end of a proof is indicated by a small box: □.
- Scattered throughout the book are "Activities". These are simple exercises that serve to check the learning of a new skill or concept, and are meant to be used in class. Solutions to these Activities are posted on the website www.ams.org/bookpages/mbk-79.
- Exercises are at the end of each chapter.
- There are several notational conventions that are used throughout this book, as follows:
    - Page 16 (footnote 3): For large numbers, we use a comma to separate groups of three digits from the right.

- Page 23: The number line is horizontal, and whole numbers are marked off to the right.
- Page 28: 3×5 means $5 + 5 + 5$ rather than $3 + 3 + 3 + 3 + 3$.
- Page 45: In an expression, multiplication is performed before addition. So, the parentheses in $(mn) + (ml)$ can be removed so that it reads $mn + ml$.
- Page 140: When rounding a whole number n to the nearest 10, if two multiples of 10 are equally close to n, then always choose the bigger number.
- Page 141: Similarly to the previous item, when rounding a whole number n to the nearest 100, 1000, etc., if two multiples of 100 (resp., 1000, etc.) are equally close to n, then always choose the bigger number.
- Page 155: For clarity, we sometimes use a raised dot $\cdot$ to replace the multiplication sign $\times$.
- Page 181: When we write fractions of a particular denominator, for example, 3, we can write 0 as $\frac{0}{3}$.
- Page 185: The fraction notation $\frac{m}{n}$ or m/n automatically assumes that $n > 0$.
- Page 188: In order to keep track of, for example, the power 5 in $\frac{24}{10^5}$, three zeros are added to the left of 24 to make sure that there are 5 digits to the right of the decimal point in 0.00024. The 0 in front of the decimal point is only for the purpose of clarity, and is optional.
- Page 236: In the context of the number line, *equal parts* means *segments of equal length*.
- Page 409: The expression $-xy$ denotes $-(xy)$.
- Page 455: The writing of a single prime by itself is "a product of primes".
- Page 497: If the denominator of a decimal fraction is 10^n, then the decimal point is placed in front of the n-th digit of the numerator of the decimal fraction *counting from the last displayed digit on the right*, which could be zero.
- Page 519: In the notation for a repeating decimal, such as $\overline{523}$, at least one of the digits under the bar is nonzero.

Part 1

Whole Numbers

Part Preview

This part discusses the **whole numbers**

$$0, 1, 2, 3, 4, \ldots$$

with a view towards laying a firm foundation for the treatment of the main topics of this book, namely, fractions, decimals, and rational numbers. Notice that we include 0 among the whole numbers. We will learn how to compute with them, both precisely and approximately. In other words, how to use the arithmetic algorithms and, when appropriate, how to make estimates. Because we use the so-called Hindu-Arabic numeral system, *the most important idea of this part is the way* counting *is done in this numeral system* (see section 1.1, **How to Count**, on page 6). The precise procedure of counting lies at the very foundation of the standard algorithms as well as the techniques for making estimates.

Most of this part will be devoted to understanding the bread-and-butter topic of elementary mathematics: the **standard algorithms**. While we will go into the procedures of these algorithms with greater precision than is normally found in the existing literature, the emphasis will not be on the procedures per se but, rather, on the logical reasoning that underlies these procedures. Your students will want to know why you make them do seemingly unnatural acts to the numbers when you teach them these algorithms, and your obligation as a teacher is to provide *explanations that make sense and are mathematically correct*. This part will help you do that. Even if you already know the usual reasoning behind these algorithms, I hope there is still something new in the explanations given here to keep you interested, such as the clarification of the role of division-with-remainder in the long division algorithm (see Chapter 7 on page 95).

The chapter on estimation (Chapter 10 on page 139) is noteworthy for its emphasis on the explanation of why estimates are important and when estimates are appropriate. Such an explanation should be in every elementary textbook but, at present, it is not.

The last chapter on numbers in an arbitrary base (Chapter 11 on page 155) rounds off the discussion of our decimal numeral system in Part 1 by showing what happens when the expanded form of a number is expressed in terms of successive powers of a whole number other than 10.

Chapter 1

Place Value

This chapter discusses the concept of *place value* in our numeral system, i.e., the Hindu-Arabic numeral system. In the usual presentations, place value is regarded as the overriding idea in whole number arithmetic, and also something every elementary student must embrace from the beginning of schooling. Of its importance there is no doubt, but the present failure to impress its importance on *every* child raises the question of whether asking a child to accept it as an article of faith is a good pedagogical decision. Regardless of pedagogy, however, it is essential that teachers know that, rather than a matter of faith, place value is an inevitable consequence of the way our numeral system keeps track of the counting of (whole) numbers. By making teachers aware of this fact, we hope to give them an option for how they want to approach place value in their classrooms. For this reason, we will devote considerable space to explaining how counting is done, and why a thorough understanding of counting elucidates other basic concepts related to whole numbers, such as "bigger", "smaller", $53 \times 100 = 5300$, or the expanded form of a (whole) number.

The sections are as follows:

How to Count

Place Value

The Use of Symbolic Notation

The Number Line

Comparing Numbers (Beginning)

Multiplication and the Expanded Form of a Number

All about Zero

The Hindu-Arabic Numeral System

1.1. How to Count

Counting is the beginning of mathematics. Counting up to a small number (two or three, for example) can be done without any fanfare, and indeed has been done even in the most primitive tribes. Counting up to a large number, such as the number of soldiers in an army or the number of sheep in an enormous flock, is, however, anything but simple and requires the help of some elaborate system to keep track. For about four hundred years now, counting in the West, and subsequently the world over, has been done using the so-called **Hindu-Arabic numeral system** (see section 1.8, The Hindu-Arabic Numeral System, on page 33 for more information). In this section, we go into some details of this counting procedure because it explains *place value*, the basic concept behind all whole number calculations.

There are two basic ideas in the construction of the Hindu-Arabic numeral system:

(1) The first is to make use of only ten symbols, 0, 1, 2, 3, 4, 5, 6, 7, 8, 9 and nothing else.

(2) The second is to use these ten symbols to write down all possible counting numbers by putting the ten symbols in different **places** or **positions**.

Because of (1), the Hindu-Arabic numeral system is also called the **decimal numeral system**. It is a common practice to refer to any one of these ten symbols in a specific place of a counting number, such as 2 in 4723, as **the digit in that place**. Thus **the third digit from the left** of 4723 is 2, or what is the same thing, **the digit in the third place from the left** of 4723 is 2.

For example, the number

$$11{,}732{,}976{,}646{,}254$$

uses 14 places; it was the U.S. national debt (in dollars) as of September 1, 2009, at 2 PM Greenwich Mean Time.[1] The fourth, sixth, and seventh digits from the right of this 14-digit number are all equal to 6. The one in the fourth place signifies 6 *thousand*, the one in the sixth place signifies 6 *hundred thousand*, and finally the one in the seventh place signifies 6 *million*. The usual terminology is that the **place value** of the position occupied by the first 6 (from the right) is a thousand, that of the second 6 is a hundred thousand, and that of the last 6 is a million. These three numbers then stand for 6 thousand, 6 hundred thousand, and 6 million, respectively. How these three meanings for the digit 6 come about is what we want to explain next.

---

[1] Strictly speaking, "11,732,976,646,254" is the *Hindu-Arabic numeral* which presents the number in dollars of the U.S. national debt as of September 1, 2009, but unless absolutely necessary, we avoid such arid formality.

1.1. How to Count

Let us take up the problem of counting from the beginning. Consider how to write down, systematically, all possible numbers if we are given only one place, and then the same problem if we are given two places, then three places, then four places, etc. By observing the pattern as we increase the number of places one by one, we get to understand the meaning of place value.

Suppose we only have one place to work with. Then we simply write down the first ten numbers by using the given symbols 0, 1, 2, 3, 4, 5, 6, 7, 8, 9, *in this order*. This place is traditionally called the **ones place**, and the numbers so obtained by making use of only the ones place are called the **one-digit numbers**. Thus there are only ten one-digit numbers. At this point, it becomes necessary to make explicit what we mean by **counting**. **Counting one step from 0** means going from 0 to 1, **counting two steps from 0** means going from 0 to 1 to 2, etc., and **counting nine steps from 0** lands us at 9, as shown:

$$0 \to 1 \to 2 \to 3 \to 4 \to 5 \to 6 \to 7 \to 8 \to 9.$$

Because of the way we count from 0, we say 1 is the **first number counting from 0** and 9 is the **ninth number counting from 0**, etc.

At 9, counting stops because there is nowhere to go if we are restricted to just one place and we are not allowed to have more symbols. We can of course artificially continue the counting by recycling the same ten symbols 0, 1, 2, 3, 4, 5, 6, 7, 8, 9 and placing them in successive rows, as follows:

```
0 1 2 3 4 5 6 7 8 9
0 1 2 3 4 5 6 7 8 9
0 1 2 3 4 5 6 7 8 9
⋮   ⋮   ⋮   ⋮
```

In this scheme, counting nine steps from 0 lands us at the 9 of the first row, and counting one more step lands us at the 0 of the second row. Continuing counting this way, the next step lands us at the 1 of the second row, and yet another step lands us at 2 of the second row, etc. Eventually, we are at the 9 of the second row. The next step then lands us at the 0 of the third row, and so on. This will allow us to count indefinitely for sure, but an obvious problem is that there is no way to differentiate among the rows so that, for example, going three steps from 0 and going thirteen steps from 0 will land us both times at the symbol 3 (even if it is the 3 in the first row and the 3 in the second row, respectively).

The central breakthrough of the Hindu-Arabic numeral system is the realization that *if we allow ourselves two places instead of one*, then we can distinguish among ten of these rows by putting these ten symbols 0, 1, 2, etc., in succession in the place to the left of each of the ten numbers.[2] This new place is called the **tens place**. Precisely, in the first row, 0, 1, 2, 3, 4, 5, 6, 7, 8, 9, we put 0 in the tens place of each number to obtain 00, 01, 02, 03, 04, 05, 06, 07, 08, 09. For the next row of numbers, we register the fact that this is a new row by putting a 1 in the tens place of each of these numbers instead, therefore obtaining 10, 11, 12, 13, 14, 15, 16, 17, 18, 19. Then the next row produces the numbers 20, 21, 22, 23, 24, 25, 26, 27, 28, 29. Yet another row yields 30, 31, 32, 33, 34, 35, 36, 37, 38, 39, etc. By the time we get to the tenth row, we put 9 in the tens place of each number. After that we have exhausted the ten symbols in the new place, and what we obtain is then the following rectangular array.

$$
\begin{array}{cccccccccc}
00 & 01 & 02 & 03 & 04 & 05 & 06 & 07 & 08 & 09 \\
10 & 11 & 12 & 13 & 14 & 15 & 16 & 17 & 18 & 19 \\
20 & 21 & 22 & 23 & 24 & 25 & 26 & 27 & 28 & 29 \\
\vdots & & \vdots & & \vdots & & \vdots & & \vdots & \\
90 & 91 & 92 & 93 & 94 & 95 & 96 & 97 & 98 & 99
\end{array}
$$

Therefore, by the use of two places, we can now distinguish among the ten rows of ten numbers. Therefore, we can now count unambiguously from 0 up to 99.

These numbers, from 00 to 99, will be all the numbers we can write down if we restrict ourselves to two places. We call these the **numbers up to two digits**, and the reason for the terminology is that the numbers in the first row, 00, 01, 02, 03, 04, 05, 06, 07, 08, 09, are nothing but the original one-digit numbers with a 0 attached to the left to indicate that we are now using two places. For this reason, it is traditional to rewrite them simply as 0, 1, 2, 3, 4, 5, 6, 7, 8, 9, so that, among numbers up to two digits, only those coming after 09, i.e., 10, 11, 12, ..., 19, 20, ..., 97, 98, 99, are called **two-digit numbers**. The usual way of writing the numbers up to two digits is therefore the following.

$$
\begin{array}{cccccccccc}
0 & 1 & 2 & 3 & 4 & 5 & 6 & 7 & 8 & 9 \\
10 & 11 & 12 & 13 & 14 & 15 & 16 & 17 & 18 & 19 \\
20 & 21 & 22 & 23 & 24 & 25 & 26 & 27 & 28 & 29 \\
\vdots & & \vdots & & \vdots & & \vdots & & \vdots & \\
90 & 91 & 92 & 93 & 94 & 95 & 96 & 97 & 98 & 99
\end{array}
$$

If we limit ourselves to two places, then we can count only up to 99. But notice that after 9 comes 10. So 10 is the first two-digit number. Notice also that when the first 100 numbers are written this way, the 1 on the left of each number of the second row signifies the *first* time the ten symbols 0, 1, 2, ..., 9 are repeated, the 2 on the left of each number of the third row

---

[2]Note that putting the new place to the *left* of the ones place is strictly a CONVENTION and nothing more. It could have been to the *right* of the ones place.

1.1. How to Count

signifies the *second* time the ten symbols 0, 1, 2, ..., 9 are repeated, the 5 on the left of each number of the sixth row signifies the *fifth* time the ten symbols 0, 1, 2, ..., 9 are repeated, etc.

This rectangular array shows clearly *why it takes ten steps to count from 0 to 10, 10 to 20, 20 to 30,* etc. Indeed, these numbers are in the left column. To go down vertically from 10 to 20, for example, we must traverse the second row from left to right, and it takes nine steps to do that because there are ten numbers per row. Then counting one more step from 19, we get to 20. This rectangular array also explains *why the left digit, such as the 3 of 38, stands not for 3 but for 30,* because

> *the 3 in 38 indicates that 38 is in the fourth row of the rectangular array, and three rows of ten numbers have to be counted before we get to 30.*

It is time to pause and reflect on the counting of the first 100 numbers. The main message of place value is that in a number such as 38, *the 3 stands not for 3 but 30, whereas the 8 is just 8.* We want students in primary school to adopt this statement as their mantra. The purpose of the preceding exposition is to lay bare the fact that if students accept the limitation of using only ten symbols for counting and go through the reasons for writing the first 100 numbers in the way described, they may come to accept the meaning of 3 as 30 on their own without being forced to memorize it as a *fait accompli*. They see that the 3 has to be placed on the left of 0, 1, 2, ..., 9 to signify that this is the *third* time we are repeating these ten symbols, and this 3 therefore carries a different meaning from, say, the 3 of 53. It is likely that even children prefer to see the reason for doing something rather than have it rammed down their throats. They may learn place value much better as a result of learning how to count in this fashion.

Once the meaning of each digit of a two-digit number is well understood, the extension of this understanding to numbers with three or more digits should not be difficult.

Thus to continue counting but still without introducing more symbols, we follow the same method as before by making use of another place, which will again be taken to be to the *left* of the tens place for consistency. In greater detail, we recycle the numbers from 00 to 99 in one row and place them in successive rows. Then as before, we can count indefinitely by counting each row from left to right and then land on the left number of the row below to continue the counting. We recognize the defect of ambiguity in this way of counting because all the rows are identical. Now we put each of the ten symbols 0, 1, 2, 3, 4, 5, 6, 7, 8, 9, in succession in the place to the left of the tens place, thereby distinguishing ten of these rows, or, what is the same thing, keeping track of the order of appearance of ten rows of numbers 00, 01, ..., 99 up to two digits. The first row is therefore recorded with a 0 in the hundreds place, thus 000, 001, 002, 003, ..., 009, 010, 011, ..., 098, 099. For the second row, we replace the 0 in the hundreds place with a 1, yielding 100, 101, 102, 103, ..., 109, 110, 111, ..., 198, 199. In like manner, the third row becomes 200, 201, 202, 203, ..., 209, 210, 211, ..., 298, 299, and so on, until we end with 900, 901, 902, 903, ..., 909, 910, 911, ..., 998, 999 in the tenth row. This is as far as we can go because we have exhausted the ten symbols in the new place. Symbolically, we have the following rectangular array of numbers.

$$
\begin{array}{cccccccccccccc}
000 & 001 & 002 & \cdots & 009 & 010 & 011 & \cdots & 019 & 020 & 021 & \cdots & \cdots & 098 & 099 \\
100 & 101 & 102 & \cdots & 109 & 110 & 111 & \cdots & 119 & 120 & 121 & \cdots & \cdots & 198 & 199 \\
200 & 201 & 202 & \cdots & 209 & 210 & 211 & \cdots & 219 & 220 & 221 & \cdots & \cdots & 298 & 299 \\
\vdots & & & & & & & & \vdots & & & & & & \vdots \\
800 & 801 & 802 & \cdots & 809 & 810 & 811 & \cdots & 819 & 820 & 821 & \cdots & \cdots & 898 & 899 \\
900 & 901 & 902 & \cdots & 909 & 910 & 911 & \cdots & 919 & 920 & 921 & \cdots & \cdots & 998 & 999 \\
\end{array}
$$

The new place is called the **hundreds place** (the third place from the right).

The counting stops at 999 if we only have ten symbols and three places to work with. These are called the **numbers up to three digits**. As before, the first row of numbers, i.e.,

000 001 002 ··· 009 010 011 ··· 019 020 021 ··· ··· 098 099

is nothing but the list of all the numbers up to two digits and are therefore just the one-digit and two-digit numbers. For this reason, they are traditionally written more simply as

0 1 2 ··· 9 10 11 ··· 19 20 21 ··· 97 98 99

and only those numbers that come after 099 are called **three-digit numbers**. As before, we notice that after 99 comes 100. Thus 100 is the first three-digit number.

We make two observations at this point. The first is that

one counts exactly 100 steps from 0 to get to 100, from 100 to get to 200, from 200 to get to 300, etc.

This is because each of 0, 100, 200, ..., is the first number of its row (they are in the left column of the rectangular array) and 400, for instance, is the first number of the fifth row and 500 is the first number of the sixth row. It takes 100 steps to get from 400 to 500 because it takes 99 steps to get from 400 to 499 (there are 100 numbers per row) and the 100th step then takes 499 to 500, as claimed. The second is that this process explains, for example, *why the 8 of 819 stands not for 8 or 80 but 800*, because

this 8 in 819 indicates that 819 is in the ninth row and therefore one must count 100 steps, 8 times, to get from 0 to 100, 100 to 200, ..., 700 to 800, rows of 100 numbers in order to get to 800.

1.1. How to Count

To continue the counting process, we need to use four places, i.e., add another place to the left of the hundreds place. Thus, in a pattern that should be all too familiar by now, we put all the numbers up to three digits, from 000 to 999, in one row and we recycle them ten times by putting them in successive rows. To distinguish among these rows, or if you like, to keep track of the order of appearance of these numbers, we put one of 0, 1, 2, 3, 4, 5, 6, 7, 8, 9 in the new place of each number in the same row. For the first row, we indicate that this is the first row by adding a 0 to the left:

0000, 0001, ..., 0009, 0010, 0011, ..., 0099, 0100, 0101, ..., 0997, 0998, 0999.

For the next row we put 1 in the new place:

1000, 1001, ..., 1009, 1010, 1011, ..., 1099, 1100, 1101, ..., 1997, 1998, 1999,

and so on. The array ends with

9000, 9001, ..., 9009, 9010, 9011, ..., 9099, 9100, 9101, ..., 9997, 9998, 9999.

The new place is called the **thousands place**.

The following rectangular array summarizes these **numbers up to four digits**.

0000	0001	0002	⋯	0009	0010	⋯	0099	0100	⋯	0998	0999
1000	1001	1002	⋯	1009	1010	⋯	1099	1100	⋯	1998	1999
⋮				⋮			⋮				⋮
9000	9001	9002	⋯	9009	9010	⋯	9099	9100	⋯	9998	9999

We note as before that the tradition in normal writing is to replace the first row by the numbers up to three digits:

0 1 2 ⋯ 9 10 11 ⋯ 98 99 100 ⋯ 998 999.

In other words, we omit the consecutive zeros from the left. Observe as before that after 999 comes 1000, and 1000 is the first **four-digit number**.

Again, we make two observations. The first is that

one counts exactly 1000 steps from 0 to 1000, from 1000 to 2000, etc.

This is because each of 0, 1000, 2000, 3000, ... is in the left column of the array, and counting from one to the other (e.g., from 2000 to 3000) requires 999 steps to traverse a whole row, and then one more step to count from 2999 to 3000. The second observation is that this process explains *why the digit in the thousands place, such as 9 of 9502, stands not for 9 or 90 or 900, but 9000*, because

the 9 of 9502 signifies that 9502 is in the tenth row of the array and one must count 1000 steps, 9 times, in order to go from 0 to 9000.

Suppose we skip-count by 10, starting with 0. Then we have already observed that it takes ten steps to go from 0 to 10, 10 to 20, etc., so that the first *nine* steps in the skip-counting by 10 are

$$0 \to 10 \to 20 \to 30 \to 40 \to 50 \to 60 \to 70 \to 80 \to 90.$$

Counting ten more steps from 90 will land us at 100 because that is the first three-digit number after 99. Thus *if we skip-count by 10, starting with 0, then the tenth step gets to 100.* This skip-counting process can be made more vivid by going down the left column of the following array for the first nine steps.

0	1	2	3	4	5	6	7	8	9
10	11	12	13	14	15	16	17	18	19
20	21	22	23	24	25	26	27	28	29
⋮		⋮			⋮			⋮	
90	91	92	93	94	95	96	97	98	99

If we skip-count by 10, but start with 100, then a similar reasoning shows that we would land at 200 after *ten* steps because we have

$$100 \to 110 \to 120 \to 130 \to 140 \to 150 \to 160 \to 170 \to 180 \to 190 \to 200.$$

Suppose we skip-count by 100, and start with 0. Then the usual reasoning based on the way counting is done shows that we have the following list of numbers with 100 steps between them:

$$0, 100, 200, 300, 400, 500, 600, 700, 800, 900, 1000, 1100, \text{etc.}$$

Therefore, *if we skip-count by 100, then in ten steps we get from 0 to 1000*:

$$0 \to 100 \to 200 \to 300 \to 400 \to 500 \to 600 \to 700 \to 800 \to 900 \to 1000.$$

At this point, the pattern should be clear as to how we can count all the numbers up to five digits, six digits, ..., or n digits, for any whole number n. For example, the same reasoning shows that if we skip-count by a thousand, starting with 0, we get to 10000 in ten steps:

$$0 \to 1000 \to 2000 \to 3000 \to 4000 \to 5000$$
$$\to 6000 \to 7000 \to 8000 \to 9000 \to 10000.$$

Activity. Assume the usual terminology that 100 is *one hundred*, 1000 is *one thousand*, etc. (a) Now make believe that you are explaining to a third grader and explain why the 3 in 352 stands for 300, the 5 stands for 50, and the 2 stands for 2. (b) Explain to this third grader why, in ten steps of skip-counting by 100, one can go from 1000 to 2000. (Solutions to all Activities are posted on the website www.ams.org/bookpages/mbk-79.)

The Hindu-Arabic numerals make use of ten symbols. Because the collection 0, 1, ..., 8, 9 is a fairly large number of symbols, the preceding discussion was not as concrete and down-to-earth as possible because it was

1.1. How to Count 13

impractical, for example, to explicitly write down all the numbers up to three digits. For this reason, we will now discuss, as an illustration, the far simpler case that uses, not ten, but only *three* symbols, 0, 1, 2, for the purpose of writing down all the counting numbers, while continuing to observe the same two basic ideas:

(1) The first is to make use of only three symbols, 0, 1, 2 and nothing else.
(2) The second is to use these three symbols to write down all possible counting numbers by putting the symbols in different *places* or *positions*.

Such a companion discussion may serve better to bring out the key ideas of the Hindu-Arabic numeral system itself.

As before, we will refer to any of these three symbols in a specific place of a counting number *in this system* as a **digit**. Limited as we are to three symbols 0, 1, 2, we only get three one-digit numbers. Let us see what numbers (in this numeral system) can be written down by the use of two places. We put these three numbers in a row: 0 1 2. Now recycle the row two more times and each time we use one of 0, 1, 2 in the place to the left to distinguish among these rows of one-digit numbers. Thus, we start off with 00, 01, 02. Then we get in succession, 10, 11, 12, 20, 21, 22, thereby obtaining the following rectangular array of **numbers up to two digits**.

$$\begin{array}{ccc} 00 & 01 & 02 \\ 10 & 11 & 12 \\ 20 & 21 & 22 \end{array}$$

Note that the initial three numbers, 00, 01, 02, are nothing but the one-digit numbers 0, 1, 2, in disguise. It is traditional to replace the former by the latter when listing all the numbers up to two digits. Thus in the normal way of writing these numbers up to two digits, we have:

$$\begin{array}{ccc} 0 & 1 & 2 \\ 10 & 11 & 12 \\ 20 & 21 & 22 \end{array}$$

The numbers after 2, i.e., 10, 11, 12, 20, 21, 22 are called **two-digit numbers**.

If we count from 0, then three steps get us to 10 ($00 \to 01 \to 02 \to 10$). So the third number in this numeral system is 10. Similarly, the sixth number is 20, and the eighth number is 22.

If we assume a familiarity with ordinary arithmetic, then there are exactly (3×3) *numbers up to two digits in this numeral system, because the three symbols* $0, 1, 2$ *are repeated three times.*

Next we list all the numbers if we avail ourselves of three places. Again, the way to do this is to put the whole collection of numbers up to two digits (which are 00, 01, 02, 10, 11, 12, 20, 21, 22) in a row, and recycle the row *three* times, and keep track of their order of appearance by putting the symbols 0, 1, 2 in succession in the third place from the right. As usual, we put these in a rectangular array:

$$\begin{array}{ccccccccc} 000 & 001 & 002 & 010 & 011 & 012 & 020 & 021 & 022 \\ 100 & 101 & 102 & 110 & 111 & 112 & 120 & 121 & 122 \\ 200 & 201 & 202 & 210 & 211 & 212 & 220 & 221 & 222 \end{array}$$

These are all the **numbers up to three digits**. Again, the numbers in the first row are nothing but the original numbers up to two digits, so it is traditional to replace them by just 0, 1, 2, 10, 11, 12, 20, 21, 22, and then continue the enumeration with 100, 101, 102, ..., 221, 222. Thus:

$$\begin{array}{ccccccccc} 0 & 1 & 2 & 10 & 11 & 12 & 20 & 21 & 22 \\ 100 & 101 & 102 & 110 & 111 & 112 & 120 & 121 & 122 \\ 200 & 201 & 202 & 210 & 211 & 212 & 220 & 221 & 222 \end{array}$$

The numbers in the second and third rows are called the **three-digit numbers.**

If we assume a familiarity with ordinary arithmetic, then there are exactly $(3\times 3\times 3)$ *numbers up to three digits, because these numbers are obtained by recycling three times the* (3×3) *numbers up to two digits.*

We see that, for example, if we count eight steps from 0 we get to 22, so that the ninth step lands us at 100, as a simple counting shows:

$$0 \to 1 \to 2 \to 10 \to 11 \to 12 \to 20 \to 21 \to 22 \to 100.$$

Thus if 1 is the first number, then 100 is the ninth and, in like manner, 200 is the 18th in this numeral system. This is because we have just seen that it takes nine steps to go from 0 to 100, and therefore another nine steps to go from 100 to 200. It also follows from this reasoning that

211 would be the 22nd number if 1 is the first,

because one must go four steps from 00 to get to 11 (00 → 01 → 02 → 10 → 11), so that one must go four steps from 200 to get to 211 (200 → 201 → 202 → 210 → 211). So if 200 is the 18th number when 1 is the first, then 211 would be four more after that, i.e., the 22nd number, as $18 + 4 = 22$ (the addition being understood to be in the usual decimal numeral system).

Put another way,

there are 22 numbers in 1, 2, 10, 11, ..., 22, 100, ..., 210, 211.

The progression to four-digit numbers should be quite clear by now. If we make use of four places, then we can list all possible **numbers up to four digits** by recycling the row of numbers up to three digits, and putting one of the symbols 0, 1, 2 in the fourth place from the right of each number

1.1. How to Count

in the same row to keep track of the order of appearance of the numbers up to three digits. Thus after

0000, 0001, 0002, 0010, 0011, 0012, ..., 0212, 0220, 0221, 0222,

we will get

1000, 1001, 1002, 1010, 1011, 1012, ..., 1212, 1220, 1221, 1222,

and then

2000, 2001, 2002, 2010, 2011, 2012, ..., 2212, 2220, 2221, 2222.

As before, the first row of numbers is traditionally replaced by

0 1 2 10 11 12 ⋯ 212 220 221 222.

Only the numbers that follow, i.e., 1000, 1001, 1002, ... 2220, 2221, 2222, are called **four-digit numbers**.

If we assume a familiarity with ordinary arithmetic, then there are exactly $(3 \times 3 \times 3 \times 3) = 81$ *numbers up to four digits, because as we have seen, we obtain these by recycling three times the* $(3 \times 3 \times 3)$ *numbers up to three digits.* We put these 81 numbers in the usual rectangular array:

0	1	2	10	11	12	20	21	22	100	101	⋯	221	222
1000	1001	1002	1010	1011	1012	1020	1021	1022	1100	1101	⋯	1221	1222
2000	2001	2002	2010	2011	2012	2020	2021	2022	2100	2101	⋯	2221	2222

We see that, for example, if we start counting with 0, the first step gets to 1 and after 27 steps we get to 1000, i.e., 1000 is the 27th number if 1 is the first because it takes 26 steps to traverse each row of 27 ($= (3 \times 3 \times 3)$) numbers, and it takes another step to go from the last number of each row to the first number of the row below. It also follows that

1211 is the 49th number in this numeral system if 1 is the first,

and the reason is the following. We break up the counting from 1 to 1211 into two steps: $1 \to 1000 \to 1211$. We have just seen that 1000 is the 27th number counting from 0. We also saw above that one has to count 22 steps to go from 000 $(= 0)$ to 211, and therefore one must count 22 steps to go from 1000 to 1211. It follows that if we start counting from 0, the 27th number is 1000, and counting another 22 steps gets us to 1211. Hence, 1211 is the 49th number counting from 0 because, in terms of the usual addition in the Hindu-Arabic numeral system, $27 + 22 = 49$.

Activity. Do counting by using only the four symbols 0, 1, 2, 3, and the same idea of using different places. (a) Write down the first 48 numbers in this numeral system if we start counting with 0. (b) What is the 35th number if we start counting from 0? Can you figure this out without looking at the list in (a)? (c) What is the 51st number if we start counting from 0? The 70th number? (Solutions to all Activities are posted on www.ams.org/bookpages/mbk-79.)

We now return to Hindu-Arabic numerals and revisit skip-counting. We begin with a general observation. We have seen that if we skip-count by 10, starting with 0, then we land at 100 after 10 steps:

$$0 \to 10 \to 20 \to 30 \to 40 \to 50 \to 60 \to 70 \to 80 \to 90 \to 100.$$

If we skip-count by 100, starting with 0, then we would land at 1000 after 10 steps:

$$0 \to 100 \to 200 \to 300 \to 400 \to 500 \to 600 \to 700 \to 800 \to 900 \to 1000.$$

If we skip-count by 1000, starting with 0, we would land at 10000 after 10 steps:

$$0 \to 1000 \to 2000 \to 3000 \to 4000 \to 5000$$
$$\to 6000 \to 7000 \to 8000 \to 9000 \to 10000.$$

Thus skip-counting by 10, 100, or 1000 in *ten* steps, and starting from 0, gets to 100, 1000, or 10000, respectively. Notice that in each case, the final destination is obtained by adding one more 0 to the right of the number representing the amount of the skip-count, and the reasoning above in each case of 10, 100, or 1000 is seen to be independent of 10, 100, or 1000. Therefore the pattern persists whether we skip-count by 10,000, 100,000, etc.[3] This is an important pattern, as we shall see by the end of this chapter, so that we should find a way to express this fact in general. Let us continue with the pattern two more times to get a better grip on it:

A skip-count by 10,000 starting at 0 lands at 100,000 after ten steps.

A skip-count by 100,000 starting at 0 lands at 1,000,000 after ten steps.

For example, the second assertion follows from

$$0 \to 100{,}000 \to 200{,}000 \to 300{,}000 \to 400{,}000 \to 500{,}000$$
$$\to 600{,}000 \to 700{,}000 \to 800{,}000 \to 900{,}000 \to 1{,}000{,}000,$$

where the critical last step in going from 900,000 to 1,000,000 is because the last six-digit number is 999,999 and the next one has to be 1,000,000, and the passage from 900,000 to 1,000,000 (like the passage from 800,000 to 900,000) traverses exactly all the numbers up to six digits (exactly 100,000 of them) plus one more step.

There is clearly a need to express *all* of these facts about skip-counting once and for all in one encompassing statement. For this, we have to resort to the use of symbols:

---

[3]When the number of digits is large, we will adopt the usual convention of separating every three digits from the right by a comma. Be sure you are aware that this convention is one we use in the U.S. but is not used in Europe, for example.

1.1. How to Count

Let n be any whole number not equal to 0. If we start counting from 0 and skip-count by $1\underbrace{00\cdots 0}_{n}$ (this is the number which is 1 followed by n zeros), then we get to the number $1\underbrace{000\cdots 0}_{n+1}$ (the number which is 1 followed by $n+1$ zeros) in ten steps.

In greater detail:

$$0 \to 1\underbrace{00\cdots 0}_{n} \to 2\underbrace{00\cdots 0}_{n} \to \cdots \to 9\underbrace{00\cdots 0}_{n} \to 1\underbrace{000\cdots 0}_{n+1}.$$

In words: if we skip-count from 0 by a number which is 1 followed by n zeros (n is a nonzero whole number), then after ten steps, we will land at the number which is 1 followed by $n+1$ zeros.

To illustrate this assertion, suppose $n = 5$. The above assertion being valid for any n, it would also be valid for this value of n, i.e., 5, and the skip-counting is by an amount equal to the number 1 followed by five zeros. Also, $n + 1 = 6$. Thus, the assertion: if we skip-count by 100,000, then starting with 0, we get to 1,000,000 in ten steps. Note that this is one of the cases we have just looked at.

Finally, we introduce the concept of **addition** in terms of counting. You may *think* that adding numbers is so simple that you don't need to know anything more about it. However, if it is simple, then it must be possible to explain it so simply that even a child can understand it. This is what we do now.[4] The *addition* of 5 to 4, written as **4 + 5**, is the whole number one arrives at by **counting 5 steps from 4**. We shall refer to this process more simply as **continued counting** or **consecutive counting**.[5] Now, if we count five more steps starting at 4, we get

$$4 \to 5 \to 6 \to 7 \to 8 \to 9,$$

and the last number is 9 (be sure to check carefully that we have counted five steps). Thus, using this definition of addition, $4 + 5$ is the number 9.

We call attention to the slippery nature of the language in the definition of addition. We use the word "step" to avoid possible misunderstanding: to say we go five steps starting at 4, it unambiguously means the following five steps: $4 \to 5$, $5 \to 6$, $6 \to 7$, $7 \to 8$, $8 \to 9$, so that the last number has to be 9. If we had said "$4 + 5$ means the fifth number after 4", or "$4 + 5$ is the number one arrives at by counting five times starting at 4", then it would be defensible to think that what is meant is 4, 5, 6, 7, 8, and we

---

[4]The explicit definition of addition that follows is absolutely essential for understanding Chapter 4 on the addition algorithm.

[5]Sometimes called **counting on**.

would seem to be defining $4 + 5$ as 8. Please be aware of this confusion in the classroom.

It is common to use the **equal sign** "=" to express "$4+5$ is the number 9" as

$$4 + 5 = 9.$$

In other words, the two numbers "$4 + 5$" and "9" are the same number. Similarly, $172 + 39$ is the number one obtains by counting 39 steps starting at 172, i.e., $173 \to 174 \to 175 \to \cdots \to 210 \to 211$. (Check this by direct counting!) Thus,

$$172 + 39 = 211$$

the addition of b to a, written as $a + b$, is the number obtained by counting b steps starting at the number a.

The number $a + b$ is also called the **sum** of a and b.

Now if we have three whole numbers a, b, and c, then the definition is

*the **sum** $a + b + c$ is the number obtained by **continued counting**, namely, starting at the number a, count b steps to arrive at the number $a + b$, then count c steps starting at $a + b$ to arrive at the number $a + b + c$.*

One gives a similar definition of **continued counting** for the sum of any number of whole numbers $a + b + \cdots + z$.

Earlier, we skip-counted ten times starting at 0 to get to 100. This is of course the same as skip-counting nine times starting at *10* to get to 100. Therefore, in terms of addition and the equal sign, we have

$$100 = 10 + \underbrace{10 + 10 + \cdots + 10}_{9} = \underbrace{10 + 10 + \cdots + 10}_{10}.$$

In the same way, we obtain

$$100 = \underbrace{10 + 10 + \cdots + 10}_{10},$$

$$1000 = \underbrace{100 + 100 + \cdots + 100}_{10},$$

$$10000 = \underbrace{1000 + 1000 + \cdots + 1000}_{10}.$$

And so on. In general, if we use symbols, then we can compress all these statements into the following: Let n be any nonzero whole number; then

(1.1) $\underbrace{1000\ldots0}_{n} = \underbrace{100\ldots0}_{n-1} + \cdots + \underbrace{100\ldots0}_{n-1}$ (ten times).

1.2. Place Value

For example, if $n = 5$,
$$100000 = \underbrace{10000 + 10000 + \cdots + 10000}_{10}.$$

Activity. If you count in the usual way, what is the 200th number beyond 6490721? What about the 230th number? And the 236th number? The 5164th number beyond 6490721?

Summary. *The concept of place value is fundamental to the understanding of the mathematics of whole numbers, but it is usually presented to students as a declaration by fiat: "the digit on the right is the ones place, the one to its left is the tens place, and then the hundreds place ...". By learning how to count, however, students get to see the genesis of this concept and may therefore find the concept more accessible. Moreover, it will be seen in the following sections that, once students learn how to count, there will be no need for them to accept the concepts of "bigger than" and "add" as intuitive but inexplicable; they will be able to explain both in terms of the more fundamental concept of counting.*

1.2. Place Value

Now that we know how to count, it is relatively simple to explain the **place value** of each digit in any number of the Hindu-Arabic numeral system. Take 28, for instance. What does each of 2 and 8 stand for? We know from the rectangular array listing the enumeration of all numbers up to two digits that all numbers with a 2 in the tens place reside in the third row, as shown.

```
00  01  02  03  04  05  06  07  08  09
10  11  12  13  14  15  16  17  18  19
20  21  22  23  24  25  26  27  28  29
30  31  3 2 33  34  35  36  37  38  39
```

In terms of counting, we get to 28 by starting with 20 and count eight more steps. It follows that the 2 in 28 stands for 20, and then 8 in 28 stands for just 8 itself, so that 28 indicates that we start with 20 and then count eight more steps. In this case, we say the *place value* of 2 in 28 is 20, and the *place value* of 8 in 28 is just 8.

What about the digits in 728? Again, we recall that in the rectangular array which enumerates all numbers up to three digits, the numbers with 7 in the hundreds place reside in the eighth row. To count to the eighth row, one must first count to 700, and then take 28 more steps. Here is a schematic representation of the eighth row:

700 701 702 $\cdots$ 709 710 711 $\cdots$ 719 720 721 $\cdots$ $\cdots$ 798 799

Combined with what we said above about 28, we see that if we start with 700, and count 20 more steps to get to 720, and then count eight more steps we will get to 728. In particular, the place value of 7 in 728 is 700, the place value of 2 is 20, and the place value of 8 is 8 itself.

Suppose the number is 3728. We interpret each digit as follows. As before, the rectangular array displaying the enumeration of all numbers up to four digits contains all the numbers with a 3 in the thousands place in the fourth row. The first number of that row is 3000, and we get to 3728 by taking 728 steps from 3000. Here is the fourth row:

3000 3001 3002 $\cdots$ 3009 3010 $\cdots$ 3099 3100 $\cdots$ $\cdots$ 3998 3999

Combined with what has been said about 728, we see that to get to 3728, we start at 3000, then count 700 more steps, count 20 more steps, and finally count eight more steps. It follows that the 3 in 3728 stands for 3000, the 7 stands for 700, the 2 stands for 20, and then 8 is just 8.

Our emphasis on the process of counting now reaps additional benefits. Recall that we defined the addition of whole numbers in terms of continued counting. The preceding statement about the place values of the digits in 3728 is now seen to be a statement about continued counting: Start at 3000, count 700 more steps, then 20 more steps, and then eight more steps. In terms of addition, this says,

$$3728 = 3000 + 700 + 20 + 8.$$

Such a sum, which exhibits the place value of each digit in 3728, is called the **expanded form** of 3728.

Activity. What is the number if we count 500 more from 516234? 50000 more from 516234?

What has been said about 3728 extends immediately to any whole number, regardless of how many digits it has. For example, 52746 is the number we get to by starting at 50000 and counting 2000 more steps, then 700 more steps, then 40 more steps, and finally six more steps. Furthermore, the definition of addition allows the preceding statement to be written as the *expanded form* of 52746:

$$52746 = 50000 + 2000 + 700 + 40 + 6.$$

In case the idea of place value has become too familiar to strike you as noteworthy, let us look at a different numeral system for comparison: in Roman numerals,[6] the number 33 is represented by XXXIII. Observe then that the three "X's" are in three different places, yet each and every one of them stands for 10, not 100 or 1000. Just 10. Similarly, the three "I's" occupy different places too, but they all stand for 1, period. Thus XXXIII represents $10 + 10 + 10 + 1 + 1 + 1$. Contrast this with the numeral 111 in

---

[6]One should point out that even in the Roman numeral system, there is a *partial* place value at work. For instance "VI" is 6 while "IV" is 4. For a simple introduction to Roman numerals, see, for example, http://www.novaroma.org/via_romana/numbers.html.

our numeral system: the first 1 on the left stands for 100, the second stands for 10, and only the third stands for 1 itself. You see the difference.

Activity. Write 88 and 99 in Roman numerals; contrast the Hindu-Arabic numeral system with the Roman numeral system in this case. Do the same with the numbers 420 and 920.

Observe that in the Hindu-Arabic numeral system, even a person unfamiliar with the counting process would know (at least in some naive sense) that "101 is bigger than 88" simply because "there are more digits in 101 than in 88". There is no such easy reading with Roman numerals: CI (which is 101) definitely looks "smaller" than LXXXV (which is 85). In subsequent sections, we will make precise the comparison of sizes in Hindu-Arabic numerals in terms of the number of digits.

1.3. The Use of Symbolic Notation

You may have noticed that in the discussion of counting, we have already begun the use of symbolic notation. For example, we had equation (1.1), which reads

$$1\underbrace{00\ldots0}_{n} = 1\underbrace{00\ldots0}_{n-1} + \cdots + 1\underbrace{00\ldots0}_{n-1} \quad \text{(ten times)}.$$

What this symbolic statement does is compress an infinite number of statements into one short statement by employing the abstract symbol n, and allow n to equal, in succession, 1, 2, 3, 4, 5, ... What we have seen is that the sum of ten 1's is 10 (the case of $n = 1$), the sum of ten 10's is 100 (the case of $n = 2$), the sum of ten 100's is 1000 (the case of $n = 3$), etc., and we want to make explicit the fact that for all other values of n, the pattern is always the same: by adding ten numbers, each of which is 1 followed by a fixed number of zeros, what we get will be the number that is 1 followed by a number of zeros that exceeds the fixed number by 1. The concise way to express this clumsy statement is to use symbols, and this is equation (1.1) above.

The use of symbolic notation is therefore nothing more than the very human desire to achieve economy of expression, e.g., everybody accepts "FBI" to mean "Federal Bureau of Investigation" without a moment's thought, *and FBI is a symbol for those three words.* Because it is so important for you as a math teacher and for your students as math learners to get accustomed to the use of abstract symbols, we will steadily increase this usage as we progress. *Let it be said explicitly that, indeed, elementary school students must begin to learn to use symbols.*

There is at present some concern about the appropriateness of introducing symbolic notation to elementary students. Considerations of "developmental appropriateness" used to be offered as reasons for not teaching children topics deemed too cognitively complex, but recent research by cognitive psychologists has led to revision of such views (cf. [**Bru02**]; [**Gea06**]; [**GW00**]; [**NMP08c**, pages 4–6]. Children's capacity for abstraction is larger than what most people realize, and this fact is further confirmed by the curricula in other developed nations such as Russia. We have just taken note of the fact that symbols in mathematics are the exact counterpart of abbreviations in everyday language, and nobody has ever advocated avoiding the use of abbreviations in elementary school. Children talk about "MTV" with ease, just as they adapt to the use of "St." in place of Street, "U.S." in place of United States, "CA" in place of California, "NBA" in place of National Basketball Association, etc., without the slightest hint of difficulty.[7] Every one of these is a symbolic representation of a longer word or phrase. Obviously one should not contemplate denoting a whole number by n in kindergarten, any more than one should teach the use of "St." before kindergartners learn to spell S-t-r-e-e-t. But it is safe to say that, if by the end of the sixth grade a student is still ill at ease with the use of symbols, then this student will probably have difficulty in the eighth grade dealing with algebra. *Adequate mathematics instruction must therefore incorporate a judicious amount of symbolic notation beginning no later than, let us say, the fourth grade.*

This book, while strongly advocating the use of symbolic notation, tries not to overdo it. For example, observation (ii) on page 27 of section 1.5, Comparing Numbers (Beginning), could be more efficiently expressed by making a heavier use of symbols, but we have refrained from doing that. You might consider it an interesting exercise to rephrase that observation with the use of symbols.

1.4. The Number Line

The discussion of numbers in this book will revolve around the so-called *number line*, a line each of whose points is uniquely identified with a (real) number. We now turn to its definition.

So far, the concept of whole numbers is developed from *counting*, but it is valuable from the standpoint of mathematics learning to have a geometric realization of these numbers so that the mental process of counting can be transferred to the process of spatial visualization as well. Recently, cognitive

---

[7]Some people claim that the use of abbreviations is different from the use of symbols, because an abbreviation such as U.S. can stand for only one thing whereas a generic symbol x can stand for many things. There is a little bit of truth in this, but ultimately such a claim is false. In the latest (2008) NBA Championship series, for example, basketball fans all understood LBJ to mean, not President Lyndon B. Johnson, but LeBron James. While older people (as of 2009) react to IRA with discomfort because it stands for "Irish Republican Army", they have also come to accept it as "Individual Retirement Account". Moreover, people in reading education recognize IRA as the "International Reading Association". **Although LOL used to mean Little Old Lady exclusively, now it is used equally frequently to stand for "laughing out loud"**. And so on.

1.4. The Number Line

psychologists have also brought out the importance of the number line from a cognitive perspective ([**NMP08c**, pages 4–8]). The benefit of having a spatial, visual analogue to help with the learning of number facts is by now recognized. In this section, we identify the whole numbers 0, 1, 2, ... as certain **equi-spaced points** (i.e., the *distances* between consecutive points are all equal) on the line. Part 2 will identify the fractions on this line, and Part 3, the rational numbers.

Fix a horizontal straight line and designate a point on it as 0. For this chapter and the next, we shall mainly concentrate on the right side of 0. To the right of 0, mark off equi-spaced points and call them 1, 2, 3, 4, etc., as on a ruler. Thus the whole numbers are identified with these equally spaced points on the right side of the line. A line with the markings of the whole numbers on it is called the **number line**.

```
  0    1    2    3    4    5    etc.
  |----|----|----|----|----|----|
```

We explicitly call attention to the fact that the counting of whole numbers corresponds to the progression of marked points from the initial point 0 to its right, as shown above. The fact that we start with a *horizontal* line and not any oblique line, and the fact that we mark off whole numbers to its *right* and not to its left are entirely a matter of CONVENTION. Until Part 3, we will have no need for the part of the line to the left of 0.

The line segment from 0 to 1 will be denoted by **[0, 1]**, and it is called the **unit segment**, and the number 1 is called **the unit**. Note that

once a unit segment has been chosen on a given straight line, all the whole numbers are fixed on the line.

The fact that a different choice of the straight line and a different choice of the unit segment would lead to a different number line is understood, but the resulting mathematical discussion, insofar as it refers only to the whole numbers, will not be affected by the possible change in position of the line or the choice of 0 and 1 on the line. This is why we can afford to abuse the language and talk about *the* number line.

Given any two points a and b on the number line, with a to the left of b, we use $[a, b]$ to denote the line segment between a and b. The points a and b are the **endpoints** of $[a, b]$. If we look at the segment $[0, 12]$, then it resembles a one-foot ruler:

```
  0         3         6         9        12
  |--|--|--|--|--|--|--|--|--|--|--|--|
```

In teaching the number line to children, one may therefore present it as a ruler. It is natural to call 12 the **length** of $[0, 12]$. More generally, we call

n the **length** of the segment $[0, n]$. In particular, *the unit segment $[0, 1]$ has length 1.* We can now define the length of any line segment $[a, b]$ by using the number line as an **infinite ruler**: $[a, b]$ is said to have **length** n for a whole number n if, after sliding $[a, b]$ to the left along the number line until a rests on 0, the right endpoint b rests on n.

When there is no fear of confusion, we generally identify a segment of length n with the segment $[\mathbf{0}, \mathbf{n}]$.

The addition of whole numbers was defined in terms of counting as *continued counting*. We now show how addition can also be interpreted in terms of the length of **concatenated line segments**, i.e., line segments connected end-to-end on the same line. For example, to find the sum $4 + 7$, we start with 4 and count 7 more steps until we reach 11, and that is the sum. The corresponding geometric statement is then to start at 4 and go to the right seven more steps, where each step is of length 1. The representation of this addition on the number line is therefore given by:

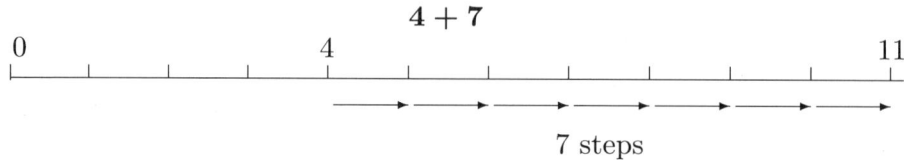

Notice that "start with 4 and count 7 more steps" is the same as concatenating the left endpoint of $[0, 7]$ to the right endpoint of $[0, 4]$ and seeing where the right endpoint of $[0, 7]$ lands. Thus using the notion of length, we can interpret the addition $4 + 7$ as the length of the concatenation of the segments of lengths 4 and 7:

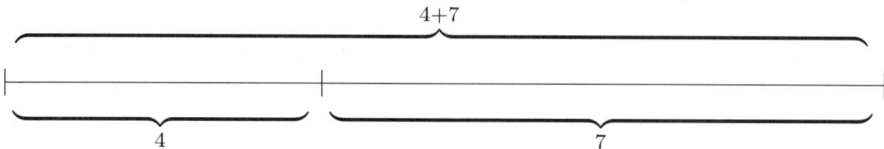

This interpretation continues to be meaningful when $4+7$ is replaced by $m+n$ for any whole numbers m and n: $\boldsymbol{m + n}$ *is the length of the concatenation of a segment of length m and a segment of length n.*

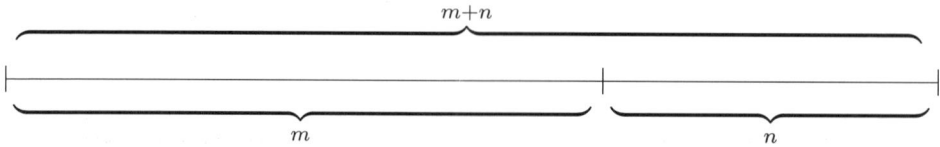

In terms of geometry then, addition is simple: *concatenate, then measure the length.*

1.5. Comparing Numbers (Beginning)

In the next two sections, we will get to see several examples of how the number line can be put to use in an essential way.

1.5. Comparing Numbers (Beginning)

Note: The section Comparing Numbers (Conclusion) starts on page 48.

We will discuss how to **compare** or **order** two whole numbers, i.e., which of two given numbers is "bigger than" the other. For example, you may think it is "obvious" that 1058 is "bigger" than 874, because a "thousand is bigger than a few hundreds", or something to this effect. But what does it mean that a thousand is *bigger* than nine hundred, and why? In order to teach mathematics well, we have to convey information as clearly and as precisely as possible. To rely on children's *intuitive feeling* that "1058 is bigger than 874" and then let them figure out on their own how to compare numbers is not good education. This is because, once we form the habit of relying solely on intuition in the teaching of mathematics, we will do the same when we get to more sophisticated topics such as fractions. Unfortunately, many people have gone down that road and the results have not been encouraging. Let us try to do better by explaining with more precision.

If there are things that you don't understand in section 1.1, **How to Count**, please go back to page 6 and review it before reading further.

Formally, given two whole numbers a and b, we define b to be **bigger than** a (or what is the same, a **smaller than** b) if, in the method of counting described in **How to Count**, a comes before b; in symbols:

$$a < b \quad \text{or} \quad b > a.$$

Thus $2 > 1$, $47 < 51$, $672 < 678$, etc. In particular, it is always the case that

if n is a nonzero whole number, then $n > 0$.

Expressions such as $a < b$ are generally referred to as **inequalities**.

If you reflect for a moment, you will agree that this definition of "bigger than" captures precisely the everyday concept of "more than".

We note immediately that for whole numbers a, b, c,

if $a < b$ and $b < c$, then $a < c$.

We shall refer to this fact as the **transitivity of** $<$.

It is also common to express $a < b$ by saying b is **greater than** a, or a is **less than** b. A whole number a so that $a > 0$ (i.e., a is nonzero) is said to be **positive**. A more complete discussion of the concept of positivity will be given in Part 3.

Given two numbers, if they are not the same, then one must come before the other in the process of counting. Therefore we have:

Trichotomy Law. *Given two whole numbers a and b, exactly one of the following possibilities holds: $a = b$, $a < b$, $b < a$.*

We now interpret the above discussion on the number line. Recall that counting corresponds to progressing to the right on the number line along the marked points that are the whole numbers. Therefore, for whole numbers a and b,

$a < b$ means exactly that the point a on the number line is to the left of b.

As to the transitivity of $<$, if $a < b$ and $b < c$, then geometrically it is clear why $a < c$:

Finally, the trichotomy law is merely the statement that two segments are either equal in length or one is longer than the other; there is no other possibility.

We now make two simple but basic observations about comparing two whole numbers. (i) *The entire discussion assumes that you know how to count*, so there will be almost no explicit mention of counting, and (ii) that *all numbers are written without leading zeros*, so that, for instance, 000218 is written as a three-digit number 218. First a few examples. We claim that $687 < 1124$. This is because, 999 being the last of the three-digit numbers, $687 < 999$. After 999 comes the first four-digit number 1000, and then of course $1000 < 1124$. By the transitivity of $<$, we have $687 < 1124$. For similar reasons, we also have $9823 < 25706$, because $9823 < 9999 < 10000 < 20000 < 25706$. Likewise, a seven-digit number such as 8,158,022 is less than a nine-digit number such as 104,057,923 because

$$8{,}158{,}022 < 9{,}999{,}999 < 10{,}000{,}000 < 100{,}000{,}000 < 104{,}057{,}923.$$

The same reasoning shows that, in general:

(i). *If a, b are whole numbers and b has more digits than a, then $a < b$.*

The arena of the local professional basketball team[8] has a rule: if the home team scores "in triple digits", then every fan gets a free pizza that night. This is a good illustration of the above observation: scores up to 99 points get you no pizza. Look for other everyday expressions that illustrate the observation.

---

[8] The Golden State Warriors.

Next, how do we compare two whole numbers with the same number of digits? Consider 7541 and 7622. We have 7541 < 7622 because
$$7541 < 7599 < 7600 < 7622.$$
Similarly 28,123,475 < 28,123,601, because
$$28{,}123{,}475 < 28{,}123{,}499 < 28{,}123{,}500 < 28{,}123{,}600 < 28{,}123{,}601.$$
In the case of the last example, the numbers 28,123,475 and 28,123,601 agree digit-by-digit as we go from left to right (the same 2, the same 8, the same 1, the same 2, and the same 3) until we come to *sixth* place from the left, when 28,123,475 has a 4 and 28,123,601 has a 6. If we review the above reasoning, we see that the inequality 4 < 6 was decisive in allowing us to conclude that 28,123,475 < 28,123,601, because 475 < 601. The same reasoning then leads to the following general observation:

(ii). *Given two whole numbers p and q with the same number of digits, suppose when going from the left, p and q agree digit-by-digit until we come to the first place where they do not. Suppose p's digit in that place < q's digit in the same place. Then p < q.*

Activity. Can you explain to a third grader why 7001 is greater than 5897? Don't forget, you are not supposed to use subtraction or any kind of computation. You have to first explain what "greater than" means, and then use that to explain 7001 > 5897.

1.6. Multiplication and the Expanded Form of a Number

We next turn our attention to the phenomenon of "too many zeros". This issue underlies our quest for a better understanding of place value, and can be explained by an example.

The U.S. population as of April 17, 2006, was 298,534,453. The expanded form of this number is therefore
$$200{,}000{,}000 + 90{,}000{,}000 + 8{,}000{,}000 + \cdots.$$
If you think the number of zeros here is dizzying, wait until you have to tackle something like the U.S. national debt, which was 8,420,415,609,294 in dollars as of April 17, 2006. Its expanded form is therefore equal to
$$8{,}000{,}000{,}000{,}000 + 400{,}000{,}000{,}000 + \cdots.$$
It is far too tedious to write out the rest of this sum. We must look for a viable alternative to this kind of writing.

To this end, we introduce the concept of **multiplication** as a shorthand for repeated addition. Precisely, we *define*
$$3 \times 5 \stackrel{\text{def}}{=} 5 + 5 + 5$$
and

$$7 \times 4 \stackrel{\text{def}}{=} 4+4+4+4+4+4+4,$$

where the symbol $\stackrel{\text{def}}{=}$ means "is defined as".

In general, if m, k are whole numbers, the definition of mk (which is the accepted *notational simplification* for $m \times k)^9$ is:

(1.2) $$mk = \begin{cases} 0 & \text{if } m = 0, \\ \underbrace{k + k + \cdots + k}_{m} & \text{if } m \neq 0. \end{cases}$$

Sometimes we refer to mk as the **product** of m and k, and call m or k a **factor** of mk. We emphasize the fact that the multiplication of k by m is *not a concept that every child is born with but is, rather, something that must be explained clearly*. This definition explicitly defines the multiplication mk as a *shorthand* for the sum of m copies of k when m is nonzero, no more and no less, and *defines 0 times any number to be 0*. Please be sure to impress this definition on your students. (See Exercise 9 on page 35.)[10]

Note that we have implicitly set up a CONVENTION in the above definition of multiplication. The product 3×5 could be defined equally well as $3 + 3 + 3 + 3 + 3$, i.e., 3 added to itself five times, but we have chosen to use the other convention instead: 5 added to itself three times. *What is important is that, once we adopt this convention, we stay with it throughout the book to avoid confusion.* The same remark applies to your teaching in the classroom.

Since addition is continued counting, the sum of two positive (whole) numbers is positive. Therefore, knowing that the product of two whole numbers is repeated addition, we conclude that *the product of two positive numbers is positive*.

For the product of three or more numbers, such as $3 \times 5 \times 8$, it will be understood unless stated to the contrary that it means the multiplications should be performed from left to right. Thus

$3 \times 5 \times 8$ means: first do 3×5 to get 15, then do 15×8 to get 120. Same for $3 \times 5 \times 8 \times 27$: it means $3 \times 5 \times 8$ times 27. Etc.

Activity. Consider the following introduction to *multiplication* taken from a third-grade textbook (the text has the goal of making sure that at the end of the third grade, students know the multiplication table of numbers up to 10):

---

[9] In this case, the notational simplification is adopted to avoid confusing "$\times$" with the letter x. Nevertheless, there are occasions when $\times$ is used on purpose.

[10] For future reference, it is worth pointing out that, once we get away from whole numbers, this meaning of multiplication as repeated addition must be modified. This remark is particularly pertinent in view of some current misconceptions about the division of fractions that result from the misconception about the multiplication of fractions as a repeated addition. See Chapter 17.

1.6. Multiplication and the Expanded Form of a Number

> Look at the 3 strips of stickers shown on the right [there is a picture of three strips of stickers]. There are 5 stickers on each strip. How can you find the number of stickers there are in all?
> You can find the total number in different ways.
>
> | You can write an ADDITION sentence. $5+5+5=15$ THINK: 3 groups of 5 $=15$. | You can write a MULTIPLICATION sentence $3 \times 5 = 15$ READ: Three times 5 equals 15. |
> |---|---|
>
> Answer: 15 stickers.

Do you think this is an ideal way to convey to third graders what *multiplication* means? Explain.

With multiplication understood as repeated addition, we can now rewrite equation (1.1) of section 1.1, **How to Count** (see page 18), as

$$(1.3) \qquad 10 \times \underbrace{100 \cdots 0}_{n-1} = \underbrace{1000 \cdots 0}_{n}$$

for every positive whole number $n = 1, 2, 3, \ldots$. Thus $10 \times 10 = 100$ in case $n = 2$ in equation (1.3), and $10 \times 100 = 1000$ in case $n = 3$, etc. Rewriting the first equation as $100 = 10 \times 10$ and substituting it into the second equation, we get

$$1000 = 10 \times 10 \times 10.$$

The case of $n = 4$ in equation (1.3) gives $10 \times 1000 = 10000$, and making use of the preceding fact concerning 1000, we get

$$10000 = 10 \times 10 \times 10 \times 10.$$

Arguing this way with $n = 5, 6, 7, \ldots$ in succession, we easily arrive at the conclusion that the number with n zeros following a 1 is equal to the product of n 10's, i.e.,

$$\underbrace{100 \cdots 0}_{n} = \underbrace{10 \times 10 \times \cdots \times 10}_{n}$$

for all $n = 1, 2, 3, \ldots$. It is standard practice in mathematics to abbreviate 10×10 as 10^2, $10 \times 10 \times 10$ as 10^3, and in general, $10 \times 10 \times \cdots \times 10$ (n times, $n > 0$) as 10^n. In other words, we define, for $n = 1, 2, 3, \ldots$

$$\mathbf{10^n} \stackrel{\text{def}}{=} \underbrace{\mathbf{10 \times 10 \times \cdots \times 10}}_{\boldsymbol{n}}.$$

We may therefore restate the equation

$$\underbrace{100 \cdots 0}_{n} = \underbrace{10 \times 10 \times \cdots \times 10}_{n}$$

as
$$\underbrace{100\cdots 0}_{n} = 10^n.$$
So far, the notation 10^n has a meaning only when n is positive. If $n = 0$, we also write
$$10^0 \stackrel{\text{def}}{=} 1$$
and, by common consent, we usually (though not always) omit the 1 in 10^1 and simply write 10.

The symbol 10^n is entirely reasonable as a labor-saving device, one that saves the day when we are confronted with the need to write too many zeros as described at the beginning of this section. The number n in 10^n is called the **power** or the **exponent** of 10, and 10^n is read as "10 to the n-th power". Thus the exponent counts the number of 10's we are multiplying together. It follows immediately that

(1.4) $$10^{m+n} = 10^m \times 10^n$$

for all whole numbers m and n.

As a final preparation for tackling the expanded form of large numbers, we note that assertions such as $3 \times 1000 = 3000$ and $7 \times 10000 = 70000$ are straightforward to verify because, for example,
$$3 \times 1000 = 1000 + 1000 + 1000 = 3000,$$
where the last equality follows from the definition of addition as continued counting. The same reasoning then shows that for any positive whole number n,
$$\underbrace{400\cdots 0}_{n} = 4 \times \underbrace{100\cdots 0}_{n} = 4 \times 10^n,$$
and the same remains true if 4 is replaced by any *single-digit* number. (We will prove at the end of the next section that when 4 is replaced by any whole number, the same assertion remains valid.)

Using this last fact, we can now revisit the expanded forms of large numbers such as the U.S. population figure at the beginning of the section. We have the following expanded form of 298,534,453 in **exponential notation** (i.e., using exponents):

$$298{,}534{,}453 = (2 \times 10^8) + (9 \times 10^7) + (8 \times 10^6) + (5 \times 10^5) + (3 \times 10^4)$$
$$+ (4 \times 10^3) + (4 \times 10^2) + (5 \times 10^1) + (3 \times 10^0),$$

where we have used **parentheses** to indicate that the operations within the parentheses are to be performed first.

Mathematically, the preferred way to write down the expanded form of a whole number is to make use of exponential notation, as above. The reason is that it displays unambiguously the place value of each digit in terms of a power of 10 regardless of how many digits the number has, e.g., 2 actually

1.6. Multiplication and the Expanded Form of a Number

stands for 2×10^8, and 9 stands for 9×10^7, while the 3 on the right stands for 3×10^0, etc. However, it is also obvious that the added clarity comes at a price: it must be said that exponential notation can be clumsy at times. For example, the ordinary expanded form of a small number such as 832,

$$832 = 800 + 30 + 2,$$

is surely simpler than one involving exponents:

$$832 = (8 \times 10^2) + (3 \times 10^1) + (2 \times 10^0).$$

Generally, we will write the expanded form of a number using exponential notation when we want to achieve maximum conceptual clarity, but we will be content to use the ordinary expanded form if there is no fear of confusion.

We will call each digit of a number (in this case, 298,534,453) the **co-efficient** of the corresponding power of 10 in an exponential form using exponents, e.g., 9 is the coefficient of the 7th power of 10, and 8 is the coefficient of the 6th.

One virtue of writing the expanded form of a number using exponents is that it makes plain the fact that the *place value* of a digit in a whole number is 10 times the *place value* of the digit to its right. For example, the 9 in 298,534,453 stands for 9×10^7 and the 8 to its right stands for 8×10^6, and

$$9 \times 10^7 = 9 \times \underbrace{10 \times 10 \times \cdots \times 10}_{7} = (9 \times \underbrace{10 \times \cdots \times 10}_{6}) \times 10 = (9 \times 10^6) \times 10,$$

and 9×10^6 would be the place value of 9 were it put in the place where 8 is at the moment.

When a number has zeros in the middle, such as 830,059, making beginners write out its expanded form using exponential notation may help them with the understanding of place value:

$$830059 = (8 \times 10^5) + (3 \times 10^4) + (0 \times 10^3)$$
$$+ (0 \times 10^2) + (5 \times 10^1) + (9 \times 10^0),$$

or, less formally with the zero terms omitted:

$$830059 = (8 \times 10^5) + (3 \times 10^4) + (5 \times 10) + 9.$$

It remains to mention that the cumbersome notation of 5×10^1 and 9×10^0 in the preceding expanded form of 830059 is occasionally preferred over the more informal 5×10 and 9, such as when absolute clarity is called for.

Activity. As an exercise in the use of exponential notation, write out the expanded form of the following numbers:

14600418, 500007009, 94009400940094.

Activity. In the expanded form of a number, the term with the highest power of 10 is called the **leading term**. (Thus 7×10^6 is the leading term of $(7 \times 10^6) + (2 \times 10^3) + (4 \times 10^2) + (1 \times 10^0)$, which is the expanded form of 7002401.) Explain why the leading term of the expanded form of any number is always larger than the sum of all the other terms in the expanded form.

1.7. All about Zero

There is a phenomenon concerning the multiplication by a power of 10 that students most likely take for granted but, just as likely, cannot explain. For example, why is 37×1000 equal to 37000? In this case, it is actually possible to give the reason without undue effort: this is the totality of 37 copies of 1000, and since we know that 10 copies of 1000 is 10000 (equation (1.3)), we see that 30 copies of 1000 is 30000. Add to it seven copies of 1000, which is 7000, we get 37000. This argument gets unbelievably tedious, however, if we try to show $375386 \times 10000 = 3753860000$. We need a comprehensive statement about what is true in general, in the following form:

$n \times 10^k =$ the whole number obtained from n by attaching k zeros to the right of the last digit of n.

The full explanation of this fact requires the distributive law, which could be dealt with right now but will be postponed to the next section for the sake of continuity. In the meantime, we will freely make use of this fact about $n \times 10^k$ when necessary.

Finally, we tie up a loose end. When we did counting in section 1.1, **How to Count**, we explained why we omitted the zeros to the left of a number symbol, e.g., 00075 is simplified to 75. We now give an alternate explanation, from the point of view of the expanded form of a number, of why it is permissible to attach any number of zeros to the left of a number without changing the number. For example, the number 830159 is the same number as 0830159, and is the same number as 000830159, etc. This is very easy to see because by the expanded form of a number,

$$000830159 = 0 \times 10^8 + 0 \times 10^7 + 0 \times 10^6 + 8 \times 10^5$$
$$+ 3 \times 10^4 + 1 \times 10^2 + 5 \times 10^1 + 9 \times 10^0$$
$$= 830159,$$

where we have used the fact that, by definition, 0 times any number is 0. This observation is conceptually important in the understanding of the various algorithms in Chapters 4–8.

1.8. The Hindu-Arabic Numeral System

It was mentioned that the name of our numeral system is the *Hindu-Arabic* numeral system. The usual reason given for the terminology is that the Arabs learned about this numeral system from the Hindus sometime before AD 800 and the translation of Arabic texts into Latin around the twelfth century introduced the Hindu numerals to the Europeans. Around 1600, this numeral system finally won universal acceptance, and hence the name. However, recent historical research indicates that this terminology is likely a misnomer.

Until recently, the history of Chinese mathematics was not known to the West mainly because of the linguistic difficulty and the lack of English translations of Chinese original sources. For example, many history of mathematics textbooks fail to mention the fact that the Chinese had a decimal numeral system since the first available record of writing dating back to at least 1000 BC. Such failure perhaps stems from the failure to recognize that the Chinese numerals, the **rod numerals** (also called the **counting board numerals**), are the equivalent of the Hindu-Arabic numerals 0, 1, 2, ..., 9. The rod numeral system was firmly in place no later than AD 200, but most likely dates to a much earlier time. It is in every respect *identical* to the Hindu-Arabic system except for the physical appearance of the ten symbols 0, 1, 2, ..., 9 themselves. Moreover, negative numbers (see Part 3) and *decimal fractions*, more commonly known as *decimals* (see Part 2), were part of the rod numeral system from the beginning. Such a numeral system has been in continuous use in China for the past two millennia. Because of the long history of contact between India and China, it may be difficult to separate what is Indian and what is Chinese in the Hindu numeral system. In recent years, the Hindu-Chinese situation is beginning to draw serious attention from Chinese scholars (cf. [**LA92**]), and confirmation of the preceding remarks is beginning to emerge, but much awaits discovery.

It should be pointed out that there have been other numeral systems in human history that also made use of place value, but they employed more than ten symbols. The Babylonians, for example, used 60 symbols;[11] this system dates back to at least 2000 BC (see [**Bur07**, Chapter 1]). These 60 symbols were constructed from two basic symbols using partial place value, and they get unwieldy. The Mayans had a place value numeral system that employed (more or less) 20 symbols, including zero;[12] this numeral system is of remote and uncertain origin (also see [**Bur07**, Chapter 1]). Again, these symbols are cumbersome. By comparison, the Hindu-Arabic system uses only *ten* easily written symbols and, not coincidentally, most of us have

---

[11]A more correct way to put it is that the Babylonian numeral system is a *base 60 place value numeral system* (see Chapter 11)

[12]The third place of the Mayan system, instead of having a value of 20^2, has the value 20×18.

ten fingers. It is hardly an accident that this has become the universally accepted numeral system today.

Exercises

There are two rules about doing the exercises in this book: (i) Unless stated to the contrary, use only what you have learned so far in the book. (ii) Every answer must come with an explanation. The explanation may be in the form of the details in a calculation, or it may be a verbal citation of particular facts used in the text. Sometimes, for emphasis, an explanation is even explicitly demanded. But whatever it is, you will have to get used to never claiming anything without giving a reason.

1. Imagine you have to explain to a fourth grader that $43 \times 100 = 4300$. How would you do it?

2. Imagine you have to explain to a fifth grader why $48 \times 500{,}000 = 24{,}000{,}000$. How would you do it?

3. What number should be added to 946,722 to get 986,722? What number should be added to 68,214,953 to get 88,214,953?

4. What number should be added to 58×10^4 to get 63×10^4? What number should be taken from 52×10^5 to get 48×10^5?

5. Which is bigger? 4873 or 12001? 4×10^5 or 3×10^6? 8×10^{32} or 2×10^{33}? 4289×10^7 or 10^{11}? 765,019,833 or 764,927,919? Explain.

6. Write each of the following numbers in expanded form:
   (a) 60,100,900,730
   (b) 2,300,000,001
   (c) 72,000,000,659

7. Explain directly (without invoking the two observations on page 26 of section 1.5, **Comparing Numbers (Beginning)**), why $872 < 1{,}304$; $100{,}002 > 99{,}817$; $803{,}429 < 804{,}021$; and $541{,}962{,}208 > 541{,}961{,}765$.

8. Explain why, for any nonzero whole number k, $10^k > m \times 10^{k-1}$ for any *single-digit* number m.

9. The following is the introduction to the concept of *multiplication* taken from a third-grade textbook. On the side of the page is the Vocabulary of the Day:

   ```
   MULTIPLICATION  an operation using at least two numbers
   to find another number, called a product.
   PRODUCT  the answer in multiplication.
   ```

   Then in the text proper, one finds:

   ```
   How many are in 4 groups of 6?
   You can use MULTIPLICATION to solve the problem.
   ```

Use cubes to model the problem and record the answer to the problem:

Number of groups	Number in each group	Product
6 ×	4 =	24

Further down:

> If Helena practices singing 3 hours each day for a week, how many hours will she practice altogether?
> Find: 3×7
> THERE IS MORE THAN ONE WAY!
> Method 1: You can use repeated addition to solve the problem.
> $$3+3+3+3+3+3+3 = 21$$
> Method 2: When the groups are equal, you can also write a MULTIPLICATION SENTENCE.
> $$7 \times 3 = 21.$$

Write down your reaction to the appropriateness of such an introduction, and compare your view with those of others in your class.

10. In the first section, **How to Count**, we have looked at the numeral system which uses only three symbols 0, 1, 2 and place value. In this numeral system, how many numbers are there from 1 to 2211? And from 1 to 12121? (Write your answers in Hindu-Arabic numerals!)

11. Use the same idea to do counting as the Hindu-Arabic numeral system, but limit yourself to the use of only four symbols: 0, 1, 2, 3. (a) What is the 121st number in this numeral system (if we start counting from 0 as usual)? The 195th number? The 254th number? The 259th number? (b) How many numbers are there from 1 to 321? And from 1 to 12321?

12. Use the same idea of counting as the Hindu-Arabic numeral system, but limit yourself to the use of only seven symbols: 0, 1, 2, 3, 4, 5, 6. Write down in this numeral system the numbers corresponding to the ordinary Hindu-Arabic numerals 7, 15, 47, 339, 352, and 1111.

13. Use the same idea of counting as the Hindu-Arabic numeral system, but use twelve symbols: 0, 1, 2, 3, 4, 5, 6, 7, 8, 9, ♡, ◇. (a) Write down in this numeral system the numbers corresponding to the ordinary Hindu-Arabic numerals 11, 15, 22, 47, 121, 145. (b) Write down in this numeral system the numbers corresponding to the ordinary Hindu-Arabic numerals 142, 1720, 1733, 8650. (c) How many numbers are there from 1 to ♡◇2?

14. Use the same idea of counting as the Hindu-Arabic numeral system, but limit yourself to the use of only two symbols: 0, 1. (a) Write down in this numeral system the numbers corresponding to the Hindu-Arabic

numerals 5, 14, 35, 59, 511, 517, 1122, 4028. (b) How many numbers are there in this numeral system from 1 to 1010? From 1 to 11101110?

Chapter 2

The Basic Laws of Operations

A major objective of Part 1 is to give the explanation of why the four standard algorithms for whole numbers are correct. It turns out that it is because of the associative and commutative laws for addition and multiplication and the distributive law.[1] This is one reason we are going to spend time in this chapter discussing these seemingly hackneyed laws.

The sections are as follows:

> The Equal Sign
> The Associative and Commutative Laws of $+$
> The Associative and Commutative Laws of $\times$
> The Distributive Law
> Comparing Numbers (Conclusion)
> An Application of the Associative and Commutative Laws of Addition

2.1. The Equal Sign

The key point of each of these basic laws is always the fact that two collections of numbers, each connected by some arithmetic operations, which look superficially distinct are in fact equal, and this equality is indicated by the ubiquitous *equal sign* "$=$", the symbol first introduced on page 18. Because the equal sign is one of the sources of confusion in elementary school mathematics, let us first elaborate on this symbol.

---

[1] These "laws" are traditionally called "properties" in school textbooks. The universally accepted terminology in mathematics is, however, the former.

The most important thing to remember is that, while the meaning of the equal sign does get more sophisticated as the mathematics gets more advanced (as we shall see later in this book), for whole numbers it is perfectly simple. Two whole numbers a and b are said to be **equal**, and we write $a = b$, if we can verify by *counting* that they are the same number. For example, $4 + 5 = 2 + 7$ because we count four objects and then five more and get nine, whereas we count two objects and then seven more and also get nine, and this is what $4 + 5 = 2 + 7$ means. Or, we can also use the number line to explain the equality of two numbers a and b. Since each number corresponds to a point on the number line, $a = b$ means exactly that *these two points on the number line coincide*.

Make an effort to explain to your students—by whatever means you can, again and again if necessary—that the equal sign between two whole numbers does *not* signify "do an operation to get an answer". It merely means:

> check the numbers on both sides of the equal sign *by counting*, or,
> by locating them on the number line to verify that they are the same point.

In order that such classroom instruction be successful, one must lay the proper groundwork. In this instance, you cannot tell your students "check the numbers on both sides of the equal sign" if they don't know what a whole number is, any more than you can ask your students to count the number of ghosts in the northeast corner of the third floor of your school building. This is why we made the effort to define whole numbers explicitly in terms of counting and, later on, as certain points on the number line. Both definitions are sufficiently specific and tangible that students will be in a position to check, when two whole numbers are presented to them, whether they are the same number.

Confusion over the equal sign can also result if a teacher sometimes inadvertently gives students the impression that *the equal sign is a command to perform a calculation*, as in "$4 + 7 = ?$". One must guard against such unintended messages, as they may contribute to students' misconceptions.

When we deal with fractions and decimals (Part 2), the equality of two fractions or two decimals will have to be more carefully explained since we will not be able to do repeated addition in the literal sense. Nevertheless, we will show that the equality in that case ultimately reduces to repeated addition (see Chapters 12 and 39).

2.2. The Associative and Commutative Laws of +

A striking application of the associativity and commutativity of addition can be found in cash registers everywhere, regardless of whether they are

2.2. The Associative and Commutative Laws of + 41

electronic, manual, fancy, clumsy, etc.; namely, the final tally of a sale is independent of the *order* in which the items are rung up![2] Imagine then that someone brings two books worth $32 and $25, a shirt worth $28, and a pair of shoes worth $55 to the cash register. Imagine also the following three of the many possible scenarios:

(1) The cashier rings up in succession the two books, the shirt of $28, and the shoes of $55. By the definition of addition in terms of consecutive counting, this is $\big((32 + 25) + 28\big) + 55$, where we have added the parentheses for emphasis.

(2) The cashier rings up the book of $25, the shoes of $55, the shirt of $28, and then the book of $32, in this order. This is $\big((25 + 55) + 28\big) + 32$.

(3) The cashier notices at a glance that the $25 book and the shoes add up neatly to $80, and the shirt and the $32 book add up to $60. So she registers 80 and then 60. Precisely, this is the sum of $(25 + 55) + (28 + 32)$.

Of course, each scenario ends up with $140 in sales.

From the point of view of addition, what this says is that

$$(2.1) \quad \big((32+25)+28\big)+55 = \big((25+55)+28\big)+32 = (25+55)+(28+32).$$

By direct computation, equation (2.1) is obviously true. (Verify this yourself!) This suggests that one can add up these four numbers in any order and the results will be the same. Indeed, something more general is true:

Theorem 2.1. *For any collection of numbers, the sums obtained by adding them up in any order are all equal.*

Since addition can be interpreted in terms of the concatenation of segments, what Theorem 2.1 says is that if we are given a finite collection of segments, then the length of their concatenation is independent of the order in which the segments are concatenated. We will see a direct application of this fact presently.

This theorem will be used in all subsequent arguments without comment. So why not just assume this and move on? Because mathematics demands simplicity, and this statement is not simple enough. It makes an assertion about the behavior of n numbers with respect to addition, for any positive whole number n, and the number of possibilities is just big enough to leave a trace of doubt. Given a choice, we would prefer to assume a little less. For this reason, mathematicians searched for simpler alternatives and came up with two disarming assertions: for any three whole numbers ℓ, m, and n, it

---

[2] I owe this observation to [**Wil02**, page 32].

is always the case that

(2.2) $$(\ell + m) + n = \ell + (m + n),$$
(2.3) $$m + n = n + m.$$

As is well known, equation (2.2) is the **associative law of addition**, and equation (2.3) is the **commutative law of addition**. The emphasis in these statements is the fact that they are valid for *any* whole numbers ℓ, m, and n. It is because of the need to express this fact that the symbols ℓ, m, and n are employed in equations (2.2) and (2.3). They are easy to believe and equally easy to verify for any explicit values of ℓ, m, and n, but we need to know that equations (2.2) and (2.3) are true no matter what ℓ, m, and n may be.

For the purpose of the present discussion, what matters is that, from these two simple laws, we can prove Theorem 2.1 concerning the addition of n numbers in any order.

Activity. Interpret equations (2.2) and (2.3) in terms of the concatenation of segments.

Unfortunately, the proof that equations (2.2) and (2.3) imply Theorem 2.1, while not difficult, is tedious and boring in the extreme; it is not a particularly instructive exercise to go through the proof. For this reason, we have put off the reasoning of why equations (2.2) and (2.3) imply equation (2.1) to the last section of this chapter on page 51 and concentrate on a discussion of the impact of Theorem 2.1 on classroom instruction. The practical consequence of Theorem 2.1 is that given any collection of numbers, say, a, b, c, ..., k, we can write down

$$a + b + c + \cdots + k$$

without any parentheses, and without any concern for their order, and there will be no ambiguity because all these sums are equal. For example,

$$a + b + c + \cdots + k = k + a + \cdots + c + b.$$

It is possible that Theorem 2.1 still impresses you as being not entirely relevant, so let us look at a simple computation: $26 + 38$. Don't forget that we only know addition as continued counting—the *addition algorithm* is yet to come. Therefore, the only way to do $26 + 38$ is to count from 26 until we stop at the 38th number. It is tedious. However, we can circumvent this tedium by a judicious but liberal application of Theorem 2.1:

$$\begin{aligned}
26 + 38 &= (10+10+6) + (10+10+10+8) \\
&= (10+10+10+10+10) + (6+8) &\text{(Theorem 2.1)} \\
&= (10+10+10+10+10) + (10+4) &\text{(Theorem 2.1)} \\
&= (10+10+10+10+10+10) + 4 &\text{(Theorem 2.1)} \\
&= 60+4 = 64.
\end{aligned}$$

If this reminds you of the addition algorithm (to be discussed in Chapter 4), then you are quite correct. We can also perform the addition geometrically as the concatenation of segments:

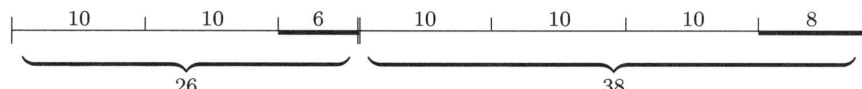

Theorem 2.1 says we may rearrange these individual segments and their concatenated length will not change:

Now one can read off the total length as 64, i.e., $26 + 38 = 64$.

Activity. $37 + 189 + 163 = ?$ $275 + 892 + 225 + 4211 + 108 = ?$

Activity. $666{,}666{,}667 + 788{,}646{,}851{,}086 + 333{,}333{,}333 = ?$

2.3. The Associative and Commutative Laws of ×

We now turn to the multiplication of whole numbers as defined in equation (1.2) on page 28. There is such a close parallel with the discussion of addition that we can be brief. The **associative law of multiplication** (equation (2.4) below) and the **commutative law of multiplication** (equation (2.5) below) are the exact analogues of the corresponding facts in addition. For any whole numbers ℓ, m, and n, the following equalities always hold:

(2.4) $$(\ell m)n = \ell(mn),$$
(2.5) $$mn = nm.$$

We have made use of the *notational convention* already mentioned in definition (1.2) of multiplication on page 28: when letters are used to stand for numbers, the multiplication sign is omitted so that "mn" stands for "$m \times n$".

Again, the same comment about the use of symbols in equations (2.2) and (2.3) applies to equations (2.4) and (2.5) as well.

Both equations (2.4) and (2.5) will again be accepted on faith. Of course, equation (2.5) is the reason why children only need to memorize about half of the 81 multiplication facts (45, to be exact) in the single-digit multiplication table. As in the last section, these two laws lead to the following.

Theorem 2.2. *Given a collection of numbers, the products obtained by multiplying them in any order are all equal.*

For example, for whole numbers a, b, c, d, e, f, it is always true that

$$\Big((e(fa))d\Big)(cb) = \Big(((ad)(bf))c\Big)e.$$

Here are two simple applications of Theorem 2.2. First, $(87169 \times 5) \times 2$ is equal to $87169 \times (5 \times 2)$, and therefore is equal to 876190 because of the fact concerning multiplication by a power of 10 stated in section 1.7, **All about Zero**, on page 32 to the effect that for any whole number N, $N \times 10^k$ equals the whole number obtained from N by attaching k zeros to the right of the last digit of N. (This fact will be proved in the next section.) Next, $10^7 \times 6572 = 6572 \times 10^7$, and therefore is equal to 65,720,000,000 by exactly the same fact.

Note that, without the commutative law, it is not so trivial to see that $10^7 \times 6572 = 65720000000$. After all, *by definition*, $10^7 \times 6572$ is the addition of 6572 to itself 10^7 times!

We have by now made use, twice, of the fact that $N \times 10^k$ equals the whole number obtained from N by attaching k zeros to the right of the last digit of N. We are now in a position to generalize this fact. Suppose we are given 6572×38000. Then

$$6572 \times 38000 = (6572 \times 38) \times 10^3,$$

and since $6572 \times 38 = 249736$, we see that $6572 \times 38000 = 249,736,000$. By using the same reasoning, one proves, in general, the following.

> $\boldsymbol{n \times (m \times 10^k)}$ *equals the whole number obtained from nm by attaching k zeros to the right of the last digit of nm.*

2.4. The Distributive Law

The **distributive law** connects addition and multiplication. It states that for any whole numbers m, n, and ℓ, the following is valid:

$$m(n + \ell) = (mn) + (m\ell).$$

2.4. The Distributive Law

Mathematicians love economy in notation (the use of notation was itself a response to the need for economy), so they devised a CONVENTION to eliminate the multiple parentheses:

> In an expression such as $\mathbf{mn} + \mathbf{m\ell}$, we always multiply the numbers $\mathbf{mn}$ and $\mathbf{m\ell}$ first before doing the addition.

With this understood, we can rewrite the above equality as follows. For any whole numbers ℓ, m, and n,

(2.6) $$m(n + \ell) = mn + m\ell.$$

Because multiplication is commutative, this could equally well be written as

$$(n + \ell)m = nm + \ell m.$$

We remark that the notational convention of omitting the parentheses is just a convention; it has no mathematical substance. In school mathematics, this convention goes under the heading of **order of operations** and, unfortunately, it sometimes assumes a central position in the curriculum. *This is wrong because, while children need to learn to observe conventions, no convention of any kind should be elevated to a position of prominence in any discipline, least of all in mathematics.*

We will need a geometric model for the distributive law when we come to fractions, so we initiate the discussion of this model for whole numbers. It requires the **area model** for multiplication. We begin with a definition of area that is adequate for the present need. A square with length 1 on each side is called a **unit square**. *The area of the unit square is by definition equal to 1.* We say a collection of rectangles $\{\mathcal{R}_j\}$ **tile** or **pave** a given rectangle R if, by combining the $\mathcal{R}_j$'s together we get the whole rectangle R, and if the $\mathcal{R}_j$'s intersect at most along their boundaries. With all this terminology in place, the **area of a general rectangle** is by definition the number of unit squares required to pave that rectangle. (Remember that we are dealing only with whole numbers at this point and therefore the lengths of all rectangles are whole numbers.) For example, 3×5 is the area of a rectangle with vertical length 3 and horizontal length 5 because there are three rows of five unit squares in the rectangle and the area of the latter is therefore $5 + 5 + 5$ (unit squares):

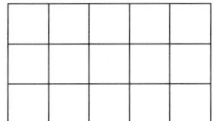

The same reasoning shows:

> *The product mn (m and n being whole numbers) is the area of a rectangle with "vertical" length m and "horizontal" length n.*

It may be mentioned that the area model of multiplication is in fact the mathematical underpinning of the *base ten blocks* manipulatives.

We now use the area model of multiplication to interpret the distributive law. This is best done through an explicit example. If $m = 3$, $n = 2$ and $\ell = 4$ in $m(n + \ell) = mn + m\ell$, then $3(2 + 4)$ is the area of the following rectangle with "vertical" length 3 and "horizontal" length 6:

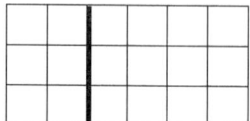

On the other hand, 3×2 is the area of the "left" rectangle and 3×4 is the area of the "right" rectangle. Thus $3 \times (2+4) = (3 \times 2) + (3 \times 4)$. Again, the essence of this picture is unchanged when 2, 3, and 4 are replaced by other triples of numbers.

The distributive law generalizes to more than three numbers. For example,

$$m(a + b + c + d) = ma + mb + mc + md$$

for any whole numbers m, a, b, c, and d. This can be seen by applying the distributive law (equation (2.6)) twice and making liberal use of Theorem 2.1, as follows.

$$\begin{aligned} m(a+b+c+d) &= m\big((a+b)+(c+d)\big) \\ &= m(a+b) + m(c+d) \\ &= (ma+mb) + (mc+md) \\ &= ma+mb+mc+md. \end{aligned}$$

Activity. One can use the distributive law to multiply a two-digit number by a one-digit number using mental math. For example, to compute 43×6, we break up 43 into $(40+3)$ so that $43 \times 6 = (40+3) \times 6 = (40 \times 6) + (3 \times 6)$, and the last is just $240 + 18 = 258$.

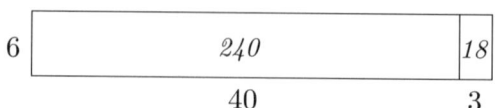

Therefore, $43 \times 6 = 258$. Following this example, use mental math to compute: (a) 24×8, (b) 53×7, (c) 39×6, (d) 79×5, (e) 94×9, (f) 47×8.

2.4. The Distributive Law

As a typical application of the distributive law, we bring closure to the assertion in section 1.7, All about Zero, on page 32 by providing an explanation for

> $\mathbf{N} \times \mathbf{10^k}$ equals the whole number obtained from N by attaching k zeros to the right of the last digit of N,

where N is any positive whole number. It suffices to consider $N = 372$ and $k = 4$ in order to simplify the exposition, as the reasoning is completely general. Thus,

$$\begin{aligned} 372 \times 10^4 &= \big((3 \times 10^2) + (7 \times 10^1) + 2\big) \times 10^4 \\ &= \big((3 \times 10^2) \times 10^4\big) + \big((7 \times 10^1) \times 10^4\big) + (2 \times 10^4) \quad \text{(dist. law)} \\ &= (3 \times 10^6) + (7 \times 10^5) + (2 \times 10^4) \quad \text{(Theorem 2.2, eq. (1.4))} \\ &= 3720000. \end{aligned}$$

We have now made the formal acquaintance of the five laws of arithmetic, but did you notice that, with the exception of the distributive law, each of them is about *one* operation alone, be it $+$ or $\times$? The distributive law is the only law that involves both $+$ and $\times$:

> *The distributive law is the glue that binds addition $+$ and multiplication $\times$ together.*

For example, if you have forgotten the definition of multiplication, the distributive law can remind you that multiplication is repeated addition, e.g., $3 \times 7 = (1 + 1 + 1) \times 7 = 7 + 7 + 7$. Yet, despite its obvious importance, it seems to be the least understood of the five laws among students. This could be partly due to the lack of insistence by teachers that students learn it. Therefore, let us begin by convincing you that the distributive law is important. Then there may be a better chance for your students to learn this law.

The distributive law (equation (2.6)) says not only that "the left side of (equation (2.6)) is equal to the right side" (i.e., $m(n + \ell)$ is equal to $mn + m\ell$), but also that "the right side of (equation (2.6)) is equal to the left side" (i.e., $mn + m\ell$ is equal to $m(n + \ell)$). In other words, while it is true that $35 \times (74 + 29)$ is equal to $(35 \times 74) + (35 \times 29)$, it is equally, if not more important to recognize that $(35 \times 74) + (35 \times 29)$ is also equal to $35 \times (74 + 29)$. As a simple demonstration, consider the straightforward computation

$$(35 \times 74) + (35 \times 29) = 2590 + 1015 = 3605.$$

However, if we use the distributive law, we can actually do this computation by mental math, as follows. We have $(35 \times 74) + (35 \times 29) = 35 \times (74 + 29)$, but $74 + 29 = 103 = (100 + 3)$. So we get $35 \times (100 + 3) = 3500 + 105 = 3605$.

But the main point of this remark is that knowing that $mn + m\ell$ is equal to $m(n+\ell)$ could be the difference between success and failure in a logical argument. See, for example, Chapter 6 when we discuss the multiplication algorithm, and Chapter 29 when we discuss the multiplication of rational numbers. It may also be mentioned that, in algebra, this way of using the distributive law underlies the skill of "collecting like terms".

Moral: Be sure you know that $mn + m\ell$ equals $m(n+\ell)$.

Activity. If a, b, c, d are whole numbers such that $a + c = b + d = 11$, what is $ba + bc + da + dc$?

2.5. Comparing Numbers (Conclusion)

The concept of *order* among whole numbers (which of two numbers is bigger?) was introduced in Chapter 1. We now conclude that discussion by relating order to addition and multiplication.

Recall that given two whole numbers a and b, the inequality $a < b$ is defined to mean that in the counting of the whole numbers starting with 0, 1, 2, ..., the number a comes before b. For convenience in logical arguments, we now recast this definition in terms of addition:

> Given two whole numbers **a** and **b**, **a** < **b** is equivalent to **b** = **a** + **c** for some positive whole number **c**.

This assertion makes it possible to achieve precision in logical arguments concerning inequalities, such as the inequalities (2.7) below. Before giving the reason for this assertion, we explain what it means to say that the two statements "$a < b$" and "$b = a + c$ for some positive whole number c" are **equivalent**. It means that both of the following implications involving these two statements are true:

> If **a** < **b**, then **b** = **a** + **c** for some nonzero whole number **c**.

> Conversely, if **b** = **a** + **c** for some nonzero whole number **c**, then **a** < **b**.

For example, $7 < 12$ means that we have to count more steps from 7 before we get to 12, and therefore $12 = 7 + 5$. Conversely, if we know $12 = 7 + 5$, then we must go five more steps from 7 before we get to 12, so $7 < 12$. The general reasoning is not much different. Given $a < b$, we know by definition that a precedes b, so that in the counting of the whole numbers, after we get to a, we need to go (let us say) c more steps before getting to b, and c is not zero. This implies, by the definition of addition in terms of counting, that $b = a + c$. Conversely, suppose $b = a + c$ is given, with $c > 0$; then after counting a objects, we have to count c more objects before we get b objects. So by the definition of "smaller than", we know $a < b$.

2.5. Comparing Numbers (Conclusion)

To summarize: If we know $a < b$, we conclude that $b = a + c$ for some positive whole number c, and if we know $b = a + c$ for some positive whole number c, then we can also conclude that $a < b$. Therefore, as far as making logical arguments is concerned, the two facts "$a < b$" and "$b = a + c$ for some positive whole number c" are interchangeable.

In terms of the number line, $a < b$ means that the segment $[0, b]$ is longer than the segment $[0, a]$ (see figure below). Knowing that addition of numbers corresponds to the concatenation of segments, we see without difficulty that the length of the segment $[a, b]$ is the number c that we seek.

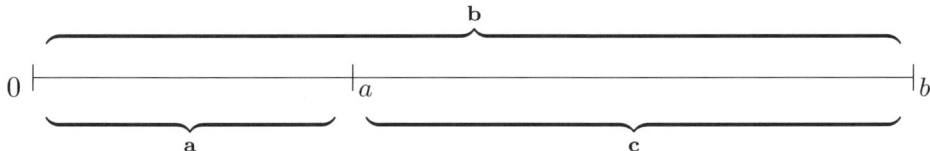

In the mathematics literature, if A and B are two mathematical statements, then alternate expressions for "**A is equivalent to B**" are:

A if and only if B (often abbreviated to **A iff B**);
A when and only when B;
a necessary and sufficient condition for A is B;
A $\iff$ B.

A less common expression, but one that is entirely correct, is "**A is true exactly when B is true**". The important thing to remember is that each time you see any of these phrases, you know that there is work ahead because you will be called upon to prove the validity of *two* implications.

Finally, we come to the main topic of this section, which is a collection of inequalities that express some simple truths about whole numbers. Let us illustrate. Knowing $22 < 29$, we know that $22 + 38 < 29 + 38$, or for that matter, $22 + 412 < 29 + 412$, simply because $60 < 67$ and $434 < 441$, respectively. Is this then *always true*, i.e., $22 + n < 29 + n$, no matter what the whole number n may be? In fact, if instead of 22 and 29, we take any two whole numbers k and ℓ so that $k < \ell$, is it *always* the case that $k + n < \ell + n$? The desire to ascertain the truth of such simple statements in general, therefore, calls for the use of symbols to express this idea correctly. Namely, we ask, for all whole numbers a, b, and c, does $b < c$ always lead to $a + b < a + c$? In like manner, does knowing $22 < 29$ imply that $n \times 22 < n \times 29$ for any whole number n? And again, what if 22 and 29 are replaced by any pair of whole numbers k and ℓ? Etc. The need for these kinds of general statements is genuine; even a casual glance at section 7.4, A Mathematical Explanation (Preliminary), on page 110 (especially footnote 9 on page 112) would testify to that. Thus, let a, b, c, d be whole numbers. Then we claim:

(2.7)
$$\begin{array}{lll} b < c & \text{is equivalent to} & a+b < a+c, \\ \text{If } a > 0, \text{ then } b < c & \text{is equivalent to} & ab < ac, \\ a < b \text{ and } c < d & \text{implies} & a+c < b+d, \\ a < b \text{ and } c < d & \text{implies} & ac < bd. \end{array}$$

One comment about the second assertion may be relevant. One should not over-read this statement: all it says is that multiplication *by a positive whole number* does not change an inequality, i.e., if $b < c$, then so long as $a > 0$, multiplying both sides of the inequality by a results in a "similar" inequality, $ab < ac$. As a teacher, it would be a good idea to introduce some caution in the teaching of this assertion and *warn students against the fixation that multiplication does not change an inequality.* Point out, for instance, that even for whole numbers, the preceding statement is true *only* when $a > 0$, because if $a = 0$, then multiplying both sides of $b < c$ by 0 would result in $0 \times b = 0 \times c \, (= 0)$. Also tell students that this assertion will be false when a is a *negative number*, which they will learn later in Chapter 31.

Activity. Assume the truth of the inequalities (2.7). Which is bigger: $86427895 \times 172945678$ or $86427963 \times 173000001$?

Activity. Without using (2.7), convince yourself that $2a < 2b$ is equivalent to $a < b$ by using the number line.

We will only give the proof of the second assertion in inequalities (2.7), since it is the most difficult of the four, and we leave the rest as (straightforward) exercises. We put the proof in small print because it is something you can come back to later.

Thus, we are given that $a > 0$. First we prove
$$b < c \quad \text{implies} \quad ab < ac.$$

For specific numbers, such as $a = 11$, $b = 7$, and $c = 9$, we know $7 < 9$ leads to $11 \times 7 < 11 \times 9$, because $77 < 99$. But without knowing the specific values of a, b, and c, and only that $b < c$, how do we conclude that $ab < ac$? We are forced to concentrate on the precise meanings (i.e., *definitions*) of $b < c$ and $ab < ac$. We know ab is the area of a rectangle $\mathcal{R}$ with vertical length a and horizontal length b, and likewise, ac is the area of a rectangle $\mathcal{S}$ with vertical length a and horizontal length c. Since $b < c$ by hypothesis, we see that while the vertical lengths of $\mathcal{R}$ and $\mathcal{S}$ are equal $(= a)$, the horizontal length of $\mathcal{R}$ is smaller than that of $\mathcal{S}$. So $\mathcal{R}$ has a smaller area than $\mathcal{S}$, and therefore $ab < ac$.

Next, we prove the converse, i.e.,
$$ab < ac \quad \text{implies} \quad b < c.$$

The reasoning of the preceding paragraph is relatively straightforward, but what follows will likely be something you cannot discover by yourself. It means that you now have the opportunity to learn something new. There are only three possibilities between b and c: $b < c$, $b = c$, and $b > c$ (this is the Trichotomy Law from page 26). If we want to conclude $b < c$, all we have to do is to show that the other two possibilities, $b = c$ and $b > c$, are not compatible with the hypothesis that $ab < ac$. There is an easy one: if $b = c$, then of course $ab = ac$. (You may even look at it this way: two rectangles with the same vertical and horizontal lengths must have the same areas.) So the possibility that $b = c$ is out, because we know $ab < ac$. What about $b > c$? Rewrite it as $c < b$. Then the preceding paragraph shows that this must imply $ac < ab$, or, what is the same thing, $ab > ac$. But again, this is incompatible with our hypothesis that $ab < ac$. So the only possibility left is $b < c$. The proof is complete.

Activity. Recall the notation of the **square** of a number: $9^2 = 9 \times 9$, $4^2 = 4 \times 4$, etc. (Of course the reason for calling, e.g., 9^2 the *square* of 9 is that it is the area of the *square* with all four sides equal to 9.) Compare 9^2 and $4^2 + 5^2$. Which is bigger? Do the same to: (i) 12^2 and $5^2 + 7^2$; (ii) 18^2 and $12^2 + 6^2$; (iii) 23^2 and $15^2 + 8^2$. Now if a and b are any two positive whole numbers, compare $(a+b)^2$ and $a^2 + b^2$ and explain why your answers to (i)–(iii) are correct.

Other applications of the assertions in inequalities (2.7) will be found throughout this book.

2.6. An Application of the Associative and Commutative Laws of Addition

We give the formal argument that the identities (2.2) and (2.3) for all numbers ℓ, m, n, lead logically to equation (2.1). We will in fact prove a little bit more, namely, that for any four numbers k, ℓ, m, n, the following holds:

(2.8) $\quad \big((k+\ell)+m\big)+n = \big((\ell+n)+m\big)+k = (\ell+n)+(m+k).$

If we let $k = 32$, $\ell = 25$, $m = 28$, and $n = 55$, then equation (2.8) becomes equation (2.1).

To prove equation (2.8), recall the statements of equations (2.2) and (2.3):

(2.2) $\qquad\qquad (\ell + m) + n = \ell + (m + n),$
(2.3) $\qquad\qquad m + n = n + m.$

Applying equation (2.2) to the three numbers $\ell + n$, m, and k, we get

$$\big((\ell+n)+m\big)+k = (\ell+n)+(m+k),$$

which is exactly the second equality in equation (2.8). To prove the first equality, we do as follows:

$$\begin{aligned}
\big((k+\ell)+m\big)+n &= (k+\ell)+(m+n) &&\text{(apply (2.2) to } k+\ell,\, m,\, n)\\
&= (k+\ell)+(n+m) &&\text{(apply (2.3) to } n,\, m)\\
&= k+\big(\ell+(n+m)\big) &&\text{(apply (2.2) to } k,\, \ell,\, n+m)\\
&= k+\big((\ell+n)+m\big) &&\text{(apply (2.2) to } \ell,\, n,\, m)\\
&= \big((\ell+n)+m\big)+k &&\text{(apply (2.3) to } k,\, \big((\ell+n)+m\big)).
\end{aligned}$$

This proves the first equality in equation (2.8), and therefore equation (2.8) itself.

This kind of formal argument is offered primarily for your benefit as a teacher or prospective teacher, and is not meant to be indiscriminately imposed on your elementary students. *Hamlet* may be great literature, but it is not the best thing to read to a toddler at bedtime. In the fifth or sixth grade, there may be occasion for you to introduce this kind of reasoning, gently. But a teacher must exercise good judgment in not overdoing anything.

Exercises

There are two rules about doing the exercises in this book: (i) Unless stated to the contrary, use only what you have learned so far in the book. (ii) Every answer must come with an explanation. The explanation may be in the form of the details in a calculation, or it may be a verbal citation of particular facts used in the text. Sometimes, for emphasis, an explanation is even explicitly demanded. But whatever it is, you will have to get used to never claiming anything without giving a reason.

1. Elaine has 11 jars, in each of which she put 16 ping pong balls. One day she decided to redistribute all her ping pong balls equally among 16 jars instead. How many balls are in each jar? Explain.

2. Before you get too comfortable with the idea that everything in this world has to be commutative, consider the following: (i) Let A_1 stand for "put socks on" and A_2 for "put shoes on", and let $A_1 \circ A_2$ be "do A_2 first, and then A_1", and similarly let $A_2 \circ A_1$ be "do A_1 first, and then A_2".[3] Convince yourself that $A_1 \circ A_2$ does not have the same effect as $A_2 \circ A_1$. (ii) For any whole number k, introduce the notation, $B_1(k)$ is the number obtained from k by adding 2 to k. Thus $B_1(7) = 7+2$, and $B_1(726) = 726 + 2$. Similarly, let $B_2(k)$ be the number obtained from k by multiplying k by 5. For example, $B_2(3) = 3 \times 5$ while $B_2(79) = 79 \times 5$. Now, show that no matter what the number n may be, $B_1(B_2(n)) \neq B_2(B_1(n))$.

3. Find shortcuts to do each of the following computations and give reasons (associative law of addition, commutative law of multiplication, etc.) for each step: (i) $833 + (5167 + 8499)$, (ii) $(54 + 69978) + 46$, (iii) $(25 \times 7687) \times 80$, (iv) $(58679 \times 762) + (58679 \times 238)$, (v) $(4 \times 4 \times 4 \times 4 \times 4) \times (5 \times 5 \times 5 \times 5 \times 5)$, (vi) 64×125, (vii) $(69 \times 127) + (873 \times 69)$, (viii) $((125 \times 24) \times 674) + ((24 \times 125) \times 326)$.

   The purpose of the last exercise is not to get you obsessed with tricks in computations. Tricks are nice to have, but they are not the main goal of a mathematics education, contrary to what some people would have you believe. What this exercise tries to do is, rather, to make you realize that the basic laws of operations discussed in this section are more than empty, abstract gestures. They have practical applications too.

4. Prove the remaining three assertions in (2.7).

---

[3] You would think that $A_1 \circ A_2$ means "do A_1 first, and then A_2", but there are powerful reasons in mathematics, related to the notation of composing functions, that make the interpretation "do A_2 first, and then A_1" more natural.

5. Introduce the symbol $\leq$ between numbers as follows: for any two numbers a and b, we say $a \leq b$ if $a < b$ or $a = b$. (This symbol is sometimes referred to as a **weak inequality**.) Prove that all the assertions in (2.7) remain true if the strict inequality symbol "$<$" is replaced by the weak inequality symbol "$\leq$".

6. True or false: "For whole numbers a and b, the fact that $a \leq b$ is equivalent to the fact that $b = a + \ell$ for some whole number ℓ." Explain.

7. What is the smallest possible area of a rectangle whose length and width are whole numbers which add up to 24? How big can the area be?

8. Let x and y be two whole numbers. (i) Explain why $(x+y)(x+y) = x(x+y) + y(x+y)$. (ii) Explain why $(x+y)(x+y) = xx + xy + yx + yy$. (iii) Explain why $(x+y)^2 = x^2 + 2xy + y^2$. (iv) Let x and y be nonzero; compare $(x+y)^2$ with $x^2 + y^2$ and explain in two ways why one is always bigger than the other—by a direct computation and by drawing a picture (cf. the Activity on page 51).

9. The following is how a fourth grade textbook introduces the *associative law of multiplication*.

    > Ramon buys yo-yos from two companies. He buys six different styles from each company and gets each style in 4 different colors. How many yo-yos does he buy in all?
    > Find 2×6×4 to solve. You can use the ASSOCIATIVE PROPERTY to multiply three factors. The grouping of the numbers does not affect the answer.
    > Step 1: Use parentheses to show grouping.
    > $$2 \times 6 \times 4 = (2 \times 6) \times 4$$
    > Step 2: Look for a known fact to multiply.
    > $$2 \times 4 \text{ is a known fact.}$$
    > Step 3: Use the COMMUTATIVE PROPERTY to change the order, if necessary.
    > $$(2 \times 6) \times 4 = (6 \times 2) \times 4$$
    > $$= 6 \times (2 \times 4)$$
    > $$= 6 \times 8 = 48$$

    Write down your reaction to such an introduction, and compare with those of others in your class.

10. Using only the laws for multiplication (equations (2.4) and (2.5) on page 43), write down an explanation of why, for four numbers ℓ, m, n, p, it is true that $\ell(m(np)) = (p\ell)(nm)$.

11. Using only the laws for addition (equations (2.2) and (2.3) on page 42), directly show that for four numbers ℓ, m, n, p, $(\ell+m) + (n+p) = (p + (\ell+n)) + m$.

Exercises

12. Let m and n be a 3-digit number and a 2-digit number, respectively. Can mn be a 4-digit number? 5-digit number? 6-digit number? 7-digit number? Explain your answer.

13. Let m and n be a k-digit number and an ℓ-digit number, respectively, where k and ℓ are positive whole numbers. How many digits can the number mn have? List all the possibilities and explain.

14. Use mental math to decide which of the following is bigger: (a) 648×427 or 649×426? (b) 207×816 or 206×819?

15. Suppose you have a calculator which displays only 8 digits (and if you have a fancy calculator, you will be allowed to use only 8 digits!), but you have to calculate 856164298×65. Discuss an efficient method to make use of the calculator to help with the computation. Explain. Do the same for 376241048×872.

16. How would you compute the square of 9,458,647,683 on a calculator with a 12-digit display?

Chapter 3

The Standard Algorithms

In the context of school mathematics, an **algorithm** is a finite sequence of explicitly defined, step-by-step computational procedures which ends in a clearly defined outcome. The purpose of this chapter is to give an overview of the so-called **standard algorithms** for the four arithmetic operations among whole numbers. The succeeding chapters will provide the mathematical explanations.

At the outset, we should make clear that there is no such thing as *the unique* standard algorithm for any of the four operations $+, -, \times, \div$, because minor variations have been incorporated into the algorithms by various countries and ethnic groups. It may also be mentioned in passing that computer programs often make use of diverse algorithms that are derived from various cultures' pencil-and-paper algorithms.[1] Such variations notwithstanding, the underlying mathematical ideas always remain the same and, to the extent that our focus is on the underlying mathematical ideas and not the explicit procedures, the nomenclature of "standard algorithms" is eminently justified. Now, to say that we will focus on the mathematical ideas is not to say that the algorithms themselves—the computational procedures—are of no interest. On the contrary, they *are*, because computational techniques are an integral part of mathematics. Furthermore, the conciseness of these algorithms, especially the multiplication algorithm and the long division algorithm, is a marvel of human invention, and one of the goals of this chapter is to make sure that you come away with a renewed respect for them. An appreciation of the standard algorithms from a slightly different perspective is given in section 11.3, Arithmetic in Base 7, (page 161).

---

[1] I am indebted to Ken Ross for this observation.

A fundamental question about these arithmetic algorithms is, *Why should you bother to learn them?* Take a simple example: What is 17 × 12? By definition, this is 12 added to itself 17 times and, in the 1990s, there was a curriculum that would have you count 17 piles of birdseed with 12 in each pile to get an answer. Indeed, if we are limited to such simple computations, one *may* be able to get away with not knowing any algorithm at all. But we are concerned with the computations for *all* whole numbers, no matter how large. For example, what about 34,609 × 549,728? Are you going to tell your students to count 34,609 piles of birdseed with 549,728 in each pile? So you see that in this case, a shortcut is clearly called for. This is where the algorithms come in: they provide a shortcut in lieu of direct counting. It takes mathematical insight into the Hindu-Arabic numeral system to arrive at these shortcuts. To learn the algorithms is to learn such insight.

The preceding discussion also explains why we would be interested in the *efficiency* of an algorithm, i.e., how to get the answer as simply and as quickly as possible. At this point you may ask, Why worry about efficiency if pushing buttons on a calculator is a very efficient way to perform a computation such as 34,609 × 549,728? There are at least two reasons why, from the point of view of mathematics education, we cannot afford to let students be completely dependent on the calculator for whole-number computations. First, without a firm grasp of the place value of our numeral system and the logical underpinning of the algorithms, it would be impossible to detect mistakes caused by pushing the wrong buttons on a calculator.[2] A more important reason is that *in mathematics learning a fact is synonymous with learning why it is true*. In the case at hand, learning the reasoning behind these efficient algorithms is a very compelling way to acquire many of the fundamental skills in mathematics, including abstract reasoning with the basic laws of operations in Chapter 2 and the ability to make deductions from precise definitions. These are skills that are absolutely essential for the understanding of fractions and decimals in the subsequent chapters and, looking further ahead, for the understanding of algebra in middle school. One can flatly state that, if students do not feel comfortable with the kind of mathematical reasoning used to justify the standard algorithms for whole numbers, then their chance of success in algebra is minimal.

More can be said along this line. If we want to expose students to mathematical reasoning early, then exposing them to the inner workings of these algorithms would be a splendid starting point. A dominant theme that runs through these algorithms is one that is also part of the basic tools of research mathematicians, namely, that whenever possible, *break down a*

---

[2]I trust that it would be unnecessary to recount the many horror stories related to a finger-on-the-wrong-button computations.

complicated task into simple subtasks. To be specific, we formulate a *leitmotif* of the standard algorithms:

> **Leitmotif.** *To perform a computation with multidigit numbers, break it down into several steps so that each step (when suitably interpreted) is a computation involving only a single digit.*

Therefore, a virtue of the standard algorithms is that, when they are properly executed, they allow students to ignore the individual numbers being computed, no matter how large, and concentrate instead on one digit at a time. This is an excellent example of the kind of abstract thinking that is critical to success in mathematics learning. If students can learn from this *leitmotif* how to break down the complex into the simple, they will gain a foothold in mastering algebra and advanced mathematics.

Ironically, it is precisely this virtue of being able to perform whole number computations by ignoring place value that has stirred up controversy in mathematics education. One objection to the teaching of the standard algorithms is that by making children focus on one digit at a time, they lose all sense of place value and consequently become prone to computational errors. The other is that the routine nature of the single-digit computations promotes the suspension of thinking, and if anything can be done without thinking, then it does not belong in a mathematics classroom. As a classroom teacher, you must confront both kinds of erroneous thinking.

The fear that teaching the standard algorithms would cause a loss of the sense of place value can only be founded on the common educational practice of teaching the standard algorithms *without also teaching the reasoning underlying these algorithms.* The fact that such harmful practices are common is well known, but less known is the fact that universities have not given prospective teachers the mathematical support they need in order to get them out of these harmful practices (see [**Wu99b**] for a more extended discussion). A main reason for the writing of the present book is, in fact, to address this very issue of mathematical support for teachers. This is why we have stressed throughout this book the importance of mathematical reasoning. Furthermore, while educators may look askance at the *routine* and *nonthinking* nature of these algorithms, the fact remains that it is the routine nature that accounts for their usefulness. To teach these algorithms without emphasizing their routine character is to miss the point of these algorithms completely, to say nothing of falsifying the mathematics.

The other objection is based on the perception that if anything can be done without thinking, then it does not belong in a mathematics classroom. *This is wrong.* If mathematicians are forced to do mathematics by having to think *every* step of the way, then little mathematics of value would ever get done and all research mathematics departments would have to close shop. What is closer to the truth is that deep understanding of a topic tends to reduce many of its sophisticated processes to simple mechanical procedures. The ease of executing these mechanical procedures then frees up mental energy to make possible the conquest of new topics through imagination and mathematical reasoning. In turn, many of these new topics will (eventually) be themselves reduced to routine or nearly routine procedures, and the process repeats itself. There is nothing to fear about the ability to execute a correct mathematical procedure without thinking unless the fluency of execution is not supported by a thorough understanding of why the procedure is valid. A teacher's charge in the classroom is therefore to promote both the facility with procedures and the ability to reason. The teaching of these algorithms must emphasize both their routine, nonthinking nature as well as the logical reasoning that lies behind the procedures.

The preceding discussion is about the kind of mathematical understanding teachers of mathematics must have in approaching the basic arithmetic algorithms. The pedagogical issue of how to introduce these algorithms to children in the early grades is something that lies outside the scope of this book and needs to be treated separately. See, however, [**Wu99a**] for a discussion of this issue from a mathematical point of view concerning the addition and multiplication algorithms.

Chapter 4

The Addition Algorithm

The sections are as follows:

> The Basic Idea of the Algorithm
> The Addition Algorithm and Its Explanation
> Essential Remarks on the Addition Algorithm

4.1. The Basic Idea of the Algorithm

Consider $263 + 4502$. By the definition of addition, we get the answer by counting 4502 steps starting at 263. This can be quite trying! However, a special feature of the Hindu-Arabic numeral system, namely, the fact that each digit of its numerals "comes prepackaged with its place value", renders such a desperate act completely unnecessary. The technique involved is called the *addition algorithm*. Let us use a simpler example to explain what is meant by the phrase in quotes. Suppose we have two sacks of potatoes, one containing 34 and the other 25 potatoes, and we want to know how many potatoes there are altogether. One way is to dump the contents of both sacks on the floor and count. But suppose upon opening the sacks, we find that in each sack, the potatoes come in bags of 10 plus some stragglers, i.e., the 34 potatoes have already been put in 3 bags of 10 each plus 4 stray ones, while the 25 potatoes in the other sack have been put in 2 bags of 10 each plus 5 stray ones. Therefore, an intelligent way to count the total number of potatoes would be to first count the total number of bags of 10's ($3 + 2$ is 5, so there are 5 bags of 10 each), and then count the stray ones separately ($4 + 5$ is 9, and so there are 9 extra). Thus the total is 5 bags of 10 each, plus 9 extra ones, which puts the total at 59. *This is exactly the idea behind the addition algorithm* because the number 3 in 34—being in the tens place—signals that there are 3 tens, and the 2 in 25 signals that there are 2 tens. Adding 2 and 3 to get 5, we know that there are 5 tens in the total. Adding 4 to 5 then rounds off the whole sum, and we get $34+25 = 59$.

There is another way to bring out this special feature of the Hindu-Arabic numeral system in the addition $34 + 25 = 59$: by the use of money. Imagine that someone has a stack of 34 one-dollar bills and another stack of 25 one-dollar bills. To find the total amount in these two stacks, she would have no choice but to do continued counting. *But*, the 34 dollars usually come in the form of

$$\begin{aligned} &3 \quad \text{ten-dollar bills, and} \\ &4 \quad \text{one-dollar bills,} \end{aligned}$$

and the 25 dollars also in the form of

$$\begin{aligned} &2 \quad \text{ten-dollar bills, and} \\ &5 \quad \text{one-dollar bills.} \end{aligned}$$

(This is exactly what is meant by 34 and 25 being "prepackaged with place value".) To find out how much money she has altogether, she only has to collect all the ten-dollar bills, and then collect all the one-dollar bills. She finds that she now has

$$\begin{aligned} &5 \, (=3+2) \quad \text{ten-dollar bills,} \\ &9 \, (=4+5) \quad \text{one-dollar bills.} \end{aligned}$$

So she has $59, exactly as before.

Activity. Suppose you want to buy twenty-five 39-cent stamps, twenty 24-cent stamps, ten 63-cent airmail stamps, and thirty 84-cent airmail stamps.[1] (a) Forget about the addition algorithm and think of the normal way you would compute the total amount of money it takes to buy this many stamps. How would you do this computation? (b) Now that you have done it, reflect on your computation and draw a parallel between it and the addition algorithm.

4.2. The Addition Algorithm and Its Explanation

The **standard addition algorithm** for two numbers is nothing more than a formal elaboration of the simple idea already explained in the preceding section. Formally, we assume that *the addition of two single-digit numbers is known*.[2] To find the sum of two arbitrary numbers, the algorithm states: Add their digits in the ones place, then add their digits in the tens place, then add their digits in the hundreds place, etc. The resulting numbers with these digits in the respective places is their sum.[3]

---

[1] Note: These are the most common postal stamps in the year 2006.

[2] This is why children have to learn how to add two single-digit numbers.

[3] In the process, you will note the ubiquitous appearance of Theorem 2.1 of Chapter 2 in the discussion.

4.2. The Addition Algorithm and Its Explanation

In the usual setup of stacking the numbers to be added vertically so that digits with the same place values are aligned, the algorithm can be restated as follows:

> The sum can be computed by adding the digits *column-wise* if the numbers are aligned so that their ones digits are in the extreme right column.

As an illustration of this well-known algorithm, consider, for example, $865 + 32$. According to the algorithm, we can add as follows:

(4.1)
$$\begin{array}{r} 8\ 6\ 5 \\ +\ \ \ \ 3\ 2 \\ \hline 8\ 9\ 7 \end{array}$$

Note that we are treating 32 as 032, so that the addition in the left column is actually $8 + 0$.

We now explain why the procedure exhibited in (4.1) gives the correct answer by making use of the expanded form of 865 and 32:

$$\begin{aligned} 865 &= 800 + 60 + 5 \\ 32 &= 0 + 30 + 2 \end{aligned}$$

Therefore, $865 + 32$ is equal to the sum of the right sides:

$$(800 + 60 + 5) + (0 + 30 + 2).$$

By Theorem 2.1 of Chapter 2, we can write this sum as

$$865 + 32 = (800 + 0) + (60 + 30) + (5 + 2).$$

But this expresses $865 + 32$ exactly as the column-wise addition of the digits in (4.1).

Activity. Compute $4502 + 273$ by the algorithm, and give an explanation to your neighbor of why it is correct.

In the example of (4.1), of course the digits in each column add up to a single-digit number and the algorithm yields the correct answer (as we have just seen). But sometimes the sum of the digits in the same column is bigger than a one-digit number, e.g., $765 + 892$, where $7 + 8 > 9$ and $6 + 9 > 9$. In that case, the algorithm requires a supplement. For the sum of two numbers, the algorithm says:

> *Do the column-wise computation from right to left. If the sum in a column exceeds 10, say 17, then enter 7 in this column, but in computing the sum of the digits in the next column (to the left of the current column), add 1.*

For example, to compute $765 + 892$, the algorithm gives:

(4.2)
$$
\begin{array}{r}
7\ 6\ 5 \\
8\ 9\ 2 \\
+\ 1\ 1 \\
\hline
1\ 6\ 5\ 7
\end{array}
$$

How we got the 1 in 1657 may be more clearly seen if we recall that $765 = 0765$ and $892 = 0892$, so that the left column is in fact the addition $0+0+1$, as in the following:

$$
\begin{array}{r}
0\ 7\ 6\ 5 \\
0\ 8\ 9\ 2 \\
+\ 1\ 1 \\
\hline
1\ 6\ 5\ 7
\end{array}
$$

Incidentally, this way of recording the carrying of the 1 to the next (left) column by entering it as a smaller number underneath is highly recommended; it reminds young kids not to forget to add the extra 1.

The explanation of the phenomenon of **carrying** in (4.2), i.e., the entering of a 1 in the column to the left, is no different. Again, the expanded forms of 765 and 892 are:

$$
\begin{aligned}
765 &= 700 + 60 + 5 \\
892 &= 800 + 90 + 2
\end{aligned}
$$

The sum $765 + 892$ is therefore equal to the sum of the right sides. Making use of Theorem 2.1 on page 41 as before, we see that the right sides add up to

$$(700 + 800) + (60 + 90) + (5 + 2).$$

The sum on the right, $5 + 2$, is just the sum in the right column of (4.2); it is 7. The sum inside the middle parentheses, $60 + 90$, is exactly the sum of the digits in the second column (from the right) of (4.2), and it is

$$60 + 90 = 100 + 50.$$

Therefore, $765 + 892$ is equal to

$$(700 + 800) + (100 + 50) + (5 + 2).$$

We use Theorem 2.1 once more to rewrite this as

$$(700 + 800 + 100) + 50 + 7.$$

The sum inside the first pair of parentheses, $700 + 800 + 100$, is exactly the sum of the digits in the third column from the right of (4.2), including the carrying of 1, and it is equal to

$$700 + 800 + 100 = 1000 + 600.$$

The "1000" in the sum $1000 + 600$ explains the carry of 1 in the left column of (4.2). Thus we have
$$765 + 892 = 1000 + 600 + 50 + 7.$$
The sum on the right is of course the expanded form of 1657, which is exactly what we see in the bottom row of (4.2).

Activity. Explain to your neighbor why the addition algorithm for $95 + 46$ works.

We have used two concrete examples to explain the underlying reasoning of the addition algorithm, at least when *two* numbers are involved (the case of more than two numbers will be dealt with in the next section). It remains to observe that the reasoning behind these two simple examples is perfectly general. The precise values of the digits in examples (4.1) and (4.2) played no role at all. In fact, the mathematical explanations of this and the subsequent standard algorithms illustrate the point that the use of abstract symbols in mathematics has its limitations: if we had used letters to represent general numbers instead of using 865, 32, 765, and 892, the explanation of the addition algorithm would have been a nightmare.

We round off the discussion by making contact with the *leitmotif* on page 59, namely, the fact that the addition algorithm reduces computations to single-digit ones. Because we chose, for the sake of simplicity, to employ the ordinary expanded form of a number rather than the expanded form using exponents in this discussion, the preceding mathematical explanation of the algorithm did not exhibit this reduction as clearly as possible. We will make amends here. Consider the first example, $865 + 32$. We saw that
$$865 + 32 = (800 + 0) + (60 + 30) + (5 + 2).$$
Using exponential notation, we have $865 = (8 \times 10^2) + (6 \times 10^1) + (5 \times 10^0)$ and $32 = (0 \times 10^2) + (3 \times 10^1) + (2 \times 10^0)$, so that the preceding equality can be rewritten as:

$$\begin{aligned}
865 + 32 &= \left((8 \times 10^2) + (0 \times 10^2)\right) + \left((6 \times 10^1) + (3 \times 10^1)\right) \\
&\quad + \left((5 \times 10^0) + (2 \times 10^0)\right) \quad\quad\quad\quad\quad\quad\quad\quad \text{(Thm. 2.1)} \\
&= (8 + 0) \times 10^2 + (6 + 3) \times 10^1 + (5 + 2) \times 10^0 \quad \text{(dist. law)}
\end{aligned}$$

The last sum now clearly displays the computation of $865 + 32$ as three separate sums of single digits: $8 + 0$, $6 + 3$, and $5 + 2$.

The same can be said of $765 + 892$, the presence of the two carries notwithstanding.

4.3. Essential Remarks on the Addition Algorithm

(i) The first remark is the obvious one: knowing how to add two single-digit numbers is a prerequisite to knowing how to execute the addition algorithm for two whole numbers. This is nothing other than a reinforcement of the *leitmotif* of Chapter 3, The Standard Algorithms.

(ii) We also see that no explanation of the addition algorithm is possible without going through place value because the algorithm itself is built on the concept of place value. Yet it is equally important to recognize a seemingly contradictory feature of this algorithm, namely, that *each step of the algorithm is strictly limited to the consideration of a single digit without regard to its place value*. (Reread the *leitmotif* on page 59 at this point.) To illustrate, consider the following two additions:

$$
\begin{array}{r} 4\ 5\ 0\ 2 \\ +\quad\ 2\ 6\ 3 \\ \hline 4\ 7\ 6\ 5 \end{array}
\qquad
\begin{array}{r} 8\ 6\ 5 \\ +\quad\ 3\ 2 \\ \hline 8\ 9\ 7 \end{array}
$$

Notice that the third column from the right of the left item and the right column of the right item are identical:

(4.3)
$$
\begin{array}{r} 5 \\ +\ \ 2 \\ \hline 7 \end{array}
$$

Yet, in terms of place value, we know that in the context of $4502+263$, the addition fact (4.3) actually stands for

$$
\begin{array}{r} 5\ 0\ 0 \\ +\ \ 2\ 0\ 0 \\ \hline 7\ 0\ 0 \end{array}
$$

because the 5 in 4502 stands for 500 and the 2 in 263 stands for 200. By contrast, in the context of $865 + 32$, the same addition fact (4.3) is literally true: it is just $5+2=7$. *As far as the algorithm is concerned*, however, the addition fact (4.3) is carried out in exactly the same way in both cases of $4502 + 263$ and $865 + 32$ without regard to this difference. This procedural simplicity, the fact that *an addition such as* (4.3) *can be carried out regardless of the place values of* 5 *and* 2, is one reason that accounts for the usefulness of this algorithm.

(iii) Looking back at the explanation of the addition algorithm, we see that the call for column-wise addition is really the call for adding the digits with the same place value. It is self-evident at this point that we may summarize the algorithm as follows:

> The addition algorithm for adding two numbers is the method of adding their digits corresponding to the same place value. Working from right to left, if the sum of two such corresponding digits is a single-digit number, record this sum as is. If, however, this sum is equal to $10 + k$ for some single-digit number k, then record k as the sum in this place but add 1 to the sum of the digits in the next place to the left.

4.3. Essential Remarks on the Addition Algorithm

(iv) So far, we have only handled the case of the addition algorithm for two numbers. Is the algorithm for more than two numbers substantively different? Not at all. Let us state the algorithm for the sum of at most ten numbers.[4]

The addition algorithm for adding n numbers ($n \leq 10$) is the method of adding the digits with the same place value. Working from right to left, if the sum of all the digits with a fixed place value is a single-digit number, record the sum as is. If, however, this sum exceeds 10, let us say it is 57, then record 7 as the sum for this place value, but add 5 to the sum of the digits in the next place value to the left.

We illustrate with $165 + 27 + 83 + 829$. Here is how the algorithm works out:

$$\begin{array}{r} 1\ 6\ 5 \\ 2\ 7 \\ 8\ 3 \\ 8\ 2\ 9 \\ +\ \ \ 1\ 2\ 2 \\ \hline 1\ 1\ 0\ 4 \end{array}$$

So we start from the right and $5 + 7 + 3 + 9 = 24$ means the ones digit of the sum would be 4, and we enter 2 in the tens column, etc.

Activity. Verify the preceding computation, and give an explanation by following the explanations of addition examples (4.1) and (4.2).

Finally, we deal with an issue that is important to the teaching of this and other standard algorithms: Would the fact that the algorithm moves from right to left make it difficult for children to learn the algorithm? Not if the need to move from right to left is explained to them. Suppose we move from *left to right* instead in the preceding addition of four numbers. It will be noted that the amount of *backtracking* becomes intolerable. So we first get $1 + 8 = 9$. Then in the next column to the right, we get $6 + 2 + 8 + 2 = 18$, which tells us we should go back to the left column and change the sum from 9 to $9 + 1 = 10$, which in turn tells us that we should have a 0 in the hundreds place and a 1 in the thousands place. Thus far we get 108, with the last column on the right yet to be added. But that turns out to be $5 + 7 + 3 + 9 = 24$, which means the ones digit of the final answer should be a 4, but that the tens column must be modified to be $(6 + 2 + 8 + 2) + 2 = 20$ on account of the 2 in 24. Now this means the tens digit of the final answer

---

[4]The reason for the restriction to ≤ 10 numbers is to avoid dealing with carrying to more than one column to the left, i.e., to make sure that the sum in a given column is < 100. This simplifies the statement of the algorithm. The principle is of course the same for any collection of numbers.

is 0 (because of the 20) and not 8, and we have to change the digit in the hundreds column too because the sum in this column is now $1+8+2=11$. So the hundreds digit of the final answer is 1 and not 0 while the thousands digit remains 1. After all that shuttling back and forth, we finally get 1104, as before.

This and other examples such as Exercise 3 below should suffice to convince your students that it is in their best interest to add from right to left.

Pedagogical Comments. *The cornerstone of the explanation of the addition algorithm is Theorem 2.1 of Chapter 2, as should be obvious from the explanations of addition examples (4.1) and (4.2). In other words, the associative and commutative laws of addition form the backbone of this algorithm. Should a teacher in teaching this and other algorithms make copious references to these laws? Few questions in education can be answered with absolute certainty, but there are at least two reasons why an* elaborate *discussion of these laws in, say, K–5 might interfere with a good mathematics education. First, such explanations tend to be tedious and students at an early age might lose interest. Second, such details might obscure the main thrust of the algorithm, which is that the addition of whole numbers is essentially the addition of single-digit numbers. As a teacher, however, you owe it to yourself and to your students to understand the importance of these laws. This is because, on the one hand, intellectual honesty demands it and, on the other hand, you must be prepared in case a precocious youngster presses you for the complete explanation.* **End of Pedagogical Comments.**

Exercises

There are two rules about doing the exercises in this book: (i) Unless stated to the contrary, use only what you have learned so far in the book. (ii) Every answer must come with an explanation. The explanation may be in the form of the details in a calculation, or it may be a verbal citation of particular facts used in the text. Sometimes, for emphasis, an explanation is even explicitly demanded. But whatever it is, you will have to get used to never claiming anything without giving a reason.

1. Explain to a fourth grader why the addition algorithm for $7032 + 845$ is correct, first using the explanation for addition example (4.1), then using money. (See section 4.1, **The Basic Idea of the Algorithm**, on page 61.)

2. Use the number line and the concatenation of segments to show that $37 + 128 = 165$. Then give an explanation of why this is correct using Theorem 2.1 of Chapter 2.

3. Do the addition $67579 + 84937$ both ways—from left to right and from right to left—and compare the work involved.

4. I have three numbers A, B, and C. A is 31986, B exceeds A by 2308, while C exceeds B by 8205. What is $B + (A + C)$?

5. Compute $123 + 69 + 528 + 4$ by the addition algorithm, and give an explanation of why the computation is correct.

6. Compute $7826+7826+7826+7826+7826$ by the addition algorithm, and give an explanation of why the computation is correct. (This exercise should whet your appetite for Chapter 6, **The Multiplication Algorithm**.)

7. Compute $172{,}993 + 90{,}008$ by the addition algorithm, and give an explanation of why the computation is correct.

8. Use the addition algorithm to compute $270{,}010{,}060{,}001 + 80{,}930{,}040$, and give a mathematical explanation. (*Hint*: You should give serious thought to using exponential notation.) Discuss how clumsy it would be to explain the working of the addition algorithm in this case by the use of money. (See section 4.1, **The Basic Idea of the Algorithm**, on page 61.)

9. (a) What is the sum of all the whole numbers from 1 to 28, i.e., $1 + 2 + 3 + \cdots + 27 + 28$? (b) What is the sum of all the whole numbers from 1 to 33? (c) Define a whole number to be **even** if it is equal to 2 times another whole number.[5] A whole number is said to be **odd** if it is equal to 1 plus an even number. If n is an odd number, what is the sum of all

---

[5] We will formally define "even" and "odd" in Part 4.

the whole numbers from 1 to n in terms of n? (d) The same question as in (c) for an even whole number n.

10. (Continuation of Exercise 9) (a) The first nine even numbers are 2, 4, 6, 8, 10, 12, 14, 16, 18. Find their sum. (b) What is the sum of the first n even numbers in terms of n? (c) The first twelve odd numbers are 1, 3, 5, 7, 9, 11, 13, 15, 17, 19, 21, 23. Find their sum. (d) Find the sum of the first n odd numbers in terms of n.

Chapter 5

The Subtraction Algorithm

Thus far we have not given a *precise* meaning of subtraction. We do so in this chapter. While subtraction is, in essence, nothing more than a different way of writing addition (see equation (5.1) on the next page), there is a subtle difference. *In the context of whole numbers,* we can add any two numbers but we cannot subtract any two numbers because there is no whole number that is equal to, for example, $3-7$. The freedom to subtract any two whole numbers only comes after we have enlarged the whole numbers to the integers; see Chapter 27. This awareness about the restrictive nature of subtraction among whole numbers will be helpful in understanding the concept of division among whole numbers (Chapter 7).

The sections are as follows:

> Definition of Subtraction
>
> The Subtraction Algorithm
>
> Explanation of the Algorithm
>
> How to Use the Number Line
>
> A Special Algorithm
>
> A Property of Subtraction

5.1. Definition of Subtraction

Let us begin with an informal discussion about subtraction. The subtraction of 15 from 37, $37-15$, is the number of steps it takes to go from 15 to 37. In terms of the number line, each step between whole numbers has unit length. Therefore, this means $(37 - 15)$ is the length of the line segment $[15, 37]$:

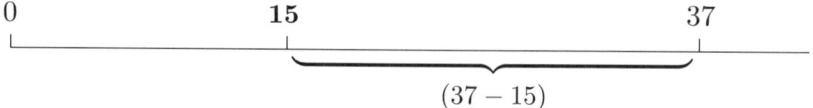

This picture exhibits the fact that the concatenation of $[0, 15]$ and a segment of length $(37-15)$ has length 37. Recalling the meaning of addition in terms of the concatenation of segments, we arrive at the fact that $37 - 15$ *is the number so that* $37 = 15+(37-15)$, or, since addition is commutative, $37-15$ *is the number so that* $37 = (37 - 15) + 15$. The geometric interpretation of the latter is the following:

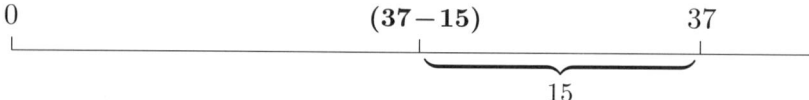

To summarize these two pictures, we see that $(37 - 15)$ *is the length of the segment left behind when a segment of length* 15 *is removed from either end of* $[0, 37]$.

This discussion gives the motivation for the following formal definition of subtraction. First, we agree to use $\boldsymbol{m \geq n}$ to denote "either $m > n$ or $m = n$". (Similarly, $\boldsymbol{m \leq n}$ means "either $m < n$ or $m = n$"; see Exercise 5 on page 54.) Now, if m, n are whole numbers and $m \geq n$, we define the **subtraction of $\boldsymbol{n}$ from $\boldsymbol{m}$**, in symbols $\boldsymbol{m - n}$, as the whole number so that when it is added to n we get back m. In symbols, for any whole numbers m, n, with $m \geq n$, if k denotes $m - n$, then by definition,

$$k + n = m.$$

Since addition is commutative, the formal definition can be equivalently rephrased as

(5.1) $\boldsymbol{m - n}$ is the whole number k that satisfies $m = k + n$.

It is common also to refer to $m - n$ as the **difference between** m and n.

The formal definition (5.1) exhibits *subtraction as an alternate but equivalent way of expressing addition*, in the sense that, if $m \geq n$, then instead of writing $m = k + n$ we can write $m - n = k$, and vice versa.

Activity. Use the formal definition of subtraction to compute and explain: $1200 - 500 = ?$; $580{,}000{,}000 - 500{,}000{,}000 = ?$; $580{,}000{,}000 - 20{,}000{,}000 = ?$; $15 \times 10^6 - 7 \times 10^6 = ?$

5.2. The Subtraction Algorithm

It follows from the definition that $m = (m-n) + n$. So in terms of the number line, $m - n$ has the following geometric meaning:

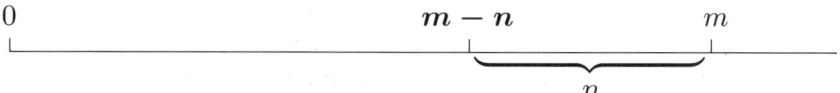

Since it is equally valid that $m = n + (m-n)$, we also have the following picture:

Therefore, an equivalent definition of $m - n$ for whole numbers m, n satisfying $m \geq n$ is that ***$m - n$ is the length of the remaining segment when a segment of length n is removed from either end of $[0, m]$***.

In Chapter 2, we discussed the distributive law. It should be mentioned that there is also a **distributive law for subtraction**: For all whole numbers k, m, n where $m \geq n$,

$$k(m - n) = km - kn.$$

If $k = 3$, $m = 6$, and $n = 2$, then this law says $3 \times (6-2) = (3 \times 6) - (3 \times 2)$. The truth of this assertion for the area model can be immediately seen from following picture.

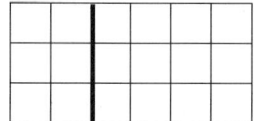

At least for the area model, the reasoning for the general case is the same; see also Exercise 15 at the end of this section.

We hasten to add that this is not a new law of operations. When we have negative numbers at our disposal in Part 3, we will show (page 411) that the distributive law for subtraction is included in the ordinary distributive law.

5.2. The Subtraction Algorithm

According to the definition of subtraction in the preceding section, the difference of two numbers, such as $1658 - 257$, is obtained by counting the number of steps it takes to go from 257 to 1658. As is the case with the addition algorithm, the purpose of the standard subtraction algorithm is to relieve the tedium of counting by offering a shortcut. The mathematics underlying this algorithm (to be introduced presently) is similar to that of the addition algorithm.

First, *assume that we know how to do single-digit subtraction.* (See the *leitmotif* on page 59) Now, let m, n be arbitrary whole numbers so that $m > n$. We want to compute $m - n$. Now put the number m *above* n, and line them up so that digits in the same column have the same place value. For example, $1658 - 257$ would be configured as:

$$\begin{array}{r} 1\ 6\ 5\ 8 \\ -\ 2\ 5\ 7 \\ \hline ?\ ?\ ?\ ? \end{array}$$

As in the case of the addition algorithm, we begin with the simplest case, where each digit of m is at least as big as the digit of n in the same column. The preceding example of $1658 - 257$, which as we know is the same as $1658 - 0257$, illustrates the simple case that we have in mind, because $1 \geq 0$, $6 \geq 2$, $5 \geq 5$, and $8 \geq 7$. In this simple case, the **subtraction algorithm** states that *the digits of $m - n$ are obtained by performing the single-digit subtraction in each column.*[1] Thus, from the four single-digit subtractions $8 - 7 = 1$, $5 - 5 = 0$, $6 - 2 = 4$, and $1 - 0 = 1$, we conclude $1658 - 257 = 1401$. Schematically:

(5.2)
$$\begin{array}{r} 1\ 6\ 5\ 8 \\ -\ 2\ 5\ 7 \\ \hline 1\ 4\ 0\ 1 \end{array}$$

Activity. Use mental math to compute $493{,}625 - 273{,}514$ and $57{,}328{,}694 - 4{,}017{,}382$.

In general, there will be columns in which the digit of m is less than the corresponding digit of n. For example, to compute $756 - 389$, we have $6 < 9$ in the ones column and $5 < 8$ in the tens column. Then the subtractions within columns, $6 - 9$ and $5 - 8$, cannot be performed using whole numbers. The subtraction algorithm now calls for **trading**, in the following sense (for notational simplicity, we state it explicitly for the subtraction $756 - 389$):

> Start from the right with the ones column: change the subtraction from $6 - 9$ to $16 - 9$ and at the same time decrease the next digit (to the left) from 5 to 4 ($= 5 - 1$). The subtraction $4 - 8$ in the tens column[2] is handled the same way: change the subtraction from $4 - 8$ to $14 - 8$ and at the same time decrease the next digit

---

[1] Recall in this context the *leitmotif* of Chapter 3 (page 59).
[2] This subtraction *was* $5 - 8$, but is now $4 - 8$ as a result of decreasing the tens digit from 5 to 4.

(to the left) from 7 to 6 ($= 7 - 1$). Finally, the subtraction in the hundreds column is now $6 - 3 = 3$, and $756 - 389 = 367$.

Schematically:

(5.3)
$$\begin{array}{r} \overset{6}{\not{7}}\ \overset{14}{\not{5}}\ \overset{16}{6} \\ -\ 3\ 8\ 9 \\ \hline 3\ 6\ 7 \end{array}$$

Activity. Use the subtraction algorithm to compute $2345 - 687$.

The possibility of trading in a given subtraction $m-n$ as described above depends on having a nonzero digit in m to the immediate left of the column where the trading is to take place. Sometimes there is no such nonzero digit. For example, to compute $50003 - 465$, the subtraction in the ones column is $3 - 5$, but the digit to the immediate left of 3 in 50003 is 0. The subtraction algorithm now takes care of this situation as follows (again, in the interest of notational simplicity, we state it explicitly for the subtraction $50003 - 465$):

> Start from the right column as usual, change the subtraction $3-5$ to $13 - 5$ and at the same time change *all* the zeros in 50003 to the immediate left of the digit 3 to nines and decrease the first nonzero digit to the left of 3 (i.e., 5) from 5 to 4 ($= 5 - 1$), and then carry out the subtraction as before.

Carrying out this algorithm on $50003 - 465$, we obtain schematically:

(5.4)
$$\begin{array}{r} \overset{4}{\not{5}}\ \overset{9}{\not{0}}\ \overset{9}{\not{0}}\ \overset{9}{\not{0}}\ \overset{13}{3} \\ -\ \ \ \ 4\ 6\ 5 \\ \hline 4\ 9\ 5\ 3\ 8 \end{array}$$

Activity. Use the subtraction algorithm to compute $300207 - 14629$.

5.3. Explanation of the Algorithm

For the explanation of the subtraction algorithm as codified in (5.2)–(5.4), the following subtraction fact is needed. In concrete terms, this fact takes the following form: if you have three piles of oranges, having 7, 8, 9 oranges, respectively, in each pile, then taking $3 + 4 + 5$ oranges away from the three piles combined would leave behind the same total number of oranges as taking successively 3 oranges from the pile of 7 oranges, 4 oranges from the pile of 8 oranges, and 5 from the pile of 9 oranges. In other words,

$$(7 + 8 + 9) - (3 + 4 + 5) = (7 - 3) + (8 - 4) + (9 - 5).$$

In general, suppose l, m, n, a, b, and c are any whole numbers so that $\ell \geq a$, $m \geq b$, and $n \geq c$. Then, we claim likewise that the following is valid:

(5.5) $\quad (\ell + m + n) - (a + b + c) = (\ell - a) + (m - b) + (n - c).$

A formal mathematical explanation of equation (5.5) can be given using the definition of subtraction; see section 5.6. A complete understanding of equation (5.5), however, must await the introduction of negative numbers in Part 3, and we will revisit this equality on page 392.

It should be remarked that equation (5.5) is valid for any number of pairs of numbers. For example, the analogue of equation (5.5) for five pairs of numbers would read: if $l \geq a$, $m \geq b$, $n \geq c$, $p \geq d$, and $q \geq e$, then

$$(\ell + m + n + p + q) - (a + b + c + d + e)$$
$$= (\ell - a) + (m - b) + (n - c) + (p - d) + (q - e).$$

The proof of this more general version is of course entirely similar to the case of three pairs of numbers.

Now that equation (5.5) is available, we can give the explanation of (5.2), the simplest form of the subtraction algorithm. We first do it schematically (note: the double arrow "$\iff$" in the following stands for "is equivalent to"):

$$\begin{array}{r} 1\ 6\ 5\ 8 \\ -\ \ \ 2\ 5\ 7 \\ \hline ?\ ?\ ?\ ? \end{array} \iff \begin{array}{r} 1000 + 600 + 50 + 8 \\ -\ \ \ \ \ 0 + 200 + 50 + 7 \\ \hline ? \end{array}$$

$$\overset{(5.5)}{\iff} \begin{array}{r} 1000 + 600 + 50 + 8 \\ -\ \ \ \ \ 0 + 200 + 50 + 7 \\ \hline 1000 + 400 + 0 + 1 \end{array} \iff \begin{array}{r} 1\ 6\ 5\ 8 \\ -\ \ \ 2\ 5\ 7 \\ \hline 1\ 4\ 0\ 1 \end{array}.$$

Notice that equation (5.5) was used in the middle "$\iff$".

We now rewrite the preceding in a more conventional manner, using the version of (5.5) for four pairs of numbers:

$$\begin{aligned} 1658 - 257 &= (1000 + 600 + 50 + 8) - (0 + 200 + 50 + 7) \\ &= (1000 - 0) + (600 - 200) + (50 - 50) + (8 - 7) \quad \text{(by (5.5))} \\ &= 1000 + 400 + 0 + 1 \\ &= 1401. \end{aligned}$$

The second line above corresponds exactly to the column-by-column subtraction in (5.2).

The explanation for (5.3) retraces the steps, run backwards, of the explanation given for the method of carrying in the addition algorithm of

5.3. Explanation of the Algorithm

Chapter 4. Again, we first do it schematically:

$$\begin{array}{r} 7\ 5\ 6 \\ -\ 3\ 8\ 9 \\ \hline ?\ ?\ ? \end{array} \iff \begin{array}{r} 700 + 50 + 6 \\ -\ 300 + 80 + 9 \\ \hline ? \end{array}$$

$$\iff \begin{array}{r} 700 + 40 + 16 \\ -\ 300 + 80 + 9 \\ \hline ? \end{array} \iff \begin{array}{r} 600 + 140 + 16 \\ -\ 300 + 80 + 9 \\ \hline ? \end{array}$$

$$\stackrel{(5.5)}{\iff} \begin{array}{r} 600 + 140 + 16 \\ -\ 300 + 80 + 9 \\ \hline 300 + 60 + 7 \end{array} \iff \begin{array}{r} 7\ 5\ 6 \\ -\ 3\ 8\ 9 \\ \hline 3\ 6\ 7 \end{array}.$$

In more conventional notation, this is expressed as follows:

$$\begin{aligned} 756 &= 700 + 50 + 6 \\ &= (600 + 100) + 50 + 6 \\ &= 600 + (100 + 50) + 6 \quad \text{(Theorem 2.1)} \\ &= 600 + 150 + 6 \\ &= 600 + (140 + 10) + 6 \\ &= 600 + 140 + (10 + 6) \quad \text{(Theorem 2.1)} \\ &= 600 + 140 + 16 \end{aligned}$$

so that

$$\begin{aligned} 756 - 389 &= (600 + 140 + 16) - (300 + 80 + 9) \\ &= (600 - 300) + (140 - 80) + (16 - 9) \quad \text{(by (5.5))} \\ &= 300 + 60 + 7 \\ &= 367. \end{aligned}$$

Note that the second line of the preceding calculation, i.e., $(600 - 300) + (140 - 80) + (16 - 9)$, corresponds exactly to the column-by-column subtractions in (5.3). Note also that we have avoided using exponents to write the numbers 756 and 389 in their expanded forms but have used the ordinary version instead, because the more complicated notation would have obscured the underlying reasoning.

Although we do not wish to over-emphasize the formalism of the laws of operations discussed in Chapter 2, it is nevertheless worthwhile to point out the *critical role played by the associative law of addition in the form of Theorem* 2.1 *in making* trading *possible in the subtraction algorithm.*

Activity. Explain the subtraction algorithm for $315 - 82$, first schematically, and then by the conventional method using (5.5).

Finally, we explain the last part of the subtraction algorithm, specifically, why it works in the case of $50003 - 465$ (see (5.4)). It is again an extended discourse on the associative law. The crucial observation may be summarized as a collection of addition facts that are a direct consequence of counting:

$$50000 = 49000 + 1000,$$
$$1000 = 900 + 100,$$
$$100 = 90 + 10,$$

so that

$$50000 = 49000 + 900 + 90 + 10.$$

Using Theorem 2.1, we get

$$\begin{aligned} 50003 &= 50000 + 3 \\ &= (49000 + 900 + 90 + 10) + 3 \\ &= 49000 + 900 + 90 + 13 \qquad \text{(Theorem 2.1)}. \end{aligned}$$

So using the version of (5.5) for four pairs of numbers, we have

$$\begin{aligned} 50003 - 465 &= (49000 + 900 + 90 + 13) - (0 + 400 + 60 + 5) \\ &= (49000 - 0) + (900 - 400) + (90 - 60) + (13 - 5) \\ &= 49000 + 500 + 30 + 8 \\ &= 49538. \end{aligned}$$

The second line corresponds to the column-by-column subtractions in (5.4).

We note that the subtraction algorithm is again one that works from right to left. Just as in the case of the addition algorithm, one can work from left to right, but the amount of backtracking needed for making corrections would be even greater here than in the case of addition, as you can experience from the following.

Activity. Do the preceding subtraction $50003 - 465$ from left to right, and compare with the computation from right to left.

It is worth repeating that there is absolutely nothing unnatural about teaching children to do something from right to left.

Finally, as in the case of the addition algorithm, we continue to take note of the *leitmotif* of Chapter 3, to the effect that the subtraction algorithm reduces subtraction to that between two single-digit numbers. Why it should be mentioned here is that, as we have seen, this is not literally true because when trading takes place, one needs to know not only subtraction between single-digit numbers, but also the subtraction of a single-digit number from numbers ≤ 18. Specifically, the following subtraction facts must be known

5.4. How to Use the Number Line

before the subtraction algorithm can be carried out.

$18 - 9$
$17 - 9$ $17 - 8$
$16 - 9$ $16 - 8$ $16 - 7$
$15 - 9$ $15 - 8$ $15 - 7$ $15 - 6$
$14 - 9$ $14 - 8$ $14 - 7$ $14 - 6$ $14 - 5$
$13 - 9$ $13 - 8$ $13 - 7$ $13 - 6$ $13 - 5$ $13 - 4$
$12 - 9$ $12 - 8$ $12 - 7$ $12 - 6$ $12 - 5$ $12 - 4$ $12 - 3$
$11 - 9$ $11 - 8$ $11 - 7$ $11 - 6$ $11 - 5$ $11 - 4$ $11 - 3$ $11 - 2$

If this shakes your confidence in the *leitmotif*, rest assured that there is nothing to worry about. The *leitmotif* continues to be true, literally, but can only be expressed with negative integers (to be taken up in Part 3); see Exercise 17 at the end of this chapter and Exercise 3 on page 402 (Chapter 28).

5.4. How to Use the Number Line

In this section, we give two examples of how one might use the number line for classroom instruction on subtraction.

The geometric description of the identity (5.5) for two pairs of whole numbers ℓ and m and a and b, i.e., $(\ell + m) - (a + b) = (\ell - a) + (m - b)$, is that if a segment **s** is the concatenation of two segments of lengths ℓ and m,

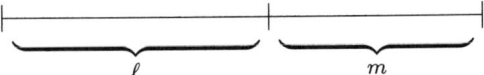

and if a segment of length $a + b$ is removed from **s**, then the length of the remaining segment is equal to the length of the concatenation of

- the remaining segment when a segment of length a is removed from the segment of length ℓ, and
- the remaining segment when a segment of length b is removed from the other segment of length m

as shown:

Therefore, $(\ell + m) - (a + b)$ is just the length of the segment that remains:

This geometric fact is easily accepted by third graders without any elaborate explanations if they are familiar with the number line. We now use this simple fact to explain the subtraction algorithm for $35 - 19$. Write this subtraction as $(20 + 15) - (10 + 9)$. We represent $20 + 15$ as the length of the following concatenated segment:

Now we take out a segment of length 10 from the left end of the left segment above, and a segment of length 9 from the right end of the right segment above, then what remains is a segment which is the concatenation of segments of lengths 10 and 6 (in the middle).

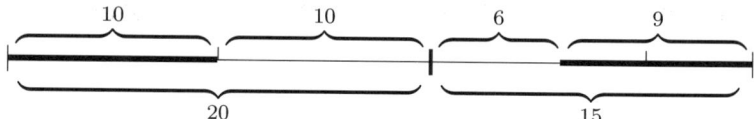

Thus we see pictorially that $35 - 19 = 16$. If we transcribe the picture into numbers,

$$35 - 19 = (20 + 15) - (10 + 9) = (20 - 10) + (15 - 9) = 10 + 6 = 16,$$

then we recognize that this is just the subtraction algorithm for $35 - 19$ with trading in the tens place.

As a second application of the number line, consider the following typical problem on subtraction:

Example. Stefanie has 15 dollars less than Linda, and the two together have 61 dollars. How much does each have?

We are going to use the number line to give a solution that is surpassingly simple. Let Stefanie have s dollars, and Linda have ℓ dollars. Note that s and ℓ are whole numbers and, by the given hypothesis, they satisfy

$$\ell + s = 61 \quad \text{and} \quad \ell - s = 15.$$

Be sure you are comfortable with this transcription from verbal information to expressions about numbers, as this is the critical step that makes the solution of the problem possible. Now the addition fact $\ell + s = 61$ can be interpreted as the concatenation of segments:

Next, $\ell - s = 15$ is equivalent to $\ell = 15 + s$. Again, this means the segment of length ℓ is the concatenation of a segment of length 15 and a segment of length s, as shown:

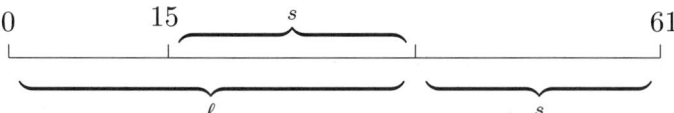

The picture clearly indicates that the segment $[15, 61]$ is the concatenation of two segments, each of length s. But $[15, 61]$ has length equal to $61 - 15 = 46$, so s being half that is equal to 23. Therefore $\ell = 15 + s = 15 + 23 = 38$. The answer is that Stefanie has 23 dollars and Linda has 38 dollars. (Check that $38 + 23 = 61$.)

Of course this assumes that you can find half of an even number, in this case, 46. In a third grade classroom, one would have to prepare students for that in advance.

5.5. A Special Algorithm

For special numbers, there are usually tricks to make computations with them much more pleasant. This applies in particular to the subtraction algorithm. Let us give one such example: the subtraction $50003 - 465$ can be done very simply by expressing the number 50003 as $49999 + 4$, so that by (5.5),

$$50003 - 465 = (49999 + 4) - (465 + 0) = (49999 - 465) + (4 - 0)$$
$$= 49534 + 4 = 49538.$$

Once the underlying reasoning is understood, one could write the preceding in a less stilted fashion, as

$$50003 - 465 = 4 + 49999 - 465 = 4 + 49534 = 49538.$$

Similarly, using (5.5) again, we have

$$30024 - 8697 = (29999 + 25) - (8697 + 0) = (29999 - 8697) + (25 - 0)$$
$$= 21302 + 25 = 21327,$$

which could be written more simply as

$$30024 - 8697 = 25 + 29999 - 8697 = 25 + 21302 = 21327.$$

The point here is that any subtraction where the first number has a row of nines can be done without trading, and can therefore be done easily by mental math. *The same trick can be used for any subtraction problem in* which the first number is only slightly bigger than a single-digit multiple of 10^n, where n is any whole number. Precisely,

> change the first number to a small number plus another one with a row of nines,

and the subtraction problem now simplifies drastically.

Activity. Compute $1004 - 758$ and $60005 - 12348$ by mental math.

5.6. A Property of Subtraction

We give a direct, formal proof of (5.5), i.e.,

$$(\ell+m+n)-(a+b+c) = (\ell-a)+(m-b)+(n-c), \text{ where } \ell \geq a, m \geq b, n \geq c,$$

using the definition of subtraction. First note that (5.5) makes sense, i.e., it is in fact true that $(\ell + m + n) \geq (a + b + c)$ so that the subtraction on the left side of (5.5) can be carried out. To see this, we make repeated use of the third assertion of (2.7) on page 50 (to the effect that $A \geq B$ and $C \geq D$ imply $A + C \geq B + D$) and use the assumption of $\ell \geq a$, $m \geq b$, and $n \geq c$ to conclude that $(\ell+m+n) \geq (a+b+c)$. Now to prove (5.5), let $x = \ell - a$, $y = m - b$, and $z = n - c$; then the right side of (5.5) becomes $x + y + z$, and we want to prove that

$$(\ell + m + n) - (a + b + c) = x + y + z.$$

According to definition (5.1) on page 72, this is equivalent to checking

(5.6) $\qquad (\ell + m + n) = (a + b + c) + (x + y + z).$

By Theorem 2.1, the right side of (5.6) is equal to $(x+a)+(y+b)+(z+c)$. By the definition of x, y, z, we have

$$\begin{aligned} x + a &= (\ell - a) + a &= \ell, \\ y + b &= (m - b) + b &= m, \\ z + c &= (n - c) + c &= n. \end{aligned}$$

Thus the right side of (5.6) is equal to $\ell + m + n$, thereby proving (5.6). Therefore, (5.5) is also proved.

Exercises

1. Give an interpretation of (5.2) in terms of money. Do the same for (5.3) and (5.4).

2. (a) Explain to a fourth grader why the subtraction algorithm for $563 - 241$ is correct, with or without the use of money. (b) Do the same with $627 - 488$.

3. Ben has 11 dollars more than Alan, but 28 dollars less than Carl. If the three together have 95 dollars, how much does each one have?

4. Consider the following problem: "Alan divided a bag of marbles between he and Ben. Alan ended up having 17 more than Ben, and Ben had 29. How many marbles were in the bag in the first place?" If you teach a third grade class, how would you make use of the number line to explain the solution?

5. Maureen has twice the number of dollars that Linda has. Nanette has 5 dollars more than Maureen, and Linda and Maureen together have 2 dollars more than Nanette. How much does each have? (Use the number line.)

6. (a) Use the subtraction algorithm to compute $2403 - 876$, and explain why it is correct. (b) Do the same with $76431 - 58914$.

7. Leon had 95 marbles in a jar. He took out some, and then put back 48 of them. He ended up with 57 marbles in the jar. How many marbles did he take out? Explain.

8. (a) Compute $60{,}013 - 58{,}325$ in two different ways. Also do it using mental math. (b) Do the same with $800{,}400 - 770{,}992$.

9. Let a, b, c be whole numbers. (a) Prove that $a < b$ and $c \leq a$ imply $a - c < b - c$. (b) Prove that $a + b < c$ is equivalent to $a < c - b$. (c) Prove that if $a \geq b$, then $a < b + c$ is equivalent to $a - b < c$.

10. Find shortcuts to compute the following: $8 \times 875 = ?$; $9996 \times 25 = ?$; $103 \times 97 = ?$; $86 \times 94 = ?$

11. Let x and y be two whole numbers and $x > y$. (i) Explain why $(x - y)(x + y) = (x - y)x + (x - y)y$. (ii) Get a simple formula for $(x - y)(x + y)$ by simplifying the right-hand side $(x - y)x + (x - y)y$.

12. (i) Compute $1^2 - 0^2$, $2^2 - 1^2$, $3^2 - 2^2$, $4^2 - 3^2$, $5^2 - 4^2$, $6^2 - 5^2$. (ii) Compute $12^2 - 11^2$, $13^2 - 12^2$, $14^2 - 13^2$. (iii) Let a be a whole number. Compute $(a + 1)^2 - a^2$ two different ways: directly, and by using the preceding exercise.

13. (Compare Exercise 10 on page 70) (i) Compute: $1 + 3$, $1 + 3 + 5$, $1 + 3 + 5 + 7$, $1 + 3 + 5 + 7 + 9$, $1 + 3 + 5 + 7 + 9 + 11$. (ii) Use Exercise 12(i) above to find the sum of the first n odd numbers $1, 3, 5, 7, \ldots, (2n - 1)$.

14. In Exercise 7 on page 54, we considered a rectangle whose length and width are whole numbers that add up to 24. Now use Exercise 11 above to *explain* why the maximum area is 144 and the minimum area is 23.

15. Prove the truth of the distributive law for subtraction on the basis of the ordinary distributive law. (*Hint*: By the definition of subtraction, $km - kn = \bigl(k(m - n)\bigr)$ means $km = kn + k(m - n)$. Now prove the latter.)

16. The following is the subtraction of two 3-digit numbers. Fill in the blanks:

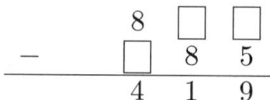

17. *This exercise assumes you know negative integers*: Here is an alternative subtraction algorithm that makes use of negative integers. (a) Consider $658 - 379$. Do single-digit subtraction of the digits with the same place value: $6 - 3 = 3$, $5 - 7 = -2$, and $8 - 9 = -1$. Explain why the number with the (generalized) expanded form, $3 \times 10^2 + (-2) \times 10^1 + (-1) \times 10^0$, is equal to $658 - 379$. (b) Use the same method to compute $3104 - 1657$, and also explain why it is correct. (c) Compute $61234 - 29876$ in two different ways.

Chapter 6

The Multiplication Algorithm

According to the definition of multiplication in equation (1.2) on page 28, 852×73 is, by definition, equal to $73+73+\cdots+73$ (852 times). We now take up the question of *how to avoid the tedium of adding* 73 *to itself* 852 *times* in order to compute 852×73. Bearing in mind the *leitmotif* on page 59, we will make use of the distributive law (see equation (2.6) on page 45) to break up the computation into a series of multiplications of *single digits*, i.e., making use of the multiplication table. This is why, in order to learn about multiplication in general, *a fluent knowledge of the multiplication table is essential.*

The sections are as follows:

The Algorithm

The Explanation

6.1. The Algorithm

The multiplication algorithm for two arbitrary whole numbers has three parts:

(i) the multiplication of two single-digit numbers,
(ii) the multiplication of an arbitrary number by a single-digit number,
(iii) the multiplication of an arbitrary number by an arbitrary number.

Part (i) is the content of the single-digit multiplication table[1] and *must be memorized*. Here we are concerned with parts (ii) and (iii).

We shall express the multiplication algorithm in terms of the specific example 852 × 73 above. Instead of computing 852 × 73 directly, we first compute the multiplications of 852 by the two single digits 7 and 3 separately (cf. (ii) above). The **First Part of the multiplication algorithm** deals with this case, i.e., multiplication of an arbitrary whole number by a single-digit number. We describe it explicitly in terms of 852 × 3. Observe how this multiplication is broken down into a series of single-digit multiplications (cf. the *leitmotif* on page 59 again).

Multiply each of the digits of 852 by 3, from right to left:

2 × 3 = 6, enter 6 in the ones column;

5 × 3 = 15, keep 5 in the tens column and enter 1 in the hundreds column;

8 × 3 = 24, keep 4 in the hundreds column and add to it the 1 that was previously entered to get 5; enter 2 in the thousands column; the thousands column now has the digit 2.

Schematically, we have:

(6.1)
$$\begin{array}{r} 8\ 5\ 2 \\ 3 \\ \times\quad 2\ 1\quad \\ \hline 2\ 5\ 5\ 6 \end{array}$$

The same algorithm, when applied to 852 × 7, yields:

(6.2)
$$\begin{array}{r} 8\ 5\ 2 \\ 7 \\ \times\quad 5\ 3\ 1 \\ \hline 5\ 9\ 6\ 4 \end{array}$$

Activity. Revisit Exercise 6 on page 69 by comparing what you get from the addition algorithm in 7826 + 7826 + 7826 + 7826 + 7826 and what you get from the multiplication algorithm in 7826 × 5.

To obtain 852×73, we now put together the two separate multiplications, 852 × 7 and 852 × 3. This is the content of the **Second Part of the multiplication algorithm**, which describes how to multiply an arbitrary

---

[1]It is only in the context of the multiplication algorithm that we come to understand the *mathematical need* to memorize the multiplication table. By the same token, we also see the frivolity of mandating the memorization the 12 × 12 multiplication table. Unhappily, some state mathematics standards (as of 2009) do mandate such frivolity.

6.2. The Explanation

whole number by a multidigit number (see (iii) above):

To obtain **852 × 73**, start with the ones digit 3 of 73: first write down the result of **852 × 3**, then write down the result of **852 × 7** below the former, *but with all the digits shifted one digit to the left of those of* **852 × 3**, and then add column-wise as in the addition algorithm:

(6.3)
$$\begin{array}{r} 852 \\ \times \quad 73 \\ \hline 2556 \\ + \quad 5964 \quad\;\; \\ \hline 62196 \end{array}$$

In case there is any confusion over this algorithm, here is a further illustrative example. Suppose instead of 852 × 73, we have 852 × 473. Then the algorithm calls for writing down the results of the three computations, 852 × 3, 852 × 7, 852 × 4, in successive rows with the columns aligned so that

the digits of **852 × 7** are shifted one digit to the left of those of **852 × 3**,

the digits of **852 × 4** are shifted two digits to the left of those of **852 × 3**,

and then add column-wise as in the addition algorithm:

(6.4)
$$\begin{array}{r} 852 \\ \times \quad 473 \\ \hline 2556 \\ 5964\;\; \\ + \quad 3408\quad\;\; \\ \hline 402996 \end{array}$$

Activity. Do the following computation using the multiplication algorithm:

$$\begin{array}{r} 527 \\ \times \quad 364 \\ \hline ? \end{array}$$

6.2. The Explanation

We have, in the main, avoided the use of exponential notation thus far in writing down the expanded form of a whole number. For the explanation of the multiplication algorithm, however, the use of exponential notation will add immeasurably to the conceptual clarity of the discussion. The slight increase in notational inconvenience is a small price to pay, all things considered.

We first explain (6.1) on 852×3 by applying the distributive law
$$852 \times 3 = \big((8 \times 10^2) + (5 \times 10) + 2\big) \times 3$$
$$= \big((\mathbf{8 \times 3}) \times \mathbf{10^2}\big) + \big((\mathbf{5 \times 3}) \times \mathbf{10}\big) + (\mathbf{2 \times 3})$$
$$= (24 \times 10^2) + (15 \times 10) + 6,$$

where we have made use of Theorem 2.2 on page 44 in the second (boldfaced) row to conclude that $(8 \times 10^2) \times 3 = (8 \times 3) \times 10^2$, etc. The underlined items correspond exactly to the single-digit computations specified in the First Part of the multiplication algorithm, and they also explain the reduction of 852×3 to a sequence of multiplications of *single*-digit numbers (in this case, 8×3, 5×3, and 2×3).

Experience with the addition and subtraction algorithms tells us that we should proceed by working from right to left. Start with the 6 in $(24 \times 10^2) + (15 \times 10) + 6$: this corresponds to the 6 in the ones column of (6.1). Next, we have
$$15 \times 10 = (10 + 5) \times 10 = 10^2 + (5 \times 10),$$
so that
$$852 \times 3 = (24 \times 10^2) + \big(10^2 + (5 \times 10)\big) + 6$$
$$= \big((24 \times 10^2) + 10^2\big) + (5 \times 10) + 6 \qquad \text{(Theorem 2.1)}$$
$$= \{(24 \times 10^2) + (1 \times 10^2)\} + (5 \times 10) + 6.$$

The last expression accounts for the 5 in the tens column in (6.1) as well as the carrying of the 1 in the hundreds column. Now, the distributive law implies that the sum within the braces { } in the last row is equal to
$$(24 \times 10^2) + (1 \times 10^2) = (24 + 1) \times 10^2 = (20 + 5) \times 10^2 = (20 \times 10^2) + (5 \times 10^2).$$
Since $20 \times 10^2 = 2 \times 10^3$ (we are using Theorem 2.2), we have
$$852 \times 3 = (2 \times 10^3) + (5 \times 10^2) + (5 \times 10) + 6.$$

This then exhibits the carrying of 2 in the thousands column of (6.1) together with the 5 in the hundreds column. The explanation of the procedures in (6.1) is now complete.

Activity. Give an analogous explanation of (6.2).

We now relate (6.1) and (6.2) to the original computation of 852×73. In the process, we will explain the Second Part of the multiplication algorithm as exhibited in (6.3). We apply the distributive law and Theorem 2.2:

(6.5)
$$852 \times 73 = 852 \times (70 + 3)$$
$$= (852 \times 70) + (852 \times 3)$$
$$= \underline{(852 \times 7)} \times 10 + \underline{(852 \times 3)}.$$

6.2. The Explanation

The underlined items clearly exhibit why the multiplication of 852 (or any whole number) by a multidigit number will be known once the multiplications of 852 by single-digit numbers are known. Now, using (6.1), (6.2), and (6.5), we obtain

$$852 \times 73 = (5964 \times 10) + 2556$$
$$= 2556 + 59640.$$

According to the addition algorithm, we can write schematically:

(6.6)
$$\begin{array}{r} 8\ 5\ 2 \\ \times \quad\ 7\ 3 \\ \hline 2\ 5\ 5\ 6 \\ +\ 5\ 9\ 6\ 4\ 0 \\ \hline 6\ 2\ 1\ 9\ 6 \end{array}$$

Because we are used to treating an empty spot as a zero (cf. the discussion of the addition algorithm immediately following (4.1) on page 63), it is customary to omit the 0 at the end of 59640. When this is done, (6.6) becomes exactly (6.3). *The shifting of 852×7 by one digit to the left is therefore explained by the fact that the 7 has place value 70, so that 852×7 is actually 852×70.*

Activity. Compute 73×852 in the form

$$\begin{array}{r} 7\ 3 \\ \times \quad 8\ 5\ 2 \\ \hline ? \end{array}$$

and give an explanation. (Note: By the commutativity of multiplication, you know in advance that the answer is 62196 ($= 852 \times 73$). So the point is to see that the multiplication algorithm gives the same answer.)

Finally, we give an explanation of the application of the multiplication algorithm to 852×473 given in (6.4). Omitting some details that have already been supplied in the explanations of (6.2) and (6.3), we have

$$852 \times 3 = 2556,$$
$$852 \times 7 = 5964,$$
$$852 \times 4 = 3408,$$

and so by the distributive law

$$852 \times 473 = 852 \times \big((4 \times 10^2) + (7 \times 10) + 3\big)$$
$$= \big(\underline{(852 \times 4)} \times 10^2\big) + \big(\underline{(852 \times 7)} \times 10\big) + \underline{(852 \times 3)}$$
$$= 340800 + 59640 + 2556.$$

Therefore, according to the addition algorithm, we have schematically:

(6.7)
$$\begin{array}{r} 852 \\ \times473 \\ \hline 2556 \\ 59640 \\ +340800 \\ \hline 402996 \end{array}$$

It is seen that (6.4) is just (6.7) with some zeros omitted.

Activity. Suppose you are presented with the following computation:

$$\begin{array}{r} 527 \\ \times3004 \\ \hline 2108 \\ +1581 \\ \hline 160208 \end{array}$$

Is this correct? Why?

We mentioned in the discussion in Chapter 3 that there are many ways of writing the standard algorithms. Here is one alternative version of the multiplication algorithm.

$$\begin{array}{r} 852 \\ \times73 \\ \hline 5964 \\ +2556 \\ \hline 62196 \end{array}$$

Procedurally, this means we run the algorithm from left to right, first multiplying by 7 before multiplying by 3. Mathematically, we do not consider such formal differences to be a difference at all. It is worth noting that even the algorithm with a single-digit multiplier can be carried out from left to right. For example, 6718×5 can be done this way:

$$\begin{array}{r} 6718 \\ \times5 \\ \hline 30 \\ 35 \\ 5 \\ +40 \\ \hline 33590 \end{array}$$

Activity. Give a precise description as well as explanation of the preceding algorithm.

6.2. The Explanation

Pedagogical Comment. *In a classroom, the most striking feature of the multiplication algorithm may well be the shifting of* 5964 *by one digit to the left in* (6.3), *or the shifting of* 3408 *by two digits to the left in* (6.4). *This phenomenon of shifting to the left should be carefully explained to students in terms of place value, in the manner of* (6.6), *i.e., we are actually looking at* 852×70 *and* 852×400, *respectively, rather than* 852×7 *and* 852×4, *so that the shifting of digits is caused by the presence of the zeros.* **End of Pedagogical Comment.**

Exercises

1. Explain to a 4th grader why the multiplication algorithm for 86×37 is correct.

2. How would you compute $32{,}897{,}546{,}126{,}349 \times 87$ with a 12-digit calculator? Describe one way you can get it done.

3. Use the multiplication algorithm to compute the following:

$$\begin{array}{r} 1\ 8 \\ \times\ \ 5\ 0\ 0\ 0\ 0\ 9 \\ \hline ? \end{array}$$

   Now apply the same algorithm to the following:

$$\begin{array}{r} 5\ 0\ 0\ 0\ 0\ 9 \\ \times\ \ \ \ \ \ \ \ \ \ \ 1\ 8 \\ \hline ? \end{array}$$

   Discuss the pros and cons of these two methods of computing 500009×18.

4. Use the expanded form of 500009 (and of course also that of 18) to explain both computations in Exercise 3 above. (Note that for a number such as 500,009, the use of exponential notation to write out its expanded form is essentially a necessity.)

5. Compute 4208×87 by the multiplication algorithm, and explain why it is correct.

6. Use mental math to do each of the following: (a) 12×45. (b) 43×22. (c) Suppose you want to buy thirty-five each of 39-cent stamps, 24-cent stamps, 63-cent stamps, and 84-cent stamps.[2] How much money do you need?

7. Ms. Wang took her five classes to visit the museum on successive Mondays. The train fare for each student was \$1.85 (which you may take to be 185 cents if you want to avoid decimals). Her classes have 28, 24, 25, 22, and 26 students. How much train fare did Ms. Wang pay for her students altogether?

8. (a) Which 2-digit number, when multiplied by 89, gives a 4-digit number that begins and ends with a 6? (b) List all the 3-digit numbers which have the following properties: the sum of the digits is 12, and when multiplied by 15 they give a 5-digit number which ends with a 5 (i.e., the ones digit is 5). (Clearly this problem can be done by guess-and-check. You are, however, asked to use reasoning to quickly dispatch it by narrowing down the choices.)

---

[2] These are the values of the most common stamps in June 2006.

9. Let a, b, m, n be whole numbers so that $a < m$ and $b < n$. Explain why $ab < mn$.

10. The following is the product of a three-digit number by a two-digit number using the multiplication algorithm. Fill in the blanks.

$$
\begin{array}{r}
\square\,8\,7 \\
\times \quad 5\,\square \\
\hline
2\,\square\,\square\,\square \\
\square\,\square\,3\,5 \\
\hline
\square\,\square\,2\,\square\,2
\end{array}
$$

Chapter 7

The Long Division Algorithm

An understanding of the long division algorithm depends on an understanding of the concept of "division" among whole numbers. For the latter, a little perspective on the general concept of an arithmetic operation is in order.

The concept of an *arithmetic operation* is a precise one in advanced mathematics,[1] but for whole numbers we can think of it as a rule that assigns another *whole number* to an ordered pair of whole numbers m and n. The fact that the outcome of an arithmetic operation, in the context of whole numbers, has to be a single whole number is important in this discussion. If the arithmetic operation in question is addition or multiplication, then indeed we get $m+n$ or mn, as the case may be. In the case of subtraction or division, however, there are complications. The main difficulty[2] is illustrated by the fact that there is no whole number equal to either $3 - 7$ or $23 \div 7$. Nevertheless, when we have a subtraction or division that does make sense, such as $7 - 5$ or $28 \div 7$, the outcome is—as expected—a whole number.

What makes any discussion of division among whole numbers confusing is the fact that there is a related concept called *division-with-remainder* of a whole number by another nonzero whole number, which associates with any ordered pair of whole numbers m and n ($n \neq 0$) a *pair* of whole numbers, the so-called quotient and remainder of the division-with-remainder of m by n. Thus, the division-with-remainder of 27 by 4 has quotient 6 and remainder 3. The fact that a division-with-remainder associates with *any* ordered pair

---

[1]Usually known as a *binary operation* or *composition law* on algebraic objects in general.

[2]We are soft-pedaling the fact that subtraction $m - n$ and division $m \div n$ cannot be defined for many ordered pairs of whole numbers m and n.

of whole numbers m and n ($n \neq 0$), not just one but *two* numbers (quotient and remainder), disqualifies it from being an arithmetic operation in the sense above. In particular, *it is distinct from the concept of division.* The failure to bring out this distinctive feature of division-with-remainder would appear to be one reason for the confusion over the concept of division in the education literature. The other reason that causes confusion is the fact that, in case m is a multiple of n, the concept of $m \div n$ coincides with the division-with-remainder of m by n when we ignore the presence of the remainder ($= 0$ in this case). Such a coincidence, while unfortunate, is all the more reason for us to keep a clear distinction between these two general concepts. *The concept of division, as an arithmetic operation, is a special case of the concept of division-with-remainder.*

In this chapter, we use the notation $m \div n$ only when it is known ahead of time that the whole number m is a multiple of n. The concept of division among whole numbers, $m \div n$, makes sense only for those ordered pairs m and n ($n \neq 0$) so that m is a multiple of n. Otherwise, we are forced to make use of division-with-remainder of m by n in any situation related to "counting how many n's there are in m". Let us illustrate. Suppose we are presented with the problem of deciding how many buses to hire in order to transport 1295 people, and each bus carries a maximum of 35 passengers. Then we have to find out how many 35's there are in 1295 *without the benefit of knowing ahead of time whether 1295 is a multiple of 35 or not.* It is division-with-remainder at work. One way to solve the problem is to patiently and methodically try out all possible whole numbers $k = 1, 2, 3, \ldots$ in $35k$ to see how soon we would get to 1295. It turns out that the 37th multiple of 35 is exactly 1295.

Obviously, we won't live very long if each time such a problem comes up, we have to go through this elaborate process of counting multiples. Just imagine being presented with the division-with-remainder of a 20-digit number by a 3-digit number! So this is where the *long division algorithm* comes in. We should have used it to get the quotient 37 and the remainder 0 because

> the long division algorithm is an efficient way to get the quotient and remainder of a division-with-remainder. It produces the quotient one digit at a time (cf. the *leitmotif* on page 59 about "one digit at a time").

What we have here is, therefore, an anomaly: whereas each of the other three standard algorithms directly addresses the arithmetic operation in question, the long division algorithm addresses division-with-remainder, which is different from the concept of division as an arithmetic operation. The long division algorithm only addresses division *accidentally*, because division happens to be the case of division-with-remainder with remainder 0.

7.1. Multiplication as Division

This is a fact that should be kept firmly in mind when we approach the long division algorithm.

It is worth pointing out that the long division algorithm is useful for locating a fraction on the number line (see section 12.6 on page 195), and is the principal algorithm for converting fractions to decimals (see the proof of Theorem 18.1 on page 301ff. and Chapter 42, Decimal Expansions of Fractions.

The sections are as follows:

Multiplication as Division

Division-with-Remainder

The Algorithm

A Mathematical Explanation (Preliminary)

A Mathematical Explanation (Final)

Essential Remarks on the Long Division Algorithm

7.1. Multiplication as Division

Division is an alternate, but equivalent, way of expressing multiplication.[3] For example, our naive understanding of division says $24 \div 3 = 8$ because if 24 apples are divided into groups of 3, there are 8 groups. Thus $24 = 3 + 3 + \cdots + 3$ (8 times). According to the definition of multiplication (equation (1.2) on page 28), $3 + 3 + \cdots + 3$ (8 times) is just 8×3. Therefore, $24 \div 3 = 8$ is exactly the same statement as $24 = 8 \times 3$.

Notice that we ask for the division $24 \div 3$, when we teach whole numbers, only because we know ahead of time that 24 is a **multiple** of 3, in the sense that there is a *whole number* k so that $24 = 3k$ (of course in this case $k = 8$).[4] With this in mind, we give the formal definition of *division among whole numbers* as follows: Let a whole number m be a multiple of another *nonzero* whole number n, let us say $m = kn$ for some whole number k. Then by definition,

(7.1) $\qquad \begin{cases} \text{the } \textbf{division of } \boldsymbol{m} \textbf{ by } \boldsymbol{n}, \; \boldsymbol{m \div n}, \text{ is} \\ \text{the whole number } k \text{ that satisfies } \; m = kn. \end{cases}$

The number k is called the **quotient** of the division $m \div n$, n the **divisor**, and m the **dividend**. Moreover, if a whole number m is the multiple of another whole number n, i.e., $m = kn$ for some whole number k, we also say **n divides m**, or **m is divisible by n**. We will go into the divisibility among whole numbers more deeply in Part 4.

---

[3] Contrary to common practice in school mathematics, this statement is not being offered as a definition of division, only as a general comment to lend some perspective to the concept of division.

[4] We will presently define "multiple" in general.

At this point, we want to explicitly call attention to the conceptual similarity between the definition of division and the definition of subtraction in (5.1) on page 72. Let us recall the latter: if m, n are whole numbers so that $m \geq n$, then

$$\boldsymbol{m - n} \quad \text{is the whole number } k \text{ that satisfies} \quad m = k + n.$$

We see that if $\div$ is replaced by $-$ and $\times$ by $+$ in (7.1), we immediately get the definition of subtraction. And of course, the analogue of the "quotient" of two numbers is the "difference" between two numbers. Please keep this analogy in mind as you learn about division, because it helps to be reminded that division is not harder than subtraction.

Because $m = kn = n + n + \cdots + n$ (k times), the division $m \div n = k$ has the following interpretation:

$\boldsymbol{m \div n}$ is the number of groups if $\boldsymbol{m}$ objects are partitioned into equal groups of $\boldsymbol{n}$ objects.

This is called the **measurement interpretation** of the division $m \div n$ among whole numbers. The reason for this terminology is clear from the following consideration. Think of a bowl of fruit punch with 48 fluid ounces. We want to measure how many cups of fruit punch are contained in this bowl. Since there are 8 fluid ounces in a cup, this is equivalent to asking how many 8 ounces there are in 48 ounces. According to the definition of division, the way to obtain the answer is to divide $48 \div 8$.

In terms of the number line, the measurement interpretation of $48 \div 8$ has a nice geometric representation. We first elaborate on the preceding definition of "multiple". Given any point on the number line, let us say 4, then there is a sequence of equi-spaced points to the right of 0 so that, going from left to right, 0 and 4 are the first two points of this sequence. (Such a sequence is easily constructed: slide the segment $[0, 4]$ to the right until the left endpoint 0 is at 4, then the new position of the right endpoint of this segment, i.e., 4, is the next point of the sequence, which is of course the point 8; next slide $[0, 4]$ to the right until 0 is at 8, then the new position of the right endpoint 4 is the next point of the sequence after 8, etc.) We call points in this sequence the **multiples of 4**. The first point to the right of 0 in this sequence (which is 4) is called the **first multiple of 4** (of course), the second point the **second multiple of 4**, the third point the **third multiple of 4**, and in general, for any whole number k, the k-th point in this sequence to the right of 0 is called the $\boldsymbol{k}$**-th multiple of 4**. It is simple to verify that the k-th multiple of 4 is just the whole number $4k$. For convenience, we agree to call 0 the **0-th multiple of 4** (which is reasonable because $0 = 0 \times 4$).

```
 0      4      8     12     16     20     24
 |------|------|------|------|------|------|
```

7.1. Multiplication as Division

In general, suppose we are given a point P to the right of 0 on the number line. Then as we saw above, there is a sequence of equi-spaced points to the right of 0 so that, going from left to right, 0 and P are the first two points of this sequence. We call this sequence the **multiples of P**, 0 the **0-th multiple of P**, P the **first multiple of P**, the next point the **second multiple of P**, and in general if k is a positive whole number, then the k-th point in this sequence to the right of 0 is called the **k-th multiple of P**. If P is a whole number, then by the construction of the multiples, the segment from 0 to the second multiple is the concatenation of two segments of length P and therefore has length $P + P = 2P$. Such being the case, the second multiple of P is by definition the point $2P$. Likewise, the k-th multiple of P is the number kP.

```
0      P      2P     3P     4P     5P     6P
|_____|_____|_____|_____|_____|_____|
```

In particular, if we start with $P = 1$, then *the multiples of 1 are exactly what we call the whole numbers*.

Now we can go back to the measurement interpretation of $48 \div 8$: by looking at the multiples of 8, it is simple to verify that it is the sixth multiple of 8 that equals 48. Thus the segment $[0, 48]$ is the concatenation of six segments of length 8. Therefore, $48 \div 8 = 6$.

```
0      8     2×8    3×8    4×8    5×8    6×8
|_____|_____|_____|_____|_____|_____|
                                          48
```

The division $m \div n$ has a second interpretation as a result of the commutativity of multiplication. Thus let $m \div n = k$ for some whole number k. This means $m = kn$ by (7.1). But $kn = nk$ and $nk = k + k + \cdots + k$ (n times). So the division $m \div n$ means also that

$$m = \underbrace{k + k + \cdots + k}_{n}.$$

Therefore, the number $k = m \div n$ can be interpreted as

> $m \div n$ is also the number of objects in each group when m objects are partitioned into n equal groups.

This is called the **partitive interpretation** of the division $m \div n$. For example, suppose you have a bowl of 48 fluid ounces of fruit punch and you want your eight guests to share the punch equally. If you wonder how much each guest gets to drink, then $48 \div 8 = 6$ fluid ounces would be the correct answer according to the partitive interpretation of $48 \div 8$. In terms of the number line, the representation of the partitive interpretation of $48 \div 8$ is the following: If you partition the segment $[0, 48]$ into eight segments of equal length, then $48 \div 8 = 6$ because each of the latter segments has length equal to 6.

Both interpretations of division are needed in problem solving, and both are commonplace in everyday life whether you are aware of it or not (see Exercises 6 to 9 at the end of the section). Furthermore, as we have seen, *the reason division has these dual interpretations is that multiplication is commutative.*

Activity. In a third grade textbook, division is introduced as follows:

> You can use counters to show two ways to think about dividing.
>
> (A) Suppose you have 18 counters and you want to make 6 equal groups. You can DIVIDE to find how many to put into each group.
>
> (B) Suppose you have 18 counters and you want to put them into equal groups, with 6 counters in each group. You can also DIVIDE to find how many groups there are altogether.

Although you can see easily in this special case that the answer to both problems is 3, discuss which of these two divisions uses the measurement interpretation and which uses the partitive interpretation. Use rectangular rows of dots to illustrate your answer.

We emphasize that, so long as we are in the context of whole numbers, our definition of division implies that *the expression $m \div n$ has meaning only when m is a multiple of n* (and of course it is understood that $n > 0$). For example, $25 \div 6$ has no meaning because there is no *whole number k* so that $25 = 6k$. When we introduce fractions in Part 2 and have extended the meaning of division *in the context of fractions*, we shall see that $25 \div 6 = \frac{25}{6}$ because $25 = \frac{25}{6} \times 6$.

The following two examples of division will be used more than once in the rest of this book.

Example 1. There is no better illustration of the two meanings of division *for whole numbers* than the study of **motion**. To this end, we give a preliminary discussion of the concept of *constant speed* (sometimes called **uniform motion**) that is suitable for the primary grades and maybe even for grade four. Contrary to popular perception, constant speed is a subtle concept. *Since we are going to use only whole numbers throughout this discussion*, however, we can approach it very simply as follows.[5] Let us

---

[5]This discussion will be continued in section 18.3, **Applications**, on page 291.

7.1. Multiplication as Division

first fix a unit of time, say, an hour, and a unit of distance, say, a mile. An object or person is said to move **at a constant speed** of v **mph** (miles per hour), where v is a whole number, if in *any* time interval *which is the length of an hour* (e.g., from the third hour to the fourth hour, or from the seventh hour to the eighth hour), the distance traveled by the object or person is v miles.

Warning to the reader: Problems involving motions of constant speed are common in school mathematics. Although such problems give the illusion of providing good examples of contextual learning, they do not in fact provide a "real world" context. Drastic oversimplifications are involved. For instance, one rarely manages to drive at a constant speed for more than a few seconds in real life unless one is on a stretch of straight freeway with no traffic and one's car has perfect cruise control (which is unlikely).

Given a motion of constant speed v as above, in a time span of t hours (t is a whole number) whether it be at the beginning of the motion, somewhere in the middle of the motion, or at the end of the motion, the total distance will always be the same, namely, tv miles.[6] This is because in t hours, the total number of miles traveled is $v + v + \cdots + v$ (t times), which is equal to tv, by the definition of multiplication (see equation (1.2) on page 28). By the same token, if the unit of time is a minute and the unit of distance is a meter, the distance traveled in t minutes by a motion of constant speed v meters per minute would be tv meters.

A typical question is then the following: *Suppose the distance between towns A and B is 264 miles and a car traveling at constant speed v mph gets from town A to town B in four hours. What is the speed v?* Each hour the car travels v miles, so in four hours, the total distance traveled is $v + v + v + v = 4v$ miles. We are given that $4v = 264$, so in terms of the number line we have the following picture:

Now we recall the meaning of division: since $4v = v \times 4$,

$$v \times 4 = 264 \quad \text{means } v \text{ satisfies} \quad v = 264 \div 4$$

So the speed is $264 \div 4$, which is 66 mph.[7]

---

[6]There is a subtle point here that will be taken up in Chapter 17, namely, the fact that here we are identifying two units: 1 in this case can be 1 hour or 1 mile. In terms of the number line, what this says is that two number lines are being identified: the number line representing the number of hours and the number line representing the number of miles.

[7]The long division algorithm to be discussed presently would provide one of several methods to get 66.

The preceding reasoning exhibits the need as well as the virtue of having clear-cut definitions, namely, those of division and constant speed. A perennial question among students is *why* does one have to *divide* distance by time to get the speed? Our reasoning shows that the use of division (rather than multiplication or subtraction or addition) is dictated by the definitions themselves.

This use of precise definitions may take awhile to get used to, but it will pay off in the long run.

Another typical question is the following: *Suppose the speed of a car is a constant* 58 *miles per hour and the distance between A and B is* 522 *miles. How many hours does it take to go from A to B?* This time around, we know that if it takes t hours to go from A to B, the total distance traveled in t hours is $t \times 58$ miles. But we are given that $t \times 58 = 522$, so $t = 522 \div 58$ hours, which turns out to be 9 hours.

Note that this discussion has a very limited scope: the numbers v and t have to be whole numbers (thus no such thing as *half* an hour is being considered), and the total distance of the motion has to be a multiple of the t and v we use. This is because up to this point, division among whole numbers makes sense only when one number is a multiple of another. Note also that, in contrast with the common practice of making students memorize the formulas "speed = distance ÷ time" and "time = distance ÷ speed" by rote, with no explanation given, we have clearly *explained* why these formulas are correct and why they should be used.

Example 2. We give a geometric interpretation of division. In Chapter 2, we introduced an area model for multiplication. According to this model, 2×3, for example, would be modeled as the area of the rectangle with vertical side equal to 2 and horizontal side equal to 3:

Now suppose we ask for $6 \div 3 = ?$ From the point of view of the area model, this means we have a rectangle with area equal to six square units and a horizontal side equal to three units, and we want to know what the length of the vertical side is:

7.2. Division-with-Remainder

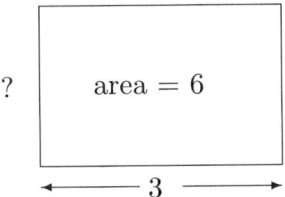

For whole numbers, such a geometric interpretation of division would seem to be nothing more than a light diversion. However, when we come to the multiplication of fractions, the geometric interpretation will acquire added significance.

Activity. If a rectangle has area equal to 84 square units and a vertical side equal to 7 units, what is the length of the horizontal side?

7.2. Division-with-Remainder

Division-with-remainder has been reduced to the role of a hackneyed topic in school mathematics, and it is usually handled as such. You will discover, however, that it is an important topic for the understanding of whole numbers and similar structures in advanced mathematics (see section 11.2, The Representation Theorem, page 158ff., and the whole of Part 4). In particular, it provides the conceptual underpinning of the long division algorithm, and we will emphasize this fact in the following sections.

Let a and d be whole numbers so that $d > 0$ (d stands for *divisor*). If a is a multiple of d, the preceding section defined the quotient q of $a \div d$ to be the whole number which satisfies $a = qd$ and gave a geometric interpretation of the quotient q as that particular multiple qd that equals a (i.e., they are the same point on the number line).

Recall that counting multiples of d in a uses the measurement interpretation of $a \div d$: it counts how many multiples of d there are in a. If a is not a multiple of d, then the measurement concept of $a \div d$ is, at least literally, no longer meaningful. But we can ask, instead, *what is the largest possible multiple of d that is smaller than a?* Or in everyday language, *how many d's are in a?* This is where the concept of division-with-remainder enters. Let us first consider a few examples. Let $a = 25$ and $d = 6$. Although no multiple of 6 equals 25, the fourth and fifth multiples of 6 are distinguished because 4×6 and 5×6 (which are 24 and 30, respectively) "trap" 25 between them. This shows that the multiple 5×6 of 6 already surpasses 25, so that

4×6 of 6 is the largest possible multiple of 6 in 25 and 4 then deserves to be called the "quotient of 25 divided by 6 in this imperfect division".

The thickened segment between 24 and 25 indicates the distance between 4×6 and the given $a = 25$. This distance $25 - (4 \times 6)$ gives what "remains behind" in this "imperfect division".

As another example, consider $a = 141$ and $d = 17$. The multiples of 17 are
$$17, 34, 51, 68, 85, 102, 119, 136, 153, \ldots.$$
Again, we see that 141 is "trapped" between the eighth multiple, $136 = 8 \times 17$, and the ninth multiple, $153 = 9 \times 17$. The "quotient of 141 divided by 17" will accordingly be taken to be 8, and the "remainder" of this "imperfect division" is 5 ($= 141 - (8 \times 17)$).

In general, suppose we have whole numbers a and d, where $d > 0$ but a is not a multiple of d. We ask, among the multiples of d,
$$0, \ d, \ 2d, \ 3d, \ \ldots, \ (q-1)d, \ qd, \ (q+1)d, \ \ldots.$$
which is the biggest that is also less than a. There is clearly such a multiple qd (just keep counting), and this q may be characterized as the **unique** whole number (i.e., one and only one q) so that the two consecutive multiples, qd and $(q+1)d$, **trap a** between them, in the following sense.
$$qd < a < (q+1)d$$

In this case, instead of having a being some multiple qd of d, what we have now is that the segment $[0, a]$ contains the first q multiples of d but not a larger multiple. The segment $[0, a]$ is then the concatenation of $[0, qd]$ and $[qd, a]$. According to the interpretation of addition as concatenation, this means
$$a = qd + (a - qd), \quad \text{where} \quad 0 < a - qd < d$$
Note that the last **double inequality** $0 < a - qd < d$ is just a different, but equivalent, way of expressing the fact that qd and $(q+1)d$ trap a between

7.2. Division-with-Remainder

them: the thickened segment $[qd, a]$ has positive length (so that $0 < a - qd$) and is contained in the segment $[qd, (q+1)d]$ of length d (so that $a - qd < d$). Therefore, using the common notation of r to denote $a - qd$, we have shown that, given whole numbers a and d with $d > 0$, then there is a unique whole number q so that

(7.2) $\qquad a = qd + r, \quad$ where r is a whole number and $\quad 0 < r < d$.

Because q is unique, the number $r = a - qd$ is also unique.

We now add one small refinement to (7.2). The requirement that a *not* be a multiple of d in such a discussion would inevitably get tedious in the long run, because a number *could* sometimes be a multiple of the given d. Let us make an allowance for this possibility in our discussion to simplify matters. To this end, observe that if a is a multiple of d, then $a = qd$ for some whole number q and the remainder is zero, i.e., $r = a - qd = 0$. It follows that, if we simply change "$0 < r < d$" to "$0 \leq r < d$", (7.2) would be valid for any whole number a, regardless of whether or not it is a multiple of the given number d ($d > 0$). We summarize this discussion in the following theorem.

Theorem 7.1. *Given any two whole numbers a and d, with $d > 0$, there are unique whole numbers q and r so that*

(7.3) $\qquad\qquad\qquad a = qd + r \qquad$ and $\qquad 0 \leq r < d$.

The statement about a, d, and q in (7.3) is called the **division-with-remainder of a by d**, and q is called the **quotient** and r the **remainder** of the division-with-remainder. As before, a is the **dividend** and d the **divisor**. Keep in mind that the remainder r is nothing more than the length of the thickened segment in the preceding picture (it could be 0, of course). Theorem 7.1 is usually called the **theorem on division-with-remainder**.[8]

Notice that whereas the division of two whole numbers as defined in the preceding section produces a single number (namely, the *quotient* of the division), in the same way that adding, subtracting, or multiplying two numbers produces a single number, the operation of *division-with-remainder* produces not one, but two numbers: the quotient *and* the remainder. Thus division-with-remainder is intrinsically different from division.

By common abuse of language, we will often refer to q as the quotient of ***a* divided by *d*** with **remainder *r***. Such an abuse is permissible only when you know the precise difference between division-with-remainder and *division*.

In case $a < d$, the quotient and remainder of the division-with-remainder of a by d are, respectively, 0 and a. This is because, if we draw the picture of

---

[8]In advanced mathematics, this theorem is (unfortunately) called the "division algorithm".

the number line, then it would be clear that a is trapped between $0 \ (= 0 \times d)$ and $d \ (= (0+1)d)$.

Activity. Without looking at the statement of Theorem 7.1 above, explain verbally to your neighbor the meaning of "the division-with-remainder of 115 by 17", first algebraically (i.e., write down the analogue of equation (7.3) for this case and explain what the numbers mean), and then by using a number line to describe the quotient and the remainder.

Activity. (a) What is the remainder of 116 divided by 6? 118 divided by 7? (b) Find the quotient and remainder of 577 divided by 23 by inspecting the multiples of 23.

We end this section with three observations. First, it is worth repeating that, except in the case of a zero remainder, division-with-remainder is *not* "division" in the sense defined in the last section. This unfortunate confusion, which is due to the presence of the word "division" in both concepts, is similar to the confusion about a sea lion not being a lion or a butterfly not being a fly.

A second observation is that, in school mathematics, the fact that the division-with-remainder of 25 by 6 has quotient 4 and remainder 1 is written as

$$25 \div 6 = 4 \text{ R } 1.$$

This notation should be struck out from all textbooks, for many reasons, one of them being that it doesn't make any sense. On the most primitive level, if we agree to write $25 \div 6 = 4$ R 1, then we have to also write $21 \div 5 = 4$ R 1. Therefore we get $25 \div 6 = 21 \div 5$ because they are both equal to 4 *R* 1. How can "four groups of 6 with 1 left over" be the same as "four groups of 5 with 1 left over"? We can analyze this notation of "4 R 1" further by concentrating on the meaning of the equal sign. We have defined the equality of two whole numbers as two points on the number line that coincide, but neither "$25 \div 6$" nor "4 R 1" is a whole number, and the equal sign between them is a travesty of symbolic notation. Even if we accept the use of fractions and real numbers in general (see Part 2, especially Chapters 12 and 21), the equality $25 \div 6 = 4$ R 1 still doesn't make any sense because "4 R 1" does not specify what number it is. The correct statement to express this division-with-remainder is "$25 = (4 \times 6) + 1$", and this is what you should bring back to your classroom.

A final observation is that, in case you haven't noticed, the main emphasis of this section has been on getting clear definitions of the concepts of *division-with-remainder, quotient,* and *remainder* in a division-with-remainder. If you wonder what this fuss is all about, consider a typical presentation of these concepts in standard textbooks:

- *division* An operation on two numbers that tells how many groups or how many in each group.
- *quotient* The answer in division.
- *remainder* The number that is left over after dividing.

Activity. How many mathematical errors can you find in the preceding passage? (There are at least three.)

The reason we need clear-cut meanings (definitions) of concepts such as division-with-remainder, quotient, and remainder, is that the **WYSIWYG** quality—**what you see is what you get**—is fundamental in mathematics. If an assumption is not clear and explicit, then it cannot be used by anybody—students or teachers. For this reason, if we are given the concepts of "division", "quotient", and "remainder" in such ambiguous language as in the preceding passage, then we would not know what is being assumed in any hypothesis involving these three terms, and it would be impossible to use such vaguely defined concepts to prove, for instance, Theorem 7.1.

7.3. The Algorithm

Given two whole numbers a, d, with $d > 0$. According to Theorem 7.1, finding the quotient in the division-with-remainder of a by d is equivalent to finding whole numbers q and r so that

$$a = qd + r \quad \text{and} \quad 0 \leq r < d.$$

Suppose we have to do the division-with-remainder of 586 by 3. Up to this point, the only way to get the quotient is to look at *all* the multiples of 3 until we hit on a whole number q so that the consecutive multiples $q \times 3$ and $(q+1) \times 3$ trap 586 between them. But this is a dreary process, and would be even more so if both the dividend and the divisor are much bigger than 3. In the long run, we want to find a method of getting the quotient in any situation that is more efficient than the monotonous checking of all the multiples of the divisor (in this case, 3). One such method is the *long division algorithm.*

What the long division algorithm does is break up the search for the quotient q of $a = qd + r$ into the simpler task of searching for each digit of q from left to right and *one digit at a time.* The fact that such a breakup is possible is remarkable. It is also appropriate to recall in this context the *leitmotif* of Chapter 3: every standard algorithm is ultimately about computing "one digit at a time".

For $a = 586$ and $d = 3$, the schematic presentation of the long division algorithm is the following:

(7.4)
$$
\begin{array}{r}
1\ 9\ 5 \\
3\,)\,\overline{5\ 8\ 6} \\
\underline{3} \\
2\ 8 \\
\underline{2\ 7} \\
1\ 6 \\
\underline{1\ 5} \\
1
\end{array}
$$

The *precise description* of how the algorithm works in this case is as follows. Each step of the algorithm is a division-with-remainder, where the divisor is always 3 but the dividend is specified step-by-step:

First Step. The dividend is the first (left) digit 5 of 586. We obtain the quotient $\underline{1}$ and remainder 2.

Second Step. The dividend is 10 times the remainder 2 of the preceding division-with-remainder plus the second digit 8 of 586, namely 28. We obtain the quotient $\underline{9}$ and remainder 1.

Third Step. The dividend is 10 times the remainder 1 of the preceding division-with-remainder plus the third digit 6 of 586, namely 16. We obtain the quotient $\underline{5}$ and remainder 1.

Final Step. The quotient of 586 divided by 3 is then the number obtained by stringing the quotients of the foregoing steps together: $\underline{195}$. The remainder of 586 divided by 3 is the remainder of the last step: 1.

It is easy to check that, indeed, 586 is $(195 \times 3) + 1$.

In general, the **Long Division Algorithm** applied to a division-with-remainder of a by d (with $d > 0$) is *a sequence of divisions-with-remainders, where the divisor is always d but the dividend is specified in each of the following steps:*

Step 1. The dividend is the first (leftmost) digit of a. Let the quotient so obtained be the single digit A, and let the remainder be r.

Step 2. The dividend is $10r +$ {**the second digit of a**}, where r is the remainder of the preceding step. Let the quotient so obtained be the single digit B, and let the remainder be s.

Step 3. The dividend is $10s +$ {**the third digit of a**}, where s is the remainder of the preceding step. Let the quotient so obtained be the single digit C, and let the remainder be t.

Repeat...

Final Step. The quotient of the division-with-remainder of a divided by d is the number whose digits starting from the left are, successively, A, B,

7.3. The Algorithm

C, ..., and the remainder is the remainder of the last division-with-remainder.

The preceding precise description *of the algorithm implicitly assumes that, at each step, the quotient (A, B, etc.) of the division-with-remainder is a* single-digit *number.* This will be proved in the last section. For now, our focus will be on the procedural aspect of the algorithm.

To make sure that the description of the algorithm is sufficiently clear, let us apply it to a second example: 1295 divided by 35.

(7.5)
$$\begin{array}{r} 0\,0\,3\,7 \\ 3\,5\,\overline{)\,1\,2\,9\,5} \\ 0 \\ \overline{1\,2} \\ 0 \\ \overline{1\,2\,9} \\ 1\,0\,5 \\ \overline{2\,4\,5} \\ 2\,4\,5 \\ \overline{0} \end{array}$$

The long division algorithm, when applied to this case, reads as follows.

First Step. Divide 1 by 35, get quotient $\underline{0}$ and remainder 1.

Second Step. Divide 12 ($= (10 \times 1) + 2$) by 35, get quotient $\underline{0}$ and remainder 12.

Third Step. Divide 129 ($= (10 \times 12) + 9$) by 35, get quotient $\underline{3}$ and remainder 24.

Fourth Step. Divide 245 ($= (10 \times 24) + 5$) by 35, get quotient $\underline{7}$ and remainder 0.

Fifth Step. The quotient of 1295 divided by 35 is therefore $\underline{0037}$, which is just 37 (see the discussion about omitting leading zeros on page 32), and the remainder is 0. Thus we have $1295 \div 35 = 37$.

Note: 1295 *is* equal to 37×35.

In the school classroom, the First Step and Second Step are usually omitted, so that the placement of the first (nonzero) digit 3 of the quotient is generally subsumed under the heading of "estimating where to place the first digit of the quotient". This suggests that when the divisor is large, the long division algorithm needs extra intervention, outside of the algorithm itself, in order to produce a quotient. Were this the case, the algorithm would be defective because a main feature of having an algorithm is to be able to perform it exactly as written with no intervention. Fortunately, the long division algorithm is perfectly sound when it is stated correctly, because as we have just seen, the algorithm itself *dictates* where to place each digit of the quotient.

Observe that the long division of 1295 by 35 had to make use of a division-with-remainder where the divisor exceeds the dividend, twice (Steps 1 and 2). This need was already anticipated in the discussion below Theorem 7.1 of the preceding section.

Activity. Find the quotient and remainder of 78645 divided by 119 *strictly* in accordance with the long division algorithm.

7.4. A Mathematical Explanation (Preliminary)

In order to explain the algorithm of the division-with-remainder of 586 by 3 as exhibited in (7.4), we must show why the mechanism described therein (see especially the description below (7.4)) produces the quotient 195 and the remainder 1 in the division-with-remainder of 586 by 3. In other words, according to Theorem 7.1 on page 105, *we must show that* (7.4) *leads to the division-with-remainder*

(7.6) $$586 = (\boxed{195} \times 3) + \boxed{1},$$

with quotient 195 and remainder 1. Now we come to a delicate issue. Any reader who takes one look at equation (7.6) is entitled to wonder what there is to show. After all, one multiplies out 195×3 to get 585, so that the right side of (7.6) is $585 + 1 = 586$, which is then equal to the left side of (7.6). End of discussion, or so it seems. *But this is not the point!* The question here is not whether (7.6) is correct, but rather whether

> the sequence of divisions-with-remainders in (7.4), each with a dividend different from 586, produces a sequence of single digits which string together to give the correct quotient of 195 and remainder 1 in (7.6).

It is not just the correctness of the final result, but also the *correctness of the procedure itself* that is at issue.

To give this comment a proper perspective, consider the following method of simplifying the fraction $\frac{16}{64}$: canceling the 6 in both the numerator and denominator, we get $\frac{1}{4}$, i.e.,

$$\frac{16}{64} = \frac{1\cancel{6}}{\cancel{6}4} = \frac{1}{4}.$$

Without a doubt, we have produced the correct answer for the problem, but is this a *correct method* of simplifying fractions? By the same token, although (7.6) is correct, we do not know if the *method* described in (7.4) that leads to (7.6) is correct. In the case of $\frac{16}{64}$, the method already fails for $\frac{35}{15}$, so we know it is wrong. With the method of (7.4), however, it seems that in other divisions-with-remainders where we put it to use, it produces the

7.4. A Mathematical Explanation (Preliminary)

correct quotients and remainders. But does this mean this method is *always* correct? In order to answer in the affirmative, the only way is to show directly how to derive equation (7.6) *by making strict use of the precise description of* (7.4) (First Step through the Fourth Step).

The failure to directly address the preceding issue in school mathematics and pre-service professional development materials is what makes the long division algorithm so notorious. It turns out that the mathematical derivation of (7.6) directly from the precise description of (7.4) is a bit sophisticated and may not be suitable for all fifth or sixth grade classrooms. For this reason, we postpone it to the next section. In the same breath, this mathematical derivation should be accessible to a sizable portion of the fifth or sixth graders. As a teacher, you owe it to yourself and to your students to make it part of your repertoire.

What we will do here is give two explanations that *can* be easily adapted to the average school classroom *and* which are both mathematically accurate. They reveal from a perspective different from the precise description of (7.4) what the procedure in (7.4) is all about and why (7.6) is correct. They both involve estimations, which will be discussed in some detail in Chapter 10. For this reason, they are interesting in their own right.

We begin by making sense of (only) the procedure in (7.4). The starting point is the rewriting of the multiplication algorithm in (6.4) (see page 87) by putting back the zeros in the empty spots, as was done in (6.7) (see page 90). There we had an example of how to restore normalcy to a seemingly strange procedure simply by filling in the missing data. We now do likewise in (7.4) by filling in the missing numbers and explaining why we do this. Let us recall (7.4):

$$
\begin{array}{r}
1\ 9\ 5 \\
3\ )\ \overline{5\ 8\ 6} \\
3 \\
\overline{\ 2\ 8} \\
2\ 7 \\
\overline{1\ 6} \\
1\ 5 \\
\overline{1}
\end{array}
$$

It is understood in the way the algorithm is taught that subtractions are performed three times, but *only in the columns where there are numbers*, and that the digits of 586 are mysteriously "pulled down" at each stage as deemed necessary (the 8 is pulled down in the fourth row and the 6 in the sixth row). What we do now is to extend the subtraction to *all the columns at each step* and, in so doing, we fill in the missing numbers to make sense of these "extended subtractions". Thus:

(7.7)
```
        1 9 5
    ┌─────────
  3 ) 5 8 6
      3 0 0   −
      ─────
      2 8 6
      2 7 0   −
      ─────
        1 6
        1 5   −
        ───
          1
```

The subtraction symbol "−" on the right makes explicit the fact that, in each case, the subtraction algorithm is being used. One easily verifies that, with the filled-in zeros (here italicized) taken into account, the subtractions as indicated are both correct and consistent with the original subtractions. *We have not altered the long division algorithm.* Next, we have to make sense of where the filled-in numbers come from and how the top row "195" fits in. Again, in the way this algorithm is usually taught, the hundreds digit 1 of 195 is supposed to multiply the 3 to get the *single digit* 3. We observe, however, that the 1 of 195 is actually 100, so that it is not 1×3 here but 100×3, which accounts for the 300 in the third row. Similarly, because the 9 in 195 is actually 90, it is not $9 \times 3 = 27$ in the fifth row but $90 \times 3 = 270$, which accounts for the 270 of the fifth row. Since the 5 of 195 is just 5, the rest of the long division algorithm is as is. So (7.7) at least shows that the long division algorithm in (7.4) is not strange at all, but is just a collection of ordinary multiplications and subtractions.

Now, we tackle the more substantive questions of why the 195 of (7.7) is the quotient of the division-with-remainder of 586 by 3 and, more pertinently, *what are these steps all about?* To answer these questions, we go back to the beginning: we are looking for whole numbers q and r so that

$$586 = (q \times 3) + r \quad \text{where} \quad 0 \leq r < 3;$$

see (7.3) on page 105. Recall that q is the largest multiple of 3 which is less than or equal to 586 (see section 7.2, **Division-with-Remainder**, on page 103). To facilitate this choice of q, we first limit the possibilities. Can q be a number with four or more digits? Now the smallest four-digit number is 1000, so if $q \geq 1000$, then $q \times 3 \geq 3000$, and since $r \geq 0$, for sure $(q \times 3) + r \geq 3000 > 586$, so that the equality $586 = (q \times 3) + r$ cannot possibly hold.[9] Hence q is at most a three-digit number. From the discussion in section 1.5, **Comparing Numbers (Beginning)**, especially assertion (*ii*) on page 27, we see that to maximize the choice of q, we should maximize, first, its hundreds digit, then its tens digit, and then its ones digit. So

---

[9] We have used, and will continue to use, the inequalities in (2.7) on page 50 without making explicit references to them so as not to interrupt the exposition.

7.4. A Mathematical Explanation (Preliminary)

let its hundreds, tens, and ones digits be a, b, and c, respectively. Then $q = (a \times 100) + (b \times 10) + c$, and therefore,

(7.8) $\quad 586 = \bigl((a \times 300) + (b \times 30) + (c \times 3)\bigr) \;+\; r \quad \text{where} \quad 0 \le r < 3.$

We first maximize the choice of the hundreds digit a. Now a cannot be too big, because $a \times 300$ cannot exceed 586. This follows from (7.8), which says 586 is the sum of $a \times 300$ and some other whole numbers, and is therefore at least as big as $a \times 300$. This is the same as saying $a \times 300 \le 586$, as claimed. Now we put this to use: Since the value of $a \times 300$ is

$$0 \ (a = 0), \quad 300 \ (a = 1), \quad 600 \ (a = 2), \quad 900 \ (a = 3), \ \ldots,$$

we see that a cannot be $2, 3, \ldots$, but a could be either 0 or 1. Therefore, we choose $a = 1$, and (7.8) can now be written as

$$586 = 300 \;+\; \bigl((b \times 30) + (c \times 3)\bigr) + r.$$

From the definition of subtraction (5.1) on page 72, we have

$$\underline{586 - 300} \;=\; \bigl((b \times 30) + (c \times 3)\bigr) + r.$$

The left side of this equality (underlined) is exactly the subtraction in the second and third rows of (7.7).

Since $586 - 300 = 286$ (see the fourth row of (7.7)), we now have

$$286 = \bigl((b \times 30) + (c \times 3)\bigr) + r \quad \text{where} \quad 0 \le r < 3.$$

We apply the same reasoning to maximize the choice of b: Since the right side is the sum of $(b \times 30)$ and two other whole numbers, clearly

$$b \times 30 \;\le\; 286.$$

Now, it is easy to see that this inequality is satisfied for all values of b $(= 0, 1, \ldots, 8, 9)$, and therefore we take the biggest possible value $b = 9$. We now obtain $286 = (9 \times 30) + (c \times 3) \;+\; r$, or,

$$286 = 270 \;+\; (c \times 3) + r.$$

Again by the definition of subtraction (5.1), we have

$$\underline{286 - 270} = (c \times 3) + r.$$

The left side of this equality (underlined) is the subtraction in the fourth and fifth rows of (7.7).

Since $286 - 270 = 16$, we are now left with

$$16 = (c \times 3) + r \quad \text{where} \quad 0 \le r < 3.$$

But this is just the standard division-with-remainder of 16 by 3, so we easily obtain the quotient $c = 5$ and remainder $r = 1$. In other words, $16 = (5 \times 3) + 1$, or,

$$\underline{16 - 15 = 1}.$$

The underlined left side is exactly the last subtraction in the sixth and seventh rows of (7.7).

We have at this point explained every computation that appears in (7.7), together with its reason for being there. We bring closure by noting that, because $a = 1$, $b = 9$, $c = 5$, and $r = 1$, (7.8) gives

$$586 = \bigl((100 + 90 + 5) \times 3\bigr) + 1 = (195 \times 3) + 1,$$

which is exactly (7.6). Our explanation of the long division algorithm (7.4) is complete.

The preceding explanation can be adapted to *any* division-with-remainder, and we will presently apply it to a division-with-remainder of 1308 by 35. But we want to first present another way of looking at this same division-with-remainder of 586 by 3 that uses money (cf. section 4.1, The Basic Idea of the Algorithm, on page 61). The advantage of using money is that it explains, directly, the long division algorithm (7.4) without the intervention of (7.7). The disadvantage is that it changes the meaning of division-with-remainder as a "measurement division" into a "partitive division", as we shall see. Moreover, it is an effective explanation only for numbers up to four digits because students cannot react too well to "ten-thousand dollar bills" and beyond. But it might be argued that for the elementary classroom, numbers up to four digits are good enough.

Suppose then we have a stack of cash which is worth 586 dollars and which consists of:

> 5 hundred-dollar bills,
> 8 ten-dollar bills,
> 6 one-dollar bills.

Needless to say, this is another way of presenting the expanded form of 586. We are trying to find out how many 3's there are in 586 (which is after all the meaning of division-with-remainder). Let us say that there are q 3's in 586, with 0 or 1 or 2 left over. This is equivalent to saying, as usual, that for some whole numbers q and r,

$$586 = (q \times 3) + r \quad \text{where} \quad 0 \leq r < 3.$$

Because multiplication is commutative, we rewrite it as

$$586 = 3q + r \quad \text{where} \quad 0 \leq r < 3$$

(by tradition, $3 \times q$ is written as $3q$). But $3q = q + q + q$ can be interpreted as 3 stacks of cash with q dollars in each stack. We may therefore rephrase our problem as one of dividing 586 dollars into 3 stacks, with the same q dollars in each stack, so that the sum of these $3q$ dollars together with the leftover 0, 1, or 2 dollars would add up to 586 dollars. We can find out what q is by creating 3 equal stacks in the following way. Before doing that, it

7.4. A Mathematical Explanation (Preliminary)

would be convenient to again recall the original long division (7.4):

(7.4)
$$
\begin{array}{r}
1\ 9\ 5 \\
3\ )\ \overline{5\ 8\ 6} \\
\underline{3} \\
2\ 8 \\
\underline{2\ 7} \\
1\ 6 \\
\underline{1\ 5} \\
1
\end{array}
$$

First, try to distribute the 5 one-hundred-dollar bills equally into these 3 stacks. In each stack we put in 1 one-hundred-dollar bill, and there are 2 left over. This corresponds to the first step in (7.4):

$$
\begin{array}{r}
1 \\
3\ )\ \overline{5\ 8\ 6} \\
\underline{3} \\
2
\end{array}
$$

Next, we convert the 2 one-hundred-dollar bills into 20 ten-dollar bills, so that (together with the original 8 ten-dollar bills already there) we now have 28 ten-dollar bills. These can be distributed into the 3 stacks equally with 9 in each stack and 1 left over. This corresponds to the second step in (7.4):

$$
\begin{array}{r}
1\ 9 \\
3\ )\ \overline{5\ 8\ 6} \\
\underline{3} \\
2\ 8 \\
\underline{2\ 7} \\
1
\end{array}
$$

Finally, we convert the 1 ten-dollar bill into 10 one-dollar bills, and we now have 16 one-dollar bills. Again, we can distribute them equally into the 3 stacks with 5 one-dollar bills in each, and 1 is left over. This corresponds to the final step of (7.4). Altogether then, the original stack of 586 dollars has been divided into 3 equal stacks, each consisting of 1 one-hundred-dollar bill, 9 ten-dollar bills, and 5 one-dollar bills, with 1 one-dollar bill left over. This is exactly the complete content of (7.4).

Activity. Write out the long division of 235 by 4, and interpret it in terms of money.

Finally, give an explanation of the division-with-remainder of 1308 by 35. The long division algorithm is carried out in standard fashion as follows:

```
              3 7
    35 ) 1 3 0 8
         1 0 5
           2 5 8
           2 4 5
             1 3
```

We fill in the missing 0:

(7.9)
```
              3 7
    35 ) 1 3 0 8
         1 0 5 0    —
           2 5 8
           2 4 5    —
             1 3
```

Because we have already gone through the division-with-remainder of 586 by 3, we will be more brief this time around. The goal is to find whole numbers q and r so that

$$1308 = (q \times 35) + r \quad \text{where } r \text{ satisfies} \quad 0 \leq r < 35.$$

Recall that this q is the largest multiple of 35 which is less than or equal to 1308. So we choose the largest such q possible. First we decide how many digits q may have. Can it have three digits? The smallest three-digit number is 100, so if q has three digits,

$$1308 \geq 100 \times 35 = 3500.$$

Impossible. This q is therefore at most a two-digit number, and if we let its tens and ones digits be a and b, respectively, then we first choose the largest possible a and then the largest possible b. Thus $q = (a \times 10) + b$, and $1308 = ((a \times 10) \times 35) + (b \times 35) + r$, so that

$$1308 = (a \times 350) + (b \times 35) + r \quad \text{where } r \text{ satisfies} \quad 0 \leq r < 35.$$

For $a = 3$ and 4, $a \times 350$ is, respectively, 1050 and 1400. Since $1400 > 1308$, a cannot be 4 or bigger. Thus $a = 3$ and

$$1308 = (30 \times 35) + (b \times 35) + r$$

so that

$$1308 - 1050 = (b \times 35) + r.$$

We see the subtraction of $1308 - 1050$ on the left side in the second and third rows of (7.9). Now $1308 - 1050 = 258$, so we have

$$258 = (b \times 35) + r \quad \text{where } r \text{ satisfies} \quad 0 \leq r < 35.$$

This is the standard division-with-remainder of 258 by 35; the quotient is therefore 7 with remainder 13. Thus $258 = (7 \times 35) + 13$ and therefore,

$$258 - 245 = 13.$$

This is the subtraction in the fourth and fifth rows of (7.9). Our explanation is complete.

7.5. A Mathematical Explanation (Final)

We now revisit the division-with-remainder of 586 by 3, and our goal is to derive (7.6) *solely* on the basis of the precise description of the long division algorithm given on page 108 following (7.4). First we recall (7.6):

(7.6) $$586 = (\boxed{195} \times 3) + \boxed{1}$$

To anticipate what comes next, recall that the long division algorithm is expected to generate the dividend 195 one digit at a time (see the *leitmotif* on page 59). Also recall that the expanded form of a number is the mechanism that isolates each digit of a number. Putting these two ideas together, we begin by rewriting (7.6) using the expanded form of 195:

$$586 = \big((1 \times 10^2) + (9 \times 10) + 5\big) \times 3 + 1.$$

Using the distributive law, we get

$$586 = \big((1 \times 10^2) \times 3\big) + \big((9 \times 10) \times 3\big) + (5 \times 3) + 1.$$

We will prove (7.6) in this form below.

According to Step 1, with dividend 5 and divisor 3, the division-with-remainder gives

$$5 = (\boxed{1} \times 3) + \boxed{2}.$$

According to Step 2, the division-with-remainder has dividend $20 + 8 = 28$ and divisor 3:

$$28 = (\boxed{9} \times 3) + \boxed{1}.$$

We are now on automatic pilot: Step 3 is a division-with-remainder with dividend $10 + 6 = 16$ and divisor 3, so that

$$16 = (\boxed{5} \times 3) + \boxed{1}.$$

Therefore, the division of 586 by 3, according to the long division algorithm, is entirely encoded in the following three (simple) divisions-with-remainders:

(7.10)
$$\begin{aligned} 5 &= (\boxed{1} \times 3) + \boxed{2}, \\ 28 &= (\boxed{9} \times 3) + \boxed{1}, \\ 16 &= (\boxed{5} \times 3) + \boxed{1}. \end{aligned}$$

We now show how to derive (7.6) by using only the collection of equations in (7.10). Using the first equation, we get

$$\begin{aligned} 586 &= (5 \times 10^2) + (8 \times 10) + 6 \\ &= ((\boxed{1} \times 3 + 2) \times 10^2) + (8 \times 10) + 6. \end{aligned}$$

Applying the distributive law, we get $(\boxed{1} \times 3 + 2) \times 10^2 = (\boxed{1} \times 3 \times 10^2) + (2 \times 10^2)$, so that

$$\begin{aligned} 586 &= ((\boxed{1} \times 10^2) \times 3) + (2 \times 10^2) + (8 \times 10) + 6 \\ &= ((\boxed{1} \times 10^2) \times 3) + (28 \times 10) + 6. \end{aligned}$$

Using the second equation in (7.10), we get

$$586 = ((\boxed{1} \times 10^2) \times 3) + ((\boxed{9} \times 3) + 1) \times 10 + 6.$$

Now, the distributive law gives $((9 \times 3) + 1) \times 10 = ((9 \times 3) \times 10) + (1 \times 10)$, so that

$$586 = ((\boxed{1} \times 10^2) \times 3) + ((\boxed{9} \times 10) \times 3) + (10 + 6).$$

Applying the last equation of (7.10) to $(10 + 6) = 16$, we get

$$586 = ((\boxed{1} \times 10^2) \times 3) + (\boxed{9} \times 10) \times 3) + (\boxed{5} \times 3) + 1,$$

which is exactly the form of equation (7.6) we derived above.

Observe that the quotient of 586 divided by 3 is displayed vertically down the first digits of the right side of (7.10), and the remainder is the remainder 1 of the last term of the last equation in (7.10).

To further firm up these ideas, we now explain the division of 1295 by 35 in (7.5). Because this is the second time around, we can be more brief. The division algorithm in this situation is captured by the following collection of divisions-with-remainders:

(7.11)
$$\begin{aligned} 1 &= (\boxed{0} \times 35) + \boxed{1}, \\ 12 &= (\boxed{0} \times 35) + \boxed{12}, \\ 129 &= (\boxed{3} \times 35) + \boxed{24}, \\ 245 &= (\boxed{7} \times 35) + \boxed{0}. \end{aligned}$$

As before, what we need to show is that, by using (7.11) alone, we can prove

$$1295 = 37 \times 35.$$

Again, as before, we expect to end up with proving

$$1295 = ((3 \times 10) + 7) \times 35.$$

We can adopt the methodical approach of last time, but we see clearly that the first two equations of (7.11) have nothing to say and so will be ignored. Now the third equation of (7.11) has 129 on the left and obviously $1295 = (129 \times 10) + 5$. Thus

$$\begin{aligned} 1295 &= (129 \times 10) + 5 \\ &= ((\boxed{3} \times 35 + \boxed{24}) \times 10) + 5. \end{aligned}$$

Using the distributive law, we simplify the last expression and guide it towards the fourth equation of (7.11),

$$(((\boxed{3} \times 35 + \boxed{24}) \times 10) + 5 = (\boxed{3} \times 35 \times 10) + (\boxed{24} \times 10) + 5$$
$$= (\boxed{3} \times 10 \times 35) + 245$$
$$= (\boxed{3} \times 10 \times 35) + (\boxed{7} \times 35),$$

where the last step uses the fourth equation in (7.11). Therefore,

$$1295 = (\boxed{3} \times 10 \times 35) + (\boxed{7} \times 35)$$
$$= ((\boxed{3} \times 10) + \boxed{7}) \times 35$$

as desired.

Activity. Perform the long division of 1162 divided by 19, and then give the explanation of why it is correct in the way we explained why (7.6) is correct.

7.6. Essential Remarks on the Long Division Algorithm

We have thus far concentrated mainly on explaining the most striking aspect of the long division algorithm, which is that, by breaking up a division problem into a sequence of simpler divisions-with-remainders, one can get the quotient and remainder of the original division simply and *mechanically*. In so doing, we have left out (on purpose) the explanation of a crucial feature of the algorithm:

> *Each division-with-remainder in the algorithm has a single-digit quotient.*

Our first remark is to explain why this assertion is true.

The explanation will become clear once we have verified it in a special case, such as the case of 586 divided by 3 (see (7.4)). Consider (7.10), the mathematical transcription of (7.4), and focus on one of the steps, say, the second:

$$28 = (9 \times 3) + 1.$$

Here the dividend is 28 and the quotient is 9. Is perhaps the fact that 9 is a single digit an accident? Not at all, because we now show that the tenth multiple of 3 (i.e., 10×3) *must* exceed the dividend 28, so that by our definition of the quotient as the largest multiple of the divisor *not to exceed the dividend*, the quotient must be smaller than 10. In other words, a single digit. The reason is this: By the statement of the Long Division Algorithm, the dividend of the division-with-remainder is 10 times the remainder 2 of the preceding step plus a single-digit number (which is 8 in this case). Because we are dividing by 3, the remainder has to be < 3 and therefore either 0, 1, or 2. Thus, 10 times the remainder can be at most 20, and if we add to this a single-digit number, then the result is at most 29. Therefore 10×3 exceeds 29. This is the conclusion we seek.

To make sure that the generality of this argument comes through, let us look at the division of 1295 by 35 in (7.5). Consider (7.11), the mathematical transcription of (7.5). Again, we isolate a single step, say the third:

$$129 = (3 \times 35) + 24.$$

The remainder from the preceding step is 12, and therefore the dividend of the third step has to be $(10 \times 12) + 9 = 129$, and the dividend of 129 divided by 35 is again a single-digit number, namely, 3. Once again, let us try to understand why this quotient has to be a single digit. If not 129, what could this dividend be? According to the statement of the Long Division Algorithm, the dividend could be 10 times the remainder from the preceding step plus a single-digit number. Since the divisor is 35, the remainder is at most 34. Therefore the dividend, being a number which is at most 340 plus a single-digit number, is at most 349. But the tenth multiple of 35 would be 350, which exceeds 349. The quotient of the resulting division-with-remainder, being the *largest multiple of* 35 *not to exceed the dividend*, must be smaller than 10. Thus, once again, it has to be a single-digit number, as we set out to show.

The general argument is basically the same. We have a divided by d and we want to isolate a division-with-remainder in the long division algorithm and see why it must have quotient equal to 9 or less. According to the algorithm, the dividend is $10R$ plus one of the digits of a, where R is the remainder of the preceding step. Recall that in a division-with-remainder, the remainder is less than the divisor d, so R is at most $(d-1)$ and the dividend is therefore at most $10(d-1) + 9 = 10d - 1$, with 9 being the maximum among single digits. The tenth multiple of d is $10d$, which is bigger than $10d - 1$. Since the quotient is the *largest multiple of* d *not to exceed the dividend* $10d - 1$, the quotient must be smaller than 10. Hence it is a single-digit number, and this settles the general case.

Our second remark is to drive home the point that *the long division algorithm is strictly a digit-by-digit procedure without regard to the place value of any of the digits*. To this end, we shall compare the division of 586 by 3 in (7.4) with the division of 58671 by 3, where the dividend 58671 has been chosen on purpose to have the same three digits 5, 8, and 6 on the left.

7.6. Essential Remarks on the Long Division Algorithm

The usual schematic display of the long division is as follows:

```
        1 9 5 5 7
    3 ) 5 8 6 7 1
        3
        ‾‾‾
        2 8
        2 7
        ‾‾‾
          1 6
          1 5
          ‾‾‾
            1 7
            1 5
            ‾‾‾
              2 1
              2 1
              ‾‾‾
                0
```

The precise description of the algorithm is then

(7.12)
$$\begin{aligned} 5 &= (\boxed{1} \times 3) + \boxed{2}, \\ 28 &= (\boxed{9} \times 3) + \boxed{1}, \\ 16 &= (\boxed{5} \times 3) + \boxed{1}, \\ 17 &= (\boxed{5} \times 3) + \boxed{2}, \\ 21 &= (\boxed{7} \times 3) + \boxed{0}. \end{aligned}$$

Now compare the first three steps of both (7.10), the precise description of (7.4), and (7.12). *They are identical.* For definiteness of discussion, let us concentrate on the third step in both. The 6 in 586 is the ones digit, which is very different from the 6 in 58671, which is the hundreds digit. In other words, while the third step of (7.10) is exactly what it says it is, $16 = (5 \times 3) + 1$, the third step in (7.12) is actually the division-with-remainder of 1600 by 3, which is therefore

$$1600 = (500 \times 3) + 100$$

if place value is taken into consideration. The point we wish to emphasize is that, *as far as the long division algorithm itself is concerned, the place value of the digit in question is irrelevant.*

Our third and final remark rounds out the earlier comment about the use of money to explain the long division algorithm in school classrooms. On the one hand, it is valuable to have such an explanation because it is one that fifth or sixth graders can relate to. On the other hand, it is important to acquire some perspective on this issue. There are some who consider the use of money to interpret the long division algorithm to be the epitome of conceptual understanding. Having seen the *mathematical* explanation of this algorithm, however, we are in a position to make explicit in what sense this naive interpretation fails to capture the essence of the algorithm.

First, the reasoning with money does not bring out the fact that this algorithm is strictly a digit-by-digit procedure without regard to the place value of each digit. Rather, the money interpretation is tied down, at each step, to the place value of the digit under consideration. Second, the money interpretation does not bring out the critical feature of the algorithm, namely, the fact that it is the synthesis of a sequence of divisions-with-remainder with the same divisor but with a far simpler dividend, so that each of these divisions-with-remainder has a *single-digit* quotient and is therefore easier to carry out. This understanding of the underlying structure of the long division algorithm is important, not only for its own sake, but also for the conversion of fractions to decimals in (42.2) and (42.6) of Chapter 42 and for the division algorithm for polynomials in algebra.

Exercises

1. Is 24 the quotient of 687 divided by 27? Could 15 be the quotient of 944 divided by 48? Explain. (No calculator is allowed.)

2. Is 6977 the remainder of 124968752 divided by 6843? Why? (No calculator is allowed.)

3. *Without using the long division algorithm,* find the quotient and remainder in each of the following cases: 964 divided by 31; 517 divided by 19; 6854 divided by 731; 1234,5497,2086 divided by 873; and 1234,5497,2086 divided by 8026,5937. You may use a four-function calculator for the last two.
   Pedagogical Comments. *The last two items furnish a good example of how to use calculators in the elementary classroom to advantage. In both cases, though, it would be instructive to first ask for a ballpark figure of the quotient without the use of a calculator. This is a good exercise in making estimates (see Chapter 10). It should also be mentioned that there is an effective way to use the (four-function) calculator to get the quotient and remainder without any trial and error. How this is done and why it is true should lead to an interesting classroom discussion (this way of using the calculator requires a knowledge of decimals, see Part 2 and Part 5).* **End of Pedagogical Comments.**

4. Explain why, if a motion has the property that for some fixed number v the distance it travels in any time interval of t units (seconds, minutes, hours, etc.) is tv units of distance (miles, meters, feet, etc.), then this motion is one of constant speed v.

5. Let r be the remainder of a divided by d for whole numbers a and d with $d \neq 0$. Suppose $a = mA$ and $d = mB$ for some whole numbers m, A, and B. Let R be the remainder of A divided by B. What is the relationship between R and r? Give a detailed explanation of your answer. (*Caution*: This problem is deceptive because it seems almost trivial, but the explanation is actually quite subtle and it requires the use of Theorem 7.1 on page 105.)

6. You give your fifth grade class a problem:
   > A faucet fills a bucket with water in 30 seconds, and the capacity of the bucket is 12 gallons. It may be assumed that the same number of gallons of water flows out of the faucet each second. How long would it take the same faucet to fill a vat with a capacity of 66 gallons?

   How would you *explain* to your class how to do this problem?

7. Consider the following two problems: (a) If you try to put 234 gallons of liquid into 9 vats, with an equal amount in each vat, how much liquid

is in each vat? (b) If you try to pour 234 gallons of liquid into buckets each with a capacity of 9 gallons, what is the minimum number of such buckets you need in order to hold these 234 gallons? Get the answer to both, and explain in each case whether you are using the partitive or the measurement interpretation of division.

8. If you have 176 marbles and want to distribute them equally into 11 jars, how many marbles does each jar get? Suppose instead you want to partition the 176 marbles into equal groups of 8 each. How many such groups are there? (Compare problem 1 on page 53.)

9. An eccentric millionaire wants to run exactly 3 miles each day and he wants to build a circular running track so that, after 16 laps around the track, he would be done with running. How long should the circumference of this circular track be (1 mile = 5280 feet)? Do you use partitive division or measurement division to do this problem?

10. The same millionaire as in the preceding problem finishes his daily run in 30 minutes. Assuming that he runs at constant speed, how many seconds does it take him to run 1056 feet?

11. Alexandra has some oranges. When she puts them into plastic bags that hold 5 each, she fills a certain number of bags and has 2 left over. She changes to bags that hold 8 each; then she fills some bags again but has 5 left over. It is known that she has no more than 50 oranges. Exactly how many does she have?

12. At a banquet, if the guests are seated 12 to a table, there will be one table with only 1 guest. If however 11 are seated to a table, then there will be one table with only 8 guests. There are fewer than 100 guests. How many guests are there exactly?

13. (a) One lap of a certain running track measures 465 meters. A person training for the marathon wants to run 20,000 meters every day on this track. How many laps of this track does he have to run? (b) Explain to a sixth grader how to do this problem, and *why* it is this way. (In other words, why is the long division algorithm the correct method to use, and how is this algorithm used correctly vis-à-vis (7.10)?)

14. Explain to a sixth grader why the long division algorithm for the division of 652 by 8 is correct.

15. Find the quotient and remainder of the division-with-remainder of 1850 by 43 using the long division algorithm. Write out the "procedural description" of this long division in the sense of (7.10) or (7.11), and use it to prove that your result is correct.

16. Do the long division of 50050 by 65 to find the quotient and remainder, describe the algorithm as a sequence of divisions-with-remainders as in (7.12), and use these to show why your quotient and remainder are correct.

Exercises

17. How many digits does the quotient of the division of 567,104,982 by 8759 have? Compute the first two digits (from the left) of this quotient. (No calculators, of course.)

18. Explain why the following *distributive law for division* is true using the definition of division. Let k, m, n, be whole numbers. Let $n > 0$ and let k, m be multiples of n. Then
$$(m \div n) + (k \div n) = (m + k) \div n.$$
(A full understanding of this problem has to wait for Part 2; see Exercise 17 on page 281.)

19. (a) An infinite sequence of letters repeats the pattern of rainbow colors indefinitely,

    ROYGBIVROYGBIVROYGBIVROYGBIVROYGBIV ...

    What is the 118th letter? (b) An infinite decimal begins with the three digits 175 and then repeats the pattern 841359 indefinitely,

    0.175841359841359841359841359841359...

    What is the 116th digit to the right of the decimal point?

Chapter 8

The Number Line and the Four Operations Revisited

We have thus far concentrated on the computational aspect of whole numbers. Now is the time to step back and survey the four arithmetic operations from the perspective of the number line.

The sections are as follows:

> The Number Line Redux, and Addition and Subtraction
> Importance of the Unit
> Multiplication
> Division
> A Short History of the Concept of Multiplication

The last section briefly discusses the evolution of the concept of multiplication between numbers.

8.1. The Number Line Redux, and Addition and Subtraction

We now pause to take stock of what we have done so far. We started with counting and agreed on a numeral system to record the whole numbers. Then we put the whole numbers on the number line (tacitly assumed to be horizontal), and identified the whole numbers with a collection of equi-spaced points to the right of a point we designated as 0 on the number line. Having done that, we are going to change our viewpoint here. *Instead of starting with some preconceived notion of a whole number and using a set of marked points on a line to represent the whole numbers, we will start with a line with equi-spaced points and define a whole number to be one of these equi-spaced points on the line.* The reason for this change of perspective will be given in the next chapter and in Chapter 12. For now, we concentrate on describing precisely the new perspective.

> Take a straight line and mark off a point as 0 (zero). Then fix a segment to the right of 0 and call it the **unit segment**. Mark the right endpoint of this segment on the line, thereby generating the first marked point. The *multiples* of this marked point (in the sense of the definition on page 99) then form a sequence of equi-spaced points to the right of 0. By reflecting this sequence to the left of 0, we now obtain a sequence of equi-spaced points on both sides of 0.

Notice that up to this point, there is no mention of whole numbers. The next step is to formalize the introduction of whole numbers by adopting the following definition.

Definition. *A **whole number** is one of the marked points on the line to the right of 0, so that, starting with the initial number 0, the next one (to the right of 0) is 1, the one after that is 2, etc., and we continue the naming of the marked points in the same way we did the counting of the whole numbers in Chapter 1. This line with the whole numbers on it is called the **number line**. A **real number**, or sometimes just a **number** when there is no danger of confusion, is by definition any point on the number line.*

As far as whole numbers are concerned, what we have defined is at least consistent with everything we have done up to this point. But by starting with points on the number line first and then naming them the whole numbers, we have made a whole number something very *concrete and explicit*:

> A whole number is one of the marked points to the right of 0 on the number line.

As in Chapter 1, we continue to treat the number line as an "infinite ruler", except now it is an infinite ruler in both directions. We also identify a whole number n with the length of the line segment $[0, n]$ from 0 to n. Until the end of Part 2, we shall be concerned exclusively with the part of the number line to the right of 0.

In the following, we shall always refer to *the* number line, and this terminology has to be understood in the following sense. The fact that a different choice of the straight line and a different choice of the unit segment would lead to a different number line is understood, but the resulting mathematical discussion, insofar as it refers only to the whole numbers, will not be affected by the possible change in position of the line or where 0 and 1 are placed on the line.[1]

The four arithmetic operations have already been interpreted in terms of the number line. Here is a quick pictorial review of addition and subtraction.

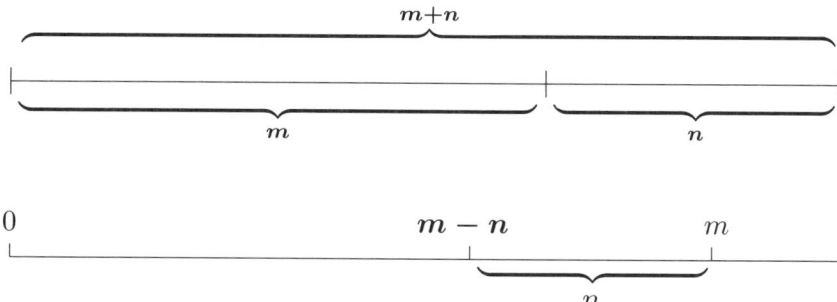

8.2. Importance of the Unit

We wish to go into some of the fine points of the number line, especially regarding the addition and subtraction of whole numbers.

The fact that the positions of the whole numbers on the number line depend on the choice of 1 on the line (as noted in the preceding section, 0 is assumed to be fixed) means that *every* whole number in a given discussion refers to the same unit 1. In terms of everyday experiences, this means, for example, that $3 + 4 = 7$ could mean "3 oranges combine with 4 oranges to make 7 oranges" (in this case, 1 is *one orange*) or "3 stars and 4 stars make 7 stars" (in this case 1 is *one star*), but never "3 stars and 4 oranges make 7 oranges". To drive home this point, consider the following scenario:

> You are going to squeeze apple juice. You worry whether you have enough apples and also wonder about how many apples will make one cup (8 fluid ounces) of juice. If you scribble $13 + 15$ on a piece of paper, which of the following is a plausible interpretation? (i) You are making a calculation of 13 apples added to 15 fluid ounces. (ii) You are adding 13 cups of juice to 15 apples. (iii) You are counting apples: 13 apples + 15 apples. (iv) You are counting how many fluid ounces of juice you have made: 13 fluid ounces added to 15 fluid ounces.

This example further illustrates the need to refer to the *same unit* in doing arithmetic operations, in this case, addition. In the example, (iii) and (iv) are the only possible interpretations of $13 + 15$ in context. The

---

[1]In mathematical terminology, this says we identify all complete ordered fields and speak of *the* real numbers.

interpretation in (i), to the effect that "13 apples + 15 fluid ounces of juice", would mean the 13 is taken from a number line where the unit 1 has been chosen to be *one apple* but 15 is taken from another line with the unit 1 chosen to be *one fluid ounce of juice*. For the same reason, "13 cups of juice + 15 apples" makes no sense.

There are many everyday examples to illustrate the confusion when addition or subtraction is done using different units. For example, inattention to the proper use of the unit can lead to statements such as $19 + 17 = 3$, ostensibly because "19 eggs plus 17 eggs is 3 *dozen* eggs". The precise definition of addition in Chapter 1 eliminates this kind of confusion by requiring all the whole numbers in the addition to refer to the same unit.[2]

Activity. Each of the following addition and subtraction statements is illegal, but can you make sense of them anyway? $9 - 2 = 1$, $8 + 16 = 2$, and $206 + 82 = 2$. (*Hint*: In no particular order, think of counting in terms of dozens, measuring area, and counting the number of days.)

We pursue the discussion of the need to refer to a fixed unit by looking at an important special case. Suppose we let the unit 1 stand for the *area* of the *unit square*. Recall that the meaning of "unit square" was earlier defined on page 45 in connection with the distributive law: it is the square whose sides all have the length of the unit segment $[0, 1]$. The number 5 means "one can count 5 such areas", and is therefore the total combined *area* of five such unit squares. For convenience, we can stack these squares together and think of 5 as the area of a rectangle whose width is a unit segment and whose length is the concatenation of 5 unit segments. The area of this particular rectangle therefore corresponds to the segment $[0, 5]$:

If there is no fear of confusion, we would simply say *this is a rectangle of width 1 and length 5*. In like manner, the number n is the area of a rectangle whose width is the unit segment and whose length is the segment $[0, n]$.

Still with 1 as the area of the unit square, what is $2 + 3$? It is the total area obtained by first counting 2 unit-square areas and then 3 more, and therefore the same as the area of the following rectangle:

(8.1)

---

[2] In mathematical terminology, what we are saying is this: Suppose a ring $\mathbf{S}$ is isomorphic to the integers $\mathbf{Z}$, and suppose under the given isomorphism that $\bar{n} \in \mathbf{S}$ corresponds to $n \in \mathbf{Z}$. Then although both $\bar{2} + \bar{5}$ and $2 + 5$ make perfect sense, we cannot perform the addition $\bar{2} + 5$ or $2 + \bar{5}$.

We will sometimes speak of this way of combining rectangles as the **concatenation of rectangles**.

Caution. In textbooks and the education literature, one often comes across statements of the following type: *"Let the unit 1 be the square"* or *"Let the unit 1 be the hexagon"* (both possibly in the context of looking at Pattern Blocks). These seem to be similar to the present discussion in making use of a two-dimensional object as a unit, but in fact there is an enormous chasm between what is done here and what is conveyed by the above statements. To us, the unit chosen is the *area* of the unit square so that in order to check what whole number is represented by a given geometric shape, what needs to be done is entirely unambiguous: *measure its area* regardless of what its shape may be. On the other hand, the statement *"Let the unit 1 be the square"* often misleads students into believing that the unit 1 is *the shape* of the square, and shape is unfortunately not a measurable quantity. Moreover, the deceptive simplicity of the statement *"Let the unit 1 be the square"* stops students from inquiring further what its true meaning is, so that in their minds the ambiguity of the concept of a number takes up permanent residence. While such ambiguity does not do students much harm in the context of learning about whole numbers, it becomes an enormous handicap when they confront fractions. In the next chapter, as well as in Part 2, we will hammer home the need for precision in the number concept.

8.3. Multiplication

The concept of multiplication is more subtle than we have let on thus far. It is time to touch on some of these subtleties (compare the last section of this chapter). In Chapter 2, we interpreted the multiplication of whole numbers as area in connection with the distributive law. In the context of the number line, some clarification of the concept of multiplication is necessary. If, for example, $2 \times 3 = 6$, then this 6 has to be the length of a segment as every whole number n is the length of a segment, e.g., $[0, n]$. Yet, how can 6 be also an area? (Recall the meaning of the equal sign "$=$" given in section 2.1, The Equal Sign, on page 39.)

Let a number line be fixed; thus a unit segment has already been chosen. The presence of a unit segment allows us to define the unit square. Now **we also identify the unit 1 of the original number line with the area of the unit square.** It is through this identification that the preceding paradox is resolved. Thus for $2 \times 3 = 6$, this 6 is the area of the obvious rectangle:

(8.2)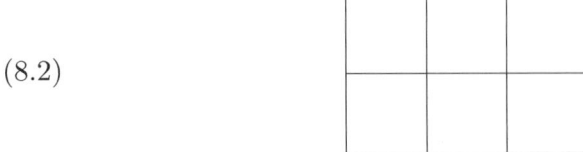

On the other hand, $2 \times 3 = 3 + 3$ by definition. Therefore 6 is also the length of the concatenated segment of two segments of length 3. But through the above identification, these two 6's are the same point on the number line.

Thus for every discussion of multiplication, the number line has to do double duty: *its unit 1 is both the length of a side of the unit square as well as the area of the unit square.* Let this be always kept in mind.

As another example to illustrate this identification, we give one possible interpretation of $5 + (2 \times 3) = 11$: it means that the area obtained by combining the rectangle of (8.1) and the rectangle of (8.2) is the same as the area of the rectangle whose width is 1 and whose length is 11. We shall be discussing the area interpretation of numbers extensively in Part 2.

8.4. Division

Finally, division. For whole numbers a, d, with d always assumed to be nonzero, let us assume for now that a is a multiple of d, say $a = qd$ for some whole number q. Then recall that

(8.3) $\quad a \div d \;$ is the whole number $\; q \;$ satisfying $\; a = qd$.

This is the place to tie up a loose end that arose in the discussion of division in Chapter 7, namely, *why one cannot divide by 0*. In order to fully understand this discussion, the most important thing is to remember the message of Chapter 7: *division is just another way of writing multiplication.* So suppose division by zero makes sense for a particular *nonzero* whole number n, say $n \div 0 = 3$. We must now find the multiplicative statement to which this division statement is equivalent. What could it be? Using (8.3) as a guide, we have to conclude that $n \div 0 = 3$ is another way of expressing $n = 3 \times 0$. But $n = 3 \times 0$ means $n = 0$, which contradicts our starting assumption that n is a nonzero number. The conclusion would still be the same if 3 is replaced by any whole number. Therefore if $n \neq 0$, $n \div 0 = k$ for *any* whole number k is *not* equivalent to any multiplication statement that makes sense. We express this fact by saying that *if n is a nonzero whole number, $n \div 0$ cannot be defined.*

Next, what about $0 \div 0$? Suppose $0 \div 0$ is a legitimate operation. Then we can start with a multiplication statement about 0 and reformulate in

8.4. Division

terms of $0 \div 0$. Thus, since $0 = 0 \times 1$, we express it according to (8.3) as a division statement $0 \div 0 = 1$. This gives one answer to the question "$0 \div 0 = ?$". But it is also true that $0 = 0 \times 2$, so also $0 \div 0 = 2$. We have therefore shown that if the division $0 \div 0$ can be given a meaning, then it must follow that $1 = 2$ because 1 and 2 would both be equal to $0 \div 0$. This is absurd. We were therefore wrong to accord $0 \div 0$ a legitimate status, and $0 \div 0$ must be excluded from our vocabulary. In summary, we have shown that

> *for any whole number n, the division $n \div 0$ cannot be meaningfully defined.*

In the school classroom, it may be instructive to students if a less formal argument is also offered, one that makes use of the partitive and measurement interpretations of division to explain why $n \div 0$ cannot be defined for a nonzero whole number n. Suppose it were definable. By the partitive interpretation, $n \div 0$ would mean the number of objects in a part when n objects are partitioned into 0 equal parts (see section 7.1, Multiplication as Division, on page 97). Because we cannot partition anything into 0 equal parts, this has no meaning. Now suppose $n \div 0$ were meaningful in the measurement sense. Then it is the number of parts when n objects are partitioned into different parts so that each part has exactly 0 objects. But if each part has no object, the partition of the n objects cannot be done. So again, this interpretation has no meaning.

The fact that division undoes multiplication (always understood in the sense of (8.3)) leads to a geometric interpretation of division that has already been mentioned in Chapter 7. Assuming that a is a multiple of d, then $a \div d$ is the other side of a rectangle whose area is a and one of whose sides is equal to d:

(8.4)

$$a \div d \quad \boxed{ a } \\ d$$

To conclude this discussion of division, it remains to point out that although the restriction that a be a multiple of d for the validity of (8.4) seems too severe, it will be seen when we come to the discussion of the division of fractions in Chapter 18 of Part 2 that, once fractions are available, this geometric interpretation will become valid *verbatim* regardless of what a and d may be. The equivalence of division and multiplication as described in (8.3) will be seen to be the key to the understanding of division in general.

8.5. A Short History of the Concept of Multiplication

The representation of the product of two numbers as the area of a rectangle has a long history behind it. Until the time of Descartes (1596–1650), this was the *only* way to understand the multiplication of numbers. In the most influential mathematics textbook of all time, Euclid's *Elements* (circa 300 BC), there was never any mention of multiplying two numbers m and n; see [**Euc56**]. Each time Euclid wanted to express that idea, he would say, "the rectangle contained by the line m and the line n". (Translation: "the rectangle" in Euclid means "the area of the rectangle", "contained by" means "having for its sides", and "the line m" means "the line segment of length m".) For this same reason, a product of three numbers, such as $12 \times 7 \times 9$, had to be interpreted as volume and therefore the product of four or more numbers was almost never considered until perhaps the third century AD, principally by Diophantus. By 1600, when Descartes argued that multiplication could also be regarded as an abstract concept independent of geometry, multiplication began to be universally accepted purely as an operation between numbers. Nowadays, a (good) college course on number systems will develop all number concepts in an abstract algebraic setting without reference to geometry. For this book, however, the geometric interpretation of multiplication not only is convenient for our purpose, but has the added advantage that it is sufficiently similar to the common manipulative base ten blocks to make beginners feel at ease. The purely algebraic approach to multiplication will be briefly discussed in Chapter 29.

Chapter 9

What Is a Number?

We now take up the issue of why we want to define a whole number as a point on the number line.

Consider the question of what the number 5 is. Here we are not talking about a description of our intuitive feelings about "5", but rather a precise definition in the sense of the definition of a **triangle** as three noncollinear points together with the line segments joining them. We know "five fingers". We also know "five books", "five days", or five of anything we see or touch because we can count. But how do we define **five** itself without reference to any concrete object? So you see that it is difficult. Do not be discouraged, however, because the general concept of a number, in the sense defined above as a point on the number line, baffled the human race for over two thousand years before it was finally pinned down in the late nineteenth century. What we need for elementary mathematics is fortunately nothing very sophisticated, just the whole numbers, integers, fractions, and negative fractions, with a few irrational numbers thrown in now and then. In other words, we will not scrutinize *every* point on the number line, only a small portion of those points. This definition of a number as a point on the number line is not ideal, but it serves our pedagogical needs admirably, in the sense that it is accessible and it lends itself to a reasonable treatment of rational numbers and decimals. See Parts 2, 3, and 5.

A precise definition of whole numbers is not strictly necessary if all we ever do in mathematics is to stay within the realm of whole numbers. For example, even if we cannot define precisely what 5 is, we can communicate the essence of it by putting up one hand with the fingers outstretched; that should be enough to communicate any kind of "fiveness" needed for conceptual understanding. This is the advantage of whole numbers: each has (at least in principle) a concrete manifestation, such as outstretched fingers, that almost renders abstract considerations about whole numbers superfluous in elementary school. But we cannot stay with whole numbers forever, because the next topic is fractions. What concrete image can one conjure up for $\frac{13}{7}$ or $\frac{119}{872}$? Children need answers to this question in the most urgent way imaginable because *they need something to anchor the many concepts related to fractions in the same way that they use a handful of fingers to anchor any discussion about* 5 *or any whole number*. Amazingly, school mathematics in our country has contrived to never answer this question. When adults abrogate their basic responsibility by ducking such basic questions, the result is entirely predictable: children become the victims. The generic nonlearning of fractions among children has become part of our national folklore, so much so that you can find many references to it in the comic strips of *Peanuts* and *FoxTrot*. We want to change this dismal scenario by adopting the down-to-earth approach of

> *giving direct answers to direct questions.*

We will define fractions, decimals, and any concept we ever take up.

Now, you may ask, why not just define fractions as precisely as we wish but leave a simple concept like whole numbers alone. The answer is that in order for children to understand fractions, fractions cannot be suddenly thrust on them out of the blue. Learning is a gradual process with each step firmly rooted in prior experiences. If we can convince children that fractions are nothing more than a natural extension of the whole numbers, then our chances of success in teaching fractions should increase immeasurably. For then, children will gain the confidence that they can use what they know about whole numbers as a guide in this new venture. At the moment, however, most (perhaps all) of the school textbooks and professional development materials would have you believe that whole numbers are simple, but fractions are a completely different breed of animal. Therefore, whole numbers are taught one way and fractions another. Sadly, this falsifies mathematics, thereby creating learning difficulties because, as mathematical concepts, whole numbers and fractions are on equal footing. They are part of the same family, the *real numbers*, and the number line provides the natural platform to showcase both. In the next part of this book, we will add fractions to this line and do unto fractions what we have been doing unto whole numbers: we will show how to add, subtract, multiply, and divide fractions in essentially the same way we do with whole numbers. For this reason, we begin with a definition of whole numbers on the number line in order to lay the groundwork for fractions.

Let us consider the philosophical question of why something as natural as a whole number should be made into something as cold and formal as "a point on the number line". The overriding fact is that we must put the whole numbers in a framework so flexible and yet so precise that it can accommodate the inclusion of fractions. This is the price we have to pay for achieving a deeper understanding of numbers. It is in the nature of human affairs that each time we try to achieve excellence in any endeavor, *doing what comes naturally is rarely good enough.* Take running, for instance. This is about as natural an activity as we are ever going to get. In fact, had our ancestors of eons ago been less good at it, all of them would have been hunted down by the predators on the African savannas and there wouldn't have been any *Homo sapiens* left to talk about fractions today. So running is entirely natural. Yet, when you talk to the Olympic sprinters, what you hear from them about running must strike you as extremely unnatural, if not downright bizarre.

If you believe an Olympic sprint event is just "getting down in the blocks, get set, and run", then you had better think again. First there is the matter of each runner adjusting the precise positions of the starting blocks in the lane and setting them up at the correct angle. This is a delicate business, and computer software has been created to optimize a runner's performance in this phase of running. Then there are about ten items on a checklist that runners should observe when they are on their marks. There are another six or seven items to check right before they are set to start, and an additional six or seven items after they have started to run, e.g., exhale, drive the arms hard, etc. In the so-called acceleration phase, the runners have another checklist of more than ten items before they can concentrate on the "stride phase", which is what we normally call "running".

Doesn't this unnatural and calculated approach to running remind you of looking at a whole number as a point on the number line?

But let's not lose our perspective. Whatever the Olympic sprinters do in a race, it is highly unlikely that they think about "keeping the face, mouth, neck, and shoulders relaxed" each time they run to catch a plane. In the same way, you need not fixate on the number line every time you count oranges in the supermarket. But, as shown on the previous pages, any thorough understanding of fractions, including decimals, must be based on a fluent use of the number line. So please try to rise to the occasion when there is a need to regard a whole number as a point on the number line and please be willing to work with your colleagues and your students on this crucial concept. If you can accept this reality, then you have already won half the battle.

Chapter 10

Some Comments on Estimation

> *Visitors to a museum were told by the curator that a certain artifact was approximately 500,013 years old. When they asked him how the age was determined so precisely, he answered "When I came to work here I was told that it was approximately half a million years old, and I've been working here for 13 years!"*

Students in primary school learn to round off numbers to the nearest ten, hundred, and thousand. They are taught the *procedure* of rounding without being told what rounding means, why they would want to round off, or when they should round off. As a result, students have no idea what estimation is about (see the museum curator above). The purpose of this chapter is to give a brief discussion of these issues. The basic lesson here is that mathematics is precise, and it remains so even when it gives out imprecise information, i.e., makes estimates.

The first section gives a precise definition of rounding, and then shows how the usual procedure taught in school follows logically from the definition. It may be observed by now that giving a precise definition of a concept before deriving the procedure associated with the concept is a recurring theme throughout this book. In addition to "rounding", other examples are division-with-remainder, the sum of two fractions, the product of two fractions, the sum of two (positive or negative) rational numbers, etc.

The exposition of this chapter breaks precedent with what we have done so far, which is to be self-contained *in the sense that we explain every concept before we put it to use.* The consideration of estimation requires that we make use of *percent* and *decimals*, concepts we will not take up until Part 2. If we insist on being self-contained, then we would have to postpone this discussion until after Part 2. But estimation, as remarked, makes its appearance in primary school, and a good deal of it makes sense as soon as whole numbers are introduced. For this reason, we take up this topic here, and compromise by asking the reader to skip part of this section if necessary, and revisit it after reading Part 2.

The sections are as follows:

Rounding

Absolute and Relative Errors

Why Make Estimates?

A Short History of the Meter

In the last section, we trace the history of the length of a meter, which is surprisingly relevant to the topic of estimation.

10.1. Rounding

We begin with a precise

Definition. *To* **round a whole number n to the nearest ten** *means to replace n by the multiple of* 10 *which is closest to n. If two multiples of* 10 *are equally close to n, the* CONVENTION *is to always choose the bigger number.*

For example, to round 248 to the nearest ten, we look at all multiples of 10 in the neighborhood of 248,

..., 210, 220, 230, 240, 250, 260, 270, ...,

and it is relatively easy to see that 250 is the number we seek. If the number is 244, then the rounded number would be 240. For 245, both 240 and 250 are equally close to 245 and 250 is the bigger of the two, so 245 rounds to 250.

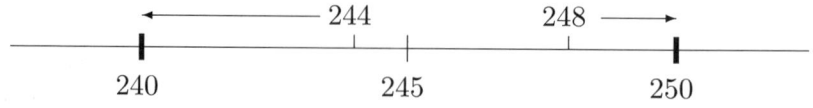

The convention to round to the bigger number in cases like 245 is an American one. Some countries adopt a different convention.

10.1. Rounding

In standard textbooks, students are usually taught to round a number n to the nearest ten by the following algorithm: If the ones digit of n is ≤ 4, change it to 0 and leave the other digits unchanged. But if the ones digit is ≥ 5, then change it to 0 but also increase the tens digit by 1 and leave other digits unchanged. This is correct in most cases, but collapses completely in the case of a number such as 12996. By the definition above, however, it is not difficult to carry out this rounding: since among 12980, 12990, 13000, 13010, ..., the multiple of 10 closest to 12996 is 13000, so the rounded number is 13000.

Nevertheless, there is a correct formulation of the algorithm to round a whole number n to the nearest ten. We can see it from an example such as 12996. We have $12996 = 12990 + 6$, and the rounding to the nearest 10 has nothing to do with 12990 but everything to do with 6: 6 itself rounds to 10, and it follows that 12996 rounds to $12990 + 10 = 13000$, as before. The same reasoning leads to the following general algorithm:

> Write n as $N + \overline{n}$, where $\overline{n}$ is the single-digit number equal to the ones digit of n (and hence N is the whole number obtained from n by replacing its ones digit with 0). Then rounding n to the nearest ten yields the number which is equal to N if $\overline{n} < 5$, and equal to $N + 10$ if $\overline{n} \geq 5$.

Rounding a whole number n to the nearest hundred (resp., thousand, etc.) can be similarly defined. For example, **rounding a whole number n to the nearest thousand** means to replace n by the multiple of 1000 which is closest to n. If two multiples of 1000 are equally close to n, the CONVENTION is to always choose the bigger number.

The corresponding algorithm is then:

> Write n as $N + \overline{n}$, where $\overline{n}$ is the three-digit number equal to the last three digits (to the right) of n. Then rounding n to the nearest thousand yields the number which is equal to N if the left digit of $\overline{n}$ is < 5, and equal to $N + 1000$ if the left digit of $\overline{n}$ is ≥ 5.

Activity. Explain to the person sitting nearest you why this algorithm is correct.

We illustrate rounding to the nearest thousand with some examples. Note that the N above is the whole number obtained from n by replacing its last three digits on the right with 000. Note also that the left digit of $\overline{n}$ is the hundreds digit of n. So to round 45,297 to the nearest thousand, we write it (according to the algorithm) as $45000 + 297$. Since the left digit 2 of 297 is < 5, the rounded number has to be 45,000. On the other hand, to round 49,501 to the nearest thousand, we write it as $49000 + 501$. Since the left digit 5 of 501 is ≥ 5, the rounded number is $49000 + 1000 = 50000$. Similarly, rounding 729,998 to the nearest thousand gives 730,000.

Activity. Round 20,245,386 to the nearest ten, nearest hundred, nearest thousand, nearest ten-thousand, and nearest million, respectively. Do the same to 59,399,248.

It remains to point out that the concept of rounding, which has been applied only to whole numbers thus far, is perfectly applicable to decimals as well. We recall a concept from page 99: given a number n (which need not be a whole number), the number kn where k is a *whole number* is called a **multiple** of n, or **whole-number multiple** for emphasis. Now, if a decimal such as 26.8741 is given, then **rounding it to the nearest hundredth (0.01)** means replacing it by the whole-number multiple of 0.01 that is closest to 26.8741.

To find this rounded number, we concentrate on the multiples of 0.01 that are near 26.8741:

$$\ldots, 26.85,\ 26.86,\ 26.87,\ 26.88,\ 26.89,\ \ldots.$$

The answer is 26.87, by inspection. More precisely, $26.88 - 26.8741 = 0.0059$, while $26.8741 - 26.87 = 0.0041$. Because $0.0059 > 0.0041$, the rounded number has to be 26.87.

We can also formulate a precise algorithm for this purpose. We need to fix another piece of terminology. Given a decimal such as 26.8741 above, its **first decimal digit** is 8, its **second decimal digit** is 7, its **third decimal digit** is 4, etc. Then

> To round a decimal m to the nearest hundredth, write it as $m = M + \overline{m}$, where M is the decimal whose digits (starting from the left) agree with those of m up to and including the second decimal digit, and are 0 elsewhere. Then the rounded decimal is M if the third decimal digit of $\overline{m}$ is < 5, and it is equal to $M + 0.01$ if the third decimal digit of $\overline{m}$ is ≥ 5.

The decimal $\overline{m}$ is the one obtained from m by replacing every digit of m to the left of its third decimal digit by 0.

Activity. Show that this procedure is correct.

For example, to round 26.8741 to the nearest hundredth, we write it as $26.87 + 0.0041$. Since the third decimal digit 4 of 0.0041 is < 5, the rounded number is 26.87, which confirms the previous answer. As another example, to round 59.99725 to the nearest hundredth, we write it as $59.99 + 0.00725$. Since the third decimal digit 7 of 0.00725 is > 5, the rounded number is $59.99 + 0.01 = 60$.

10.1. Rounding

The formulation of the definition of rounding a decimal to the nearest $0.0\cdots01$, where the 1 is the kth decimal digit, is similar to the case of whole numbers and will be left as an exercise.

Activity. Round 1.70995 to the nearest 10^{-k}, for $k = 1, 2, 3, 4$.

Now we take up the question of why anyone would ever want to round off any number. Consider the population of Berkeley (California) which, according to the city website, is 102,743 (Census 2000). One could raise a legitimate question about how meaningful such a figure is. It is not just the variability of a city's population over time (needless to say, the census was taken sometime before 2000), but even at the moment the census was taken, such a figure had to be inherently inaccurate. Given the unpredictability of death, the mobility of the population (especially considering the large student population), and the acknowledged inability to get all the residents to participate in the census, a conservative estimate is that at least the last three digits, 743, are meaningless. One way the city of Berkeley could have indicated the uncertain nature of this figure would be to avoid any reference to 743 by rounding 102,743 to the nearest thousand. As we have seen, this rounding yields the figure of 103,000. Thus, one can say that the population of Berkeley is about 103,000, and the three zeros at the end of this figure are usually taken to be due to rounding. As a matter of common sense, one would consider "about 103,000" to be a more reasonable description, in context, of the population of Berkeley than "102,743". This is one example of why one wants to round off figures.

One can make a case that even the thousands digit of 102,743 (i.e., 2) is suspect. In that case, we want to eliminate any reference to 2,743 altogether by rounding 102,743 to the nearest ten-thousand. (At the end of the next section, **Absolute and Relative Errors**, we will explain in a quantitative way why we can afford to round off this way.) Thus to the nearest ten-thousand, the population of Berkeley is about 100,000. Now in everyday conversation, you are more likely to hear "The population of Berkeley is about a hundred thousand" than "The population of Berkeley is about a hundred and three thousand". So the decision to round to the nearest ten-thousand rather than the nearest thousand, while giving a cruder figure, better serves the purpose of "giving a ballpark figure" in everyday life.

As another example of the need for rounding, imagine that you are the leader of a research team and members of your team came up with the result of 58.41672 cm for a particular measurement. But you know that the instrument used in that measurement, while capable of giving readings containing many digits past the decimal point, is accurate only up to the second decimal digit, i.e., only 58.41 is a reliable figure. In this case, you have no choice but to round off to the nearest hundredth to avoid giving out misinformation. In your research report, you therefore give the measurement as 58.42, and the scientific community would understand, as a matter of convention, that the last digit (i.e., 2) involves rounding. This is then another situation where rounding is called for.

10.2. Absolute and Relative Errors

Rounding introduces error. It is therefore imperative that you be aware of the error that comes with the rounding (to the nearest hundred? nearest thousand?) and whether or not the magnitude of the error is something you can live with.

The need of this awareness is well illustrated by the following example. A fourth grade textbook illustrates the usefulness of rounding by looking at an addition problem: $127 + 284 = ?$ After computing directly to get the answer 411, it proceeds to use rounding to check the correctness of 411. It says 127 rounds down to 100 while 284 rounds up to 300, so $127 + 284$ is roughly $100 + 300 = 400$. Since 400 is close to 411, according to this textbook, 411 is a reasonable answer.

The mathematics of the preceding paragraph is extremely flawed and serves as an object lesson on how *not* to teach the usefulness of rounding. To see this, consider the following two questions. First, in saying 411 is close to 400, what is meant by being "close"? Second, should students expect similar closeness each time they estimate the sum of two 3-digit numbers? We begin with the second question.

Suppose we have to add $149 + 147$. Rounding to the nearest hundred gives $100 + 100 = 200$, but the exact answer is of course 296. Right away we see that, no matter how "closeness" is defined, 200 is not close to 296 the same way that 400 is close to 411. Rounding to the nearest hundred is therefore not a good way to check the correctness of additions of 3-digit numbers.

Next, to quantify "closeness", we introduce two standard concepts. The difference between the correct value and an estimated value (always taken to be a positive number) is called the **absolute error**. The ratio

$$\frac{absolute\ error}{correct\ value},$$

which is commonly expressed in percent, is called the **relative error** of the estimation. What the relative error does is to put the absolute error in perspective. Let us explain the last sentence by way of an analogy. Suppose you get two problems wrong on each of two exams, and the two exams have three questions and 15 questions, respectively. Then the number 2 by itself does not tell the full story of your performance; you need to consider that you got $\frac{2}{3}$ of the first exam wrong and but only $\frac{2}{15}$ of the second wrong before you see the difference. This idea of putting the number 2 in the

10.2. Absolute and Relative Errors

proper context is a basic motivation for introducing the concept of relative error.

For $127 + 284$, the absolute error is $411 - 400 = 11$, and the relative error is

$$\frac{11}{411} \approx 3\%,$$

where "$\approx$" means "approximately equal to" and we have rounded off the percent to the nearest one. On the other hand, the estimation of $149 + 147$ as 200 has an absolute error of $296 - 200 = 96$, so that its relative error is

$$\frac{96}{296} \approx 32\%.$$

Although there is no universally accepted definition of the concept of "closeness"—and it must be admitted that what is close or not close depends on the particular situation at hand—one can nevertheless assert that a relative error of more than 10% *generally* would not qualify as "close". The textbook's mistake is not to have forewarned students about the pitfalls of the proposed method. This kind of instruction on the use of estimation is therefore unacceptable.

We now show how one can make use of estimation in a positive way by showing that rounding to the nearest hundred is an effective way to check the addition of 4-digit numbers. Take $4257 + 3461$. Rounding to the nearest hundred yields an estimate of $4300 + 3500 = 7800$. Since the correct answer is 7718, the absolute error is $7800 - 7718 = 82$, and the relative error is

$$\frac{82}{7718} \approx 1\%.$$

This shows that the estimation is close to the correct answer, by any definition of "close". We can argue in general. Suppose we have two 4-digit numbers m and n. If we round both numbers to the nearest hundred and add, we introduce an absolute error of at most 100 to the addition, for the following reason. Rounding m alone leads to an absolute error of at most 50 because in choosing among all multiples of 100 for the rounding (see the definition of rounding in the preceding section), m is never more than 50 from a multiple of 100. Same for n. The combined error in the addition is therefore never more than 100. In the case of $4257 + 3461$, we found the absolute error to be 82; one can see easily that for $4250 + 3450$, the absolute error becomes exactly 100. Now to return to the task at hand, both m and n being 4-digit numbers, the **leading digit** (i.e., the left digit) of m or n is at least 1, so that the leading digit of $m + n$ is at least 2. *Now the leading digit of a 4-digit number is the thousands digit, and therefore $m + n$ is at least 2000.* Thus

$$(m+n) \geq 2000 \quad \text{and therefore} \quad \frac{1}{m+n} \leq \frac{1}{2000}.$$

(We will prove in Chapter 15 that if two whole numbers m, n satisfy $m \geq n > 0$, then $\frac{1}{m} \leq \frac{1}{n}$.) The relative error of the estimation of $m + n$ by rounding m and n to the nearest hundred is then at most

$$(10.1) \qquad \frac{100}{m+n} \leq \frac{100}{2000} = 5\%.$$

The conclusion is that

if we estimate the sum of two 4-digit numbers by rounding to the nearest hundred, the relative error is at most **5%**.

As we said, a relative error of 5% gives us confidence that the estimate is reasonable.

In the last argument, we saw the importance of the place value of the leading digit of a number. If the numbers had been 3-digit ones, then all we could have said was that $m + n \geq 200$ and the computation in (10.1) above would have been

$$\frac{100}{m+n} \leq \frac{100}{200} = 50\%.$$

Of course, to have a relative error which might be as great as 50% is nothing to write home about, and this is consistent with the earlier comment that for 3-digit numbers, rounding to the nearest hundred is no way to check the accuracy of addition.

In making estimates, it is critically important to be aware of the relative error of each estimate.

It is time to point out that the place value of the leading digit of a whole number is called the **order of magnitude** of a number. Recall that if the order of magnitude of a whole number is 10^n, then the number has $n + 1$ digits. So one may think of order of magnitude as a different indication of "the number of digits" of a number. The order of magnitude of a number is the most basic statement about the size of a number and, not infrequently, the only thing that one cares to know about a number. For example, the national debt as of April 22, 2006, is \$8,379,388,245,684.45, which is of course computed according to a fixed formula and therefore is given down to two decimal places.[1] The order of magnitude is therefore 10^{12}. Eight *trillion* plus. The fact that it is given down to two decimal places is obviously not to be taken seriously: it is a constantly evolving figure that changes with every minute and we as a nation could not care less whether we owe 95 cents more or less. It would be fair to say that "about 8 trillion" is all that matters to an average citizen concerned with this figure.

Let us look at the national debt more closely. How much information do we lose if we only know that it is "about 8 trillion"? Now "8 trillion" implies that *we are rounding to the nearest* 10^{12}. So 8 trillion could be as

---

[1] It is nevertheless sobering to note that our national debt as of April 23, 2004, was only \$7,141,602,592,641.44.

high as $8,499,999,999,999 or as low as $7,500,000,000. The absolute error is at most 5×10^{11}, and the relative error is then at most

$$\frac{5 \times 10^{11}}{7.5 \times 10^{12}} \approx 6.7\%.$$

Thus *by clinging to the simplistic view that the national debt is 8 trillion, which amounts to throwing away 12 digits of a 13-digit number, we are off by only* 6.7%. This makes the trade-off of precision for simplicity altogether worthwhile.

Activity. The area of the U.S.A. is 9,629,091 square kilometers. Round it to the nearest thousand, ten-thousand, hundred-thousand, and million. In everyday conversation, which of these rounded figures do you think would be most useful? What is the relative error in that case?

Finally, let us bring closure by returning to our initial example of the population of the city of Berkeley, which is 102,743. The reason that the last four digits 2743 are unreliable has been explained earlier. Let us estimate the relative error of rounding this figure to the nearest 10^4, which then gives a rounded figure of 100,000. Now, although there is no such thing as "the correct value" of Berkeley's population, we can at least make a stab at it. Believing the last four digits of 102,743 to be unreliable, we may assume that the population is, *at the least*, off by 5,000, so that 97,743 is the smallest possible correct population figure for Berkeley. The biggest possible "correct population" of Berkeley is $102743 + 5000 = 107743$, so that the biggest absolute error would be $107743 - 100000 = 7743$. Therefore the relative error of the estimation of the population as 100,000 does not exceed

$$\frac{7743}{97743} \approx 8\%.$$

An error of at most 8% in an everyday context is altogether tolerable.

10.3. Why Make Estimates?

Building on the examples given in the preceding sections, we now give a comprehensive summary of some of the reasons why one would use estimations in place of precise numbers. There are at least three.

(I) *Precision is unattainable or unavailable.* Practically all measurements made in the sciences are approximations. This is because every instrument has built-in limitations to its accuracy. The simplest illustration of this fact is to try to measure the length of your desk by an ordinary metric ruler which has clear markings of millimeters but nothing smaller. Thus the instrument you use for this measurement is only reliable up to one millimeter, i.e., $\frac{1}{10^3}$ of a meter. Therefore, even assuming that your ruler is

100% accurate,[2] your final measurement, given in meters, will be accurate only up to the third decimal digit. If you give a figure with four decimal digits, then the fourth decimal digit could only be an estimate and should be rounded off.

It may be of some interest to note that there is one measurable quantity that is *completely* precise, but that is because of a **fundamental hypothesis** in physics: The speed of light[3] is *exactly* 299,792,458 meters per second, and it is not an estimate because the General Conference on Weights and Measures (CGPM) decided in 1983 to *define* the length of a meter to be 1/299,792,458 of the distance traveled by light in one second. See the last section, A Short History of the Meter on page 150.

Another class of examples of numbers with built-in inaccuracies are astronomical measurements of the distances of stars. You are probably so used to reading about such-and-such a star is x light years away that you do not realize how fantastic it is to be able to make estimates of such astronomical distances. Here we are talking about stars other than the sun, the nearest of which is about 4.3 light years away. Thus if we send a signal to the nearest star other than the sun, it would take 4.3 years to get there. Any thought of a direct measurement is therefore out the window. What estimated distances we come up with have to be the results of extremely sophisticated and indirect inferences. Therefore measurements in an astronomical context are generally estimates and not exact.

Another kind of need for estimations rather than precise values arises from our imprecise use of language. We speak of the "distance" we travel to go to work, for example. We also speak of "the temperature" of a city on a given day. We will leave as an exercise (Exercise 13 at the end of this section) to explain why these are inherently imprecise concepts.

As final examples of the unattainability of precision, we recall the notion of a national debt and the population of a city or state, or for that matter, the nation. These are clearly numbers that have been reluctantly extracted from a sea of ambiguities.

(II) *Precision is unnecessary.* Next time you enter an elevator, look for a sign that says "Capacity x lbs." This x may be 4000, or it may be 2500, but it is hardly likely that if a load of 4001 lbs. is put in the elevator in the former case, the elevator would immediately drop to the bottom of the building. So the main purpose of this number x is more to give a general warning against overloading than a precise determination of carrying capacity. More likely the true capacity has been rounded down from something like 4500 lbs. to 4000 for reasons of safety.

---

[2] This is of course impossible.

[3] Strictly speaking, it is the speed of light *in a vacuum*.

10.3. Why Make Estimates?

Consider now the case of weather reporting. Sometimes one hears during freezing winter months announcements about the day's "balmy weather in the sixties". If one insists on precision, such an announcement would be abominable. But precision is hardly called for when the overriding concern is whether you have to go out bundled up in a Siberian overcoat.

Next, we return to the case of the national debt. Consider the following data:

On April 22, 2006, it was \$8,379,388,245,684.45.
On February 23, 1996, it was \$5,017,056,630,040.53.

With these precise numbers in the background, now think about the alternative, *imprecise* rendering of these facts by the simple statement that "our national debt ballooned from 5 trillion to 8 trillion in a span of ten years from 1996 to 2006". Some would say that the latter conveys as much information as a statement giving the precise figures. In a similar vein, it can be argued that for everyday purposes, knowing that the world's population is "about 6 billion people in 2004" is as good as knowing the census figure of 6,610,401,734 people worldwide in April of 2006.

The moral of all these examples is that, sometimes, less is more. Less precision may serve one's needs as well as the full story.

(III) *Estimation is used as an aid to achieve precision.* The most obvious example here is division-with-remainder (Chapter 7), which begins with an estimate—given whole numbers a and b with $b > 0$—of the number q so that $qb \leq a$ but $(q+1)b > a$. Then we obtain the precise value of the remainder of the division as $a - qb$. Note the parallel between estimating the quotient q and rounding a number a to the nearest hundred or nearest thousand (see section 7.4, A Mathematical Explanation (Preliminary), on page 110): both involve trapping a given whole number between two successive multiples of a fixed number, the multiples of b in the case of division-with-remainder and the multiples of 100 or 1000 in the case of rounding to the nearest hundred or nearest thousand. The long division algorithm (Chapter 7) then carries this process of estimation one step further by requiring the estimate of a quotient at each step of the algorithm.

In the last section, we saw how to use rounding *properly* to partially confirm an addition ($4257 + 3461 = 7718$). Of course estimation can be used to check the other arithmetic operations as well. Let us give a crude example: could $78 \times 86 = 5608$ be correct? No, because $70 \times 80 = 5600$, and 78×86 should be quite a bit bigger than 70×80. Estimations are most powerful for checking order of magnitude. For example, 285×461 cannot possibly be equal to 87385, because rounding 285 down to 250 and 461 to 400 shows that $285 \times 461 > 250 \times 400 = 100,000$, so that 285×461 should be at least a 6-digit number.

The need for estimation also arises in other contexts. Suppose you are in a supermarket and you have put seven items in the basket but you only have $15 with you. It suddenly occurs to you that maybe you do not have enough money. You must make a quick estimate that the seven items do not cost more than $15. You have neither pen and paper nor calculator with you and you must do the calculations in your head. The price tags read: $1.25, $3.25, $1.39, $0.99, $1.49, $2.42, $2.79. You round everything to the nearest dollar and add: $1 + 3 + 1 + 1 + 1 + 2 + 3 = 12$. *But you also know that each rounding introduces an error of $0.50*, so that if the errors were cumulative, the estimation of the cost could be off by $7 \times 0.5 = 3.5$ dollars. The total cost of the seven items could therefore be as high as $12 + 3.5 = 15.5$ dollars. If this were the case, you would be in trouble. This calls for a double-check: the error could be 3.50 only if you have rounded *down* each time, but you actually rounded *up* twice: $0.99 and $2.79. Therefore the error is at most $5 \times 0.5 = 2.5$, in which case the total cost is at most $12 + 2.5 = 14.5$ dollars. So you are safe. At the cash register, you have your estimate confirmed: the total comes to $13.58.

The preceding example gives meaning to the following statement made at the beginning of the last section. We use it to conclude this discussion of estimation:

> It is imperative that you be aware of the error that comes with any rounding and whether or not the magnitude of the error is something you can live with.

10.4. A Short History of the Meter

The meter was promulgated to be the unit of length in 1795 under Napoleon's order, and was defined to be 1/10,000,000 (one ten-millionth) of a quarter of the circumference of the earth as measured by the meridian passing through Paris (and the North Pole). This was a grandiloquent decision not grounded in common sense because, given the unevenness of the surface of the earth and the fact that such a meridian is not really a circle (it is more-or-less an ellipse; see Exercise 10 below), this circumference does not lend itself to an easy, precise measurement. Consequently, the length of the meter was, from the very beginning, a fictitious absolute standard at best. In fact, modern measurements yield a value of this quarter circumference about two thousand meters longer than the putative ten million meters. For this reason, the exact value of the meter was revised several times in the past two hundred years.

In the meantime, the question of whether the speed of light in a vacuum is a constant or a variable quantity was heavily debated towards the end of the nineteenth century. The fact that it is constant was first verified to within acceptable limits of accuracy by the Michelson–Morley experiment in

10.4. A Short History of the Meter

1887, and was adopted as a fundamental hypothesis in physics by Einstein for his theory of special relativity in 1905. When in 1975, the speed of light was measured by lasers to be 299,792,458 meters per second, with an accuracy up to 1.2 meters per second, the radical idea of *recalibrating the meter in terms of the speed of light* began to take hold. The subsequent definition of the meter by the General Conference on Weights and Measures ten years later as 1/299,792,458 of the distance traveled by light in a vacuum in one second was the result.

Exercises

1. Give a precise definition of **rounding a whole number n to the nearest 10^k** (for a whole number k), and also give the precise algorithm of the rounding.

2. Use **rounding to the nearest 50** (i.e., find the multiple of 50 that is closest to the given number) to check the addition of $127 + 284 = 411$. What is the absolute error, and what is the relative error? Also discuss in general the absolute error and relative error if we use rounding to the nearest 50 to check the addition of two 3-digit numbers. What about the addition of two 4-digit numbers?

3. Use rounding to the nearest ten to check $127 + 284 = 411$. Again, what is the absolute error, and what is the relative error? Do the same for the addition of two 3-digit numbers in general.

4. Round 61,499,995 to the nearest ten, nearest 10^5, and nearest 10^6.

5. Round 950,249,936 to the nearest 10^k, with $k = 1, \ldots, 8$.

6. Give a precise definition of **rounding a decimal m to the nearest $\frac{1}{10^k} = 0.0\cdots01$** ($(k-1)$ zeros after the decimal point), where k is a positive whole number. Also give the precise algorithm of the rounding.

7. Round 1.70995 to the nearest $\frac{1}{10^k}$, for $k = 1, 2, 3, 4$. Do the same for 0.0028394.

8. A student wants to purchase a hi-fi system with a price tag of $3285. He makes $165 a week. Assuming that he can save all his weekly earnings for this noble goal, use mental math—and rounding to the nearest 50— to estimate how many weeks it would take him to save enough money to make the purchase. Compare your estimate with the correct answer.

9. The earth (like all planets in the solar system) revolves around the sun in an elliptical orbit, with a maximum distance of 152,007,016 km and a minimum distance of 147,000,830 km from the sun. Yet this orbit is presented in everyday life as well as in most atlases as circular.[4] Can you think of any reason why this over-simplified representation of the orbit is not entirely unreasonable? (*Hint:* Think of relative error.)

10. The earth is not a sphere, but rather an *ellipsoid* which results from rotating an ellipse around the line joining the poles. The equatorial diameter is 12,756.8 km while the distance between the poles is 12,713.8 km. Is it reasonable to present the earth as a sphere for most purposes?

11. Australia is often referred to in tour books as being "roughly the same size" as the U.S. The total area of Australia is 7,617,930 sq km and

---

[4]Unfortunately, an assessment item in a state standardized test on mathematics blandly asked for the radius of the earth's orbit, and that is not so good.

the total area of the *contiguous* United States is 9,158,960 sq km. Is it reasonable to say they are roughly the same size?

12. The population of the U.S. is 292,922,990, according to the census of 2000. What would it be if it were rounded to the nearest 10^7? What is the relative error of the rounding? And if rounded to the nearest 10^8? Would it be reasonable to just say the population is about 300 million?

13. Discuss why it is impossible to discuss with total precision the concepts of "the distance from one's house to one's school" and "the temperature of the day".

14. In March of 2004, the media reported that the two Martian rovers, Spirit and Opportunity, were sending back from 106 million miles away images of a rock with rippled layering, showing wave motion of water eons ago. The figure of "106 million miles" clearly indicated rounding to the nearest million (10^6). What is the maximum relative error in making such a rounding?

Chapter 11

Numbers in Base b

The purpose of this chapter is to give a brief discussion of numbers in base b for an arbitrary positive whole number b, $b > 1$. Our numeral system corresponds to the case of $b = 10$. This discussion complements section 1.1, How to Count, on page 6, in that instead of approaching numbers in base b from the vantage point of counting, we now approach it from the vantage point of the expanded form of a number. The hope is that such a discussion will put the arithmetic properties of our (Hindu-Arabic) numeral system in a new perspective.

The sections are as follows:

Basic Definitions

The Representation Theorem

Arithmetic in Base 7

Binary Arithmetic

11.1. Basic Definitions

In Chapter 1 we introduced the concept of the *expanded form* of a number. For instance, 36402 is equal to

$$(11.1) \qquad 3 \cdot (10^4) + 6 \cdot (10^3) + 4 \cdot (10^2) + 0 \cdot (10^1) + 2 \cdot (10^0),$$

where, for the particular needs of this section, we have adopted the *algebraic convention* of using a raised dot $\cdot$ to replace the multiplication sign $\times$. Expression (11.1) is a sum of multiples of decreasing powers of 10, and only single-digit multiples are employed. Recall also that these multiples of powers of 10, namely, $\{3, 6, 4, 0, 2\}$ are called the *coefficients* of the expansion.

We now approach the expression (11.1) from a different angle. We first extend the concepts of "exponent" and "power" to numbers other than 10. If b is any positive number and n is a whole number, then by definition,

$$b^0 = 1, \quad \text{and} \quad b^n = \underbrace{bb\cdots b}_{n} \quad \text{if} \quad n > 0.$$

Thus $1296^0 = 1$, while $17^5 = 17 \times 17 \times 17 \times 17 \times 17$, and $5^3 = 5 \times 5 \times 5$. As before, the n in b^n is called the **power** or **exponent** of b. We also note that, as in equation (1.4) on page 30, we have

$$b^m b^n = b^{m+n}$$

for any whole numbers m and n. This is one of the so-called **laws of exponents**, and it will be used without explicit mention in the following.

We can express 36402 as a sum of multiples of decreasing powers of any positive whole number b, *using only multiples* $< b$. (This is not at all obvious, but will be explained in the next section.) For example, with $b = 7$, and $b = 12$, we have

(11.2) $36402 = 2 \cdot (7^5) + 1 \cdot (7^4) + 1 \cdot (7^3) + 0 \cdot (7^2) + 6 \cdot (7^1) + 2 \cdot (7^0)$

and

(11.3) $\quad 36402 = 1 \cdot (12^4) + 9 \cdot (12^3) + 0 \cdot (12^2) + 9 \cdot (12^1) + 6 \cdot (12^0).$

Expressions (11.2) and (11.3) are called, respectively, the **base 7** and **base 12 expansions** (or **representations**) of 36402. With a calculator available, it is painless to verify that (11.2) and (11.3) are correct. The symbols **(211062)$_7$** and **(19096)$_{12}$** will be used to denote the expansions (11.2) and (11.3), respectively, and the rationale of these symbols is easily explained. For example, the subscript 7 of $(211062)_7$ indicates that this is a base 7 expansion, and the sequence 211062 records the so-called *coefficients* of the respective powers of 7 from left to right of (11.2).

In general, if b is a positive whole number and n is any whole number, then a sum of multiples of decreasing powers of b,

(11.4) $\quad\quad\quad\quad a_n b^n + a_{n-1} b^{n-1} + \cdots + a_1 b^1 + a_0 b^0$

so that each a_n, a_{n-1}, ..., a_0 is a whole number $< b$, will be called a **base b expansion**, or a **base b representation**, and will be denoted by $(a_n a_{n-1} \cdots a_1 a_0)_b$. The a_i for each i is called a **coefficient** of the expansion.[1] For example, with $b = 12$ and $a_4 = 1$, $a_3 = 9$, $a_2 = 0$, $a_1 = 9$, $a_0 = 6$, while $a_i = 0$ for any other whole number i, we get $(19096)_{12}$, and (11.4) becomes (11.3) for these values of b and a_j for all whole numbers j. From this point of view, the expanded form (11.1) is nothing other than a **base 10 expansion**, or **decimal expansion**, of what we normally write

---

[1] We are forced to employ symbols with a subscript here, because there is no other way to represent n whole numbers whose values are unspecified when n can be arbitrarily large. Please read section 1.3, The Use of Symbolic Notation, on page 21 again.

11.1. Basic Definitions

as 36402. Of course, 36402 should have been written as $(36402)_{10}$, but the primacy of the decimal (Hindu-Arabic) numeral system in our culture decrees that the simpler notation of 36402 suffices.

Activity. (a) Write out in base 2, 5, and 7 the first 20 numbers starting with 1 (rather than 0). (b) What is the base 3 expansion of 79?

We make a few comments to round out the picture. First of all, the definition of a base 5 expansion says that the expression
$$3 \cdot (5^4) + 6 \cdot (5^3) + 4 \cdot (5^2) + 2 \cdot (5^0)$$
is not a number in base 5 because the second coefficient 6 exceeds 5. This is analogous to the fact that the expression
$$19 \cdot (10^3) + 7 \cdot (10^2) + 2 \cdot (10^1) + 36 \cdot (10^0)$$
is not the decimal expansion (i.e., the expanded form) of a whole number. A second comment concerns notation. An expression such as
$$11 \cdot (12^6) + 8 \cdot (12^4) + 10 \cdot (12^1) + 5 \cdot (12^0)$$
is a base 12 expansion, and we will denote it by
$$(\underline{11}\,0800\,\underline{10}\,5)_{12},$$
where the bars under 11 and 10 indicate that 10 and 11 are the coefficients of the base 12 expansion. Otherwise, without the bars, $(110800105)_{12}$ could mean
$$1 \cdot (12^8) + 1 \cdot (12^7) + 8 \cdot (12^5) + 1 \cdot (12^2) + 5 \cdot (12^0).$$

Activity. (a) What are the following numbers in our decimal numeral system? $(6507)_8$, $(101110101)_2$, $(\underline{58}\,\underline{20}\,48\,\underline{16})_{60}$. (b) What is the base 5 expansion of 98?

Two special values of the base b should be singled out: $b = 2$ and $b = 60$. In case $b = 2$, it is customary to say **binary expansion** (or **representation**) and **binary numbers** in place of "base 2 expansion" and "numbers in base 2". This is the preferred numeral system in computer science because the computer stores its data in **bits**, and each bit can be in only one of two states, *on* or *off*, hence 1 or 0. On the other hand, each coefficient of a binary expansion, being a whole number smaller than 2, can only be 0 or 1. Therefore binary numbers are a perfect fit for the needs of the digital computer. (Notice that on most computers, the switches are 0 or 1 instead of "off" or "on".) We will devote the last section of this chapter to a discussion of binary numbers.

The case of $b = 60$ was used by the Babylonians some 4000 years ago; numbers in this system are called **sexagesimal expansions** (or **representations**) and **sexagesimal numbers**. It is easy to explain why you cannot ignore the sexagesimal numeral system: "$(2\,\underline{52}\,\underline{15})_{60}$ seconds" is what is known in the modern world as "2 hours 52 minutes and 15 seconds", and "$(60)_{60}$ degrees" is what we call "360 degrees". (Please confirm both for yourself!) A sexagesimal number can be deceptively large. For example, a 4-digit sexagesimal number could be bigger than ten million in our numeral system:

$$\begin{aligned}(\underline{48}\ \underline{17}\ 31)_{60} &= 48 \cdot (60^3) + 17 \cdot (60^2) + 3 \cdot (60^1) + 1 \cdot (60^0) \\ &= 10{,}429{,}381.\end{aligned}$$

11.2. The Representation Theorem

We will give a proof in this section of the

Representation Theorem. *If b is a whole number > 1, then every whole number has one and only one base b representation.*

In the interest of clarity, we will prove this theorem only for the special case where $b = 7$ and the given whole number is 3644. It will be clear that the reasoning used is perfectly general and will be applicable to any base and any number. We give two different proofs, and both arguments will be seen to be perfectly applicable to any number other than 3644, and any base other than 7. What both proofs have in common is their heavy reliance on division-with-remainder (Chapter 7). Because the second proof is being offered as an option, we will leave it in a smaller print.

The first proof is nothing more than a repeated application of division-with-remainder to 3644 with 7 as the divisor. Let us begin with a simple case. If the number is 18, then from $18 = 2 \cdot 7 + 4$, we immediately conclude that $18 = (24)_7$. If the number is 123, then applying division-with-remainder once gives $123 = 17 \cdot 7 + 4$. Unfortunately, we cannot conclude from this that $123 = (\underline{17}\,4)_7$, because $(\underline{17}\,4)_7$ is not a number in base 7 due to the fact that $17 > 7$. This hurdle can be easily overcome, however, by applying division-with-remainder one more time to 17 to get $17 = 2 \cdot 7 + 3$. Thus by substituting $17 = 2 \cdot 7 + 3$ into $123 = 17 \cdot 7 + 4$, we get

$$\begin{aligned}123 &= (2 \cdot 7 + 3) \cdot 7 + 4 \\ &= 2 \cdot (7^2) + 3 \cdot (7^1) + 4 \cdot (7^0) \\ &= (234)_7.\end{aligned}$$

We now use the same idea to derive the base 7 representation of 3644. We have

$$3644 = 520 \cdot 7 + 4.$$

11.2. The Representation Theorem

Repeated applications of division-with-remainder to the coefficient of 7 lead to

$$3644 = 520 \cdot 7 + 4,$$
$$520 = 74 \cdot 7 + 2,$$
$$74 = 10 \cdot 7 + 4,$$
$$10 = \underline{1} \cdot 7 + 3.$$

Because $1 < 7$ in the last equation, the process stops. Also observe that the remainder in each division-with-remainder with 7 as divisor (i.e., 4, 2, 4, 3) is a whole number smaller than 7. The fact that each of these numbers is smaller than 7 will be seen to guarantee that we get the coefficients of the base 7 representation of 3644.

As in the case of 123, we replace 520 in the first equation by the expression $74 \cdot 7 + 2$ in the second equation, then we replace 74 by the expression $10 \cdot 7 + 4$ in the third equation, and finally we replace 10 by the expression $1 \cdot 7 + 3$ in the last equation. Putting these substitutions together, we obtain

$$\begin{aligned} 3644 &= 520 \cdot 7 + 4 \\ &= (74 \cdot 7 + 2) \cdot 7 + 4 \\ &= 74 \cdot 7^2 + 2 \cdot 7 + 4 \\ &= (10 \cdot 7 + 4) \cdot 7^2 + 2 \cdot 7 + 4 \\ &= 10 \cdot 7^3 + 4 \cdot 7^2 + 2 \cdot 7 + 4 \\ &= (1 \cdot 7 + 3) \cdot 7^3 + 4 \cdot 7^2 + 2 \cdot 7 + 4 \\ &= 1 \cdot 7^4 + 3 \cdot 7^3 + 4 \cdot 7^2 + 2 \cdot 7 + 4 \\ &= (13424)_7. \end{aligned}$$

Notice that among the coefficients 1, 3, 4, 2, 4 of the base 7 representation of 3644, $\{3,4,2,4\}$ are exactly the remainders of the successive divisions-with-remainders.

Activity. (a) Find the base 7 expansion of 280. (b) Find the binary expansion of 67.

> We now give the second proof that 3644 has a base 7 representation. Consider the successive powers of 7:
> $$7 < 7^2 \,(= 49) < 7^3 \,(= 343) < 7^4 \,(= 2401) < 7^5 \,(= 16807).$$
> We see that 3644 falls between 7^4 and 7^5. We therefore apply division-with-remainder to the dividend 3644 and the divisor 7^4 ($= 2401$) to get
> $$3644 = 1 \cdot 7^4 + 1243.$$
> Now the remainder 1243 falls between 7^3 and 7^4, so with 7^3 ($= 343$) as divisor and 1243 as dividend, division-with-remainder gives
> $$1243 = 3 \cdot 7^3 + 214.$$

Repeating the process on the remainder 214:

$$214 = 4 \cdot 7^2 + 18,$$
$$18 = 2 \cdot 7 + 4.$$

The last remainder is of course smaller than 7 (the divisor being 7). We can now replace the remainder in each division-with-remainder with the expression in the next division-with-remainder to get

$$\begin{aligned}
3644 &= 1 \cdot 7^4 + 1243 \\
&= 1 \cdot 7^4 + (3 \cdot 7^3 + 214) \\
&= 1 \cdot 7^4 + 3 \cdot 7^3 + 214 \\
&= 1 \cdot 7^4 + 3 \cdot 7^3 + (4 \cdot 7^2 + 18) \\
&= 1 \cdot 7^4 + 3 \cdot 7^3 + 4 \cdot 7^2 + 18 \\
&= 1 \cdot 7^4 + 3 \cdot 7^3 + 4 \cdot 7^2 + (2 \cdot 7 + 4) \\
&= 1 \cdot 7^4 + 3 \cdot 7^3 + 4 \cdot 7^2 + 2 \cdot 7 + 4 \\
&= (13424)_7,
\end{aligned}$$

which gives the same expansion as before.

Activity. Derive the base 3 representation of 279.

It remains to show that the representation for 3644 in base 7 is the only one possible for 3644. So let a, b, c, d, e be whole numbers ranging from 0 to 6 and suppose $(abcde)_7 = (13424)_7 = 3644$.[2] We have

$$3644 = a \cdot 7^4 + b \cdot 7^3 + c \cdot 7^2 + d \cdot 7 + e = 1 \cdot 7^4 + 3 \cdot 7^3 + 4 \cdot 7^2 + 2 \cdot 7 + 4$$

so that

(11.5) $\qquad 3644 = (a \cdot 7^3 + b \cdot 7^2 + c \cdot 7 + d) \cdot 7 + e,$

(11.6) $\qquad 3644 = (1 \cdot 7^3 + 3 \cdot 7^2 + 4 \cdot 7 + 2) \cdot 7 + 4.$

Because both e and 4 are less than 7, both (11.5) and (11.6) are divisions-with-remainders of 3644 by 7. According to Theorem 7.1 on page 105, the quotient and remainder of a division-with-remainder are both unique. Since e and 4 are both remainders, the uniqueness means $e = 4$. Furthermore, the uniqueness of the quotient means that

(11.7) $\qquad a \cdot 7^3 + b \cdot 7^2 + c \cdot 7 + d = 1 \cdot 7^3 + 3 \cdot 7^2 + 4 \cdot 7 + 2.$

Now denote the number common to both sides of (11.7) by k. Then we have

$$k = (a \cdot 7^2 + b \cdot 7 + c) \cdot 7 + d,$$
$$k = (1 \cdot 7^2 + 3 \cdot 7 + 4) \cdot 7 + 2.$$

---

[2] Strictly speaking, one should also allow for the fact that the putative representation $(abcde)_7$ has more than five digits, but the argument is the same in that case.

Again, since both d and 2 are less than 7, these are two divisions-with-remainders of k by 7. We are now in a position to repeat the same argument and conclude that $d = 2$ and

$$a \cdot 7^2 + b \cdot 7 \ + \ c \ = \ 1 \cdot 7^2 + 3 \cdot 7 \ + \ 4.$$

It is now clear how to proceed to finish the proof that $c = 4$, $b = 3$, and $a = 1$. Thus 3644 has only one base 7 representation.

11.3. Arithmetic in Base 7

The purpose of this section is to lend clarity and perspective to the standard algorithms of Chapters 3–7 by doing arithmetic in an arbitrary base. We will use base 7 for definiteness, but the reasoning generalizes to any base.

Given two numbers $(103)_7$ and $(242)_7$ in base 7, we want their sum *as numbers in base* 7. We can do that by first converting them back to base 10, doing the addition, and then converting them back to base 7 to obtain

$$(103)_7 + (242)_7 = (345)_7.$$

This is a valid way to do base 7 addition, but it is not the simplest because one can in fact add numbers directly in base 7 in a way that is identical to the addition algorithm of base 10 given in Chapter 4:

(11.8)
$$
\begin{array}{r}
1\ 0\ 3 \\
+\ \ 2\ 4\ 2 \\
\hline
3\ 4\ 5
\end{array}
$$

The reasoning is the same as in Chapter 4. We make use of the associative law and commutative law of addition (precisely, Theorem 2.1 on page 41) and the distributive law as explained in Chapter 2 to get

$$
\begin{aligned}
(103)_7 + (242)_7 &= (1 \cdot 7^2 + 0 \cdot 7 + 3) + (2 \cdot 7^2 + 4 \cdot 7 + 2) \\
&= (1 \cdot 7^2 + 2 \cdot 7^2) + (0 \cdot 7 + 4 \cdot 7) + (3 + 2) \\
&= (1 + 2) \cdot 7^2 + (0 + 4) \cdot 7 + (3 + 2).
\end{aligned}
$$

The last line shows clearly why the addition $(103)_7 + (242)_7$ could be done as column-by-column addition in (11.8).

The case where *carrying* takes place in an addition problem does not present new difficulties, but we should first construct an **addition table** for single-digit numbers in base 7:

+	0	1	2	3	4	5	6
0	0	1	2	3	4	5	6
1	1	2	3	4	5	6	10
2	2	3	4	5	6	10	11
3	3	4	5	6	10	11	12
4	4	5	6	10	11	12	13
5	5	6	10	11	12	13	14
6	6	10	11	12	13	14	15

Activity. Verify that the preceding table is correct.

What this table shows is how adding two single-digit numbers in base 7 leads to a double-digit number; see the lower right corner of the table, e.g., $(5)_7 + (6)_7 = 5 + 6 = 11 = 1 \cdot 7 + 4 = (14)_7$. Armed with this table, we can add, e.g., $(64)_7 + (25)_7$:

$$
\begin{array}{r}
6\ 4 \\
+\ \ \ 2\ 5 \\
\hline
{}^1\ {}^1\ \ \ \\
1\ 2\ 2
\end{array}
$$

This is because

$$
\begin{aligned}
(64)_7 + (25)_7 &= (6 \cdot 7 + 4) + (2 \cdot 7 + 5) \\
&= (6 \cdot 7 + 2 \cdot 7) + (4 + 5) \\
&= 8 \cdot 7 + 9 \\
&= (1 \cdot 7 + 1) \cdot 7 + (1 \cdot 7 + 2) \\
&= \mathbf{1 \cdot 7^2 + 1 \cdot 7 + 1 \cdot 7 + 2} \\
&= 1 \cdot 7^2 + 2 \cdot 7 + 2.
\end{aligned}
$$

In the next-to-the-last line (boldfaced) of the preceding calculation, we see carrying at work in the middle column of the preceding schematic presentation.

Activity. $(66)_7 + (1)_7 = ?$ $(2666)_7 + (1)_7 = ?$ $(266660)_7 + (10)_7 = ?$

Subtraction can be done in exactly the same way as in addition if the definition of subtraction in equation (5.1) on page 72 is kept in mind. For example, $(502)_7 - (213)_7$ is computed as:

$$
\begin{array}{r}
{}^4\ {}^6\ \ \\
\not{5}\ \not{0}\ 2 \\
-\ \ 2\ 1\ 3 \\
\hline
2\ 5\ 6
\end{array}
$$

Activity. Give an explanation of the preceding subtraction.

11.3. Arithmetic in Base 7

We conclude with a discussion of multiplication in base 7. As in the case of base 10, the first thing is to construct the **multiplication table** for single-digit numbers in base 7:

×	1	2	3	4	5	6
1	1	2	3	4	5	6
2	2	4	6	11	13	15
3	3	6	12	15	21	24
4	4	11	15	22	26	33
5	5	13	21	26	34	42
6	6	15	24	33	42	51

Activity. Check that each entry in the multiplication table is correct.

There is no difference between base 10 and base 7 as far as multiplication is concerned. We follow Chapter 6 and begin with the multiplication of an arbitrary number by a single-digit number, e.g., $(265)_7 \times (4)_7$. Keeping the multiplication table in mind, we get:

(11.9)
$$\begin{array}{r} 2\ 6\ 5 \\ 4 \\ \times \quad 1\ 3\ 2 \\ \hline 1\ 4\ 5\ 6 \end{array}$$

The reason is

$$\begin{aligned}(265)_7 \times (4)_7 &= (2 \cdot 7^2 + 6 \cdot 7 + 5) \cdot 4 \\ &= 8 \cdot 7^2 + 24 \cdot 7 + 20 \\ &= (\mathbf{\underline{1} \cdot 7 + 1}) \cdot \mathbf{7^2} + (\mathbf{\underline{3} \cdot 7 + 3}) \cdot \mathbf{7} + (\mathbf{\underline{2} \cdot 7 + 6}) \\ &= 1 \cdot 7^3 + (1+3) \cdot 7^2 + (3+2) \cdot 7 + 6.\end{aligned}$$

In the next-to-the-last line (boldfaced), the underlined numbers correspond to the numbers entered in the third row of (11.9). We see carrying at work.

Activity. $(345)_7 \times (10)_7 = ?$ $(345)_7 \times (100)_7 = ?$ $(345)_7 \times (10000)_7 = ?$

We now come to the main feature of the multiplication algorithm in base 7, which is that when we multiply two multidigit numbers together, there is a *shifting to the left* in each successive row, in the following sense. Let us look at $(265)_7 \times (34)_7$. The algorithm calls for separate multiplications of $(265)_7$ by the two digits of $(34)_7$, namely 3 and 4:

$$(265)_7 \times (4)_7 = (1456)_7,$$
$$(265)_7 \times (3)_7 = (1161)_7.$$

Then the product of $(265)_7$ and $(34)_7$ is obtained by adding $(265)_7 \times (4)_7$ and $(265)_7 \times (3)_7$ together, but *with the latter shifted one column to the left*, as shown.

$$\begin{array}{r} 2\ 6\ 5 \\ \times \quad\quad 3\ 4 \\ \hline 1\ 4\ 5\ 6 \\ +\ 1\ 1\ 6\ 1\quad \\ \hline 1\ 3\ 3\ 6\ 6 \end{array}$$

The main observation is that *it is the same algorithm as the decimal case* explained in Chapter 6.

Here is the explanation in terms of the distributive law:

$$(265)_7 \times (34)_7 = (265)_7 \times ((30)_7 + (4)_7)$$
$$= ((265)_7 \times (30)_7) + ((265)_7 \times (4)_7).$$

Now observe that since $(30)_7 = (3)_7 \times (10)_7$, we have

$$(265)_7 \times (30)_7 = ((265)_7 \times (3)_7) \times (10)_7;$$

therefore,

$$(265)_7 \times (34)_7 = ((265)_7 \times (3)_7) \times (10)_7 + ((265)_7 \times (4)_7).$$

This shows first of all that we have reduced the multiplication of $(265)_7$ by a 2-digit number to the multiplication of $(265)_7$ by single-digit numbers. Moreover, since $((265)_7 \times (3)_7) \times (10)_7 = (11610)_7$, comparing it with $(265)_7 \times (3)_7 = (1161)_7$ allows us to see the reason for the shifting by one column to the left in the preceding schematic presentation.

Activity. Use the multiplication algorithm for base 7 to directly compute $(540)_7 \times (26)_7$. Check the result by converting both numbers in base 7 to decimal numbers, multiplying them, and then converting the result back to base 7.

Once multiplication is understood, long division in base 7 can be done exactly as in Chapter 7. Because of space considerations, we will leave this topic to the reader.

Summing Up. The key point of this section is that the standard algorithms are *universal* in the sense that they are the same algorithms for place-valued numeral systems in *any* base. Therefore, far from a collection of isolated skills, the standard algorithms are broadly applicable skills.

11.4. Binary Arithmetic

It was remarked earlier that the binary numeral system is used exclusively in computer science. The purpose of this section is to work out a few examples in binary arithmetic for the purpose of illustration. We intentionally tackle

11.4. Binary Arithmetic

this topic last because, *without a good understanding of the arithmetic in an arbitrary base, binary arithmetic is very confusing.*

Recall that, because we are forced to use whole numbers smaller than 2, we have only two symbols to work with: 0 and 1. Following the method of counting given in section 1.1, **How to Count**, on page 6, we list the first 20 binary numbers (we omit the notation "$(\ )_2$" for simplicity):

$$0, 1, 10, 11, 100, 101, 110, 111, 1000, 1001, 1010, 1011, 1100,$$
$$1101, 1110, 1111, 10000, 10001, 10010, 10011.$$

Activity. Verify that these are indeed the first 20 numbers.

The following binary **addition** and **multiplication tables** are easily verified:

+	0	1
0	0	1
1	1	10

×	0	1
0	0	0
1	0	1

The fact (from the addition table) that $(1)_2 + (1)_2 = (10)_2$ has a generalization. We first note some special cases:

$$(10)_2 + (10)_2 = (100)_2$$

because the left side is $2 + 2 = 2 \times 2 = 2^2 = (100)_2$. Also,

$$(100)_2 + (100)_2 = (1000)_2$$

because the left side is equal to $2^2 + 2^2 = 2 \times 2^2 = 2^3 = (1000)_2$. More generally, we have

$$(1\underbrace{00\cdots0}_{m})_2 + (1\underbrace{00\cdots0}_{m})_2 = (1\underbrace{00\cdots0}_{m+1})_2$$

for any whole number m. This is because $(1\underbrace{00\cdots0}_{m})_2 = 2^m$ by definition, and so by the definition of multiplication:

$$2^m + 2^m = 2 \times 2^m = 2^{m+1} = (1\underbrace{00\cdots0}_{m+1})_2.$$

We now give the full generalization:

$$\underbrace{(1\overbrace{00\cdots0}^{m})_2 + \cdots + (1\overbrace{00\cdots0}^{m})_2}_{2^n} = (1\overbrace{00\cdots0}^{m+n})_2$$

for any whole numbers m and n. The computation is entirely similar. For $m = 0$ and $n = 2$, we have the following special case:

$$(1)_2 + (1)_2 + (1)_2 + (1)_2 = (100)_2.$$

Similarly, with $m = 1$ and $n = 2$, we get
$$(10)_2 + (10)_2 + (10)_2 + (10)_2 = (1000)_2.$$
Consequently,
$$\begin{aligned}(11)_2 + (11)_2 + (11)_2 + (11)_2 &= \underbrace{\{(10)_2 + (1)_2\} + \cdots + \{(10)_2 + (1)_2\}}_{4} \\ &= \underbrace{(10)_2 + \cdots + (10)_2}_{4} + \underbrace{(1)_2 + \cdots + (1)_2}_{4} \\ &= (1000)_2 + (100)_2 \\ &= (1100)_2.\end{aligned}$$

Activity. Do the addition $(11)_2 + (11)_2 + (11)_2 + (11)_2$ vertically, as in ordinary addition, and use the addition algorithm. What do you notice about carrying?

Activity. Can you think of an analogue of the equation $(11)_2 + (11)_2 + (11)_2 + (11)_2 = (1100)_2$ in base 10?

Another addition fact of interest is
$$(\underbrace{11\cdots 1}_{m})_2 + (1)_2 = (1\underbrace{00\cdots 0}_{m})$$
for any whole number m. If one is willing to just use the addition algorithm, then it is obvious. Let us show this for $m = 4$; the proof in general is the same.

```
        1 1 1 1
              1
   +  1 1 1
   ─────────
      1 0 0 0 0
```

It may also be instructive to verify this directly:
$$\begin{aligned}(1111)_2 + (1)_2 &= (2^3 + 2^2 + 2 + 1) + 1 \\ &= 2^3 + 2^2 + 2 + 2 = 2^3 + 2^2 + 2 \cdot 2 \\ &= 2^3 + 2^2 + 2^2 = 2^3 + 2 \cdot 2^2 \\ &= 2^3 + 2^3 = 2 \cdot 2^3 \\ &= 2^4 \\ &= (10000)_2.\end{aligned}$$

Compare Exercise 6(c) at the end of the chapter to see where this discussion leads.

The following multiplication fact is of interest:
$$(1\underbrace{00\cdots 0}_{m})_2 \times (1\underbrace{00\cdots 0}_{n})_2 = (1\underbrace{00\cdots 0}_{m+n})_2$$

11.4. Binary Arithmetic

for any whole numbers m and n. This is because the left side is $2^m \cdot 2^n$, which is equal to 2^{m+n}, which is exactly the right side. (Compare Exercise 8(a) below.)

Finally, we illustrate the multiplication algorithm by showing that
$$(1111)_2 \times (11)_2 = (101101)_2.$$

$$
\begin{array}{r}
1\ 1\ 1\ 1 \\
\times \qquad\quad 1\ 1 \\
\hline
1\ 1\ 1\ 1 \\
1\ 1\ 1\ 1 \\
+ \quad 1\ 1\ 1 \\
\hline
1\ 0\ 1\ 1\ 0\ 1
\end{array}
$$

Exercises

1. In section 1.1, **How to Count**, on page 6, we showed how to build up the decimal (Hindu-Arabic) numeral system by the counting process, using only the ten symbols 0, 1, ..., 9, and the concept of place value. Suppose you are now limited to the use of only the first seven symbols, 0, 1, 2, 3, 4, 5, 6, and you are asked to build up a numeral system by imitating the process described in **How to Count**. What is the corresponding *expanded form* of a number in this system? What is the 50th number in this system? the 98th? the 343rd? How is this related to numbers in base 7?

2. Take an arbitrary whole number, 8704, for example. Apply the method used to prove the Representation Theorem to get the base 10 representation of 8704. Is this something you have encountered before?

In Exercises 3–7, you are expected to perform all the computations in the given base rather than converting those numbers back to base 10, doing the needed computations, and converting back to the original base.

3. Write 793 as a number in base 4, in base 5, in base 8, in base 60, and in base 2. (Use a scientific calculator.)

4. Write $(3021)_4$ as a number in base 7, in base 8, in base 12, and in base 60. Do the same with $(15\ 11\ 3)_{20}$. (Use a scientific calculator.)

5. (a) $(1234)_5 + (4213)_5$.   (b) $(51\ 42\ 0\ 25)_{60} + (43\ 9\ 36\ 59\ 40)_{60}$.
   (c) $(2\ 45)_{60} \times (56\ 22\ 3)_{60}$.   (d) $(10011011)_2 \times (1100010)_2$.
   (e) $(11\ 8\ 10)_{12} \times (76)_{12}$. (f) $(111111111)_2 + (10)_2$. (g) $(111111111)_2 + (101)_2$. (h) $(4213)_5 - (1234)_5$. (i) $(738)_{12} - (5\ 11\ 10)_{12}$. (j) $(123456)_7 \times (10000)_7$. (k) $(123456)_8 \times (10000)_8$. (l) $(100000)_7 - (6543)_7$.

6. (a) Explain why $1 + 2^2 + 2^3 + 2^4 + 2^5 = 2^6 - 1$. (b) How is this related to $(111111)_2 + (1)_2 = (1000000)_2$? (c) Explain why for any whole number $n > 0$, $1 + 2^2 + 2^3 + \cdots + 2^n = 2^{n+1} - 1$.

7. $(111)_2 + (111)_2 + (111)_2 + (111)_2 = ?$

8. For any whole number $b > 1$, compute the following.
   (a) $(1\underbrace{00\cdots0}_{m})_b \times (1\underbrace{00\cdots0}_{n})_b$ for any whole numbers m and n.
   (b) $(a_1 a_2 \cdots a_n)_b \times (1\underbrace{00\cdots0}_{m})_b$.
   (c) $(k0k)_b \times (11)_b$ for $0 \le k \le (b-1)$.
   (d) $\big((b-1)(b-1)\cdots(b-1)\big)_b + (1)_b$.

9. In the last two sections, it is implicitly assumed that addition and multiplication of numbers in base b ($b > 1$) are associative, commutative, and distributive. Explain why this is true.

Part 2

Fractions

Part Preview

*The understanding of this part requires a knowledge of at least Chapters 8 and 9 in Part 1, **Whole Numbers**, especially the concept of the number line. It is also strongly recommended that you read Chapter 24 starting on page 367 before embarking on a study of this part in order to get a general idea of where this part is headed, and why.*

The following is a new approach to the teaching of fractions. It is not new in the sense of introducing new concepts; the subject is too old for that. Rather, it is new in the conceptual way the various skills and concepts are introduced and developed in order to satisfy the dual requirement that the presentation be usable in the elementary classroom, especially in grade 5 and up, and that it be consistent with the minimum standards of mathematics in terms of reasoning, coherence, and precise definitions (cf. [**Wu06**]). For example, the traditional presentation asks you to believe that a fraction is a piece of pizza, parts of a whole, a ratio, a division, and an operator. Not only is it difficult to accept these *multiple meanings* on faith, but also none of these meanings even remotely hints at the fact that a fraction is above all a *number*. All the while, we ask students to compute with fractions as numbers. *The almost unbridgeable chasm between what we teach students and what we require them to do goes a long way towards explaining the non-learning of fractions* (for a fuller discussion, see [**Wu10b**]). Recall that in Part 1, we have already discussed the concept of a number as a point on the number line. This chapter builds on this conceptual framework by defining a fraction as a point on the number line constructed in a well-determined manner. Once you accept this definition, you can use logical reasoning to explain *all* other meanings of this concept. Moreover, you also add, subtract, multiply and divide fractions in the same way you do whole numbers; the strategies are identical and the differences lie only in the details. Every skill or concept that appears in the usual presentation of fractions will make its appearance in due course in this coherent development of the subject. *Everything* will be explained, and nothing will be taken for granted.

The finite decimals form a special class of fractions. The fact that *a decimal is by definition a fraction* and that *a thorough understanding of finite decimals must go through fractions* does not seem to be generally recognized in school mathematics education, but this book hopes to make a strong case for this recognition. The basic properties of finite decimals and their arithmetic will be briefly explained in this chapter. Although there are some fine points about decimals that deserve to be treated separately, such as infinite decimals (see Part 5), the importance of finite decimals in everyday life demands that they be included in any discussion of fractions.

We emphasize the centrality of the number line and the logical coherence in this presentation of fractions, and ask that when you teach fractions, you do likewise. The reason why this subject is the bane of elementary school students is, without a doubt, the fact that *the concept of a fraction, unlike that of whole numbers, is an abstraction*. Whereas the learning of whole numbers can be grounded on the counting of fingers, there is no obvious replacement of one's fingers for the learning of fractions. Pieces of a pizza clearly do not get the job done. A main reason for the introduction of the number line in the study of fractions is precisely to offer such a replacement. We will see that this line will make it possible to carry out the arithmetic operations on fractions visually. In addition, it is in the nature of an abstract concept that, because it is off the beaten track of the concrete and the intuitive, its learning needs the support of precise descriptions (in the form of definitions) and transparent reasoning. The purpose of a logically coherent presentation of fractions is to provide this support.

Precise definitions and logical reasoning, the *sine qua non* of mathematics, are for the most part not found in the mathematics education literature, especially the literature on fractions, whether it be school texts, professional development materials, or research articles (see [**Wu08**, pages 1–4 and 33–42], for example). We have come to accept the mystical and mathematically incoherent presentations of fractions to students and teachers alike as a *fait accompli*, but it may be time for us to ask why. We can do better, and this book shows one way of doing that.

We are going to develop the whole subject of fractions from the beginning. They will be treated as a direct continuation of whole numbers, *which they are*. At least from a logical standpoint, we will assume nothing from your previous knowledge of fractions. We hope, now that you have acquainted yourself with the precision and the unremitting emphasis on reasoning of Part 1, you will encounter none of the usual discontinuities in the transition from whole numbers to fractions. Most of the procedures, such as identifying equivalent fractions or the multiplication of fractions, should be familiar to you, but it is likely that you will find something new in their explanations nonetheless. For example, we will ask you to remember, again and again, that a fraction is a point on the number line (see Chapter 8 in Part 1), and we use this fact as the basis of all explanations. The idea that a fraction is a point on the number line and not a piece of pizza will take some time before it sinks in, because old habits die hard. However, we believe you will find it worthwhile in the long run.

This presentation of fractions can be imported directly into a fifth or sixth grade classroom with only the requisite modifications that must accompany any successful school lesson. In year 2010, however, there still remains considerable resistance to the very idea of *defining* a fraction as a point on the number line because, as the argument goes, the concept of a fraction is so complex that students should not be limited to the number line but should be exposed to the many "personalities" of a fraction[1] from the beginning. Our response to this objection is twofold. First, the definition of a fraction as a point on the number line is superior to the use of a piece of pizza as a starting point, because we will be able to reason directly with the number line to get to all the basic facts about fractions whereas one cannot do that with pizzas; e.g., there is no way that students can learn to multiply two pieces of pizza to arrive at another piece of pizza without invoking some facts that have not been discussed. Indeed, students' frustration with doing things in ways they know not why is a main obstacle to the learning of fractions. In addition, by using the definition of a fraction as a point on the number line, we will use reasoning to arrive at all the "personalities" of a fraction[2] without making students learn these personalities by rote. In this way, having a precise definition of a fraction opens students' eyes to the fact that, *in mathematics*, reasoning—and not rote learning—is the rule of the day.

For the primary grades, the issue of formally defining a fraction as a point on the number line needs some clarification. School mathematics for the primary grades is essentially the mathematics of whole numbers, and there is no need for the introduction of abstraction on a large scale.[3] By the fifth grade, however, abstraction becomes a necessity, not only because the systematic treatment of fractions demands it, but because it is around this time that school mathematics begins to prepare students for algebra (cf. [**Wu01**], [**NMP08a**], [**NMP08b**], and [**Wu09a**]). One may paraphrase the situation by saying that the primary grades are a laboratory for students'

---

[1] In the technical sense of the research literature.

[2] In other words, we *prove* as theorems that a fraction has all these personalities.

[3] One has to put this statement in context: The very use of the Arabic numerals 1, 2, 3, ..., to represent the whole numbers is, by itself, a significant abstraction. But by and large, the presence of abstraction in the primary grades is not obtrusive.

mathematical explorations, but that beginning about the fifth grade, the scientific method of organizing the data and establishing causal relationship must be brought to bear on these explorations.

In the primary grades, it is therefore not a matter of urgent necessity that a precise definition of a fraction be introduced. The belief is, however, that having seen a logical development of fractions such as the one in this book, a teacher would be immeasurably better equipped to provide primary students the mathematical foundation they need in the later grades. For example, a teacher familiar with such a development would recognize the futility of cutting pizzas exclusively for the purpose of teaching fractions. Working on the number line is conceptually *and pictorially* far simpler and mathematically more to the point.

One can explain, from a different perspective, the need for teachers in the primary grades to be knowledgeable about a mathematically correct treatment of fractions. Think of the analogous situation in an infant's attempt to learn English. Although the English used in baby talk is truly basic, would it not be preferable that a baby be exposed to someone who speaks correct normal English rather than to someone who has a limited vocabulary and a faulty command of grammar? By the same token, although one would never *prove* in the primary grades the theorem that a fraction $\frac{a}{b}$ is equal to *a divided by b*, this so-called "division interpretation of a fraction" is informally discussed in the primary grades. A primary teacher who understands this theorem would be much more likely to open a child's eyes to the mathematical reasoning embedded in the theorem than one who rigidly clings to the notion that one must make primary students learn, *by rote*, the division interpretation of a fraction in order to promote the "conceptual understanding of a fraction". For this and similar reasons, the content of Part 2 should be part of the education of every teacher in elementary school as long as they are all required to teach mathematics. (For a different proposal on this issue, see [**Wu09b**].)

As I tried to explain at the beginning of this book (see **To the Reader** on page xxv), learning mathematics requires more work than is commonly acknowledged in the education literature, especially that of the past two decades. But if the material you read contains mathematics that is *incorrectly* presented, then no amount of hard work can help you learn it. The goal of this part is to present fractions in a way that makes sense not only pedagogically, but mathematically as well. I believe that such a presentation is at least worthy of your hard work.

Chapter 12

Definitions of Fraction and Decimal

Recall that a number is a point on the number line (see Chapter 8). In Part 1, we discussed a special collection of these points, the whole numbers. In this chapter, we are going to focus our attention on a larger collection of points that includes the whole numbers as a subcollection; these are the fractions.

As we have emphasized again and again: definitions are important in mathematics because mathematics is a precise discipline and we need to know *exactly* what we are talking about. In the subject of fractions, however, there is now a tradition that students should learn how to compute with fractions and do word problems with them, not by being told what a fraction is and how to make logical deductions therefrom, but by exposing them to discussions using *only metaphors and analogies* (see [**Wu10b**]). For example, the most common explanation of a fraction is that it is a piece of pizza. Now, a student cannot possibly take this explanation seriously because when she computes $\frac{2}{3}$ of 55 miles, for example, she would not be thinking of a piece of pizza with the cheese dripping over a straight line 55 miles long. So she learns quickly that the pizza reference can only be taken with a grain of salt. In other words, *she cannot trust what she reads or hears*. Is this quite the right frame of mind for a student learning mathematics?

A few better textbooks for professional development do make an attempt at a definition, and typically what we find is the following:

```
A fraction has three distinct meanings.
    PART-WHOLE. The part-whole interpretation of a
    fraction such as 2/3 indicates that a whole has been
```

partitioned into three equal parts and two of those parts are being considered.

QUOTIENT. The fraction $\frac{2}{3}$ may also be considered as a quotient, $2 \div 3$. This interpretation also arises from a partitioning situation. Suppose you have some big cookies to give to three people. You could give each person one cookie, then another, and so on until you had distributed the same amount to each. If you have six cookies, then you could represent this process mathematically by $6 \div 3$, and each person would get two cookies. But if you only have two cookies, one way to solve the problem is to divide each cookie into three equal parts and give each person $\frac{1}{3}$ of each cookie so that at the end, each person gets $\frac{1}{3} + \frac{1}{3}$ or $\frac{2}{3}$ cookies. So $2 \div 3 = \frac{2}{3}$.

RATIO. The fraction $\frac{2}{3}$ may also represent a ratio situation, such as there are two boys for every three girls.

This, too, is not a satisfactory definition of what a fraction is, for several reasons. To say that something you try to get to know is three disparate things simultaneously (i.e., part-whole, quotient, and ratio) strains one's credulity. For instance, if I tell you I have discovered a substance that is as hard as steel, as light as air, and as transparent as glass, would you believe it? Another reason for objection is that a fraction is being explained in terms of a "ratio", but most people do not know what a ratio is.[1] In addition, when students approach fractions, they carry with them their experience from whole numbers, and for whole numbers, the idea of a division $a \div b$ makes sense only when a is a multiple of b (see Chapter 7). They do not as yet know what $2 \div 3$ means. For example, they cannot imagine dividing 2 marbles into 3 equal parts. Therefore to use the concept of "2 divided by 3" to explain the meaning of $\frac{2}{3}$ is mathematically illegitimate and pedagogically a disaster. This "definition" of a fraction is therefore not one that can be used to advantage in a school classroom. Moreover, we anticipate that fractions would be added, subtracted, multiplied, and divided, and it is not clear how one goes about adding, subtracting, multiplying, and dividing part-wholes, quotients, or ratios.

We need a new beginning, and this is what we set out to do.

The sections in this chapter are as follows:

Prologue

The Basic Definitions

---

[1] We will precisely define "ratio" in Chapter 22.

Decimals

Importance of the Unit

The Area Model

Locating Fractions on the Number Line

Issues to Consider

12.1. Prologue

This section gives an *informal* discussion of the motivation behind our approach to fractions. *Suppose we know what a fraction is.* If we want to put all the fractions on the number line, how should we proceed? For definiteness, let us start with fractions whose denominators are all equal to 3.

As usual, the number 1 is our unit, and the "whole", in the context of the number line, will be taken to be the unit segment $[0, 1]$ from 0 to 1. Then a fraction such as $\frac{1}{3}$ would be, by common consent, one part when $[0, 1]$ is divided into three equal parts. But common consent notwithstanding, more precision regarding the meaning of the "whole" and "equal parts" is needed if our discussion is to make sense. To explain what is at issue, consider the following situation: We have a triangle ABC, and D is the midpoint of the side BC. Let the "whole" in this case be $\triangle ABC$; thus $\triangle ABC$ is 1. With $\triangle ADC$ as part of the whole, would $\triangle ADC$ be $\frac{1}{2}$?

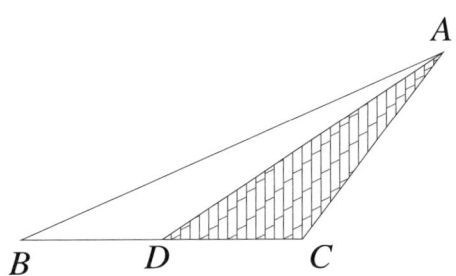

Note that $\triangle ABD$ and $\triangle ADC$ have bases of equal length (D is the midpoint of BC by hypothesis) and evidently the same height relative to the base BD or DC. Therefore they have the same area. It would be natural to believe that $\triangle ADC$ is $\frac{1}{2}$. But if 1 stands for the *triangle ABC*, which is a geometric entity, then one would also believe that $\frac{1}{2}$ stands for one of the regions when $\triangle ABC$ is divided into two *congruent* regions (the "equal parts"). But $\triangle ABC$ is divided into the two triangles ABD and ADC, and they are visibly *not* congruent to each other. Examples like this make us aware that our conception of what a "whole" is and what "equal parts" means need greater precision. The "whole", the number 1, is therefore *not* the geometric entity $\triangle ABC$ itself, but rather the *area of* $\triangle ABC$, and "equal" does not mean congruent but *equal in area*. A little reflection would reveal that, indeed, such precision is needed in any consideration concerning the part-whole concept of a fraction.

To go back to $[0, 1]$, let us try to clean up our language. The **whole**, which is 1, is now the *length* of the unit segment and is *not* the segment itself. When we say $[0, 1]$ is divided into **equal parts**, we actually mean to say that $[0, 1]$ is divided into *segments of equal length*. The fraction $\frac{1}{3}$ would therefore be the *length* of any segment so that three segments of the same length, when pieced together on the number line, would form a segment of length 1. Since all segments between consecutive whole numbers have length 1, when we likewise divide each of the segments $[0,1]$, $[1,2]$, $[2,3]$, ..., into three segments of equal length, the *length* of every one of these segments is also $\frac{1}{3}$. In particular, each of the following thickened segments has length $\frac{1}{3}$ and is therefore a legitimate representation of $\frac{1}{3}$:

Now concentrate on the thickened segment on the left. The distance of its right endpoint from 0 is naturally $\frac{1}{3}$. Since all the whole numbers on the number line indicate their distance from 0, we proceed to label the right endpoint of this segment by the fraction $\frac{1}{3}$, and we call this segment the **standard representation of** $\frac{1}{3}$ and denote this thickened segment by $[\mathbf{0}, \frac{\mathbf{1}}{\mathbf{3}}]$, because the notation clearly exhibits the left endpoint as 0 and the right endpoint as $\frac{1}{3}$.

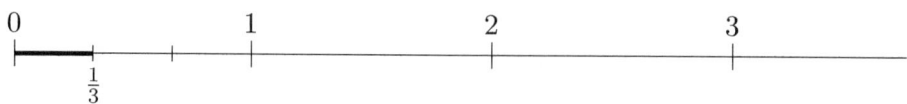

In like manner, each of the following individual collections of thickened segments can be thought of as $\frac{5}{3}$ since each collection has five parts when a segment of length 1 is divided into 3 parts of equal length.

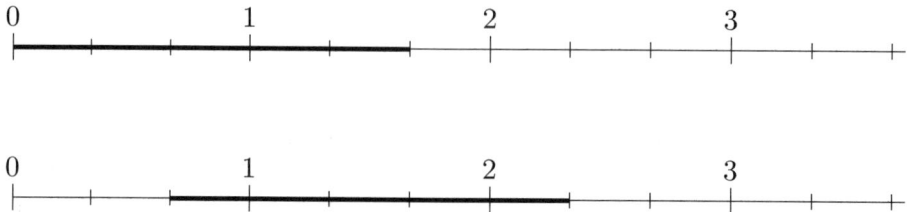

12.1. Prologue

Again, the right endpoint of the segment in the first of the preceding three pictures is at a distance of $\frac{5}{3}$ from 0. We therefore label the right endpoint of this segment by $\frac{5}{3}$ and call this segment the **standard representation of** $\frac{5}{3}$, to be denoted by $[0, \frac{5}{3}]$.

Once again, we emphasize that a fraction such as $\frac{5}{3}$ is represented by many segments which can all legitimately claim to be "the total length of five parts when a segment of length 1 is divided into 3 parts of equal length", but its standard representation possesses the virtue that its right endpoint is at a distance of exactly $\frac{5}{3}$ from 0. *We may therefore identify $\frac{5}{3}$ with the right endpoint of its standard representation.* Here are the first few fractions with denominator equal to 3. By convention, we agree to let 0 be written as $\frac{0}{3}$.

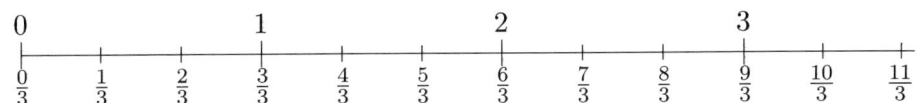

We make two observations. First, these fractions are just *multiples* of $\frac{1}{3}$ as defined on page 86 of Chapter 7. Second, the point $\frac{3}{3}$ is the same point as 1, $\frac{6}{3}$ is the same point as 2, $\frac{9}{3}$ is the same point as 3, etc.

Of course the consideration of fractions with denominator equal to 3 extends to any other fraction. For example, replacing 3 by 5 in $\frac{8}{3}$, we would get a typical fraction such as $\frac{8}{5}$. It would be the label of the right endpoint of the following thickened segment, because its distance from 0 is 8 times a fifth of the distance from 0 to 1.

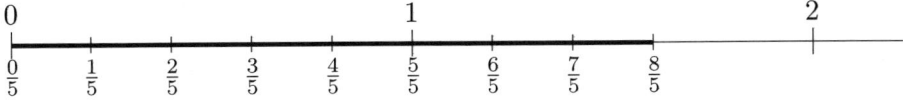

We summarize this discussion as follows:

(a) Once we fix a denominator, say 3, then each of the fractions $\frac{0}{3}, \frac{1}{3}, \frac{2}{3}, \frac{3}{3}, \frac{4}{3}, \ldots$ is the right endpoint of a segment whose left endpoint is always 0 and whose length is, respectively, 0, 1, 2, 3, 4, ... times a third of the whole.

(b) The fractions $\frac{3}{3}, \frac{2\times 3}{3}, \frac{3\times 3}{3}, \ldots$ are labels for the points already labeled by the whole numbers 1, 2, 3, ..., respectively.

(c) Each of the segments $[0, \frac{1}{3}], [0, \frac{2}{3}], [0, \frac{3}{3}], \ldots$ is, respectively, completely identified with the right endpoint $\frac{1}{3}, \frac{2}{3}, \frac{3}{3}, \ldots$. *Knowing the segment is equivalent to knowing the right endpoint.*

Now mathematics tries to say everything as concisely as possible, even if the conciseness is sometimes achieved at the expense of immediacy or intuition. In the long run though, conciseness wins out, because the alternative is to drag along an unwieldy collection of descriptive statements. In the present situation, item (c) above says that, in order to describe parts-of-a-whole, *we can forget about segments and just concentrate on their endpoints*. Thus, the collection of segments $[0, \frac{1}{3}], [0, \frac{2}{3}], [0, \frac{3}{3}], [0, \frac{4}{3}]$, etc., will be completely replaced by their right endpoints, $\frac{1}{3}, \frac{2}{3}, \frac{3}{3}, \frac{4}{3}$, etc. In turn, the latter, instead of being described in terms of parts-of-a-whole, may be simply described as *multiples* of $\frac{1}{3}$:

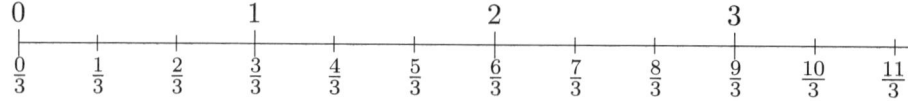

In a sense, identifying a fraction such as $\frac{1}{3}$ as a point on the number line rather than just "a segment of length $\frac{1}{3}$" is nothing new. The whole numbers 1, 2, 3, ... are already identified as points on the number line rather than just segments of lengths 1, 2, 3, ... For example, 2 is simply the point to the right of 0 and of distance 2 from 0.

So far we have only looked at fractions with a fixed denominator, such as 3. We may now replace 3 systematically by each of the positive whole numbers. Fractions with denominator 1 are of course just the whole numbers. We can then consider fractions with denominator 2, and 3, and 4, etc. Each whole number n therefore generates an infinite collection of equi-spaced points (corresponding to fractions with denominator equal to n). The fractions, when put on the number line, are therefore exactly *the totality of all these infinite collections* as n runs through 1, 2, 3, ...

The foregoing discussion assumes that we know what a fraction is, and all we did was to put fractions on the number line. *But do we know what a fraction is?* The answer is no. See the introduction of this chapter and

Chapter 24, On the Teaching of Fractions in Elementary School, on page 367. The fact that we do not know what a fraction is now puts the above discussion on very shaky grounds, because we cannot make a fictitious quantity real simply by putting it on the number line, any more than we can pretend that a unicorn is a real animal by giving it a name. On the other hand, if we accept the fact that we have as yet no definition of a fraction, then *the collections of infinite equi-spaced points on the number line can serve as a definition of fractions*. This will in fact be our point of departure. A fraction in this book will simply be one of the numbers on the number line *constructed in the precise manner described above*. This is the easy part. The hard part, which is where mathematics comes in, is to *explain in a logical manner everything we know about fractions on the basis of this definition, in a manner that is accessible to elementary school students*. The rest of this chapter will attempt to do this.

12.2. The Basic Definitions

This section starts the mathematical discussion by giving a formal definition of fractions. We will not make use of any facts in the preceding section for the mathematical reasoning that follows, but will be guided in spirit by that discussion.

We begin with the *number line* on which a collection of equi-spaced points have been chosen, one of them being designated by 0. The point 0 together with the subcollection of equi-spaced points to the right of 0 are the *whole numbers* (see Chapter 8 in Part 1): $\{0, 1, 2, 3, \ldots\}$.

Recall from Chapter 8 that a *number* is just a point on the number line, and that for any two numbers a and b, the line segment between a and b is denoted by $[a, b]$. Then a and b are called the **endpoints** of $[a, b]$, and the segment $[0, 1]$ is called the **unit segment**. Also recall that the concept of the *multiple* of a point was introduced on page 99. Then:

Definition. *The **fractions** are the points on the number line defined in the following manner: Fixing a whole number $n > 0$, we divide the unit segment into n parts of equal length. Then the first division point to the right of 0 will be denoted by $\frac{1}{n}$. The multiples of $\frac{1}{n}$ then form an equi-spaced sequence associated with n. The totality of all the points in these sequences as n runs through $1, 2, 3, \ldots$ is by definition the collection of all the fractions.*

In greater detail, consider the case of $n = 3$, and we will describe the points generated by 3. Divide the unit segment $[0, 1]$ into three parts of equal length. Of the two division points in $[0,1]$, the one next to 0 is denoted by $\frac{1}{3}$. We now give names to each multiple of $\frac{1}{3}$: 0 is the zeroth multiple and will be denoted by $\frac{0}{3}$ in this sequence, $\frac{1}{3}$ has already been designated as the first multiple, $\frac{2}{3}$ is the second multiple, $\frac{3}{3}$ the third multiple, and in general, the m-th multiple of $\frac{1}{3}$ will be denoted by $\frac{m}{3}$. Here is the picture:

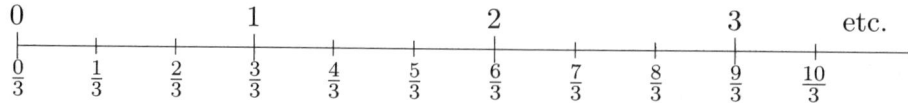

Note that the way we have just introduced the multiples of $\frac{1}{3}$ on the number line is exactly the same way that the multiples of 1 (i.e., the whole numbers) were introduced on the number line in Chapter 8. In both cases, we start with a fixed point (in Chapter 8 it was "1", and here it is "$\frac{1}{3}$"), and then we take its multiples. One could paraphrase this situation by saying that the multiples of $\frac{1}{3}$ are the analogues of the whole numbers if $\frac{1}{3}$ is used in place of 1.

In general, the fraction denoted by $\frac{m}{n}$ (for whole numbers m and n, $n \neq 0$) is the point which is m-th multiple of the division point next to 0 when the unit segment is divided into n parts of equal length.

Activity. Describe in words without looking at the definition what $\frac{7}{5}$ is, what $\frac{12}{35}$ is, what $\frac{33}{17}$ is, and what $\frac{127}{63}$ is.

Fractions are therefore points constructed on the number line in a specific way. The whole purpose of having a definition is that, once accepted, *it has to be the starting point of all future discussions of fractions.* In your classroom, therefore, once you adopt this definition, *you as well as your students will be obligated to refer back to it for any explanation about fractions.* Likewise, this book will explain everything known about fractions using only this definition of a fraction.

A word about the use of symbols in the study of fractions. In the above process of introducing the multiples of $\frac{1}{3}$, we also gave each whole number (a point on the number line!) another name: 1 is now also $\frac{3}{3}$, 2 is $\frac{6}{3}$, 3 is $\frac{9}{3}$, and by extension, 4 is $\frac{12}{3}$, 5 is $\frac{15}{3}$, etc. If we allow ourselves the use of symbolic notation, then we can succinctly summarize all these assertions in one brief symbolic statement:

In general, $m = \frac{3m}{3}$ for every whole number m.

The emphasis is of course on the phrase, "for *every* nonzero whole number m". Thus when m takes on all the values 1, 2, 3, 4, ... in succession, we get an infinite number of statements compressed into one symbolic statement. There is naturally no need to favor the number 3. If we take any whole

12.2. The Basic Definitions

number $\ell > 0$, it is evident for exactly the same reason that $\frac{\ell}{\ell} = 1$, $\frac{2\ell}{\ell} = 2$, $\frac{3\ell}{\ell} = 3$, $\frac{4\ell}{\ell} = 4$, and more generally,

$$(12.1) \qquad \frac{\ell m}{\ell} = m, \quad \text{for all whole numbers } m, \ell, \text{ where } \ell > 0.$$

In particular, if we take in succession $\ell = 1$ and $m = 1$, then we obtain

$$\frac{m}{1} = m \quad \text{and} \quad \frac{\ell}{\ell} = 1$$

for any whole number $m > 0$ and $\ell > 0$. This way of making use of symbols to achieve both precision *and* conciseness will be used extensively in the subject of fractions, and we suggest that you do likewise in your classroom, *gradually*, starting with at least the fifth grade; see [**Wu01**] and [**Wu09a**].

The statement in (12.1) points to the fact that *a whole number is a fraction*, and also that *each fraction is denoted by many symbols*, e.g., $2 = \frac{4}{2} = \frac{6}{3} = \frac{8}{4}$. This is nothing new. In Chapters 1 and 11 in Part 1, it has already been noted, for example, that the symbol we call 9 in the Hindu-Arabic numeral system represents the number that is represented by 100 in base 3, 21 in base 4, 14 in base 5, 13 in base 6, etc.

The number m in $\frac{m}{n}$ is called the **numerator of the fraction** $\frac{m}{n}$, and the number n its **denominator**. This is the accepted terminology. Nevertheless, an *abuse of language* has been built into this terminology, and we want to discuss it briefly. A fraction is not a symbol but, rather, *a point on the number line* and, as noted in connection with (12.1), it can be denoted by many symbols. Therefore, it is, strictly speaking, not correct to say "m is the numerator of the fraction $\frac{m}{n}$". Rather, one should say

> The number m in $\frac{m}{n}$ is called the **numerator of the fraction symbol** $\frac{m}{n}$ representing the m-th multiple of the fraction represented by $\frac{1}{n}$.

Same for the denominator. However, clarity is one thing, and reasonable normal usage is another. Nobody[2] ever talks about fractions in this stilted language in normal mathematical communications. There will always be abuse of language in mathematics, but if the abuse is in the service of simplicity, we accept it and move on.

Of course, we reserve the right to use the precise language when absolutely necessary.

For typographical reasons, a fraction $\frac{m}{n}$ is sometimes written as m/n. In common language, we say $\frac{m}{n}$ is "m-nths", e.g., $\frac{2}{7}$ is two-sevenths and $\frac{4}{5}$ is four-fifths. Also, we adopt for convenience the CONVENTION that

> the fraction notation $\frac{m}{n}$ or $\boldsymbol{m/n}$ automatically assumes that $\boldsymbol{n > 0}$.

---

[2] Except pedants.

It is common to call $\frac{m}{n}$ a **proper** fraction if $m < n$, and **improper** if $m \geq n$, but by the way we define a fraction, a fraction is just a point on the number line and *we normally do not call attention to whether a fraction is proper or improper.*

We can now define the lengths of more line segments than in Chapter 1, where every length is a whole number. Let $\frac{m}{n}$ be a fraction. We say a line segment from x to y on the number line, denoted by $[x, y]$, has **length** $\frac{m}{n}$ if, after sliding $[x, y]$ to the left until x rests on 0, the right endpoint y rests on $\frac{m}{n}$. Observe that if $n = 1$, then $\frac{m}{n} = m$ and this definition of length coincides with the one given in Chapter 1. Moreover, it follows from the definition of $\frac{m}{n}$ that

$\frac{m}{n}$ is the length of the concatenation of m segments each of length $\frac{1}{n}$.

For brevity, we shall also agree to express the preceding sentence as

$$\frac{m}{n} \text{ is } m \text{ copies of } \frac{1}{n}.$$

This is the terminology that will be used often in the rest of the chapter.

We give two more examples of fractions. The first is the collection of the multiples of $\frac{1}{5}$:

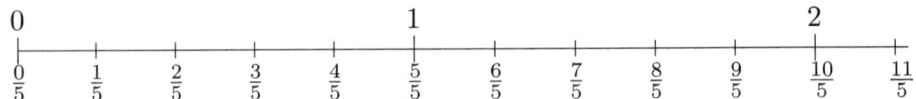

The next is the collection of the multiples of $\frac{1}{8}$:

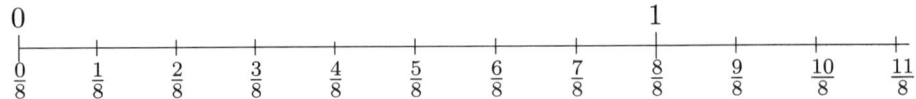

You see that there are many fractions on the number line, but it should not be assumed that every number (i.e., point) to the right of 0 is a fraction. For example, if $[0, c]$ is the line segment with the same length as the diagonal of the unit square, then it will be shown in section 36.3 that c is not a fraction. In fact, "most" numbers are not fractions.

You may not be completely at ease with this definition of a fraction yet. Here is an activity that could help you get there.

Activity. Using the preceding examples as models, describe in words where each of the fractions is on the number line and also draw a rough picture to show its location. (a) $\frac{7}{9}$. (b) $\frac{6}{11}$. (c) $\frac{9}{4}$. (d) $\frac{17}{5}$. (e) $\frac{17}{3}$. (f) $\frac{k}{5}$, where k is a whole number satisfying $11 \leq k \leq 14$. (g) $\frac{k}{6}$, where k is a whole number satisfying $25 \leq k \leq 29$.

12.3. Decimals

Finally, it is to be remarked that there is at present some confusion in the use of the word "fraction" in education literature. Some writers define a fraction to be $\frac{x}{y}$, where x and y can be any real numbers. Thus $\frac{\sqrt{2}}{5}$ is a fraction according to the latter, but not here.

In this book, a fraction is a multiple of $\frac{1}{n}$ for some nonzero whole number n, so that the m and n in a fraction symbol $\frac{m}{n}$ are always *whole numbers*.

One has to be careful about this conflicting use of the term.

12.3. Decimals

There is a special class of fractions that deserves to be singled out at the outset: those fractions whose denominators are equal to positive powers of 10, e.g.,
$$\frac{1489}{100}, \quad \frac{24}{100000}, \quad \frac{58900}{10000}.$$

These are called **decimal fractions**, but they are better known in a different notation and under a slightly different name. It has been recognized since 1593 by the German Jesuit astronomer C. Clavius (see [**Gin28**]) that a decimal fraction is easier to write if we abandon the fraction symbol: just use the numerator and then keep track of the number of zeros in the denominator (two in the first decimal fraction, five in the second, and four in the third) by the use of a so-called **decimal point**, thus:

$$14.89, \quad 0.00024, \quad 5.8900,$$

respectively. A number expressed in this notation of the decimal point is called a **finite** or **terminating decimal**. (Regrettably, this terminology is very confusing.) The rationale of the notation is clear: the number of **decimal digits**, i.e., the number of digits to the right of the decimal point, corresponds to the number of zeros in the respective denominators, two in 14.89, five in 0.00024, and four in 5.8900. In particular,

we regard 5.8900 *as having four decimal digits.*

If we rewrite the above decimal fractions using exponents (see Chapter 1), as we will do from now on, then they become

$$\frac{1489}{10^2}, \quad \frac{24}{10^5}, \quad \frac{58900}{10^4}.$$

The rule of the decimal point then states that

> *the number of decimal digits in a decimal is equal to the exponent of* 10 *in the denominator of the corresponding decimal fraction, e.g., two in* 14.89, *five in* 0.00024, *and four in* 5.8900.

When it is clear from context that only finite decimals are being discussed, we usually omit any mention of "finite" or "terminating" and just say **decimals**. Notice the CONVENTION that, in order to keep track of the power 5 in $\frac{24}{10^5}$, three zeros are added to the left of 24 to make sure that there are five digits to the right of the decimal point in 0.00024. The 0 in front of the decimal point is only for the purpose of clarity, and is optional.

You may be struck by the strange-looking number 5.8900 because you are used to deleting the zeros at the right end of 5.8900 and writing it as 5.89. Why do you think you can do that? Do you just delete the zero at the end of a whole number such as 100? So clearly there is something here *that must be proved*. We will do that in the next section.

Activity. (a) Express $\frac{163079}{10^8}$ and $\frac{230000}{10^2}$ in decimal notation. (b) Express 10000.2001 and 0.000000071008000 as fractions.

> *This definition of a decimal may present the first serious hurdle for the reader in this book in terms of mathematics learning. Let us repeat the main message: the meaning of a decimal such as 0.0938 is that it is a fraction, namely,*
>
> $$\frac{938}{10^4}.$$
>
> *You may likely consider this particular "interpretation" of a decimal mildly entertaining, pause briefly to take note, and proceed to forget all about it. However, what this definition does is to ask you to do a wholesale re-evaluation of your existing knowledge of decimals by rethinking everything you know about decimals and reorganizing your knowledge from scratch using this definition as a starting point. This is not an easy thing to do because you are probably used to treating 5.89 as 5 and 8 tenths and 9 hundredths without worrying about what that means and the kind of trouble such a conception of a decimal gets you into when you do computations with decimals. We understand the effort it takes to start over again with a new definition (which is no different from learning a new language) and will give you every support possible in subsequent sections. Nevertheless, we ask you to make the effort because your knowledge of decimals will continue to be problematic until you do.*

12.4. Importance of the Unit

We wish to underscore the important role of a unit disguised as the number 1 in the definition of a fraction. Compare the discussion in section 8.2.

We have so far limited ourselves to looking at the unit 1 abstractly as a point on the number line. Now suppose we interpret 1 as the *weight* of a

12.4. Importance of the Unit

piece of ham that weighs three pounds. Then the number 1 stands for three pounds, and the number 4, being $1+1+1+1$, will now be interpreted as the weight of four pieces of ham of the same weight (therefore twelve pounds). Now what would $\frac{1}{3}$ represent? According to our definition, we divide our unit (three pounds) into three parts of equal weight[3] (each part therefore weighs one pound), and one of these parts is $\frac{1}{3}$. In this context, $\frac{1}{3}$ represents one pound. In ordinary language, $\frac{1}{3}$ is *"the weight of a third of this piece of ham"*. More generally, the same reasoning tells us that, if we decide on using the weight of an object X as our unit, then $\frac{5}{7}$ (say) would represent the weight of five parts of X after X has been partitioned into seven parts of equal weight. Therefore in this setting, $\frac{5}{7}$ is what we usually refer to as "the weight of five-sevenths of X". (Compare the more formal treatment of this phrase in Chapter 15, page 245 below.) Thus depending on what the unit 1 is, a fraction can have many interpretations. A fraction $\frac{5}{7}$ could be the "volume of five-sevenths of a bucket of water", the "volume of five-sevenths of a pie",[4] "five-sevenths of your life savings in dollars", etc., depending on what the unit is.

An example of what might lead to misconceptions is the use of pattern blocks as so-called "exploration of fractions with shapes". Recall that there are six pieces in the pattern blocks set: a triangle, a square, a large blue rhombus which is the combination of two triangles, a small white rhombus ("Rhombus B"), a hexagon which is the combination of six triangles, and a trapezoid which is half of the hexagon. Suppose we take the unit to be the *area* of the triangle. Then the trapezoid would be represented by 3, the large rhombus by 2, and the hexagon by 6. The square, according to the description in the pattern blocks literature, would be represented (roughly) by $\frac{23}{10}$, because its *area* is about $\frac{23}{10}$ of the area of the triangle if the latter is taken to be 1. Thus, contrary to what you may have been led to believe,

> pattern blocks are not manipulatives that explore fractions with *shapes*, but rather, are explorations of fractions with *area*.

(See Exercise 10 at the end of this chapter.) On the other hand, if we take 1 to be, *not area*, but *the total number of noncongruent polygons in the pattern blocks collection*, then with respect to this unit, the hexagon, the square, the triangle, the two rhombi, and the trapezoid would each be represented by

---

[3]Notice that we did not say "divide our unit into three equal parts", but said instead *"three parts of equal weight"*. The reason is that, insofar as we are dealing with a piece of ham, the former statement might be misinterpreted to mean "equal" in volume, or "equal" (congruent) in shape.

[4]In textbooks, imprecision in the description of the unit often leads to unfortunate misconceptions; see Exercises 9 and 10 at the end of this chapter. The usual statement is to just say "cut the pie into seven equal parts and take five". It would be more clear to say instead, "cut the circle into regions of equal area". If "equal parts of the whole" are never made explicit to students once and for all, there will be misconceptions about "equal division" and therefore about fractions.

$\frac{1}{6}$, and the collection of the two (different) rhombi would be represented by $\frac{2}{6}$.

Activity. Suppose the unit 1 represents the (value of a) dollar. What numbers would represent a penny, a nickel, a dime, and a quarter? If on the other hand, the unit represents (the value of) a quarter, what numbers would represent a penny, a nickel, a dime, and a dollar? Still with a quarter as your unit, how many dollars would $\frac{13}{5}$ represent?

In terms of a fixed unit, we sometimes informally *paraphrase* the definition of a fraction as follows when there is no fear of confusion:

> Let k, ℓ be whole numbers with $\ell > 0$. Then $\frac{1}{\ell}$ is by definition one part when the unit is divided into ℓ equal parts, and $\frac{k}{\ell}$ is by definition (the totality of) k of these parts.

Pedagogical Comments. *One should take note that the preceding passage does not give the precise meaning of a fraction because "one part" and "divided into ℓ equal parts" are vague statements.[5] If the unit represents the length of a certain segment, then "equal parts" would mean "segments of the same length". If the unit represents the volume of a fixed set, then "equal parts" would mean "subsets of equal volume", and so on. If the unit in question represents the length of a segment, then the preceding paraphrase gives essentially the original definition of a fraction.[6] In the classroom, this way of introducing a fraction in terms of a fixed unit may be initially more acceptable to some students, but if it is used, do not forget to constantly remind the students about the presence of the unit.* **End of Pedagogical Comments.**

12.5. The Area Model

Other than the length of a unit interval, the most interesting and the most common unit is the area of a unit square. We single it out for an extended discussion because of its importance in the development of the theory of fractions in this book.

Recall from Chapter 2 that **unit square** refers to a square with each side of length 1 (a certain unit being assumed to have been chosen on the number line).

---

[5] See preceding footnote.
[6] What is missing from this paraphrase is the clarity of a fraction as a definite point on the number line. Here, a "part" must be left to the imagination.

12.5. The Area Model

Because we will have to use the concept of **area** in a more elaborate fashion, let us first give a more detailed discussion of the basic properties of area. The basic facts about area that we need are rather mundane and are summarized below.

(a) The area of a planar region is always a number.

(b) The area of the unit square is by definition the number 1.

(c) If two regions are *congruent*, then their areas are equal.

(d) If two regions have at most (part of) their boundaries in common, then the area of the region obtained by combining the two is the sum of their individual areas.

We shall not define "congruent regions" precisely except to use the intuitive meaning that congruent regions have the "same shape and same size", or that one can check congruence by sliding, rotating, and reflecting one region to see if it can be made to coincide completely with the other. Thus the following two regions A and B are congruent:

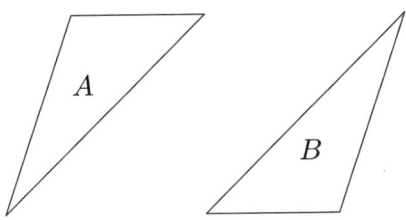

The meaning of item (d) is illustrated by the following: the area of the combined region of C and D below is the sum of the areas of C and D.

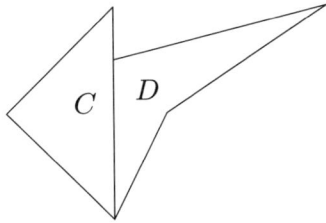

Let us show, for example, that *each of the following four triangular regions in the unit square has area* $\frac{1}{4}$.

Here is the explanation. By item (b) above, the area of the unit square is 1, which we now regard as the unit of the number line. Let us accept as self-evident that these triangular regions are congruent to each other so that, by item (c), these four regions have equal area. Therefore the unit (= area of unit square) has been divided into four equal parts (each part being the area of a triangular region). By the definition of $\frac{1}{4}$ relative to this unit, each triangular region now represents $\frac{1}{4}$. Or more precisely, since the unit is the unit area, the fraction $\frac{1}{4}$ represents an area of $\frac{1}{4}$.

We pause to address the issue of why we bother with something you consider to be completely obvious. There are at least three reasons.

(i) It is obvious to you because you have more or less seen it before, but is it obvious to a student in grades K–3 who is coming across it for the first time? Probably not. So you have to learn how to explain this fact.

(ii) We will need this fact and others similar to it in Chapter 17 below, so we should be able to supply a reason.

(iii) We want to illustrate what it means to have a definition, i.e., *we can explain any assertion related to the definition on the basis of the definition itself.* This is the attitude we take about definitions.

On a practical level, should you make a big deal of this explanation in your classroom? No. The explanation of such "obvious" facts is not easy for children. It may be sufficient that you impress on them that there *is* an explanation and move on. For your own growth as a teacher of mathematics, however, it is good to learn how to give such an explanation in case you need it.

This kind of argument can be carried out in like manner in similar situations, and $\frac{1}{4}$ can be seen to have many pictorially distinct representations. It can be a part of any division of the unit square into four parts of equal area. We give some examples. In the following pictures, each square is assumed to be the unit square. It is understood that the unit 1 is the area of the square. Moreover, the division of each side of the unit square is always understood to be an equi-division. Then the area of each of the following shaded regions in the respective squares is $\frac{1}{4}$:

12.5. The Area Model

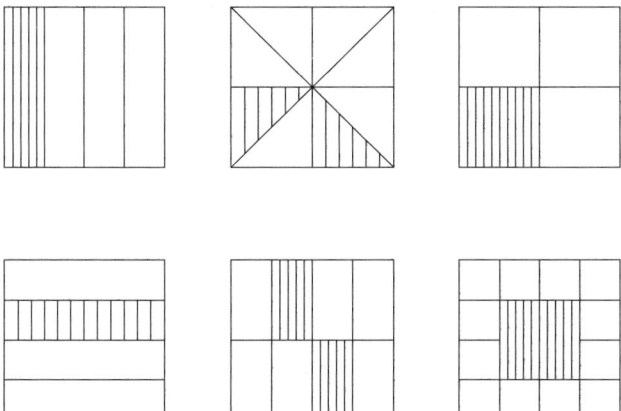

Activity. In the following picture, suppose the unit 1 is the area of the whole square.

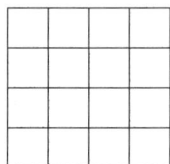

Shade two different regions so that each represents $\frac{3}{8}$; do the same for $\frac{7}{16}$. Can you tell by visual inspection which of $\frac{3}{8}$ and $\frac{7}{16}$ has more area?

If it helps you to think of a fraction as some kind of a pictorial object—part of a pie, part of a square (such as above), or a collection of dots—by all means do so. *In mathematics, do whatever it takes to help you learn something—including the use of manipulatives, metaphors, and analogies—provided you do not lose sight of the precise definitions or skills.* In the case of fractions, it means you may use any pictorial image you want to process your thoughts on fractions, but *at the end, you should be able to formulate logical arguments in terms of the original definition of a fraction as a point on the number line.*

One word of caution about the use of pictures: even in informal reasoning, we should try not to damage students' intuitive grasp of the basic mathematics. Therefore, if the area of a square or a circle is used as a unit, what we should be careful about is to keep the size of the unit the same (or as much as hand-drawing allows!) under all circumstances. Here are some examples of how *not* to do it.

EXAMPLE 1. Tell students that the following shaded area represents $\frac{3}{2}$:

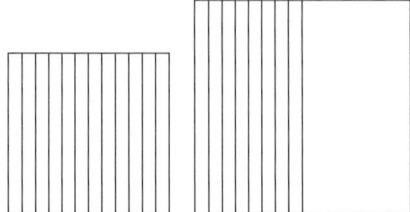

What is wrong is that, if the area of the left square is implicitly taken as the unit, then the area of the right square (which is visibly bigger) would be bigger than 1 and consequently the total shaded area would represent more than $\frac{3}{2}$. Or, if the area of the right square is taken as the unit 1, then the total shaded area would be smaller than $\frac{3}{2}$.

EXAMPLE 2. Tell students that the following shaded area represents $\frac{3}{2}$:

 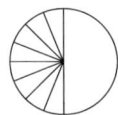

This is exactly the same visual misrepresentation of a fraction as the preceding example, but a pizza has replaced the square. If the area of the left pizza is taken as a unit, then the right pizza represents a number bigger than 1, so that the total shaded area would be more than $\frac{3}{2}$. *Because the pizza representation of a fraction is so popular, it is hoped that, by bringing this problematic issue to the forefront, such misleading representations of fractions will disappear from the classroom.* An additional remark is that many teachers manage to avoid this misrepresentation because they only work with proper fractions, in which case only one pizza is used at all times. Such a practice is not pedagogically sound because students must get used to seeing all kinds of fractions, proper and improper.

EXAMPLE 3. "What fraction is represented by the following shaded area?"

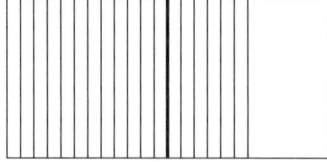

The problem here is that the unit is not clearly specified, so that it would be perfectly legitimate to assume that the area of the whole rectangle is the unit 1, in which case the shaded area would be $\frac{3}{4}$ instead of $\frac{3}{2}$. It would be a good idea to avoid this kind of ambiguity right from the beginning, by emphasizing the important role of a unit.

12.6. Locating Fractions on the Number Line

We now give some examples on how to locate fractions, approximately, on the number line. Observe how division-with-remainder comes in naturally. In a sense, this is a natural continuation of Chapter 10 on estimation: roughly how big is a given fraction?

For example, on the following line, between which whole numbers should the fraction $\frac{16}{3}$ be placed?

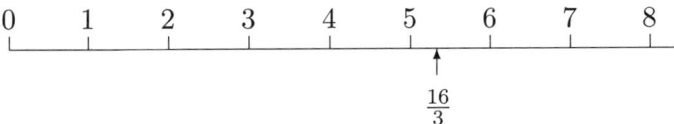

For this simple case, we can do it by a simple mental calculation. We know $15 = 5 \times 3$ and $18 = 6 \times 3$. Therefore $\frac{15}{3} = 5$ and $\frac{18}{3} = 6$, by virtue of equation (12.1), so that inasmuch as the multiples of $\frac{1}{3}$ propagate to the right as $\ldots, \frac{15}{3}, \frac{16}{3}, \frac{17}{3}, \frac{18}{3}, \ldots$, the fraction $\frac{16}{3}$ must be somewhere between $5 \,(= \frac{15}{3})$ and $6 \,(= \frac{18}{3})$, and closer to 5 than to 6, as shown below.

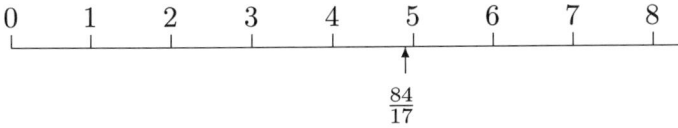

In general, when simple mental calculation does not come as easily, it will be necessary to use division-with-remainder (Chapter 7). To illustrate, consider the problem of where to put $\frac{84}{17}$ on the same number line. Now look at the multiples of 17, namely 0, 17, 34, 51, 68, 85, ... Clearly 84 lies between 68 $(= 4 \times 17)$ and 85 $(= 5 \times 17)$, and its exact position can be obtained from the division-with-remainder: $84 = (4 \times 17) + 16$. So

$$\frac{84}{17} = \frac{(4 \times 17) + 16}{17}.$$

and therefore $\frac{84}{17}$ should be the point on the number line which is $\frac{16}{17}$ to the right of the number 4. So if each step we take is of length $\frac{1}{17}$, we go to the right of 4 another 16 steps. If we go 17 more steps, we would get to 5. Therefore $\frac{84}{17}$ should be quite near 5, as shown below.

Activity. Give the approximate locations of the following fractions on the number line. (a) $\frac{29}{100}$. (b) $\frac{255}{101}$. (c) $\frac{1234}{2467}$. (d) $\frac{49}{5}$. (e) $\frac{73}{12}$.

12.7. Issues to Consider

Certain issues that arise in the preceding discussion are quite subtle and deserve to be examined at some length.

First and foremost, it is imperative for our purpose that we have an *explicit definition* of a fraction, for two reasons. In this book, every mathematical statement is supported by reasoning, and mathematical reasoning must be grounded in exact and precise information. So the definitions have to be precise in order to give the necessary support. A second reason is that pieces of a pizza cannot lead to a clear and logical development of fractions. For example, nobody knows how to divide one piece of pizza by another. Since the definition of a fraction as a certain point on the number line has proven to be the only one that is mathematically viable thus far, this is the definition we will use.

A second issue is that there is some uneasiness among students concerning the strange notation $\frac{k}{\ell}$ for a fraction; they wonder why it takes *two* whole numbers k and ℓ and a bar between them to denote a single object. We believe this notation would be strange only when the concept of a fraction is a mystery, so that *any* notation used to denote a mysterious object would seem strange. But we know better: fractions are a definite collection of points on the number line, most of which lie *between* whole numbers. To locate the point denoted by $\frac{5}{3}$, for example, ask your students if there is any other way to specify the location of this point between 1 and 2 using only *one* whole number. Obviously not. Furthermore, explain to them that the bar between 5 and 3 is strictly for clarity and nothing else. If we eliminate the bar, then $\frac{5}{3}$ would be written as $\genfrac{}{}{0pt}{}{5}{3}$, and how long would it be before this symbol becomes 35 or 53 in the hands of sloppy writers? Better to put the bar back in! When such explanations are supplied, the notation $\frac{5}{3}$ will look much less strange.

A third issue is that you may have some concern about the idea of dividing the unit segment into any number of equal parts, for example, 7 or 11. Thus, in order to place the fractions with denominator equal to 11 on the number line, we have to divide the unit segment into 11 parts of equal length. In practical terms, this concern is entirely reasonable because most of us have trouble dividing anything into equal parts other than halves or fourths. However, for the understanding of the definition of fractions, there is no need to fret about the *practical* means of doing this but only about the *theoretical* possibility, and the latter is not in doubt. Furthermore, be aware that in the school classroom, one can cheat a little bit with this kind of division: put together 11 short segments of the same length on a line and

12.7. Issues to Consider

declare *that* to be the unit segment. Then the equal division of the unit segment into 11 equal parts would be built-in.

A fourth issue that merits discussion is this: *what does the equality of two fractions mean* in general?[7] For example, we encountered in equation (12.1) the phenomenon that $\frac{n\ell}{\ell} = n$ for any whole numbers $n, \ell > 0$, so

$$\text{what does } \text{``} \frac{n\ell}{\ell} = n \text{''} \text{ mean?}$$

This is a continuation of the discussion of the equality of two whole numbers begun in Chapter 2, page 39. For whole numbers, it was either to check by counting, or to check by looking at the number line to see if both whole numbers are the same point. For fractions, counting is out of the question. But *because we have a precise definition of a fraction*, we are still in a position to enunciate unambiguously what it means for two fractions to be equal.

Definition. *The **equality** $\frac{a}{b} = \frac{m}{n}$ means that the two points denoted by the fraction symbols $\frac{a}{b}$ and $\frac{m}{n}$ are the same point on the number line. Two fractions that are equal are also said to be **equivalent fractions**.*

A more correct terminology for "equivalent fractions" is **equivalent fraction symbols**, but as usual, we simply follow tradition when it is more or less harmless to do so. In subsequent sections, we will often be called upon to verify that two fractions are equal.

The discussion of the equality of fractions should be put in the more general context of **comparing** or **ordering fractions**. In Chapter 1, we expressed the concept of $A < B$ for two whole numbers A and B in terms of their relative positions on the number line, namely, $A < B$ if A is to the left of B. We now define the same concept for fractions in exactly the same way. Given two fractions A and B, we say $A < B$ (A **is less than** B, or B **is greater than** A) if A is to the left of B as points on the number line:

This is the same as saying that the segment $[0, A]$ is shorter than the segment $[0, B]$.

A final remark on the definition of a fraction is that, from a mathematical standpoint as well as in our present approach to fractions, there is no difference between "big" fractions and "small" fractions (in the sense of the numerators or denominators being big and small numbers), because on the number line, all points (numbers) are on equal footing. You will never need

---

[7]The meaning of the equal sign has attracted the attention of educational researchers in algebra. The fundamental issue here is whether students are taught what *equality* means from the very beginning of their mathematics study.

to favor fractions with single-digit numerators or denominators, as is the common practice in elementary classrooms.

Exercises

There are two rules about doing the exercises in this book: (i) Unless stated to the contrary, use only what you have learned so far in the book. (ii) Every answer must come with an explanation. The explanation may be in the form of the details in a calculation, or it may be a verbal citation of particular facts used in the text. Sometimes, for emphasis, an explanation is even explicitly demanded. But whatever it is, you will have to get used to never claiming anything without giving a reason.

1. With the area of a unit square as the unit 1, draw two distinct pictorial representations of each of: (a) $\frac{5}{6}$. (b) $\frac{7}{4}$. (c) $\frac{9}{4}$.

2. Suppose the unit 1 on the number line is the area of the following shaded region in a division of the given square into four parts of equal area:

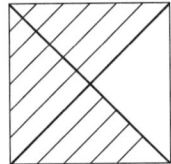

Write down the fraction of that unit representing the shaded area of each of the following, and *give a brief explanation of your answer* (you may assume that the division of the sides of the square in the middle figure is a division into four parts of equal length, and that the right figure consists of two squares, with the right vertical line dividing the side of the right square into two halves):

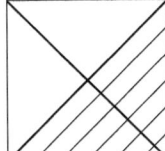 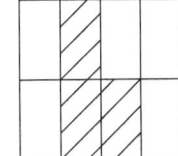 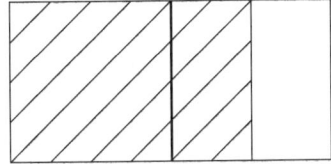

3. Repeat the preceding problem if the unit 1 is now the area of the following shaded region:

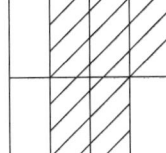

4. With the unit as in Exercise 2 above, write down the fraction representing the area of the following shaded region (assume that the top and

bottom sides of the square are each divided into three segments of equal length):

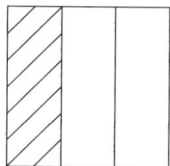

5. Take a pair of opposite sides of a unit square and divide each side into 48 equal parts. Join the corresponding points of division to obtain 48 thin rectangles (we will assume that these are rectangles). For the remaining pair of opposite sides, divide each into 203 equal parts and also join the corresponding points of division; these lines are perpendicular to the other 48 lines. The intersections of these 48 and 203 lines create 48 × 203 small rectangles which are congruent to each other (we will assume that too). What is the area of each such small rectangle, *and why*? (This problem is important for Chapter 17 below.)

6. Indicate the approximate position of each of the following on the number line, and briefly explain. (a) $\frac{102}{1003}$. (b) $\frac{65}{16}$. (c) 2.51. (d) 4.26. (e) $\frac{459}{23}$. (f) $\frac{1502}{24}$. (g) $\frac{9}{28}$. (h) $\frac{17}{84}$.

7. *Review section 12.4, Importance of the Unit, on page 188 before doing this problem. Also make sure that you do it by a careful use of the definition of a fraction rather than by some transcendental intuition you possess which cannot be explained to your students.*
   (a) After driving 150 miles, we have done only two-thirds of the driving for the day. How many miles did we plan to drive for the day? Explain.
   (b) After reading 200 pages of a book, I am exactly four-fifths of the way through. How many pages are in the book? Explain.
   (c) Helena was three-quarters of the way to school after having walked 0.9 miles from home. How far is her home from school?

8. Three segments (thickened) are on the number line, as shown:

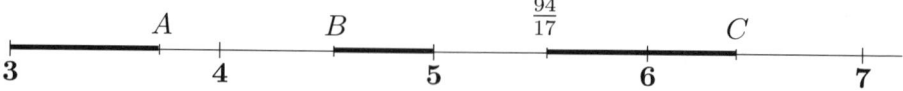

It is known that the length of the left segment is $\frac{9}{14}$, that of the middle segment is $\frac{7}{15}$, and that of the right segment is $\frac{15}{17}$. What are the fractions A, B, and C? (*Caution:* Remember that you have to explain your answers, and that you know nothing about "mixed numbers" until we come to this concept in Chapter 14.)

9. (a) I have a friend who earns two dollars for every three times she walks her parents' dog. She knows that this week she will walk the dog twelve times. How much will she earn?

   (b) Suppose your friend tells you that he taught his fifth grade class to do the problem in part (a) by using fractions and "setting up a proportion"
   $$\frac{2}{3} = \frac{?}{12},$$
   and he wonders why his class didn't "get it". How would you straighten him out to help him?

10. A text on professional development claims that students' conception of "equal parts" is fragile and is prone to errors. As an example, it says that when a circle is presented this way to students:

    they have no trouble shading $\frac{2}{3}$, but when these same students
    ```
    are asked to construct their own picture of 2/3, we often
    see them create pictures with unequal pieces, such as
    the following:
    ```

    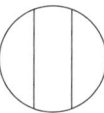

    (a) What kind of faulty mathematical instruction might have promoted this kind of misunderstanding on the part of students? (Hint: Note the phrase "unequal pieces" above, and see footnotes 3 and 4 on pages 189 and 189.) (b) What would you do to correct this kind of mistake by students?

11. *You may need to look up the case of* Two Green Triangles *on p. 86 of the Case Book* [**BGJ94**] *for this exercise.* In a teacher's attempt to teach fractions, she made use of Pattern Blocks and tried to get her students to "figure out the fractional relationships of the different colored blocks". She held up two green triangles and said, "If yellow equals 1 then how much is this?" Recall that "yellow" is the big hexagon, and six green (equilateral) triangles pieced together in the usual manner form a hexagon congruent to the yellow hexagon. The teacher expected her students to say "two sixths", but they kept saying "two" instead. Explain which aspect of the teacher's teaching might lead to such misconception on the part of the students.

12. The following is found in a certain third grade workbook:

> Each of the following figures represents a fraction:
>
>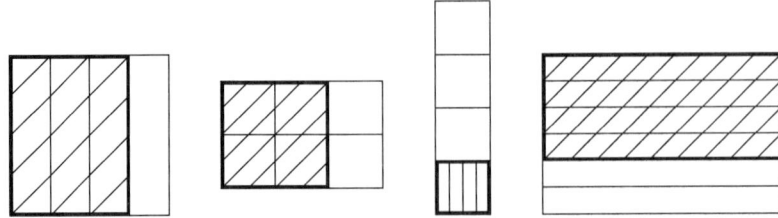
>
> Point to two figures that have the same fractions shaded.

If you are the third grade teacher teaching from this workbook, how would you change this problem to make it suitable for classroom use?

Chapter 13

Equivalent Fractions and FFFP

We are all familiar with the fact that $\frac{1}{2}$ is equal to $\frac{2}{4}$, because "taking two parts out of four is the same as taking one part out of two". Similarly, $\frac{1}{3}$ is equal to $\frac{2}{6}$. And so on. According to the definition on page 197, what we have is that $\frac{1}{2}$ and $\frac{2}{4}$ are equivalent fractions, and that $\frac{1}{3}$ and $\frac{2}{6}$ are also equivalent fractions.

The main purpose of this section is to try to come to grips with equivalent fractions by answering two questions:

Is there a general reason that underlies the equivalence of $\frac{1}{2}$ and $\frac{2}{4}$, and $\frac{1}{3}$ and $\frac{2}{6}$?

How do we recognize whether two fractions are equivalent?

If the usual examples such as $\frac{1}{2}$ and $\frac{2}{4}$ and $\frac{1}{3}$ and $\frac{2}{6}$ lead you to believe that everything along this line can be done by visual inspection and that there is nothing to learn here, ask yourself whether $\frac{161}{91}$ and $\frac{253}{143}$ are equivalent and, if so, how would you explain it to your students? You will be able to do that by the end of this chapter (see Exercise 1 on page 218).

The sections in this chapter are the following.

Theorem on Equivalent Fractions (Cancellation Law)

Applications to Decimals

Proof of Theorem 13.1

FFFP

The Cross-Multiplication Algorithm

Why FFFP?

In the last section, we explain the general reasoning behind the Fundamental Fact of Fraction Pairs (FFFP).

13.1. Theorem on Equivalent Fractions (Cancellation Law)

At the beginning of the study of fractions, students are taught the rote skill that given a fraction, such as $\frac{6}{8}$, we would get the same fraction if we multiply both the numerator and denominator by the same whole number. Thus,
$$\frac{6}{8} = \frac{5 \times 6}{5 \times 8} \left(= \frac{30}{40} \right).$$
Another part of this rote skill is that if we divide the numerator and denominator by a common divisor, we also get the same fraction. Thus,
$$\frac{6}{8} = \frac{6 \div 2}{8 \div 2} \left(= \frac{3}{4} \right).$$
We now give the reason for both in a general setting in the form of a theorem:

Theorem 13.1. *Given two fractions $\frac{m}{n}$ and $\frac{k}{\ell}$, suppose there is a nonzero whole number c so that*
$$k = cm \quad \text{and} \quad \ell = cn.$$
Then
$$\frac{m}{n} = \frac{k}{\ell}.$$

Recall from our convention in Chapter 12 that n and ℓ, being the denominators of the fractions $\frac{m}{n}$ and $\frac{k}{\ell}$, respectively, are automatically assumed to be nonzero. This theorem is usually stated more briefly in the form

(13.1) $$\frac{m}{n} = \frac{cm}{cn}$$

for all fractions $\frac{m}{n}$ and for all nonzero whole numbers c. If $n = 1$ in equation (13.1), then the resulting equation, $m = \frac{cm}{c}$ for any whole number m and any nonzero whole number c, is just the equation (12.1) in Chapter 12. If, however, m and n are fixed and the c in equation (13.1) runs through $1, 2, 3, \ldots$, then we get an infinite number of fractions equivalent to $\frac{m}{n}$. Thus, let $m = 2$, $n = 3$, and let $c = 2, 3, \ldots, 9$, we obtain
$$\frac{2}{3} = \frac{4}{6} = \frac{6}{9} = \frac{8}{12} = \frac{10}{15} = \frac{12}{18} = \frac{14}{21} = \frac{16}{24} = \frac{18}{27}.$$
To say that one *knows* Theorem 13.1 implies that one recognizes, for example, that the fractions $\frac{6}{9}$ and $\frac{18}{27}$ are equal, and can explain why they are equal.

Theorem 13.1 is usually called the **theorem on equivalent fractions**, as it gives a sufficient condition for two fractions to be equivalent. Equation

13.1. Theorem on Equivalent Fractions (Cancellation Law)

(13.1) is sometimes called the **cancellation law** of fractions (you "cancel" the c from both the top and bottom of $\frac{cm}{cn}$ in equation (13.1) to get $\frac{m}{n}$).

Theorem 13.1 may be regarded as the *fundamental fact in the subject of fractions.*

We will present the proof of Theorem 13.1 in the section after next. For now, we will assume that it has been proved and concentrate on exploring some of its ramifications. The remaining chapters in Part 2 will be seen to be nothing more than applications, direct or indirect, of this theorem. Conceptually, the message of the theorem is clear: the essence of the fraction symbol $\frac{m}{n}$ lies not so much in the individual numbers m and n in the symbol as in their "relative sizes" because $\frac{m}{n} = \frac{2m}{2n} = \frac{3m}{3n} = \cdots$.

The following corollary rephrases the cancellation law (13.1) in terms of division.

Corollary 13.1.1. *Given a fraction $\frac{k}{\ell}$. If a whole number c divides whole numbers k and ℓ, then*

$$\text{(13.2)} \qquad \frac{k}{\ell} = \frac{k \div c}{\ell \div c}.$$

Before proving the corollary, we should mention that the passage from $\frac{k}{\ell}$ to $\frac{k \div c}{\ell \div c}$ is called **reducing the fraction** $\frac{k}{\ell}$ to $\frac{k \div c}{\ell \div c}$.

You will note that the effectiveness of using the cancellation law for the purpose of reducing a fraction depends entirely on finding a whole number (different from 1) that divides both numerator and denominator. Sometimes it may not be possible to guess such a number. For instance, it is not obvious that $\frac{171}{285}$ is equal to $\frac{3}{5}$, because 57 does not jump out at us as a number that divides both 171 and 285. Other striking examples abound, such as the following two: $\frac{253}{161} = \frac{11}{7}$ because 23 is a common divisor of 253 and 161, and $\frac{1651}{762} = \frac{13}{6}$. Fortunately, there is an algorithm that produces such a common divisor, and this is the Euclidean Algorithm; see Chapter 35.

Activity. Prove $\frac{1651}{762} = \frac{13}{6}$.

For the proof of the corollary, let us first look at an example. Given $\frac{38}{57}$. Suppose we know that 19 divides both 38 and 57; then $38 \div 19 = 2$ and $57 \div 19 = 3$ and the claim (according to the corollary) is that $\frac{38}{57} = \frac{2}{3}$. This is so because we also have $38 = 19 \times 2$ and $57 = 19 \times 3$, so that

$$\frac{38}{57} = \frac{19 \times 2}{19 \times 3} = \frac{2}{3}$$

by equation (13.1) with $c = 19$. The reasoning is the same in the general case.

Proof. Indeed, by the definition of division on page 97, if c divides k, let $k = cm$ and if c divides ℓ, let $\ell = cn$ for some whole numbers m and n; thus

we have $m = k \div c$ and $n = \ell \div c$. Therefore, using equation (13.1), we get
$$\frac{k}{\ell} = \frac{cm}{cn} = \frac{m}{n} = \frac{k \div c}{\ell \div c},$$
which is exactly equation (13.2). The corollary is proved. □

A fraction $\frac{k}{\ell}$ is said to be **reduced**, or **in lowest terms**, if there is no whole number $c > 1$ so that c divides both k and ℓ. It is a fact that every fraction is equal to one and only one fraction in reduced form. For example, $\frac{2}{3}$ is the reduced form of $\frac{18}{27}$ and $\frac{13}{6}$ is the reduced form of the fraction $1651/762$. This fact about the reduced form of a fraction is plausible, but its proof (to be given on page 476) is not so trivial as it requires something like the Euclidean Algorithm. However, it is good to keep this fact in mind because it justifies our occasional reference to *the* reduced form of a given fraction.

Pedagogical Comments. *It seems to be a tradition in school mathematics to regard nonreduced fractions as something "illegitimate", and students usually get points deducted if they give an answer in terms of non-reduced fractions. This is perhaps the right place to give this issue some perspective. On the one hand, a fraction is a fraction is a fraction. Nowhere does the definition in Chapter 12 say that some fractions are "better" than others. So from a* mathematical *standpoint, all fractions are on the same footing. The fraction $\frac{5}{10}$ is as good as $\frac{1}{2}$, even if we prefer the latter. Moreover, it is sometimes difficult to justify our preference even under the most generous conditions. If we insist that every fraction be in lowest terms, should we accept $\frac{38}{57}$ from a fifth grader? After all, few adults would recognize that it is not in lowest terms. And what about an extreme case like $\frac{1333}{2279}$? (It is actually equal to $\frac{31}{53}$.) A pedantic insistence on having everything in lowest terms is difficult, if not impossible, to defend.*

On the other hand, it does get annoying if students get into the habit of never simplifying fractions such as $\frac{4}{2}$ or $\frac{3}{9}$. Some common sense is thus called for in dealing with this situation. One way is to make it a policy that if the answer is a fraction with single-digit numerator and denominator, then it must be reduced. The teacher can also teach students the special skill of reducing fractions to lowest terms and, on exams, explicitly require that certain answers be reduced. She can of course lead by example if she always simplifies some obviously reducible fractions at the board, e.g., $\frac{15}{20} = \frac{3}{4}$. Beyond that, a teacher should make allowances for unreduced fractions.
End of Pedagogical Comments.

It may not be superfluous to point out that the preceding discussions only make use of the statement of Theorem 13.1 with seemingly *no reasoning* involved. Is this learning by rote? *Absolutely not.* Learning how to use a tool (in this case Theorem 13.1 or the cancellation law (13.1)) is an important part of learning mathematics. The preceding discussions are designed to get you used to applying Theorem 13.1 *without having to repeat any part of its proof.*

Learning how to make effective use of the tools at one's disposal is an important part of achieving mathematical proficiency. Of course we also insist that you learn why the tools are mathematically valid. The proof of Theorem 13.1, given in the section after next, is an integral part of the learning.

13.2. Applications to Decimals

As an application of Theorem 13.1, we bring closure to the discussion in the section 12.3, Decimals, about why the decimals 5.8900 and 5.89 are equal. Recall that we had, by definition,
$$\frac{58900}{10^4} = 5.8900,$$
so that
$$5.8900 = \frac{58900}{10^4} = \frac{589 \times 10^2}{10^2 \times 10^2} = \frac{589}{10^2} = 5.89,$$
where the next-to-the-last equality makes use of Theorem 13.1. We observe that the same reasoning proves something more general:

> *one can add or delete zeros to the right end of the decimal point without changing the decimal.*

To drive home this point, let us show that $12.70000 = 12.7$:
$$12.7 = \frac{127}{10} = \frac{127 \times 10^4}{10 \times 10^4} = \frac{1270000}{10^5} = 12.70000.$$

Now we make use of Theorem 13.1 to do something different with decimals: we show how to convert some fractions to finite decimals. Such a conversion is usually taught in schools in three steps:

(i) Add any number of zeros to the right of the numerator and then divide the resulting number by the denominator using the long division algorithm.

(ii) Obtain a decimal by the judicious placement of a decimal point in the quotient.

(iii) Equate this decimal with the original fraction.

No explanation is given. The lack of any explanation for why the resulting decimal is *equal* to the original fraction contributes to the erosion of students' understanding of the equal sign.

It turns out that the three-step procedure (i)–(iii) is correct, but the reasoning is more subtle than is commonly realized. We will discuss this in section 18.4 and more fully in Chapter 42. Here, in the most naive way possible, and *using only the theorem on equivalent fractions*, we show why certain simple fractions are equal to finite decimals. An overriding fact is that a *reduced* fraction is equal to a finite decimal when, and only when, its denominator is a product of 2's and 5's (it is understood that either 2 or 5 could be absent). This theorem will be proved in section 36.2; it will not be explicitly invoked in the present discussion, but it does explain why the fractions in the ensuing discussion are what they are.

The most common conversions from fractions to decimals are these:

$$\frac{1}{2} = 0.5, \quad \frac{1}{4} = 0.25, \quad \frac{1}{5} = 0.2.$$

They can be quickly explained *if we remember the definition of a finite decimal*:

$$\frac{1}{2} = \frac{5 \times 1}{5 \times 2} = \frac{5}{10} = 0.5,$$

$$\frac{1}{4} = \frac{25 \times 1}{25 \times 4} = \frac{25}{100} = 0.25,$$

$$\frac{1}{5} = \frac{2 \times 1}{2 \times 5} = \frac{2}{10} = 0.2.$$

The idea here is that we want the denominator to be a power of 10, and we use equivalent fractions to "bring the denominator up" to a power of 10 by multiplying it with 2's and 5's when necessary (10 is the product of only 2 and 5). This method always works if the denominator is a product of 2's and 5's. With this in mind, we understand why $\frac{3}{8} = 0.375$, because

$$\frac{3}{8} = \frac{3}{2^3} = \frac{5^3 \times 3}{5^3 \times 2^3} = \frac{375}{10^3} = 0.375,$$

where we write $\mathbf{2^3}$ for $2 \times 2 \times 2$, $\mathbf{5^3}$ for $5 \times 5 \times 5$, etc.; see page 156. Similarly, $\frac{18}{625} = 0.0288$, because

$$\frac{18}{625} = \frac{18}{5^4} = \frac{2^4 \times 18}{2^4 \times 5^4} = \frac{288}{10^4} = 0.0288.$$

It is clear from the preceding examples why the exponent of 2 or 5 in the denominator (3 in the case of 8 and 4 in the case of 625) accounts for the number of decimal digits in each answer (3 in the case of 0.375 and 4 in the case of 0.0288).

13.3. Proof of Theorem 13.1

We now turn to the proof of Theorem 13.1. First, let us see how it is usually handled in school textbooks. A typical explanation is the following:

> Different fractions which name the same amount are called EQUIVALENT FRACTIONS.
>
> We will practice a method for making equivalent fractions. If we multiply a number by a fraction that is equivalent to 1, the answer will be a different name for the same number. Thus
> $$\frac{1}{2} \times 1 = \frac{1}{2} \times \frac{2}{2} = \frac{2}{4}, \qquad \frac{1}{2} \times 1 = \frac{1}{2} \times \frac{3}{3} = \frac{3}{6}.$$
> The fractions $\frac{1}{2}$, $\frac{2}{4}$, and $\frac{3}{6}$ are equivalent fractions.

Let us assume that students understand what is meant by "name the same amount". The most worrisome feature of an explanation of this kind is that the explanation of the validity of equation (13.1) is made to depend on the concept of fraction multiplication. As we shall see in Chapter 17 below, the latter is by no means a simple concept and needs the kind of careful definition that is mostly missing in school texts. Therefore, such an explanation of equivalent fractions, being dependent on another concept which is even more difficult to explain, is not acceptable (see the opening essay of this book, **To the Reader**). A correct explanation of equation (13.1) turns out to be both simple and intuitive, as we now proceed to demonstrate.

We will first prove Theorem 13.1 for the special case of
$$\frac{1}{2} = \frac{3}{6} \ \left(= \frac{3 \times 1}{3 \times 2} \right).$$

This example is actually *too* simple to help with the proof in the general case. However, it does give a good introduction to the basic ideas involved. The reasoning is a good illustration of why we need precise definitions. According to the definition of a fraction, $\frac{3}{6}$ is three copies of $\frac{1}{6}$ (recall from Chapter 12, this means that $\frac{3}{6}$ is the length of the concatenation of three segments each of length $\frac{1}{6}$), and we want to know why this is equal to $\frac{1}{2}$. Let us first look at this from the intuitive point of view of cutting pies. Of course the pies are represented two-dimensionally by circles and we try to cut the circles into congruent circular sectors. So $\frac{1}{2}$ is represented by half of a circle that represents 1.

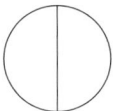

Now we cut each half into three congruent pieces, thereby obtaining a division of the pie into six congruent smaller pieces. Each of these smaller pieces is therefore $\frac{1}{6}$.

Look at, say, the left half of the circle, which is $\frac{1}{2}$. It is now divided into three congruent pieces, each being $\frac{1}{6}$. This is then the statement that $\frac{1}{2}$ is three copies of $\frac{1}{6}$; i.e., $\frac{1}{2}$ is $\frac{3}{6}$.

The intuitive idea of the preceding argument can be easily translated into a formal argument using the number line. In the interest of brevity,

> we shall henceforth write **equal parts** for *segments of equal length* in the context of the number line.

Now divide $[0, 1]$ into two equal parts, the point of division (i.e., $\frac{1}{2}$) being indicated by the vertical arrow below the line segment:

Next, divide each of the segments of length $\frac{1}{2}$ into three equal parts (the "three" is because we are looking at $\frac{3\times 1}{3\times 2}$). Now $[0, 1]$ is divided into six equal parts, and each of these parts therefore has length $\frac{1}{6}$. The picture makes it clear that $\frac{1}{2}$ is three copies of $\frac{1}{6}$. Knowing that three copies of $\frac{1}{6}$ is equal to $\frac{3}{6}$, we see that $\frac{1}{2}$ is $\frac{3}{6}$.

Let us look at another example:
$$\frac{4}{3} = \frac{20}{15}.$$

The argument in this example fully illustrates the complexities of the general situation, but observe how this argument elaborates on—rather than departs from—the simple ideas of the preceding number line argument for $\frac{1}{2} = \frac{3}{6}$. We begin with the left side: $\frac{4}{3}$ is four copies of $\frac{1}{3}$, and we want to know why it is also 20 copies of $\frac{1}{15}$. On the number line, consider all multiples of $\frac{1}{3}$.

13.3. Proof of Theorem 13.1

The segments between consecutive multiples are all of length $\frac{1}{3}$. These are the segments between the vertical arrows below the number line:

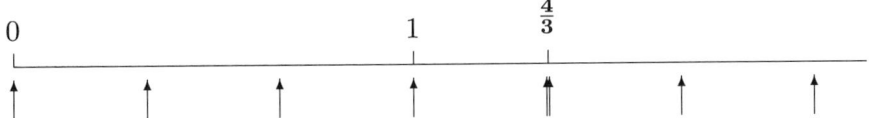

Now divide each of these segments of length $\frac{1}{3}$ into five equal parts (the "five" is because we are looking at $\frac{20}{15} = \frac{5 \times 4}{5 \times 3}$); call these parts *small segments*. All small segments have the same length. Here is a picture of these *small segments*:

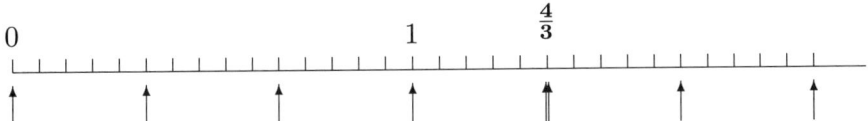

What is the length of each *small segment*? The unit interval $[0, 1]$ is the concatenation of three copies of $\frac{1}{3}$'s, and each of the latter is the concatenation of five *small segments*. Thus $[0, 1]$ is the concatenation of $3 \times 5 = 15$ *small segments* all of which have the same length. Thus each *small segment* has length $\frac{1}{15}$, and since each segment of length $\frac{1}{3}$ is the concatenation of five *small segments*, we see that $\frac{1}{3}$ is five copies of $\frac{1}{15}$. Since $\frac{4}{3}$ is four copies of $\frac{1}{3}$ and $\frac{1}{3}$ is five copies of $\frac{1}{15}$, it follows that $\frac{5}{3}$ is $5 \times 4 = 20$ copies of $\frac{1}{15}$. In other words, $\frac{4}{3}$ is $\frac{20}{15}$.

The **proof of Theorem 13.1** is quite straightforward by now. Given fraction $\frac{m}{n}$, we want to prove that for any nonzero whole number c,

$$\frac{m}{n} = \frac{cm}{cn}.$$

We know that $\frac{m}{n}$ is m copies of $\frac{1}{n}$, and we want to prove that it is also cm copies of $\frac{1}{cn}$. On the number line, consider, first, all the multiples of $\frac{1}{n}$, and then all the multiples of $\frac{1}{cn}$. Call the segments between consecutive multiples of $\frac{1}{cn}$ *small segments*. Then for each segment lying between consecutive multiple of $\frac{1}{n}$, the *small segments* divide it into c equal parts.

Observe that each segment between consecutive multiples of $\frac{1}{n}$ has length $\frac{1}{n}$ and, likewise, each *small segment* has length $\frac{1}{cn}$. But the former is the concatenation of c of the latter, so $\frac{1}{n}$ is c copies of $\frac{1}{cn}$. Since $\frac{m}{n}$ is m copies of $\frac{1}{n}$ and $\frac{1}{n}$ is c copies of $\frac{1}{cn}$, it follows that $\frac{m}{n}$ is cm copies of $\frac{1}{cn}$. But since

$\frac{cm}{cn}$ is also cm copies of $\frac{1}{cn}$, we have proved that $\frac{m}{n}$ is equal to $\frac{cm}{cn}$. This completes the proof of Theorem 13.1.

When the unit on the number line is the area of the unit square, there is a very intuitive proof of Theorem 13.1 that may make the theorem more accessible to elementary school students. We present it here. We can illustrate the idea of the proof with a concrete example. Let us see why

$$\frac{5}{2} = \frac{15}{6}.$$

So we let the unit 1 be the area of a unit square. Then $\frac{5}{2}$ is the area of five half-squares, as indicated by the shaded region in the following picture.

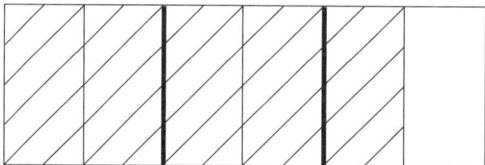

Prompted by the fact that $15 = 3 \times 5$ and $6 = 3 \times 2$, we divide each unit square *horizontally* into equal thirds, thereby producing six congruent small rectangles in each unit square:

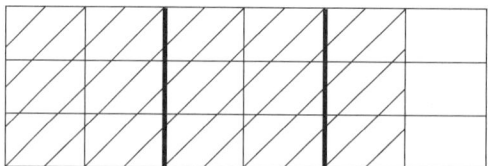

Now we see that this new shaded area consists of $3 \times 5 = 15$ small rectangles, all congruent to each other. But six of these rectangles pave a unit square, so the usual argument using the basic properties of area (see (a)–(c) on page 191 in section 12.5, **The Area Model**, and subsequent discussion) shows that the area of a small rectangle is $\frac{1}{6}$, so that the area of the shaded region is $\frac{15}{6}$. Therefore, $\frac{15}{6} = \frac{5}{2}$.

The general proof of Theorem 13.1 using the area model will be left as an exercise.

13.4. FFFP

From the perspective of the conceptual development of fractions, the essential content of Theorem 13.1 is that any two fractions may be put **on equal footing**, in the sense that they can be represented by equivalent fractions

13.4. FFFP

(or more correctly, by equivalent fraction symbols) *with the same denominator*. Consider, for example, two fractions such as $\frac{4}{3}$ and $\frac{9}{7}$. How can we find fractions equivalent to these two so that these fractions have the same denominator? Given that the denominators are 3 and 7, it is simplest to think of 3×7 as the common denominator because, by equation (13.1), we have
$$\frac{4}{3} = \frac{4 \times 7}{3 \times 7} = \frac{28}{21} \quad \text{and} \quad \frac{9}{7} = \frac{9 \times 3}{7 \times 3} = \frac{27}{21}.$$

So instead of $\frac{4}{3}$ and $\frac{9}{7}$, we can use $\frac{28}{21}$ and $\frac{27}{21}$ which have the same denominator 21 to represent these fractions. In many situations, having such alternative representations of the same fractions is a decisive advantage. For example, we see that $\frac{4}{3}$, being the 28-th multiple of $\frac{1}{21}$, is to the right of $\frac{9}{7}$ on the number line as the latter is merely the 27-th multiple of the same fraction $\frac{1}{21}$. Therefore by the definition of "<", we have $\frac{9}{7} < \frac{4}{3}$. This is hardly an obvious fact.

In general, we have the following **Fundamental Fact of Fraction-Pairs (FFFP)**:

Any two fractions can be denoted by fraction symbols which have the same denominator. For example, if the given fractions are $\frac{k}{\ell}$ and $\frac{m}{n}$, where k, ℓ, m, n are any whole numbers, then they are, respectively, also equal to

$$\frac{kn}{\ell n} \quad \text{and} \quad \frac{\ell m}{\ell n}.$$

The reason for FFFP is of course that, from equation (13.1), we know $\frac{k}{\ell} = \frac{kn}{\ell n}$ and $\frac{m}{n} = \frac{\ell m}{\ell n}$. For a discussion of the general conceptual background behind FFFP, see section 13.6, **Why FFFP?**

Activity. Rewrite $\frac{15}{13}$ and $\frac{2}{17}$ as two fractions with equal denominator. Do the same for $\frac{8}{9}$ and $\frac{17}{11}$, also for $\frac{15}{4}$ and $\frac{13}{25}$.

It should not be inferred from the statement of FFFP that changing $\frac{k}{\ell}$ and $\frac{m}{n}$ to
$$\frac{kn}{\ell n} \quad \text{and} \quad \frac{\ell m}{\ell n}$$
is the only way to put the original fractions on equal footing. For example, $\frac{3}{4}$ and $\frac{1}{6}$ can be put on equal footing by replacing them with
$$\frac{9}{12} \quad \text{and} \quad \frac{2}{12}$$
rather than
$$\frac{18}{24} \quad \text{and} \quad \frac{4}{24}.$$

Therefore, what FFFP does is to provide at least *one* fail-safe way of achieving this goal.

There will be numerous applications of FFFP in subsequent sections.

There is also **an analogue of FFFP for decimals**:

> *Any two decimals may be rewritten as two decimals with the same number of decimal digits.*

For example, suppose 0.021 and 6.12295 are given, with three and five decimal digits, respectively. But we know that $0.021 = 0.02100$. So 0.02100 and 6.12295 are now both decimals with five decimal digits. We hasten to add that the FFFP for decimals is a consequence of the FFFP for fractions. Indeed, using the same example, the decimals 0.021 and 6.122959 are, respectively,

$$\frac{21}{10^3} \quad \text{and} \quad \frac{612295}{10^5}.$$

It is obvious that one way to put these fractions on equal footing is to use 10^5 as common denominator:

$$\frac{21 \times 10^2}{10^3 \times 10^2} \quad \text{and} \quad \frac{612295}{10^5}.$$

By definition, these are the decimals 0.02100 and 6.12295, respectively, which are exactly the same decimals as before.

In general, if decimal A has more decimal digits than decimal B, then both A and B can be rewritten as two decimals with the same number of decimal digits as A by adding the appropriate number of zeros to the right end of B.

The FFFP for decimals will be especially useful when we deal with the addition and subtraction of decimals.

13.5. The Cross-Multiplication Algorithm

Given two fractions $\frac{m}{n}$ and $\frac{k}{\ell}$, Theorem 13.1 tells us that if $k = cm$ and $\ell = cn$ for some nonzero whole number c, then $\frac{m}{n} = \frac{k}{\ell}$. Now suppose $\frac{m}{n} = \frac{k}{\ell}$. Then does the *converse* hold; i.e., is it true that

> if $\frac{m}{n} = \frac{k}{\ell}$, then $k = cm$ and $\ell = cn$ for some nonzero whole number c?

The example $\frac{6}{9} = \frac{16}{24}$ (both are equal to $\frac{2}{3}$) shows that the answer is no. However, a nontrivial statement *can* nevertheless be made about the whole numbers m, n, k, and ℓ when $\frac{m}{n} = \frac{k}{\ell}$; it will in fact completely answer the question, "How can we tell if two fractions (or more correctly, two fraction symbols) are equivalent?"

13.5. The Cross-Multiplication Algorithm

Theorem 13.2. (Cross-Multiplication Algorithm) *We are given two fractions $\frac{m}{n}$ and $\frac{k}{\ell}$. Then*

$$\text{(13.3)} \qquad \frac{m}{n} = \frac{k}{\ell} \iff m\ell = nk.$$

Here we have again used the symbol "$\iff$" to stand for "is equivalent to"; see page 49 for a general discussion of the meaning of "is equivalent to".

Before giving the proof, we first revisit the preceding example, $\frac{6}{9} = \frac{16}{24}$. Theorem 13.2 implies that $6 \times 24 = 16 \times 9$, and this is easily verified directly. But how do these "cross-products" come up in the first place, and is there any a priori reason why they should be equal? To answer these questions, first observe that the equality $\frac{6}{9} = \frac{16}{24}$ says six copies of $\frac{1}{9}$ is equal to 16 copies of $\frac{1}{24}$; this is certainly not intuitively obvious. But knowing FFFP, we know how to remedy the situation by re-expressing these two fractions as fractions with equal denominators,

$$\frac{6}{9} = \frac{24 \times 6}{24 \times 9} \quad \text{and} \quad \frac{16}{24} = \frac{9 \times 16}{9 \times 24}.$$

We now see how the products 6×24 and 16×9 arise naturally, and also why they must be equal. As there is nothing special about the numbers 6, 24, 16, and 9 in this reasoning, we know that we have essentially proved one half of Theorem 13.2, the part that says "$\frac{m}{n} = \frac{k}{\ell}$ implies $m\ell = nk$". More formally, we have

Proof of Theorem 13.2. First, suppose $\frac{m}{n} = \frac{k}{\ell}$. Using FFFP, we rewrite this as

$$\frac{m\ell}{n\ell} = \frac{nk}{n\ell}.$$

Thus the $m\ell$-th multiple of $\frac{1}{n\ell}$ is equal to the nk-th multiple of the same fraction $\frac{1}{n\ell}$. This is possible only if $m\ell = nk$.

Conversely, suppose $m\ell = nk$. Then

$$\frac{m\ell}{n\ell} = \frac{nk}{n\ell}.$$

By Theorem 13.1, the left side is $\frac{m}{n}$ and the right side is $\frac{k}{\ell}$. Thus we have $\frac{m}{n} = \frac{k}{\ell}$. The theorem is proved. $\square$

The cross-multiplication algorithm is an algorithm that is as basic in fractions as the standard algorithms are in whole numbers. Just as the standard algorithms, the cross-multiplication algorithm was often taught as a rote skill with no reasoning, and just as the standard algorithms, it has recently been erroneously suppressed or de-emphasized in textbooks. Here are some examples to illustrate the utility of this algorithm.

EXAMPLE 1. Compare $\frac{84}{119}$ and $\frac{228}{323}$. Since $84\times 323 = 27132 = 119\times 228$, the fractions are equal, by Theorem 13.2. In fact, both fractions turn out to be equal to $\frac{12}{17}$. Without the cross-multiplication algorithm, it would be difficult to do this problem.

EXAMPLE 2. I have a whole number x with the property that the fraction $\frac{39}{x}$ equals $\frac{63}{105}$. What is x?

Solution: We are given $\frac{39}{x} = \frac{63}{105}$. By Theorem 13.2, we have $39 \times 105 = 63x$, so that $63x = 4095$. Therefore $x = 4095 \div 63 = 65$.

EXAMPLE 3. If $\frac{a}{b} = \frac{c}{d}$ for nonzero whole numbers a, b, c, d, then $\frac{a}{c} = \frac{b}{d}$. This is because, by Theorem 13.2, the truth of either equality $\iff ad = bc$.

13.6. Why FFFP?

The basic idea behind FFFP is to put any two fractions on equal footing. Consider, for example, the two fractions $\frac{4}{7}$ and $\frac{3}{5}$. Suppose we want to know which of the two is bigger. It is difficult to tell because, by definition,

$\frac{4}{7}$ is four copies of $\frac{1}{7}$,

$\frac{3}{5}$ is three copies of $\frac{1}{5}$,

and most of us cannot compare four copies of one thing with three copies of another: the two "units" $\frac{1}{7}$ and $\frac{1}{5}$ are different. To make this point clear, imagine for a moment that we already know that

$\frac{4}{7}$ is 20 copies of $\frac{1}{35}$,

$\frac{3}{5}$ is 21 copies of $\frac{1}{35}$.

Then we would be able to immediately conclude that $\frac{4}{7}$ is to the left of $\frac{3}{5}$ because, among the multiples of $\frac{1}{35}$, $\frac{4}{7}$ is one fewer than $\frac{3}{5}$ and is therefore to the left of $\frac{3}{5}$. Therefore, this suggests that if we can further express $\frac{1}{7}$ and $\frac{1}{5}$ in terms of a common "unit", then our job will be done. The main thrust of FFFP is precisely to fill this need: it shows that with regard to the two fractions $\frac{1}{7}$ and $\frac{1}{5}$,

$\frac{1}{7}$ is the fifth multiple of $\frac{1}{35}$,

$\frac{1}{5}$ is the seventh multiple of $\frac{1}{35}$.

Thus $\frac{1}{35}$ is the common unit we sought. From this perspective, we understand now why $\frac{4}{7}$ is 20 copies of $\frac{1}{35}$ and $\frac{3}{5}$ is 21 copies of $\frac{1}{35}$.

The idea of getting a common unit for two different quantities is of course no more than common sense. For example, suppose we want to find

out whether 3500 yards or 3.2 km is longer. Clearly, we need to express both yards and km in terms of a common unit of measurement, say, a *meter*:

1 yard = 0.9144 meters
1 km = 1000 meters

Then 3500 yards = 3200.4 meters and 3.2 km = 3200 meters. Conclusion: 3500 yards is barely longer than 3.2 km.

This kind of reasoning is fundamentally no different from that of FFFP.

Exercises

1. Are the fractions $\frac{161}{91}$ and $\frac{253}{143}$ equivalent? (This answers the question raised at the beginning of this chapter on page 203.)

2. Reduce the following fractions to lowest terms. (You may use a four-function calculator to test the divisibility of the given numbers by various whole numbers.)

$$\frac{27}{126}, \quad \frac{72}{48}, \quad \frac{42}{91}, \quad \frac{52}{195}, \quad \frac{204}{85}, \quad \frac{414}{529}, \quad \frac{1197}{1273}.$$

3. Express each of the following fractions as a decimal (you may use a four-function calculator):

$$\frac{9}{4}, \quad \frac{36}{125}, \quad \frac{15}{8}, \quad \frac{81}{25}, \quad \frac{19}{64}, \quad \frac{218}{625}.$$

(*Caution*: If you say $\frac{9}{4}$ is 2.25 because you found it *by long division*, then you should ask yourself, Why is this true? We will give an explanation in Chapter 18, but not before then. In the meantime, do this problem again using only what you know up to this point.)

4. Explain each of the following by drawing pictures using the number line (e.g., imagine you are doing it for fifth grade students) *without* making use of Theorem 13.1:

$$\frac{6}{14} = \frac{3}{7}, \quad \frac{28}{24} = \frac{7}{6}, \quad \frac{30}{12} = \frac{5}{2}, \quad \text{and} \quad \frac{12}{27} = \frac{4}{9}.$$

5. (a) Use the representation of 1 as the area of a unit square to give a proof of why $\frac{6}{14} = \frac{3}{7}$ and $\frac{30}{12} = \frac{5}{2}$. (Compare the discussion starting on page 212.) (b) Write out a proof of Theorem 13.1 for the area model.

6. (a) What is the meaning (i.e., definition) of the fraction $\frac{159}{52}$? (b) Describe roughly where this fraction is on the number line. (c) What is the meaning of the fraction $\frac{n^2+1}{n}$ for a whole number n, and roughly where is it on the number line?

7. Generalize FFFP as follows: Given a finite collection of fractions, show that they can be rewritten as a collection of fractions with the same denominator.

8. (a) For which fraction $\frac{m}{n}$ is it true that $\frac{m}{n} = \frac{m+1}{n+1}$? (b) For which fraction $\frac{m}{n}$ is it true that $\frac{m}{n} = \frac{m+b}{n+b}$, where b is a positive whole number?

9. Prove that the following three statements are equivalent for any four whole numbers a, b, c, and d, with $b \neq 0$ and $d \neq 0$:
    (a) $\frac{a}{b} = \frac{c}{d}$.
    (b) $\frac{a}{a+b} = \frac{c}{c+d}$.
    (c) $\frac{a+b}{b} = \frac{c+d}{d}$.

10. Place the three fractions $\frac{13}{6}$, $\frac{11}{5}$, and $\frac{9}{4}$ on the number line and explain how they get to where they are.

11. (Simplified version of an SAT problem.) A flock of geese on a pond were being observed continuously. At noon, $\frac{1}{5}$ of the geese flew away. At 1 PM, $\frac{1}{8}$ of the geese that remained flew away. Then 56 geese remained. At no time did any geese arrive or fly away or die. How many geese were in the original flock? (*Hint:* Use the number line.)

12. We pair each whole number n with another whole number N so that $\frac{n}{N} = \frac{91}{39}$. (a) If $n = 49$, what is the corresponding N that is paired with 49? (b) Which whole number n is paired with N if $N = 33$? (c) If the whole number n is paired with N, and n' is paired with N', is there any relationship among n, n', N, and N'?

Chapter 14

Addition of Fractions and Decimals

We are now in a position to deal with the addition of fractions. For whole numbers, addition is calculated by combining two groups of objects and just counting. For fractions, we do not have that luxury. It is not possible to combine two segments, one of length $\frac{11}{13}$ and another of length $\frac{4}{7}$, and "count", or combine $\frac{7}{8}$ of a bucket of water and another $\frac{5}{11}$ of a bucket of water and "count". However, in Chapter 1 we gave a geometric definition for the addition of two whole numbers in terms of the concatenation of segments, and *this geometric definition makes sense verbatim for the addition of any two fractions.* For example, we can concatenate two segments of lengths $\frac{7}{8}$ and $\frac{5}{11}$ and measure the length of the resulting segment; the latter length would then be the *sum* of $\frac{7}{8}$ and $\frac{5}{11}$. This is in fact the correct way to define the addition of fractions in general.

After you learn how to add fractions correctly, it is hoped that you will *never again* teach your students to add fractions using the *least common denominator*.

The sections are as follows. In the last section, we give an example in which the addition of fractions has to be handled with some finesse.

Definition of Addition and Immediate Consequences

Addition of Decimals

Mixed Numbers

Refinements of the Addition Formula

Comments on the Use of Calculators

A Noteworthy Example of Adding Fractions

14.1. Definition of Addition and Immediate Consequences

Definition. *Given fractions $\frac{k}{\ell}$ and $\frac{m}{n}$, we define their **sum** by*

(14.1) $$\frac{k}{\ell} + \frac{m}{n} = \textit{the length of two concatenated segments,}$$
$$\textit{one of length } \frac{k}{\ell}, \textit{ followed by one of length } \frac{m}{n}:$$

```
           k/ℓ                              m/n
|-------------------------|----------------------|
         _____/
                    k/ℓ + m/n
```

It follows directly from this definition that if both denominators are ℓ, then

(14.2) $$\frac{k}{\ell} + \frac{m}{\ell} = \frac{k+m}{\ell}$$

because both sides are equal to the length of $m + k$ concatenated segments each of length $\frac{1}{\ell}$. This tells us how to compute the sum of two fractions with the same denominator. If we have two fractions $\frac{k}{\ell}$ and $\frac{m}{n}$, but with no prior guarantee that the denominators ℓ and n are equal (although they could be equal), then we can use FFFP of Chapter 13 to reduce this situation to that of equal denominator. Namely, we have $\frac{k}{\ell} = \frac{kn}{\ell n}$ and $\frac{m}{n} = \frac{\ell m}{\ell n}$, so that by (14.2),

$$\frac{k}{\ell} + \frac{m}{n} = \frac{kn}{\ell n} + \frac{\ell m}{\ell n} = \frac{kn + \ell m}{\ell n}.$$

In other words, by expressing both fractions as multiples of the same fraction $\frac{1}{\ell n}$, we easily get an explicit formula for their sum. We restate this general **addition formula** for fractions for future reference:

(14.3) $$\frac{k}{\ell} + \frac{m}{n} = \frac{kn + \ell m}{\ell n}.$$

We make three comments on (14.3). First, for the thinking behind this addition formula, see section 13.6, Why FFFP?, on page 216. Second, the general formula (14.3) includes the earlier formula involving the same denominator, (14.2), as a special case. Indeed, if $\ell = n$ in (14.3), then (14.3) becomes

$$\frac{k}{\ell} + \frac{m}{\ell} = \frac{k\ell + \ell m}{\ell \ell} = \frac{\ell(k+m)}{\ell \ell} = \frac{k+m}{\ell},$$

where the last step uses the cancellation law of Chapter 13 to cancel ℓ from the top and bottom. We have therefore recovered equation (14.2). Finally, this formula is different from the usual formula given in textbooks involving the *lcm* (least common multiple[1]) of the denominators ℓ and n, and we shall comment on the difference below.

We pause to take note of the fact that addition among fractions satisfies the associative and commutative laws. The associative law says
$$(A + B) + C = A + (B + C)$$
for any fractions A, B, and C. This is obvious from the following picture.

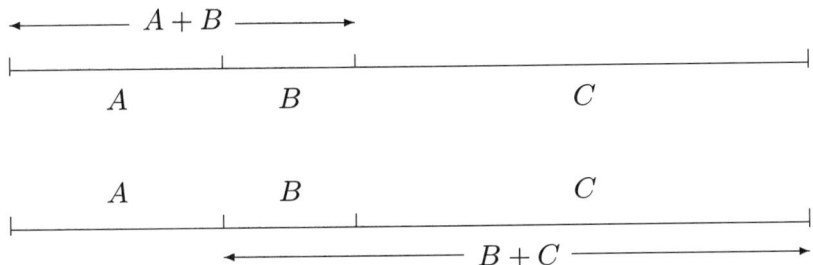

Similarly, the commutative law holds because from the picture

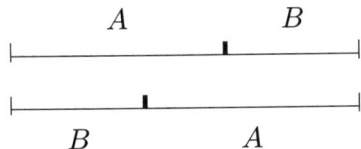

we get
$$A + B = B + A.$$
These two laws are the exact counterparts of equations (2.2) and (2.3) on page 42 in Chapter 2. The same reasoning as that which led to Theorem 2.1 in that chapter now shows that we may write
$$A + B + C + \cdots + Z$$
for an arbitrary sum of fractions A, B, ..., Z without the use of parentheses. We shall refer to this extended version of Theorem 2.1 as "Theorem 2.1 of Chapter 2 for fractions". With this understood, we have

(14.4) $$\frac{k_1}{\ell} + \frac{k_2}{\ell} + \cdots + \frac{k_n}{\ell} = \frac{k_1 + k_2 + \cdots + k_n}{\ell}$$

for any whole numbers $k_1, k_2, \ldots, k_n, \ell$ ($\ell > 0$), as both sides are equal to the length of $k_1 + k_2 + \cdots + k_n$ concatenated segments each of length $\frac{1}{\ell}$.

---

[1] A knowledge of the lcm of two whole numbers is not needed here. Any discussion involving lcm, such as that to be found later in this section, is only peripheral to our purpose. For this reason, we do not even define what lcm means here. For the definition and a detailed discussion, see Chapter 36.

We call special attention to the fact that formula (14.3) was obtained by a deductive process that is conceptually simple and entirely natural. It should go a long way towards explaining why the addition of fractions *could not* take the form of
$$\frac{k}{l} + \frac{m}{n} = \frac{k+m}{l+n}.$$
For example, let $k = m = 1$ and $l = n = 2$. Then by the definition of addition, the left side is the concatenation of two segments each of length one-half and therefore has length 1. The right side, however, is equal to $\frac{1+1}{2+2} = \frac{1}{2}$, which is not 1. The moral of the story is that, *if we give students a clear definition of what addition of fractions means,* then they will have a chance to decide for themselves what is right and what is wrong.

14.2. Addition of Decimals

The first application of fraction addition is the explanation of the addition algorithm for (finite) decimals. For example, consider
$$4.0451 + 7.28.$$
The decimal addition algorithm calls for

(i) lining up the decimal points of the two decimals,

(ii) adding the two numbers as if they are whole numbers, and

(iii) putting the decimal point back in the resulting number.

We now supply the reasoning. First of all, we use the FFFP for decimals to rewrite the two decimals with the same number of decimal digits, i.e., $4.0451 + 7.28 = 4.0451 + 7.2800$. This corresponds to (i). Then,

$$\begin{aligned} 4.0451 + 7.28 &= \frac{40451 + 72800}{10^4} \quad \text{(corresponds to (ii))} \\ &= \frac{113251}{10^4} \\ &= 11.3251 \quad \text{(corresponds to (iii))}. \end{aligned}$$

The reasoning is of course completely general and is applicable to any other situation.

A second application is to get the so-called complete expanded form of a (finite) decimal. For example, given 4.1297, we know it is the fraction
$$\frac{41297}{10^4}.$$
But the expanded form of 41297 reads
$$41297 = (4 \times 10^4) + (1 \times 10^3) + (2 \times 10^2) + (9 \times 10^1) + (7 \times 10^0).$$

We also know that $\frac{4\times 10^4}{10^4} = 4$, $\frac{1\times 10^3}{10^4} = \frac{1}{10}$, etc. Thus,

$$4.1297 = 4 + \frac{1}{10} + \frac{2}{10^2} + \frac{9}{10^3} + \frac{7}{10^4}.$$

This expression of a (finite) decimal as a sum of *single-digit* multiples of $\frac{1}{10}, \frac{1}{10^2}, \frac{1}{10^3}$, etc., where the coefficients are the successive decimal digits, is what is called the **complete expanded form** of 4.1297. In the same way, a decimal $0.d_1 d_2 \cdots d_n$, where each d_j is a single-digit number, has the following **complete expanded form**:

$$0.d_1 d_2 \cdots d_n = \frac{d_1}{10} + \frac{d_2}{10^2} + \cdots + \frac{d_n}{10^n}.$$

14.3. Mixed Numbers

We next introduce a staple in the school curriculum, the concept of a mixed number.

Recall that if we have a fraction $\frac{22}{5}$, then because $22 = (4 \times 5) + 2$, we get

$$\frac{22}{5} = \frac{(4 \times 5) + 2}{5} = \frac{4 \times 5}{5} + \frac{2}{5} = 4 + \frac{2}{5}$$

on account of equation (14.2) above. By tradition, whenever we have the sum of a whole number and a *proper* fraction, the addition sign is omitted so that

$$4 + \frac{2}{5} \stackrel{\text{def}}{=} 4\frac{2}{5}.$$

In general, for a whole number q and a *proper* fraction $\frac{r}{l}$, the notation

$$q\frac{r}{\ell} \quad \text{stands for} \quad q + \frac{r}{\ell},$$

and it is called a **mixed number**.

Please be warned that the symbolic notation for a mixed number is a confusing one, because $q\frac{r}{l}$ suggests the product of q and $\frac{r}{l}$. Our advice is to avoid using it whenever possible.

Many textbooks introduce the concept of a mixed number before they explain what the addition of two fractions means. Consequently, the relationship between a mixed number and a fraction can only be presented as a rote skill. Both practices account for the well-documented fear of mixed numbers. Of course, with the notation of a mixed number clearly explained, the conversion of an improper fraction to a mixed number, and vice versa, are almost nonissues. For instance, given $\frac{25}{7}$, we have $25 = (\underline{3} \times 7) + \underline{4}$ by division-with-remainder. Thus

$$\frac{25}{7} = \frac{(3 \times 7) + 4}{7} = \frac{3 \times 7}{7} + \frac{4}{7} = 3 + \frac{4}{7},$$

and the last item is exactly $3\frac{4}{7}$, by definition. Conversely, given $5\frac{1}{3}$, we know from the definition that
$$5\frac{1}{3} = 5 + \frac{1}{3} = \frac{15}{3} + \frac{1}{3} = \frac{16}{3}.$$

Activity. Convert each of the following improper fractions to a mixed number, and vice versa:
$$77\frac{5}{6}, \quad 4\frac{5}{7}, \quad 6\frac{1}{7}, \quad 13\frac{4}{5}, \quad \frac{32}{7}, \quad \frac{148}{9}, \quad \frac{166}{15}.$$

EXAMPLE. Compute $2\frac{5}{9} + \frac{7}{8}$ and $15\frac{4}{17} + 16\frac{12}{13}$.

Using the associative and commutative laws of addition without comment, we have
$$2\frac{5}{9} + \frac{7}{8} = 2 + \left(\frac{5}{9} + \frac{7}{8}\right) = 2 + \frac{103}{72} = 2 + 1 + \frac{31}{72} = 3\frac{31}{72},$$
$$15\frac{4}{17} + 16\frac{12}{13} = (15 + 16) + \frac{256}{221} = 31 + 1 + \frac{35}{221} = 32\frac{35}{221}.$$

However, we could have carried out the additions differently:
$$2\frac{5}{9} + \frac{7}{8} = \frac{18+5}{9} + \frac{7}{8} = \frac{23}{9} + \frac{7}{8} = \frac{(23 \times 8) + (7 \times 9)}{9 \times 8} = \frac{247}{72},$$

$$15\frac{4}{17} + 16\frac{12}{13} = \frac{(15 \times 17) + 4}{17} + \frac{(16 \times 13) + 12}{13}$$
$$= \frac{259}{17} + \frac{220}{13} = \frac{(259 \times 13) + (220 \times 17)}{17 \times 13}$$
$$= \frac{7107}{221}.$$

The fact that both answers are equally acceptable in each case is something we already had occasion to emphasize. The fact that the two answers are actually equal to each other in each case is something for you to verify!

Activity. Verify that the two answers above are indeed the same.

Although the tradition in school mathematics is to insist that every improper fraction be automatically converted to a mixed number, *there is no mathematical reason for this practice*. This tradition seems to be closely related to the one which insists that every answer in fractions must be in reduced form (see the Pedagogical Comments on page 206).

Given the general confusion about mixed numbers in school mathematics, there is a strong case for the suggestion that this concept be used sparingly in the school curriculum at present. One reason for entertaining mixed numbers at all is, for example, the fact that the fraction $\frac{22}{5}$ is between 4 and 5 is made more explicit by writing it as $4\frac{2}{5}$. However, the same fact is better displayed in the decimal notation,[2] i.e., $\frac{22}{5} = \frac{44}{10} = 4.4$. The universality of decimals in our daily life would seem to justify this suggestion.

14.4. Refinements of the Addition Formula

We should now address some of the fine points of formula (14.3). The salient feature of this formula is its simplicity—not only its *formal* simplicity, but also the simplicity of the reasoning behind its derivation. Another noteworthy feature is its generality: it is valid under all circumstances. It provides an easy-to-use formula for the addition of fractions for all occasions. This is a "security blanket" that is invaluable to many students.

As a rule, however, the generality of a formula works against it in special situations where there may be cute tricks to provide shortcuts. Formula (14.3) is no exception. For example, in the case of fractions with equal denominators, we have already observed that (14.2) is simpler than (14.3), although (14.3) includes (14.2) as a special case in principle. Another example is when one denominator is a multiple of another, then one should just use the bigger denominator:

$$\frac{2}{9} + \frac{5}{36} = \frac{8}{36} + \frac{5}{36} = \frac{13}{36}.$$

More generally,

$$\frac{m}{n\ell} + \frac{k}{\ell} = \frac{m}{n\ell} + \frac{nk}{n\ell} = \frac{m+nk}{n\ell},$$

whereas by comparison, equation (14.3) would give us

$$\frac{m}{n\ell} + \frac{k}{\ell} = \frac{m\ell + nk\ell}{n\ell^2}.$$

We point out again that this answer is nevertheless correct because, using equation (13.1) of Chapter 13,

$$\frac{m\ell + nk\ell}{n\ell^2} = \frac{(m+nk)\ell}{n\ell\ell} = \frac{m+nk}{n\ell}.$$

The next special case of (14.3) is worthy of a more elaborate discussion. As motivation, first consider the following example.

EXAMPLE. Compute $\frac{3}{4} + \frac{5}{6}$.

$$\frac{3}{4} + \frac{5}{6} = \frac{18 + 20}{24} = \frac{38}{24} = \frac{19}{12},$$

which could also be written as $1\frac{7}{12}$. However, in this case one sees that it is not necessary to go to $\frac{1}{24}$ as the common unit of measurement of the fractions $\frac{1}{4}$ and $\frac{1}{6}$, because by visual inspection, $\frac{1}{4} = \frac{3}{12}$ and $\frac{1}{6} = \frac{2}{12}$, so $\frac{1}{12}$ would be a

---

[2]The conversion of a fraction to a finite decimal is taken up in Chapter 18 below, and the case of possibly infinite decimals will be taken up in complete generality in Part 5.

simpler common unit for $\frac{1}{4}$ and $\frac{1}{6}$. Hence we could have computed the sum this way:
$$\frac{3}{4} + \frac{5}{6} = \frac{3 \times 3}{12} + \frac{2 \times 5}{12} = \frac{19}{12} = 1\frac{7}{12}.$$

This example brings us to the consideration of the usual formula for adding fractions. So suppose $\frac{k}{\ell}$ and $\frac{m}{n}$ are given, and suppose we know that there is a whole number A which could be different from ℓn but which is nevertheless a multiple of both n and ℓ. For example, if
$$\frac{k}{\ell} = \frac{3}{4} \quad \text{and} \quad \frac{m}{n} = \frac{5}{6},$$
then
$$\ell = 4 \quad \text{and} \quad n = 6, \quad \text{and we may let} \quad A = 12.$$
Then we have $A = nN = \ell L$ for some whole numbers L and N, so that
$$\frac{k}{\ell} = \frac{kL}{\ell L} = \frac{kL}{A} \quad \text{and} \quad \frac{m}{n} = \frac{mN}{nN} = \frac{mN}{A}.$$
It follows that

(14.5) $$\frac{k}{\ell} + \frac{m}{n} = \frac{kL + mN}{A}, \quad \text{where} \quad A = nN = \ell L.$$

As remarked elsewhere,[3] if A is taken to be the lcm of n and ℓ, then this is the formula used in most textbooks to teach students how to add fractions. Thus, for the example above, (14.5) would yield the previous computation of $\frac{3}{4} + \frac{5}{6}$ if we let $\ell = 4$, $n = 6$, and $A = 12$.

Using formula (14.5) to add fractions is a useful skill that all students of fractions should learn (see section 14.6). However, we should also add a strong word of caution against using (14.5)—with A equal to the *lcm* of n and ℓ—as the *definition* of the addition of the two fractions $\frac{k}{\ell}$ and $\frac{m}{n}$. There are reasons from advanced mathematics as to why such a definition is conceptually the *wrong* way to define the addition of fractions,[4] but for our purposes, it is enough to point out that formula (14.5) is a pedagogical disaster when used as the definition for the addition of fractions for the following reasons. When (14.5) is used as a definition of fraction addition, students' perception is that the addition of fractions bears no resemblance to the way two whole numbers are added. Therefore they conclude that "fractions are a different kind of number". It may be noted that this book makes a great effort to dispel this erroneous perception. An additional problem with using (14.5) as the definition of fraction addition is that students'

---

[3] Again, see Chapter 36 for a definition as well as discussion of lcm.

[4] In technical language, this would imply that the definition of addition in the field of quotients of an integral domain requires the domain to be something like a unique factorization domain (UFD).

generic confusion of the *lcm* and *gcd*[5] of two numbers immediately makes the addition of fractions a difficult skill to learn.

The example below illustrates the difficulty with the adding of fractions when the lcm must be used.

EXAMPLE. Compute $\frac{2}{323} + \frac{3}{493}$.

According to formula (14.3),

$$\frac{2}{323} + \frac{3}{493} = \frac{(2 \times 493) + (3 \times 323)}{323 \times 493} = \frac{1955}{159239}.$$

(A calculator was used in the whole-number computations.) Now suppose a sixth grader is taught to add fractions only by using (14.5) with A as the lcm of ℓ and n. She would have a difficult time finding the lcm of 323 and 493 and may therefore give up doing the problem altogether. But we have just seen that there is *nothing* at all difficult with such a routine problem: $\frac{1955}{159239}$ is the answer (with the help of a four-function calculator).

To bring this discussion to closure, we observe that the lcm of 323 and 493 is in fact $17 \times 19 \times 29$, because $323 = 17 \times 19$ and $493 = 17 \times 29$, so that according to (14.5),

$$\frac{2}{323} + \frac{3}{493} = \frac{2 \times 29 + 3 \times 19}{17 \times 19 \times 29} = \frac{115}{9367}.$$

Of course, $\frac{1955}{159239} = \frac{115}{9367}$, but in both practical and theoretical terms, there is little advantage in having $\frac{115}{9367}$ as the answer instead of $\frac{1955}{159239}$.

Activity. Check that the equality $\frac{1955}{159239} = \frac{115}{9367}$ is correct.

The preceding remarks must be understood in the context of the definition of addition in (14.1). We have seen how simple and natural it is to explain the addition formula (14.3) on the basis of definition (14.1). We can only hope that

> *formula (14.5) will never be used again in school classrooms as the definition of the addition of fractions.*

14.5. Comments on the Use of Calculators

The increasing use of large numbers thus far in both the main text as well as in the exercises is intentional. It is designed to get everyone out of the habit of working only with single-digit numerators and denominators. This habit seems to be linked to several common classroom practices which obstruct both the teaching and learning of fractions. We list some of these practices and indicate how the routine use of large numbers may provide the needed corrective action.

---

[5] See Chapter 36.

(i) *The over-reliance on drawing pictures of pies in every phase of the learning of fractions.* Students see no need to acquire a more abstract understanding of what a fraction is, thereby retarding their acquisition of a basic disposition towards algebra. If, however, fractions such as $\frac{159}{68}$ and $\frac{21}{825}$ appear often, then the pressing need of coming to grips with the fraction concept and all its associated operations will be self-evident.

(ii) *The exclusive reliance on (14.5) (for the case of A as the lcm of the denominators) as a way of adding fractions.* As soon as large numbers are used, the ineffectiveness of (14.5) as the definition of the addition of fractions will be exposed. (See the Example on the preceding page.)

(iii) *The failure to acquire the needed computational fluency with fractions.* So long as the fractions that students encounter in their classroom always have single digit numerators and denominators, there will be no need to remember—or indeed, to understand—the formulas for adding, subtracting, and dividing fractions. For these simple fractions, it is common practice to draw pictures of pies to do fraction computations. If large numbers are never used, the default way of computing with fractions will continue to be the drawing of pictures. If all that students know about fractions can be described by the drawing of pictures, it would be a safe bet that they will have excessive difficulty in learning algebra.

This book therefore strongly advocates the routine use of large numbers in teaching fractions and, for this reason, also the frequent use of a four-function calculator to alleviate the tedium of computing with large numbers.

There is an additional pragmatic issue to address. *When calculator use is explicitly allowed, the teacher should make sure that all intermediate steps of a computation are clearly displayed so that the calculator does not short-circuit students' need to remember the computational algorithms.* We recall an earlier calculation as an example:

$$\frac{2}{323} + \frac{3}{493} = \frac{(2 \times 493) + (3 \times 323)}{323 \times 493} = \frac{1955}{159239}.$$

In this instance, the calculator enters only in the last step, but in a way that is invisible. There is no way of telling whether the arithmetic computations are done by hand or the calculator. At the level of grades 4–6, it may be a good rule of thumb that *if the presence of the calculator is invisible in students' work, then the calculator is not a distraction in students' learning.*

14.6. A Noteworthy Example of Adding Fractions

If n is a whole number, we define $n!$ (read: **n factorial**) to be the product of all the whole numbers from 1 through n. Thus $5! = 1 \times 2 \times 3 \times 4 \times 5$.

14.6. A Noteworthy Example of Adding Fractions

We also define the **binomial coefficients** $\binom{n}{k}$ for any whole number k satisfying $1 \leq k \leq n$ as
$$\binom{n}{k} = \frac{n!}{(n-k)!\, k!}$$
For example,
$$\binom{5}{3} = \frac{5!}{2!\, 3!} = \frac{5 \times 4 \times 3 \times 2 \times 1}{(2 \times 1)(3 \times 2 \times 1)} = 10.$$
Then a very useful formula which, in a special case, says
$$\binom{5}{3} = \binom{4}{3} + \binom{4}{2},$$
can be checked directly:
$$10 = \frac{4 \times 3 \times 2 \times 1}{1 \times (3 \times 2 \times 1)} + \frac{4 \times 3 \times 2 \times 1}{(2 \times 1)(2 \times 1)}.$$
In general, we will prove the following identity:

(14.6) $$\binom{n}{k} = \binom{n-1}{k} + \binom{n-1}{k-1}$$

If you are familiar with *Pascal's Triangle*, then (14.6) is the description of how Pascal's Triangle is constructed. However, our concern here is to use (14.5) to prove (14.6). We shall start from the right side of (14.6) and show that the addition of these two fractions gives the left side. Thus

$$\binom{n-1}{k} + \binom{n-1}{k-1} = \frac{(n-1)!}{(n-1-k)!\, k!} + \frac{(n-1)!}{((n-1)-(k-1))!\, (k-1)!}$$
$$= \frac{(n-1)!}{(n-k-1)!\, k!} + \frac{(n-1)!}{(n-k)!\, (k-1)!}.$$

Observe now that $(n-k)!\, k!$ is a multiple of both denominators, since
$$(n-k)!\, k! = (n-k) \cdot \{(n-k-1)!\, k!\},$$
$$(n-k)!\, k! = k \cdot \{(n-k)!\, (k-1)!\}.$$
So using (14.5) with A as $(n-k)!\, k!$, we have
$$\frac{(n-1)!}{(n-k-1)!\, k!} + \frac{(n-1)!}{(n-k)!\, (k-1)!} = \frac{(n-1)!\, (n-k) + (n-1)!\, k}{(n-k)!\, k!}.$$
By the distributive law, the numerator can be simplified:
$$(n-1)!\, (n-k) + (n-1)!\, k = (n-1)!\, \{(n-k) + k\} = (n-1)!\, n = n!$$
Thus,
$$\frac{(n-1)!}{(n-k-1)!\, k!} + \frac{(n-1)!}{(n-k)!\, (k-1)!} = \frac{n!}{(n-k)!\, k!} = \binom{n}{k}.$$
So identity (14.6) is proved.

Exercises

1. (a) Explain to a fifth grader why $\frac{2}{5} + \frac{5}{2} = \frac{29}{10}$. (b) Explain to a seventh grader why $\frac{b}{a} + \frac{a}{b} = \frac{a^2+b^2}{ab}$ for nonzero whole numbers a, b.

2. Children are usually told that, for example, $4.27 = 4 + 0.27$ as a matter of course. You are asked to explain this carefully to a fourth grader on the basis of the definitions we use, i.e., the definition of a decimal as a fraction and the definition of the addition of fractions in (14.1).

3. (a) Compute: $5.09 + 7.9287 = ?$ How would you explain this to a sixth-grader? (b) $0.57 + 14.3 + 27.0802 = ?$

4. (a) $\frac{17}{50} + \frac{4}{3} = ?$ (b) $\frac{8}{5} + \frac{16}{11} + \frac{4}{15} = ?$ (c) $(3\frac{1}{5} + 2\frac{7}{8}) + \frac{3}{4} = ?$ (d) $3\frac{1}{5} + (2\frac{7}{8} + \frac{3}{4}) = ?$ (e) $\frac{4}{9} + 0.37 = ?$ (f) $2.7 + 1\frac{1}{4} + 45.08 + \frac{25}{16} = ?$

5. You may use a four-function calculator to do the whole-number calculations: (a) $81\frac{25}{31} + 145\frac{11}{12} = ?$ (b) $78\frac{23}{54} + \frac{67}{14} = ?$

6. (a) Let $\frac{a}{bc} = \frac{d}{e}$ for some whole numbers $a, \ldots, e$, where $b > 0, c > 0, e > 0$. Is it true that $\frac{a}{c} = \frac{bd}{e}$? (b) Given that $\frac{4953}{6604} = \frac{51}{68}$, is it true that $\frac{6604}{4953} = \frac{68}{51}$?

7. If a, b, c are nonzero whole numbers, what is $\frac{1}{ab} + \frac{1}{bc} + \frac{1}{ac}$? Simplify your answer as much as possible.

8. Which *whole number* is closest to the following sum: $3\frac{5}{6} + \frac{9}{5} + 1.9$? (*Caution:* Be sure your explanation is complete.)

9. Which *whole number* is closest to the following sum: $\frac{12987}{13005} + \frac{104}{51}$? How would you explain this to a fifth grader?

10. Given that $3 \times 2392 = 552 \times 13$, place $\frac{2392}{552}$ on the number line.

11. The following points on the number line have the property that the thickened segments $[A, 1], [B, 1.7], [2, C], [D, 3], [4\frac{1}{3}, E]$, all have the same length:

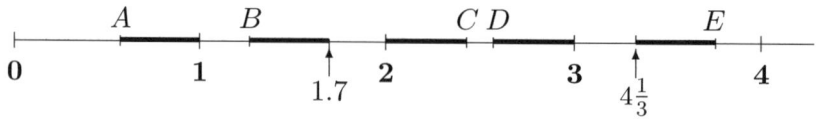

If $A = \frac{5}{9}$, what are the values of B, C, D, E?

12. For any point A on the number line, consider the following rule for moving A to a new position: A is moved a distance of a third of the length of $[0, A]$ to the right of A:

Using this rule, we move the number 1 four times in succession. Between which two whole numbers will the new position of 1 be? (*Caution*: Only use fraction addition to do this problem.)

Chapter 15

Equivalent Fractions: Further Applications

We pick up the thread from Chapter 13 and explore further the ramifications of Theorems 13.1 and 13.2 of that section. It is quite astonishing how much can be said about fractions just by knowing these two theorems.

The interpretation of a fraction as a division is a key result in the subject of fractions. Unfortunately, there are severe pedagogical problems that accompany the teaching of this interpretation in the average classroom. If the meaning of a fraction is not clearly put forth, and if the meaning of "division" among whole numbers (such as $5 \div 7$) remains vague, it is impossible for students to make any sense of the transition from one unclear concept (fractions) to another (division among arbitrary whole numbers[1]). A main purpose of this section is to give precise meaning to "division between any two whole numbers" and to explain why a fraction is a division in this sense.

After some simple observations about the conversion of fractions to decimals, we discuss in detail how to compare fractions. But the most substantive part of this section is the in-depth discussion of a common expression such as "$\frac{2}{3}$ of $4\frac{1}{2}$". This discussion introduces a technique that will be used many times over in the rest of this book.

The sections are as follows:

A Different View of a Fraction

A New Look at Whole Number Divisions

Comparing Fractions

---

[1]From section 7.1, we know that in order for $m \div n$ to make sense between two whole numbers, m has to be a multiple of n.

The Concept of $\frac{m}{n}$ of $\frac{k}{\ell}$

15.1. A Different View of a Fraction

We now introduce a new meaning for a fraction. We claim that, for any fraction $\frac{m}{n}$ on the number line,

$$(15.1) \quad \frac{m}{n} = \text{the length of one part when a segment of length } m \text{ is divided into } n \text{ equal parts.}$$

(Recall the convention from Chapter 13: *equal parts* in the context of the number line means *segments of equal length*.) In other words, divide the segment $[0, m]$ into n equal parts, and the division point to the right of 0 is exactly $\frac{m}{n}$.

This gives a completely different understanding of $\frac{m}{n}$, i.e., a completely different way of locating the point $\frac{m}{n}$ on the number line. To understand what this means, we must remember the *definition*[2] of the fraction $\frac{m}{n}$:

partition $[0, 1]$ into n equal parts, then the m-th multiple of the first division point to the right of 0 is $\frac{m}{n}$.

This definition says that to locate $\frac{m}{n}$ on the number line, all we need to do is to look at the unit segment $[0, 1]$. There is no need to consider the segment $[0, m]$ at all. On the other hand, the prescription on the right side of (15.1) for locating $\frac{m}{n}$ on the number line involves the *whole segment* $[0, m]$ *from the beginning*, because we have to divide this (presumably long) segment into n equal parts. There is no a priori reason why these two operations are related, much less why they would produce the same point on the number line.

Let us first see why (15.1) is true in a special case. We will show that

$$\frac{4}{5} = \text{the length of a part when the segment } [0, 4]$$

is partitioned into five equal parts.

For the proof, observe that the difficulty here is that $[0, 4]$ is naturally divided into four equal parts, namely $[0, 1]$, $[1, 2]$, $[2, 3]$, and $[3, 4]$. The question is how to divide it into *five* equal parts. Now equation (12.1) on page 185 enters in a rather surprising way. If we think of 4 (the length of $[0, 4]$) as $4 = \frac{5 \times 4}{5}$, then 4 is 20 copies of $\frac{1}{5}$. Now think of $\frac{1}{5}$ as one object, then 4 is a collection of 20 such objects. But it is not difficult at all to divide 20 equal objects into five equal groups: just put four such objects into each group. Thus if we want to divide 4 into five equal parts, one part is just four copies

---

[2]At the risk of overstating our case, we reiterate the fundamental importance of knowing the definitions of the basic concepts.

15.1. A Different View of a Fraction

of $\frac{1}{5}$, which is then precisely the fraction $\frac{4}{5}$. The special case of (15.1) is proved.

We can easily translate the preceding reasoning into geometric language. The multiples of $\frac{1}{5}$ divide $[0,4]$ into 20 ($= 5 \times 4$) equal parts, and therefore groups of four copies of $\frac{1}{5}$ would partition $[0,4]$ into five equal parts. Thus the positions of the 4-th, 8-th, 12-th, 16-th multiples, shown by the vertical arrows below the number line in the picture below, provide a partition of $[0,4]$ into 5 equal parts.

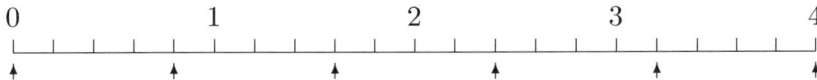

Then the length of the first segment (from the left) is visibly $\frac{4}{5}$.

It remains to point out that the preceding algebraic reasoning is perfectly general and is in no way dependent on the specific numbers 4 and 5. So replacing 4 by m and 5 by n in the preceding argument will result in a valid proof of (15.1). Precisely, let us divide the length m into n equal parts in the following way. Using equation (13.1) (in fact (12.1) is enough), we have

$$m = \frac{nm}{n}$$

so that $[0, m]$ is exhibited as nm copies of $\frac{1}{n}$. Thus the segments, so that each consists of *m copies of* $\frac{1}{n}$, divide $[0, m]$ into n equal parts; it follows that each of these n parts is m copies of $\frac{1}{n}$, i.e., $\frac{m}{n}$, as desired.

We can get a firmer grasp of the message of (15.1) if we apply it to a concrete situation. Take a collection of 15 pencils and choose our unit 1 to be the number of pencils in this collection (i.e., 15). Therefore, the number 2 now represents the number of pencils in two such collections, i.e., 30 pencils; the number 3 represents the number of pencils in three collections, i.e., 45 pencils; and so on. What is $\frac{4}{5}$? By the definition in Chapter 12, we first partition the unit (corresponding to $[0, 1]$ on the number line) into five equal parts (i.e., equal number of pencils), so each part consists of three pencils. Now we aggregate four of these parts (corresponding to concatenating four segments each of length $\frac{1}{5}$), thereby obtaining 12 pencils. So far we have only made use of the definition of the fraction $\frac{4}{5}$. Now we look at what (15.1) says about another way to get at the fraction $\frac{4}{5}$. We put together four collections of these pencils (corresponding to $[0, 4]$) and divide the totality (of $4 \times 15 = 60$ pencils) into five equal parts. Then the size of a part in terms of the number of pencils, which is $60 \div 5 = 12$ pencils, coincides with what is represented by the fraction $\frac{4}{5}$.

You see that the two numbers at the end of these two processes are the same (i.e., 12 pencils), but the intermediate steps of the processes look completely different.

Activity. I bought 4 lbs. of ice cream, and I have to distribute it equally to 25 children. How much ice cream does each child get? Explain.

15.2. A New Look at Whole Number Divisions

Let m and n be whole numbers, and let us put ourselves for a moment back in the situation of Part 1, where *only whole numbers are considered*. Suppose we are given m objects, and m *is a multiple of* n, with n always assumed to be nonzero in this discussion. Recall the partitive interpretation of $m \div n$ (see page 99):

> when m is a multiple of n, $m \div n$ is the number of objects in each group when the collection of m objects is partitioned into n equal groups.

We now use (15.1) to relate $m \div n$ to $\frac{m}{n}$.

Let us begin with an example. If $m = 12$ and $n = 3$, then $12 \div 3 = 4$. Now, the fraction $\frac{12}{3}$, when interpreted as in (15.1), is the length of one part when $[0, 12]$ is divided into three parts of equal lengths. Since $[0, 12]$ is the concatenation of the three segments $[0, 4]$, $[4, 8]$, and $[8, 12]$, we see that $\frac{12}{3}$ is the length of $[0, 4]$, which is of course 4. Conclusion: $12 \div 3 = \frac{12}{3}$.

In general, if m is a multiple of n, so that $m = nk$ for some whole number k, then $m \div n = k$. On the other hand, $[0, m]$ is the concatenation of the following n segments of length k:

$$[0,k], \quad [k,2k], \quad \ldots, \quad [(n-2)k,(n-1)k], \quad [(n-1)k,m].$$

This gives an explicit partition of $[0, m]$ into n segments of equal length k. According to (15.1), k is just $\frac{m}{n}$. Hence, if m is a multiple of n, $n > 0$, then

$$\frac{m}{n} = m \div n.$$

If m and n are arbitrary whole numbers (in particular, m may no longer be a multiple of n) and *if only whole numbers are considered*, then the notation "$m \div n$" has no meaning in this context.[3] For example, one cannot divide four bowling balls into seven equal groups, or three airplanes into five equal groups. But if we put ourselves *in the context of fractions* so that the unit 1 can be divided further into equal parts (e.g., segments of equal length),

---

[3]This fact is consistently overlooked in standard textbooks.

then we can *extend the meaning of division among ALL whole numbers* in the following way:

Definition. *For whole numbers m and n with $n > 0$, $m \div n$ is by definition the length of one part when a segment of length m is divided into n equal parts.*

With this definition in place, the full significance of (15.1) can now be displayed in one symbolic statement: for *any* whole numbers m and n with $n > 0$, where m is no longer required to be a multiple of n, we have

$$\frac{m}{n} = m \div n.$$

For this reason, we will write $\frac{m}{n}$ in place of $m \div n$ for all whole numbers m and n, $n > 0$. *We shall henceforth retire the symbol $\div$ from all symbolic computations and instead use fractions to denote division.*

The preceding idea of extending the meaning of a concept to make it more inclusive, e.g., in this case, giving meaning to **m divided by n** regardless of whether m is a multiple of n or not, will be a recurrent one in this book.

Let us summarize what we have done thus far: We gave a clear definition of a fraction as a point on the number line in terms of part-whole (cf. Chapter 12), and we demonstrated why on the basis of this definition alone, $\frac{m}{n}$ is equal to $m \div n$ when m is a multiple of n. When m is not a multiple of n, we also proved in (15.1) that $\frac{m}{n}$ has the *formal* property of a division in the partitive sense. Modifying the partitive interpretation of division, we *defined* "m divided by n" in general without regard to whether m is a multiple of n or not. Then we concluded that, on account of (15.1), a fraction $\frac{m}{n}$ as defined in Chapter 12 furnishes the correct notion of "m divided by n" for arbitrary whole numbers m and $n > 0$.

It remains to note that this discussion of fraction-as-division is not yet complete, but will be complemented by the general discussion of the division of fractions in general; see the end of section 18.2.

15.3. Comparing Fractions

In Chapter 13, we used the cross-multiplication algorithm to decide if two fractions are equivalent. We now expand the use of this algorithm to the *comparison* of two fractions, i.e., which is bigger? Recall from page 197 that we say a fraction A is *less than* another fraction B if A is to the left of B as points on the number line. It follows from the definition that $0 < A$ for every nonzero fraction A, and that the following two facts, valid for whole numbers, continue to hold for fractions:

Transitivity: *If A, B, C are fractions so that $A < B$ and $B < C$, then $A < C$.*

Trichotomy law: *Given two fractions A and B, exactly one of the following possibilities holds: $A = B$, $A < B$, and $A > B$.*

As with whole numbers, the weak inequality $\mathbf{A \leq B}$ between fractions A and B means A is less than or equal to B.

If two fractions $\frac{k}{\ell}$, $\frac{m}{\ell}$ have the same denominator ℓ, then ordering them is easy:

$$\frac{k}{\ell} < \frac{m}{\ell} \text{ is equivalent to } k < m; \quad \frac{k}{\ell} = \frac{m}{\ell} \text{ is equivalent to } k = m$$

This is because $\frac{k}{\ell}$ is the k-th multiple of $\frac{1}{\ell}$ and $\frac{m}{\ell}$ is the m-th multiple of the same $\frac{1}{\ell}$, and since the multiples of a number (first, second, third, ...) progress from left to right, we see that the k-th multiple is to the left of the m-th multiple if and only if $k < m$, and the two multiples are the same point if and only if $k = m$.

There is one aspect of the preceding passage that should be singled out: the fact that we took pains to explain why $k < m$ implies $\frac{k}{\ell} < \frac{m}{\ell}$. The reason given was that this is because *the multiples of $\frac{1}{\ell}$ progress from left to right* so that the k-th multiple (i.e., $\frac{k}{\ell}$) is to the left of the m-th multiple (i.e., $\frac{m}{\ell}$) since $k < m$. In your classroom, you may wish to stress this fact as otherwise some students would vaguely conclude that the inequality $\frac{k}{\ell} < \frac{m}{\ell}$ between fractions is the result of a corresponding inequality between numerators *or* denominators. In other words, these students would believe that $\frac{2}{9} < \frac{2}{11}$ because $9 < 11$.

Since we have already dealt with the case of equality in Chapter 13, we concentrate on the case of inequality here. Let $\frac{k}{\ell}$, $\frac{m}{n}$ have different denominators ℓ and n. Then FFFP of Chapter 13 says that *they may as well have the same denominator*, in the sense that we can write

$$\frac{k}{\ell} = \frac{kn}{\ell n} \quad \text{and} \quad \frac{m}{n} = \frac{\ell m}{\ell n}.$$

Therefore the preceding case of two fractions with the same denominator implies the following theorem, also called the cross-multiplication algorithm.

Theorem 15.1. (Cross-Multiplication Algorithm) *Given any fractions $\frac{m}{n}$ and $\frac{k}{\ell}$,*

(15.2) $$\frac{m}{n} < \frac{k}{\ell} \iff m\ell < nk.$$

Like Theorem 13.2 of Chapter 13, this theorem has been mistakenly identified as a rote procedure in recent years and therefore banished from

15.3. Comparing Fractions

many recent texts. One reason this happens is that textbooks and professional development materials usually do not define *explicitly* what it means for a fraction to be greater than another. The absence of a definition then renders any reasoning in support of Theorem 15.1 impossible; consequently, Theorem 13.2 can only be a rote skill. *Now that you know how to prove this theorem and that this is not a rote procedure, be sure to keep it in mind every time you try to compare fractions.*

In the education literature, the way to compare $\frac{k}{\ell}$ and $\frac{m}{n}$ is, first, to convert both to fractions with a common denominator, $\frac{kn}{\ell n}$ and $\frac{\ell m}{\ell n}$, and then say that $\frac{k}{\ell} < \frac{m}{n}$ if $\frac{kn}{\ell n} < \frac{\ell m}{\ell n}$. One therefore infers that, *by definition*, one fraction is bigger than the other if, *after* they have been rewritten as fractions with a common denominator, the numerator of one of them is bigger than the numerator of the other. What is wrong with this de facto definition? First, instead of explaining what it means for one fraction to be bigger than another, it tells you to follow a rote procedure, and then tells you how to identify the bigger fraction after the fact. As we have repeatedly emphasized, *it is conceptually wrong to teach fractions this way.* Furthermore, such a definition says that, in order to compare $121\frac{5}{6}$ with $122\frac{1}{897}$, one has to first change both to fractions with a common denominator.[4] Thus, the mixed numbers become $\frac{731}{6}$ and $\frac{109435}{897}$, and then

$$\frac{731 \times 897}{6 \times 897} \quad \text{and} \quad \frac{6 \times 109435}{6 \times 897},$$

so that, from $731 \times 897 < 6 \times 109435$, we conclude $121\frac{5}{6} < 122\frac{1}{897}$. This is not only Byzantine, it is absurd! By contrast, the definition of $A < B$ above says right away that $121\frac{5}{6} < 122\frac{1}{897}$, because $121\frac{5}{6}$ is to the left of 122 while $122\frac{1}{897}$ is to the right of 122 on the number line.

Proof of Theorem 15.1. We must prove (again, see page 49 for a discussion of the meaning of "is equivalent to") the following.

If $\frac{m}{n} < \frac{k}{\ell}$ is true, then so is $m\ell < nk$.
If $m\ell < nk$ is true, then so is $\frac{m}{n} < \frac{k}{\ell}$.

We now prove both carefully. If $\frac{m}{n} < \frac{k}{\ell}$, then the point $\frac{m}{n}$ on the number line is to the left of $\frac{k}{\ell}$. Because

$\frac{m}{n}$ (being equal to $\frac{m\ell}{n\ell}$) is the $m\ell$-th multiple of $\frac{1}{n\ell}$, and

$\frac{k}{\ell}$ (being equal to $\frac{nk}{n\ell}$) is the nk-th multiple of $\frac{1}{n\ell}$,

---

[4]Don't forget: Once a definition is given, it must be followed literally. A common misconception is that *a definition is something one uses when it suits one's purpose*. This cavalier attitude toward definitions (and mathematics in general) is what we are trying hard to eliminate.

it follows that the $m\ell$-th multiple of $\frac{1}{n\ell}$ is to the left of the nk-th multiple of $\frac{1}{n\ell}$. Since multiples of a point progress from left to right, we see that $m\ell < nk$. This proves the first statement.

Conversely, suppose $m\ell < nk$. Then the $m\ell$-th multiple of $\frac{1}{n\ell}$ is to the left of the nk-th multiple of the same number $\frac{1}{n\ell}$. Thus by the definition of one fraction being less than another,
$$\frac{m\ell}{n\ell} < \frac{nk}{n\ell}.$$
By the theorem on equivalent fractions, this implies $\frac{m}{n} < \frac{k}{\ell}$. This completes the proof. □

Because decimals are fractions, in principle there is nothing more to be said about the comparison of decimals once we learn how to compare fractions. However, the following observation is useful:

> The comparison of decimals is reduced immediately to the comparison of whole numbers by the FFFP for decimals.

For example, which is larger: 0.093 or 0.10018? We may rewrite these as 0.09300 and 0.10018, and so it becomes the comparison between
$$\frac{9300}{10^5} \text{ and } \frac{10018}{10^5}.$$
This is a comparison between 9300 and 10018 because we are asking whether 9300 copies of $\frac{1}{10^5}$ or 10018 copies thereof is bigger. It is 10018. So 0.10018 is larger.

There are some subtle aspects of the comparison between decimals in the context of *scientific notation*; see Part 5, Chapter 40.

Activity. Compare the following pairs of fractions (you may use a four-function calculator for the last two pairs):

$$\tfrac{5}{6} \text{ and } \tfrac{4}{5}; \quad \tfrac{6}{7} \text{ and } \tfrac{8}{9}; \quad \tfrac{9}{51} \text{ and } \tfrac{51}{289}; \quad \tfrac{49}{448} \text{ and } \tfrac{56}{512}.$$

EXAMPLE 1. Compare $\frac{9}{16}$ and $\frac{14}{25}$. Because $14 \times 16 = 224 < 225 = 9 \times 25$, equation (15.2) implies that $\frac{14}{25} < \frac{9}{16}$.

EXAMPLE 2. Which is bigger: $\frac{1}{24}$ or $\frac{1}{25}$? It is intuitively clear that if you divide something into a larger number of equal parts, each part gets smaller. (In a classroom, one can get this kind of intuition by doing a hands-on activity on dividing a line segment into two equal parts, three equal parts, four equal parts, etc.) So you would say $\frac{1}{25}$ is smaller than $\frac{1}{24}$. But can you always convince a fourth or fifth grader of this fact? Better yet, can you *convince yourself* that $\frac{1}{12345678}$ is smaller than $\frac{1}{12345677}$? So you see that it would be better to have available a valid reason for your belief that if $a > b$ for whole numbers a and b, then $\frac{1}{a} < \frac{1}{b}$. You will have a chance to do this

15.3. Comparing Fractions

in a problem at the end of this section, but let us deal directly with $\frac{1}{24}$ and $\frac{1}{25}$ here.

One way is to start with the point $\frac{1}{24}$ on the number line. Its 24th multiple is 1. But if you start with $\frac{1}{25}$, its 24th multiple is only $\frac{24}{25}$, which is short of 1. So $\frac{1}{25}$ is smaller. Another way is to use the cross-multiplication algorithm. However, it would be better not to use it explicitly in a school classroom right after this algorithm is taught because, if the student is not yet entirely convinced by the algorithm,[5] an explanation that makes use of it would not be persuasive. Pedagogically, it would be a better strategy to argue directly by making use of the *idea* behind the algorithm. So, using the theorem on equivalent fractions, $\frac{1}{25} = \frac{24}{24 \times 25}$ while $\frac{1}{24} = \frac{25}{24 \times 25}$. Since $24 < 25$, we conclude that $\frac{1}{25} < \frac{1}{24}$.

EXAMPLE 3. Which is greater: $\frac{23}{24}$ or $\frac{24}{25}$? Do it both with and without computation.

Using the cross-multiplication algorithm, we see that $\frac{23}{24} < \frac{24}{25}$ because $23 \times 25 = 575 < 576 = 24 \times 24$. However, this result could also be done *by inspection*! Both fractions are points on the unit segment $[0, 1]$:

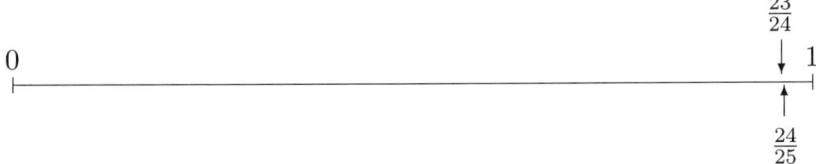

It suffices therefore to decide which of the following is the longer segment: the one between $\frac{23}{24}$ and 1, or the one between $\frac{24}{25}$ and 1. But the first segment has length $\frac{1}{24}$ while the latter has length $\frac{1}{25}$; we have just seen that the latter is shorter. So $\frac{23}{24} < \frac{24}{25}$.

EXAMPLE 4. Prove that for any nonzero fractions $\frac{a}{b}$, $\frac{c}{d}$,

$$\text{if} \quad \frac{a}{b} < \frac{c}{d}, \text{ then } \quad \frac{b}{a} > \frac{d}{c},$$

assuming that a, b, c, and d are all nonzero.

This must be done carefully. By equation (15.2), $\frac{a}{b} < \frac{c}{d}$ implies $ad < bc$, which may be rewritten as $da < cb$. By (15.2) again, the latter implies $\frac{d}{c} < \frac{b}{a}$, which is equivalent to $\frac{b}{a} > \frac{d}{c}$.

Remark. Given a fraction $\frac{a}{b}$ (both a and b nonzero), the fraction $\frac{b}{a}$ is called its **reciprocal**. What Example 4 says is that taking reciprocals reverses an inequality between fractions. This would be more intuitive if we look at the simplest case of whole numbers. Start with $3 < 5$. Then because $3 = \frac{3}{1}$ (by (12.1)), the reciprocal of 3 is $\frac{1}{3}$. Similarly, the reciprocal of 5 is $\frac{1}{5}$.

---

[5]For most students, it takes time before they really buy into what they are taught.

But clearly $\frac{1}{3} > \frac{1}{5}$ (if this is not "clear" to you, please review Example 2 on the preceding page), so the inequality-reversal upon taking reciprocals is understandable in this case. You may also observe that $\frac{1}{7} < \frac{1}{3}$ is equivalent to $7 > 3$. And so on.

EXAMPLE 5. We want to make some red liquid. One method is to mix 18 fluid ounces of liquid red dye in a pail of 230 fluid ounces of water, and the other method is to mix 12 fluid ounces of red dye in a smaller pail of 160 fluid ounces of water. The question: which method do you think would produce a redder liquid?

In the first method, we have 18 parts of red dye out of $230 + 18 = 248$ parts of liquid. In the second method, we have 12 parts of red dye out of $160+12 = 172$ fluid ounces of liquid. The liquid corresponding to the greater of the two, $\frac{18}{248}$ and $\frac{12}{172}$, would produce the redder liquid—if common sense prevails. Now
$$12 \times 248 = 2976 < 3096 = 18 \times 172.$$
Thus by Example 4, $\frac{18}{248} > \frac{12}{172}$. So we know that the first method gives a redder liquid.

We can also think about the problem in a different way. In the first method, we distribute 18 fluid ounces of red dye among 230 parts of water. We now use the division interpretation of a fraction: by (15.1) on page 236, each part of water gets $\frac{18}{230}$ fluid ounces of red dye. Similarly, in the second method, each part of water gets $\frac{12}{160}$ fluid ounces of red dye. We now compare $\frac{18}{230}$ and $\frac{12}{160}$. We have
$$12 \times 230 = 2760 < 2880 = 18 \times 160.$$
By equation (15.2) again, $\frac{12}{160} < \frac{18}{230}$, and so we get the same result that the first method would provide a redder liquid.

In the last example, both ways of doing the problem end up with the same conclusion. Is this just luck, or is something deeper involved here? How are the two inequalities
$$\frac{12}{172} < \frac{18}{248} \quad \text{and} \quad \frac{12}{160} < \frac{18}{230}$$
related? Two obvious relationships stand out:
$$\frac{12}{172} = \frac{12}{160+12} \quad \text{and} \quad \frac{18}{248} = \frac{18}{230+18}.$$

Activity. Use the following theorem to verify that no luck was involved when both methods in Example 5 led to the same answer.

Theorem 15.2. *The following three statements are equivalent for any four whole numbers a, b, c, and d, with $b \neq 0$ and $d \neq 0$:*

(a) $\dfrac{a}{b} < \dfrac{c}{d}$.

(b) $\dfrac{a}{a+b} < \dfrac{c}{c+d}$.

(c) $\dfrac{a+b}{b} < \dfrac{c+d}{d}$.

Proof. Because we are mainly interested in (a) and (b) being equivalent, we will only prove this part. The rest of the proof we leave to an exercise.

Why (a) implies (b): If (a) is true, then by equation (15.2) above, $ad < bc$. Adding ac to both sides gives $ac + ad < ac + bc$ (by inequalities (2.7) on page 50), which is equivalent to $a(c+d) < (a+b)c$, by the distributive law. By equation (15.2) again, this implies $\frac{a}{a+b} < \frac{c}{c+d}$.

Why (b) implies (a): If (b) is true, then by virtue of equation (15.2), $a(c+d) < c(a+b)$, which is equivalent to $ac + ad < ac + bc$, which in turn is equivalent to $ad < bc$ (again by inequalities (2.7)). Now equation (15.2) says $\frac{a}{b} < \frac{c}{d}$, as desired. $\square$

15.4. The Concept of $\frac{m}{n}$ of $\frac{k}{\ell}$

We now come to what is arguably the most significant application of the theorem on equivalent fractions. First, we give a precise meaning to a common expression, "two-thirds of something", or more generally, "$\frac{m}{n}$ of something". What is meant by, for example, "I ate two-thirds of a pie"? We are asking this question in the realm of everyday life, so we can afford to go by our normal beliefs without worrying about precision. (We should point out retroactively that we made use of the everyday meaning of this phrase in Exercises 7 and 8 on page 200; there was no fear of confusion in that situation.) With this understood, it would appear that, if we look at the pie as a circular disk and ignore its depth, and we cut it into three parts of equal *area*, then the statement "I ate two-thirds of a pie" should mean I ate two of these parts. Another example: what is meant by "he gave three-fifths of a bag of rice to his roommate"? Most likely, he measured his bag of rice by *weight* and, after dividing the bag of rice into five equal parts by weight, he gave away three parts. Now consider the statement: "I put away three-quarters of the ham". Now this leaves a lot of room for interpretation because it is not made clear how the ham was measured. If it was measured lengthwise, then the ham would have been cut into four parts of equal length. If, however, it was measured by weight, then it would have been cut into four parts of equal weight. This example illustrates that such statements can be ambiguous even in everyday life, and *the choice of the unit of measurement has to be made explicit in each situation.*

Still in the context of everyday life, *suppose a unit of measurement has been made explicit*; then $\frac{m}{n}$ **of something of size** x would generally mean the total length of m parts when the segment $[0, x]$ is divided into n equal parts, where x is put on the number line whose unit 1 represents the chosen unit of measurement.

Now we leave the realm of everyday life and enter the domain of mathematics. If we want everybody to understand *precisely* what we mean by "$\frac{m}{n}$ of something", then no amount of misunderstanding can be tolerated, and the only course of action is to *define this phrase precisely*. This is what definitions are for.

Definition. $\frac{m}{n}$ **of a fraction** $\frac{k}{\ell}$ means the length of m concatenated parts when the segment $[0, \frac{k}{\ell}]$ is divided into n parts of equal length.

At the risk of harping on the obvious, by putting $\frac{k}{\ell}$ on the number line, we have implicitly fixed a unit, and $\frac{k}{\ell}$ then refers to this unit.

You may wonder, in case $\frac{m}{n}$ and $\frac{m'}{n'}$ are equivalent fractions, whether $\frac{m}{n}$ of $\frac{k}{\ell}$ is equal to $\frac{m'}{n'}$ of $\frac{k}{\ell}$. We will answer this question affirmatively below (see Corollary 15.3.1 of Theorem 15.3 on page 249).

From the perspective of this definition, what was proved in (15.1) on page 236 can now be reformulated as follows:

$$\text{any fraction } \frac{m}{n} \text{ is equal to } \frac{1}{n} \text{ of } m.$$

You would also find that the reasoning in that section resonates with the reasoning in the rest of this section.

Henceforth, if we use the phrase "$\frac{m}{n}$ of a fraction $\frac{k}{\ell}$" in the context of a *mathematical* discussion, this phrase will be understood in terms of this definition, provided a unit for $\frac{k}{\ell}$ has been clearly specified and $\frac{k}{\ell}$ is a point on the number line where 1 represents this unit.

> *We have to make explicit a general rule in the interface between the everyday context and the mathematical context. Because we have now given a precise definition of the phrase* $\frac{m}{n}$ **of a fraction** $\frac{k}{\ell}$, *you will have to understand this phrase* only *in this precise sense* each time you come across it in the rest of this book. *In particular, the least you can do is to commit the preceding definition to memory to the point of automatic recall. The phenomenon that a common phrase, once it becomes a precise mathematical term, must henceforth be understood in the precise mathematical sense will happen a few more times in this book (cf. average speed in Chapter 18 or ratio in Chapter 22).*

Activity. *Without looking at the preceding definition*, verbally explain to your neighbor the meaning of: (a) $\frac{4}{5}$ of $21\frac{1}{3}$; (b) $3\frac{3}{7}$ of $1\frac{1}{2}$; (c) $\frac{7}{3}$ of $\frac{3}{4}$. In the last item, locate the point which is $\frac{7}{3}$ of $\frac{3}{4}$.

15.4. The Concept of $\frac{m}{n}$ of $\frac{k}{\ell}$

It remains to show how to compute $\frac{m}{n}$ of $\frac{k}{\ell}$. Let us work out some examples first. What is $\frac{1}{3}$ of $\frac{9}{8}$? We recognize that $\frac{9}{8}$ is nine copies of $\frac{1}{8}$, so a third of it is just three copies of $\frac{1}{8}$, i.e., $\frac{3}{8}$.

Similarly, $\frac{1}{3}$ of $\frac{21}{5}$ is $\frac{7}{5}$, because $\frac{21}{5}$ is 21 copies of $\frac{1}{5}$ so that a third of it is seven copies of $\frac{1}{5}$.

What would $\frac{2}{3}$ of $\frac{6}{7}$ be? We have to divide the segment $[0, \frac{6}{7}]$ into three segments of equal length and then concatenate two of them. Now $\frac{6}{7}$ is six copies of $\frac{1}{7}$, and 6 consists of three equal groups of 2. Therefore $\frac{6}{7}$ is the concatenation of three segments, where each segment is itself the concatenation of two copies of $\frac{1}{7}$, as shown:

$$0 \quad \frac{2}{7} \quad \frac{4}{7} \quad \frac{6}{7} \quad 1$$

One can now read off from the picture that $\frac{2}{3}$ of $\frac{6}{7}$ is $\frac{4}{7}$.

Activity. What is $\frac{4}{5}$ of $\frac{45}{7}$?

Next, what about $\frac{1}{4}$ of $\frac{7}{9}$? This differs from the previous examples in that we are supposed to divide seven copies of $\frac{1}{9}$ into four equal groups and take one, i.e., divide $[0, \frac{7}{9}]$ into four parts of equal length and take one part. But 7 not being divisible by 4, we will have to find a way to get around this difficulty. If we remember[6] the theorem on equivalent fractions (see page 204), then we know we can "force" the numerator of a fraction to be divisible by 4, because

$$\frac{7}{9} = \frac{4 \times 7}{4 \times 9} = \frac{28}{36}$$

so that we may regard $\frac{7}{9}$ as 28 copies of $\frac{1}{36}$. Now 28 is four equal groups of 7. Thus if $[0, \frac{7}{9}]$ is partitioned into four parts of equal length, then one part is seven copies of $\frac{1}{36}$, i.e., has length $\frac{7}{36}$. We conclude that $\frac{1}{4}$ of $\frac{7}{9}$ is $\frac{7}{36}$.

EXAMPLE. After Stefanie had walked $\frac{2}{5}$ of the distance from home to the train station, there was still $\frac{7}{8}$ of a mile to go. How far is home from the train station?

We can draw the distance from home to the train station on the number line with 0 being home and T being the train station. Then it is given that, when this segment from 0 to T is partitioned into five segments of equal length, Stefanie was at the second division point after 0:

---

[6]Memory is important!

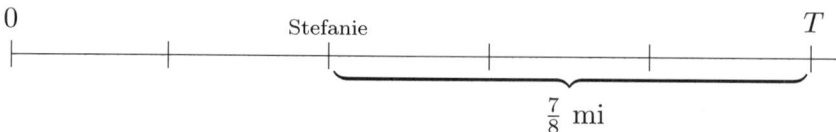

The length of any of the five segments is determined by the fact that the total distance from home to the train station T is 5 times this length. We are given that the distance from where Stefanie stands to T is $\frac{7}{8}$ of a mile, and this distance comprises three of these segments. It therefore suffices to find out how long "a third of $\frac{7}{8}$ of a mile" is. From a previous consideration, we know

$$\frac{7}{8} = \frac{3 \times 7}{3 \times 8} = \frac{21}{24},$$

so that $\frac{7}{8}$ is equal to 21 copies of $\frac{1}{24}$. It follows that

a third of $\frac{7}{8}$ is a third of 21 copies of $\frac{1}{24}$, and is thus seven copies of $\frac{1}{24}$.

Therefore the length of any one of the five segments between 0 and T, in miles, is seven copies of $\frac{1}{24}$. The total distance from 0 to T, in miles, is 5×7 copies of $\frac{1}{24}$, which is $\frac{35}{24}$ miles.

Remark. The preceding example is one of the standard word problems on fractions which is usually given after the multiplication of fractions has been introduced. The standard solution method is given out in the form of an algorithm: "flip over $(1 - \frac{2}{5})$ to multiply $\frac{7}{8}$". We now see that there is in fact no need to use multiplication of fractions for the solution, and moreover, the solution of the problem does not need any rote skills. Simple reasoning suffices.

We can now deal with *the general case*. Given a fraction $\frac{k}{\ell}$, suppose we want to find $\frac{m}{n}$ of it. Now the numerator k of $\frac{k}{\ell}$ may not be divisible by n, but by equation (13.1),

$$\frac{k}{\ell} = \frac{nk}{n\ell},$$

and the numerator nk of $\frac{nk}{n\ell}$ is now divisible by n. So an n-th of $\frac{k}{\ell}$ is an n-th of $\frac{nk}{n\ell}$, which is obviously equal to $\frac{k}{n\ell}$, i.e., k copies of $\frac{1}{n\ell}$. Therefore $\frac{m}{n}$ of $\frac{k}{\ell}$, being m copies of an n-th of $\frac{k}{\ell}$, is mk copies of $\frac{1}{n\ell}$, which is $\frac{mk}{n\ell}$.

We summarize:

Theorem 15.3. *Given nonzero fractions $\frac{m}{n}$ and $\frac{k}{\ell}$. Then $\frac{m}{n}$ of $\frac{k}{\ell}$ is equal to $\frac{mk}{n\ell}$.*

We now tie up a loose end of an earlier discussion in this section:

15.4. The Concept of $\frac{m}{n}$ of $\frac{k}{\ell}$

Corollary 15.3.1. *If $\frac{m}{n}$ and $\frac{M}{N}$ are equivalent fractions and $\frac{k}{\ell}$ and $\frac{K}{L}$ are also equivalent fractions, then $\frac{m}{n}$ of $\frac{k}{\ell}$ is equal to $\frac{M}{N}$ of $\frac{K}{L}$.*

Proof. By Theorem 15.3,
$$\frac{m}{n} \text{ of } \frac{k}{\ell} \text{ is equal to } \frac{mk}{n\ell}$$
and
$$\frac{M}{N} \text{ of } \frac{K}{L} \text{ is equal to } \frac{MK}{NL}.$$
We want to show that
$$\frac{mk}{n\ell} = \frac{MK}{NL}.$$
By Theorem 13.2 of Chapter 13, this would be the case if $mkNL = n\ell MK$. Because $\frac{m}{n}$ and $\frac{M}{N}$ are equivalent fractions by hypothesis, the cross-multiplication algorithm implies that $mN = nM$. Similarly, $kL = \ell K$. Multiplying the left sides and also the right sides of the preceding two equalities, we get exactly $mNkL = nM\ell K$. The corollary is proved. □

The reasoning we gave in the previous examples, which leads up to the proof of Theorem 15.3, will be used many more times in the rest this chapter.

Exercises

1. (a) Suppose we try to put 2710 pieces of candy into 21 bags with an equal number in each bag. What is the maximum number of pieces we can put in each bag, and how many are left over? (b) Suppose we try to divide a (straight) path of 2710 feet into 21 parts of equal length. How many feet are in each part?

2. Assume that you can cut a pie into any number of equal portions (in the sense of cutting a circle into congruent circular sectors). (a) How would you cut seven pies in order to give equal portions to 11 kids? (b) Find two different ways to cut 11 pies to give equal portions to seven kids.

3. (a) Without appealing to (15.1), show directly that $\frac{4}{7}$ is the length of a part when a segment of length 4 is divided into seven equal parts. (b) Do the same to $\frac{7}{3}$.

4. (The following problem is a fifth grade problem. You are asked to do it without the use of proportions. You are also asked to explain the solution clearly to the class.) Ballpoint pens are sold in bundles of 4. Lee bought 20 pens for 12 dollars. How much would 28 pens cost?

5. Write out a proof of (15.1) for the case where the unit on the number line represents the area of the unit square. (Compare Exercise 5 on page 218.)

6. Let the unit on the number line stand for *a dozen eggs*. (a) How many eggs are represented by the number 8 on this number line? (b) How many eggs are represented by the number $8 \div 3$? (c) How many eggs are represented by the fraction $\frac{8}{3}$? (d) What do you notice about the answers to (b) and (c)?

7. Explain to a sixth grader directly, without using Theorem 15.3, why $\frac{3}{7}$ of $\frac{5}{16}$ lb. is equal to $\frac{15}{112}$ lb.

8. (a) A wire 314 feet long is only four-fifths of the length between two posts. How far apart are the posts? (b) Helena was three-quarters of the way to school after having walked $\frac{8}{9}$ miles from home. How far is her home from school? (c) After driving 18.5 miles, I am exactly three-fifths of the way to my destination. How far away is my destination?

9. James gave a riddle to his friends: "I was on a hiking trail, and after walking $\frac{7}{12}$ of a mile, I was $\frac{5}{9}$ of the way to the end. How long is the trail?" Help his friends solve the riddle.

10. (a) Write down a fraction that is between $\frac{31}{63}$ and $\frac{32}{63}$, and one between $\frac{5}{8}$ and $\frac{8}{13}$. (b) Write down a decimal that is between 1.0356 and 1.0357. (c) Write down a decimal that is between 1.03567 and 1.03667.

11. *Explain* to your students how to do the following problem: Nine students chip in to buy a 50-pound sack of rice. They are to share the rice equally by weight. How many pounds should each person get? (If you just say, "divide 50 by 9", that will not be good enough. You must explain what is meant by "50 divided by 9", and why the answer is 5 and $\frac{5}{9}$ of a pound.)

12. (a) $\frac{3}{7}$ of a fraction is equal to $\frac{5}{6}$. What is this fraction? (b) $\frac{m}{n}$ of a fraction is equal to $\frac{k}{\ell}$. What is this fraction?

13. Show that (a) and (c) of Theorem 15.2 on page 244 are equivalent.

14. A number has the following property: $\frac{2}{5}$ of this number added to $\frac{1}{5}$ is equal to the number itself. What is this number?

15. On April 30, 2009, *Cape Cod Times* reported that in the town of Truro, MA, officials declared that voters had "narrowly approved one of four zoning amendments" by meeting the legal requirement of a two-thirds vote. It turned out that the precise vote was 136 to 70, and the officials said since the calculator gave a value of 136 to 0.66 × 206 when rounded to the nearest whole number, 136 was two-thirds of the total vote count of 206. Discuss whether the town officials were right in saying 136 is two-thirds of 206 by *only* using what we have learned thus far.

16. In baseball, a player's **batting average** is the total number of his hits divided by the total numbers of his **at bats** (the number of times he gets to face a pitcher). Suppose going into the last day of the season, both players A and B have 200 hits in 630 at bats. On the last day, player A gets two hits for two at bats whereas player B gets three hits for four at bats. Who wins the batting title?

Chapter 16

Subtraction of Fractions and Decimals

The purpose of this chapter is twofold: to give a precise definition of the subtraction of fractions, and to describe the algorithm for the subtraction of fractions and decimals. This algorithm is very similar to the addition algorithm.

The sections are as follows:

Subtraction of Fractions and Decimals

Inequalities

16.1. Subtraction of Fractions and Decimals

We now say a few words about the subtraction of fractions; the brevity of our comments is warranted by the similarity between subtraction and addition. Suppose as usual that $\frac{k}{\ell}$ and $\frac{m}{n}$ are given and that $\frac{k}{\ell} \geq \frac{m}{n}$. Imitating the definition of subtraction among whole numbers in (5.1) on page 72, we define the **difference** $\frac{k}{\ell} - \frac{m}{n}$ as the fraction, so that

$$\left(\frac{k}{\ell} - \frac{m}{n}\right) + \frac{m}{n} = \frac{k}{\ell}.$$

Geometrically, this means, in view of the definition of addition in definition (14.1),

$$\frac{k}{\ell} - \frac{m}{n} = \text{the length of the remaining segment when a}$$

$$\text{segment of length } \frac{m}{n} \text{ is removed from one}$$

$$\text{end of a segment of length } \frac{k}{\ell}$$

Using FFFP, we have

$$\frac{k}{l} - \frac{m}{n} = \frac{kn}{ln} - \frac{\ell m}{\ell n}$$
$$= \text{the length of the remaining segment when } \ell m \text{ copies}$$
$$\text{of } \frac{1}{\ell n} \text{ are removed from } kn \text{ copies of } \frac{1}{\ell n}$$
$$= \text{the length of } (kn - \ell m) \text{ copies of } \frac{1}{\ell n}$$
$$= \frac{kn - \ell m}{\ell n},$$

where the last equality is by definition of the fraction $\frac{kn-\ell m}{\ell n}$. This yields the following formula: if $\frac{k}{\ell} \geq \frac{m}{n}$, then

(16.1) $$\frac{k}{\ell} - \frac{m}{n} = \frac{kn - \ell m}{\ell n}.$$

It is of more than a little interest to take a closer look at one aspect of (16.1): in view of the fact that the subtraction between whole numbers kn and ℓm can take place only if $kn \geq \ell m$, how do we know that the subtraction $kn - \ell m$ on the right side of (16.1) makes sense? This is where the assumption $\frac{k}{\ell} \geq \frac{m}{n}$ comes in: by the cross-multiplication algorithm (Theorem 15.1 on page 240) we do have $kn \geq \ell m$.

The subtraction formula (16.1) is in particular applicable to mixed numbers and, in that context, brings out a special feature which is not particularly important[1] but which is cute nevertheless. This is the phenomenon of **trading** in the subtraction of mixed numbers.

Instead of explaining trading using symbolic notation, we shall as usual illustrate it with an example. Consider the subtraction of $17\frac{2}{5} - 7\frac{3}{4}$. There

---
[1]Except perhaps on standardized assessments!

16.1. Subtraction of Fractions and Decimals

is an obvious way to make (16.1) directly applicable to this case regardless of what the relevant numbers may be, which is to convert the mixed numbers into improper fractions:

$$17\frac{2}{5} - 7\frac{3}{4} = \frac{85+2}{5} - \frac{28+3}{4} = \frac{87}{5} - \frac{31}{4} = \frac{(87 \times 4) - (31 \times 5)}{5 \times 4} = \frac{193}{20}.$$

(We emphasize again that there is no need to convert this back to a mixed number unless of course there is an explicit instruction to do so.) However, there is another way to do the computation:

$$17\frac{2}{5} - 7\frac{3}{4} = \left(17 + \frac{2}{5}\right) - \left(7 + \frac{3}{4}\right).$$

Now we use the analogue of identity (5.5) on page 75 for fractions. In other words, we assume the validity of

$$(L + M + N) - (A + B + C) = (L - A) + (M - B) + (N - C)$$

for all fractions $A, B, \ldots, N$, so that $L \geq A$, $M \geq B$, and $N \geq C$. The proof is the same as that in section 5.6, **A Property of Subtraction**, on page 82, but we will not go into the details here because we will prove this identity in full generality in (27.8) on page 392. In any case, with $L = 17$, $M = \frac{2}{5}$, $N = 0$, $A = 7$, $B = \frac{3}{4}$, $C = 0$, we get

$$17\frac{2}{5} - 7\frac{3}{4} = (17 - 7) + \left(\frac{2}{5} - \frac{3}{4}\right) = 10 + \left(\frac{2}{5} - \frac{3}{4}\right).$$

But (16.1) is not applicable to $\frac{2}{5} - \frac{3}{4}$ as it stands because $\frac{2}{5} < \frac{3}{4}$. (Can you *prove* this inequality?) To remedy the situation, we will imitate the subtraction algorithm in Chapter 5: if in the subtraction of (for example) $82 - 57$ we find that the subtraction in the ones digit (i.e., $2 - 7$) cannot be done using whole numbers, we simply trade a 1 from the tens digit to the ones digit to convert the 2 to $10 + 2 = 12$. Likewise, since the fraction $\frac{2}{5}$ is not big enough for the subtraction $\frac{2}{5} - \frac{3}{4}$, we will trade a 1 from 17 to convert it to $\frac{5}{5} + \frac{2}{5} = \frac{7}{5}$. In more detail,

$$17\frac{2}{5} = (16 + 1) + \frac{2}{5} = 16 + \left(1 + \frac{2}{5}\right) = 16 + \left(\frac{5}{5} + \frac{2}{5}\right) = 16 + \frac{7}{5}.$$

Now we can carry out the foregoing subtraction:

$$17\frac{2}{5} - 7\frac{3}{4} = \left(16 + \frac{7}{5}\right) - \left(7 + \frac{3}{4}\right) = (16 - 7) + \left(\frac{7}{5} - \frac{3}{4}\right) = 9 + \frac{13}{20} = 9\frac{13}{20}.$$

Incidentally, $9\frac{13}{20} = \frac{193}{20}$, exactly the same as before.

Activity. (a) Compute $4\frac{1}{10} - 2\frac{3}{4}$ both ways. (b) For a simple subtraction such as $2\frac{1}{6} - 1\frac{1}{3}$, it may be a good idea to do it via pictures as well as direct computations. Do both.

In the case of decimals, *the FFFP for decimals essentially reduces subtraction among decimals to subtraction among whole numbers.* Let us explain what this means for a special case such as $14.02 - 8.4257$, but the reasoning will be seen to be perfectly general. Recall that the algorithm for the subtraction of decimals calls for:

(i) lining up the decimal points of the two decimals,

(ii) subtracting the two numbers as if they were whole numbers, and

(iii) putting the decimal point back in the resulting number.

The algorithm indeed bears out the fact that the actual subtraction is carried out only among whole numbers, and the question is why this is so. We now explain. By FFFP for decimals, we may rewrite these two decimals as 14.0200 and 8.4257, i.e., they now have the same number of decimal digits. Then,

$$\begin{aligned}14.02 - 8.4257 &= 14.0200 - 8.4257 \\ &= \frac{140200}{10^4} - \frac{84257}{10^4} \\ &= \frac{140200 - 84257}{10^4} \quad \text{(corresponds to (i))} \\ &= \frac{55943}{10^4} \quad \text{(corresponds to (ii))} \\ &= 5.5943 \quad \text{(corresponds to (iii))}.\end{aligned}$$

Activity. $0.402 - 0.0725 = ?\quad 3.14 - 1\frac{5}{8} = ?$

16.2. Inequalities

We conclude this chapter with some remarks on inequalities among fractions. With the availability of the concept of addition among fractions, the direct extensions of some of the assertions in Chapter 2 (see inequalities (2.7) on page 50) are straightforward. Let A, B, C, D be fractions. Then

(a) $A < B$ is equivalent to $A + C = B$ for a nonzero fraction C;

(b) $A < B$ is equivalent to $A + C < B + C$ for all C;

(c) $A < B$ and $C < D$ imply $A + C < B + D$.

Assertion (a) provides a different way to look at the comparison among fractions. It is worth mentioning that one can also express (a) in terms of subtraction as:

(a′) $B > A$ implies $B - A > 0$.

The counterparts of these assertions for whole numbers have been proved in section 2.5, **Comparing Numbers (Conclusion)**; see page 48. Because in terms of the number line, there is no conceptual difference between the addition of whole numbers and the addition of fractions, we can safely leave the proofs of these statements to the exercises.

Exercises

1. Compare the following pairs of fractions. $\frac{70}{105}$ and $\frac{38}{57}$, $\frac{4}{9}$ and $\frac{3}{7}$, $\frac{9}{29}$ and 0.31, $\frac{13}{17}$ and 0.76, $\frac{12}{23}$ and $\frac{53}{102}$.

2. (a) Compare the fractions $\frac{94}{95}$ and $\frac{311}{314}$ both ways, with and without using the cross-multiplication algorithm. (b) Do the same for $\frac{85}{119}$ and $\frac{227}{325}$ (compare Example 2 on page 242).

3. Use a calculator to do the whole number computations (and *only* the whole number computations) if necessary to see which is greater: $\frac{112}{234}$ and $\frac{213}{435}$, $\frac{577}{267}$ and $\frac{863}{403}$.

4. (a) How would you explain to a fifth grader that the reason the inequality $\frac{4}{9} > \frac{3}{7}$ is true is that $4 \times 7 > 3 \times 9$? (b) Explain to a fifth grader directly, without using formula (16.1), how to do the subtraction $\frac{7}{18} - \frac{5}{12}$.

5. $\frac{17}{50} - \frac{1}{3} = ?$ $(5\frac{1}{5} - 2\frac{7}{8}) - 1\frac{1}{3} = ?$ $(4\frac{2}{3} - 2.6) - \frac{6}{7} = ?$ $4\frac{2}{3} - (2.6 + \frac{6}{7}) = ?$ $25.56 - \frac{184}{11} = ?$

6. (a) Find a fraction B so that $4\frac{2}{5} - B = 1\frac{3}{4}$. (b) Find a fraction A so that $\frac{4}{7} + A = \frac{7}{8}$.

7. Without computing the exact answer, estimate which of the following is bigger: $\left(\frac{91}{624} + \frac{8}{9}\right)$ and 1. Explain how you did it.

8. Prove assertions (a)–(c) about inequalities among fractions on page 256.

9. Let A, B, C be fractions. (a) Prove that $A + B < C$ is equivalent to $A < C - B$. (b) Suppose $C < A$ and $C < B$. Prove that $A < B$ is equivalent to $A - C < B - C$ (cf. Exercise 9 on page 83).

10. Three numbers A, B, and C are given. It is known that $B = 14\frac{2}{5}$, B exceeds A by $1\frac{2}{3}$, while C exceeds A by $2\frac{1}{4}$. (a) What is $C - B$? (b) What is the sum $A + B + C$?

11. (a) If a and b are nonzero whole numbers such that $a < b$, explain why $\frac{1}{a} > \frac{1}{b}$. (b) Still with $a < b$, is it true that for any nonzero whole number c, $\frac{c}{a} > \frac{c}{b}$? Explain.

12. A bucket with a capacity of six gallons was filled with water. Someone used two-thirds of the water. Another person then took away a quarter of the rest. Later $1\frac{2}{5}$ gallons of water were poured into the bucket. How much water is in the bucket?

13. Let $\frac{a}{b}$ be a nonzero fraction. Order the following (infinite number of) fractions: $\frac{a}{b}$, $\frac{a+1}{b+1}$, $\frac{a+2}{b+2}$, $\frac{a+3}{b+3}$ .... (*Hint*: Try $\frac{a}{b} = \frac{2}{3}$, and then try $\frac{a}{b} = \frac{3}{2}$.)

14. (a) Which is the better buy: three pencils for 59 cents or ten pencils for $1.99? (b) Which is the better buy: 12 candles for $1.75 or three candles for 45 cents?

15. Alan has two books. Together they are $9\frac{1}{3}$ cm thick. One book is $1\frac{3}{4}$ cm thicker than the other. How thick is each book?

16. A pitcher of orange juice is poured into three glasses, A, B, and C. The total amount of orange juice in glasses A and C exceeds the amount in glass B by $2\frac{5}{12}$ fl. oz. Moreover, the amount in glass B exceeds the amount in glass C by $2\frac{1}{4}$ fl. oz., while the amount in glass C exceeds the amount in glass A by $\frac{7}{12}$ fl. oz. How much orange juice was in the pitcher to begin with? (*Hint:* Use the number line.)

17. An alcohol solution mixes five parts water with 23 parts alcohol. Then three parts water and 14 parts alcohol are added to the solution. Which has a higher concentration of alcohol, the old solution or the new?

18. Suppose a, b are whole numbers so that $1 < a < b$. Which is bigger: $\frac{a-1}{a}$ or $\frac{b-1}{b}$? Can you do it by inspection? What about $\frac{a+1}{a}$ and $\frac{b+1}{b}$?

19. (Problem from SAT.) A flock of geese on a pond was being observed continuously. At 1:00 PM, $\frac{1}{5}$ of the geese flew away. At 2:00 PM, $\frac{1}{8}$ of the geese that remained flew away. At 3:00 PM, three times as many geese as had flown away at 1:00 PM flew away, leaving 28 geese on the pond. At no other time did any geese arrive or fly away or die. How many geese were in the original flock?

Chapter 17

Multiplication of Fractions and Decimals

It is well known that the division of fractions is regarded as a difficult concept in school mathematics. It is also a fact that the most interesting problems in applications, such as those on percent and rate, depend on it. Less known is the fact that much of the difficulty evaporates if there is a firm foundation in multiplication. In this chapter, we will try to lay such a firm foundation.

Before we can discuss how to multiply two fractions, we must first find out what the operation means. For whole numbers, multiplication means repeated addition[1] (e.g., $3 \times 5 = 5 + 5 + 5$). Clearly, this definition is not applicable to fractions because $\frac{2}{5} \times \frac{1}{4}$ is neither adding $\frac{1}{4}$ to itself $\frac{2}{5}$ times, nor adding $\frac{2}{5}$ to itself $\frac{1}{4}$ times. Most textbooks, and education writing in general, simply duck the issue of giving any meaning to multiplying two fractions and, instead, take advantage of students' general willingness to be strung along with no explanations given and no questions asked. The following passage from a fifth grade textbook is *typical* of the effort along this line:

> In this lesson we will multiply fractions. Consider this
> multiplication problem: How much is one half of one half?
> Using our fraction manipulatives, we show one half of a
> circle. To find one half of one half, we divide the half
> in half. We see that the answer is one fourth. Written
> out, the problem looks like this.
> $$\frac{1}{2} \times \frac{1}{2} = \frac{1}{4}.$$

---

[1] See equation (1.2) on page 28.

> We find the answer to a fraction multiplication problem
> by multiplying the numerators to get the new numerator
> and multiplying the denominators to get the new
> denominator.

Notice that *nothing* is said about the meaning of multiplying two fractions. Instead, by suggesting that "half of one half" is "one fourth", the authors hope to make $\frac{1}{2} \times \frac{1}{2} = \frac{1}{4}$ palatable, and therefore get students to believe that the multiplication of fractions is nothing more than "multiplying the numerators to get the new numerator and multiplying the denominators to get the new denominator". As mathematics, this whole passage makes no sense whatsoever.

The simplicity of the *procedure* of multiplying fractions ("multiplying the numerators to get the new numerator and multiplying the denominators to get the new denominator") is the siren song that lures many in mathematics education to take the fateful step of teaching how to multiply fractions without giving a definition. What they willfully choose to ignore is the resulting nonlearning of fractions, especially the nonlearning of the division of fractions. For this reason, we will resist this temptation.

There are at least two interpretations of the multiplication of fractions that everybody must know. The first is the areas of rectangles, and the other is the concept of "$\frac{m}{n}$ of something" that was defined earlier in Chapter 15. Either can be used as a definition of the multiplication of fractions, but once adopted, the other interpretation then becomes a theorem that must be proved. We will use the geometric one, because many elementary teachers seem to welcome its conceptual affinity with base ten blocks, but *both interpretations are equally important*.

The geometric definition may be briefly described as follows. We saw in Chapter 2 that the multiplication of whole numbers can be interpreted in terms of area: $m \times n$ is the area of a rectangle of sides m and n (more precisely, of "vertical" side m and "horizontal" side n), where m and n are whole numbers. Since the area of a rectangle *does* make sense even when its sides are fractions, we will use the area of rectangles to *define* fraction multiplication. Notice the conceptual similarity between this approach to defining multiplication of fractions with the earlier approach to defining addition of fractions: both are direct descendants of the same concepts in whole numbers.

The sections are as follows:

The Definition and the Product Formula

Immediate Applications of the Product Formula

A Second Interpretation of Fraction Multiplication

Inequalities

Linguistic vs. Mathematical Issues

17.1. The Definition and the Product Formula

Fix a number line. The unit of length, being the length of the unit segment, is thus defined and, therewith, also the unit square (recall: this is the square whose side has length 1). *Let the unit 1 of the number line also stand for the area of the unit square.* (See the discussion of multiplication in Chapter 8.) Thus we are identifying two number lines: one where the unit 1 is the length of a side of a fixed unit square, and the other where the unit 1 is the area of the same square.

In the following, we shall adopt the common abuse of language by referring to a rectangle with sides of lengths $\frac{m}{n}$ and $\frac{k}{\ell}$ as **a rectangle with sides $\frac{m}{n}$ and $\frac{k}{\ell}$**. With this understood, we define the **multiplication of fractions** as follows:

$$\frac{m}{n} \times \frac{k}{\ell} = \text{the area of a rectangle with sides } \frac{m}{n} \text{ and } \frac{k}{\ell}.$$

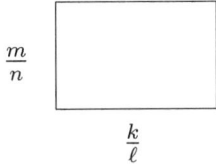

It follows immediately from this definition that if fractions A, B, C, and D satisfy $A = B$ and $C = D$, then $AC = BD$. Indeed, a rectangle with sides A and C is congruent to a rectangle with sides B and D, and consequently these rectangles have the same area. So by definition, $AC = BD$. We will make repeated use of this algebraic fact in the following.

We call attention to the fact that, *at the moment*, we do not know as yet what fraction the simple product $\frac{1}{2} \times \frac{1}{2}$ is equal to. Your conditioned reflex tells you that it is $\frac{1\times 1}{2\times 2}$ and therefore equal to $\frac{1}{4}$, but conditioned reflexes do not count in mathematics. *We have to explain why it is $\frac{1}{4}$ on the basis of the above definition and the above definition alone, and by making use of mathematical reasoning.* Thus we have to show, according to the definition of $\frac{1}{2} \times \frac{1}{2}$, why a square with side $\frac{1}{2}$ has area $\frac{1}{4}$. Fortunately, the general reasoning has already been given on page 192. Thus, in the following picture where the big square is the unit square and the sides of the four small squares all have length $\frac{1}{2}$, the area of the unit square (which is 1) is divided into four parts of equal area by the four small squares. Each small square therefore has area equal to $\frac{1}{4}$. Referring now to the shaded square in the bottom row on the left, we have the following.

$$\frac{1}{2} \times \frac{1}{2} = \left(\text{area of shaded square with side } \frac{1}{2}\right)$$
$$= \frac{1}{4}$$

Before we tackle the problem of computing $\frac{m}{n} \times \frac{k}{\ell}$ in general, let us work through some simple cases to secure our footing in this definition.

The following may be the simplest:

$$1 \times \frac{k}{\ell} = \frac{k}{\ell}.$$

We first show $1 \times \frac{1}{\ell} = \frac{1}{\ell}$, i.e., the area of a rectangle with sides 1 and $\frac{1}{\ell}$ is $\frac{1}{\ell}$. By the argument in Chapter 12 using the basic properties (a)–(c) of area, any rectangle contained in the unit square with sides of length 1 and $\frac{1}{\ell}$ will have area $\frac{1}{\ell}$. This is because ℓ such rectangles provide a partition of the unit square into ℓ congruent rectangles, and therefore provide a division of the unit (i.e., the area of the unit square) into ℓ equal parts (equal areas). For the case $\ell = 7$, one of these seven rectangles is shown below with thickened sides:

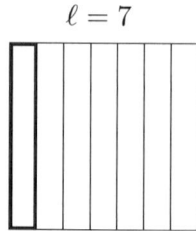

$\ell = 7$

This then proves $1 \times \frac{1}{\ell} = \frac{1}{\ell}$.

In general, $1 \times \frac{k}{\ell}$ is the area of the rectangle with sides 1 and $\frac{k}{\ell}$. If $k = 3$ and $\ell = 7$, such a rectangle is shown below with thickened sides:

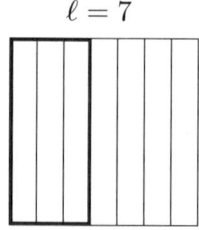

$\ell = 7$

Therefore $1 \times \frac{3}{7}$ equals the area of the preceding shaded rectangle, equals the sum of the areas of three "narrow" rectangles with sides 1 and $\frac{1}{\ell}$, equals

17.1. The Definition and the Product Formula

$\frac{1}{7} + \frac{1}{7} + \frac{1}{7} = \frac{3}{7}$. Although we have used $k = 3$ which is smaller than $\ell = 7$, the reasoning is independent of how big or small k is relative to ℓ. Therefore, for any k, $1 \times \frac{k}{\ell}$ equals the area of k such rectangles each with sides 1 and $\frac{1}{\ell}$, and is therefore equal to $\frac{1}{\ell} + \cdots + \frac{1}{\ell}$ (k times) $= \frac{k}{\ell}$.

Let us look at two more simple cases: $\frac{1}{2} \times \frac{1}{3} = ?$ and $\frac{1}{3} \times \frac{1}{6} = ?$

For $\frac{1}{2} \times \frac{1}{3}$, take a unit square and divide the vertical sides into two equal parts and divide a horizontal sides into three equal parts. Joining corresponding division points both horizontally and vertically then partitions the square into six congruent rectangles, and therefore six rectangles with equal areas, as shown.

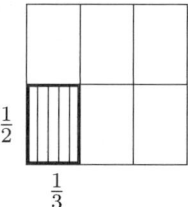

By construction, each of the 2×3 (= 6) rectangles has sides $\frac{1}{2}$ and $\frac{1}{3}$ and its area is by definition $\frac{1}{2} \times \frac{1}{3}$. However, the total area of these 2×3 rectangles is the area of the unit square, which is 1, so the shaded rectangle is one part in a partition of the unit square (i.e., the *unit*) into 2×3 rectangles of equal area. Therefore, the area of the shaded rectangle is $\frac{1}{2 \times 3}$ by the definition of a fraction in Chapter 12. Thus $\frac{1}{2} \times \frac{1}{3} = \frac{1}{2 \times 3}$.

Next, $\frac{1}{3} \times \frac{1}{6}$. Again we divide the vertical sides of a unit square into three equal parts and the horizontal sides into six equal parts. Joining the corresponding division points, both horizontally and vertically, leads to a partition of the unit square into 3×6 (= 18) congruent rectangles, and therefore 18 rectangles of equal areas:

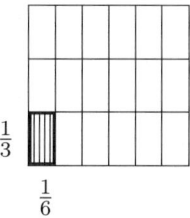

By construction, each rectangle has sides of lengths $\frac{1}{3}$ and $\frac{1}{6}$, so its area is $\frac{1}{3} \times \frac{1}{6}$, by definition. But since these 3×6 rectangles are identical, they partition the unit square (which has area equal to 1) into 3×6 rectangles of equal areas. The area of each rectangle is therefore $\frac{1}{3 \times 6}$, by definition of a fraction. Consequently, $\frac{1}{3} \times \frac{1}{6} = \frac{1}{3 \times 6}$.

Activity. Compute $\frac{1}{4} \times \frac{1}{3}$.

At this point, we are in a position to prove that for any whole numbers $\ell > 0$ and $n > 0$,

(17.1) $$\frac{1}{n} \times \frac{1}{\ell} = \frac{1}{n\ell}.$$

The proof goes as follows. Divide the two vertical sides of a unit square into n equal parts and the two horizontal sides into ℓ equal parts, respectively. Joining the corresponding division points, both horizontally and vertically, creates a partition of the unit square into $n\ell$ congruent rectangles, as shown.

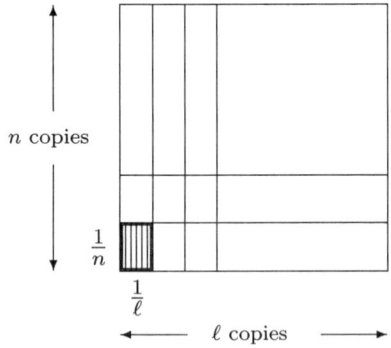

Because each of these rectangles has sides $\frac{1}{n}$ and $\frac{1}{\ell}$ by construction, its area is $\frac{1}{n} \times \frac{1}{\ell}$ by definition. Moreover, these $n\ell$ congruent rectangles partition the unit square into $n\ell$ rectangles of equal areas, so each of them has area $\frac{1}{n\ell}$. (This is the definition of $\frac{1}{n\ell}$ as one part when the unit—area of the unit square—is divided into $n\ell$ equal parts; see page 183.) Thus $\frac{1}{n} \times \frac{1}{\ell} = \frac{1}{n\ell}$, which proves equation (17.1).

Before attacking the general case of $\frac{m}{n} \times \frac{k}{\ell}$, let us again consider a concrete example: $\frac{2}{7} \times \frac{3}{4}$. This is by definition the area of a rectangle with sides $\frac{2}{7}$ and $\frac{3}{4}$. *Now we change strategy completely:* instead of partitioning the unit square, we use small rectangles of sides $\frac{1}{7}$ and $\frac{1}{4}$ to build a rectangle of sides $\frac{2}{7}$ and $\frac{3}{4}$. By the definition of $\frac{2}{7}$, it is the concatenation of two segments each of length $\frac{1}{7}$. Similarly, the side of length $\frac{3}{4}$ consists of three concatenated segments each of length $\frac{1}{4}$. Joining the corresponding points on opposite sides yields a partition of the original rectangle into 2×3 identical small rectangles each with sides $\frac{1}{7}$ and $\frac{1}{4}$, as shown below.

17.1. The Definition and the Product Formula

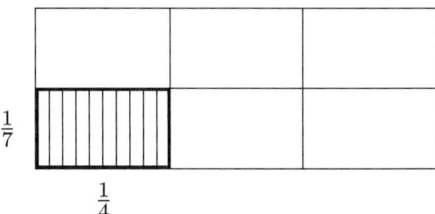

By equation (17.1), each of the small rectangles has area $\frac{1}{7\times 4}$. Since the big rectangle contains exactly 2×3 such congruent rectangles, its area (as a fraction) in terms of the unit area 1 is

$$\underbrace{\frac{1}{7\times 4}+\frac{1}{7\times 4}+\cdots+\frac{1}{7\times 4}}_{2\times 3} = \frac{2\times 3}{7\times 4}.$$

Recall that the area of the big rectangle is $\frac{2}{7}\times\frac{3}{4}$. Thus at least in this case, we have the expected product rule:

$$\frac{2}{7}\times\frac{3}{4} = \frac{2\times 3}{7\times 4}.$$

Activity. Compute $\frac{3}{5}\times\frac{4}{3}$.

Finally we prove the general **product formula** for any fractions $\frac{m}{n}$ and $\frac{k}{\ell}$:

(17.2) $$\frac{m}{n}\times\frac{k}{\ell} = \frac{mk}{n\ell}.$$

We are given a rectangle with sides $\frac{m}{n}$ and $\frac{k}{\ell}$, so that its area is $\frac{m}{n}\times\frac{k}{\ell}$, by definition. Our task is to show that its area is also equal to $\frac{mk}{n\ell}$. Now its side of length $\frac{m}{n}$ consists of m concatenated segments each of length $\frac{1}{n}$ and its side of length $\frac{k}{\ell}$ consists of k concatenated segments each of length $\frac{1}{\ell}$. Joining corresponding division points on opposite sides leads to a partition of the big rectangle into mk congruent small rectangles, as shown.

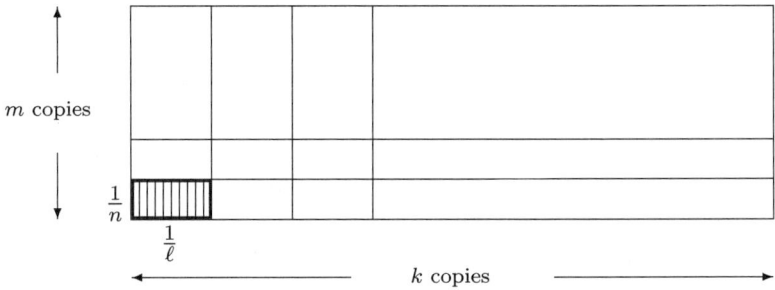

Since each of these small rectangles has sides equal to $\frac{1}{n}$ and $\frac{1}{\ell}$, its area is $\frac{1}{n\ell}$ by virtue of equation (17.1). But the big rectangle is paved by exactly

mk such small rectangles, so its area is

$$\underbrace{\frac{1}{n\ell} + \frac{1}{n\ell} + \cdots + \frac{1}{n\ell}}_{mk} = \frac{mk}{n\ell},$$

thereby proving product formula (17.2).

Observe that if we let $n = 1$ in (17.2), we would have

$$m \times \frac{k}{\ell} = \frac{m}{1} \times \frac{k}{\ell} = \frac{mk}{\ell} = \left(\frac{k}{\ell} + \frac{k}{\ell} + \cdots + \frac{k}{\ell}\right) \quad (m \text{ times}),$$

where the last is because of equation (14.4) in Chapter 14. This shows that for any fraction A and any nonzero whole number m,

(17.3) $\qquad\qquad m \times A = \underbrace{A + A + \cdots + A}_{m}.$

If A is a whole number, (17.3) is just the definition of multiplication for whole numbers (see the definition in equality (1.2) on page 28). The equality (17.3) is therefore an *extension* of the concept of multiplying a whole number by a whole number m to multiplying a fraction by m. This extension is conceptually important if we want to approach the multiplication of fractions using (17.7) below instead of rectangles (as we have done). See the discussion of (17.7).

Note that from (17.3), we recover the earlier result that $1 \times A = A$ for any fraction A.

This is the right place to note that

> *in the context of fractions, multiplication by a nonzero number can lead to a smaller number.*

For example, if we start with 15, then multiplying it by $\frac{1}{15}$ gets 1, and multiplying it by $\frac{1}{750}$ would get an even smaller $\frac{1}{50}$. More importantly, one should draw rectangles to show, *using the definition of multiplication*, why multiplying by a fraction can lead to a smaller number. It is well to note also that, even for whole numbers, multiplication by a nonzero number does not *always* result in a bigger number, e.g., multiplication by 1 gives back the original number and multiplication by 0 gives 0. The usual complaint about children's misconception that "multiplication leads to a bigger number" is therefore more likely a statement about faulty instruction rather than students' misconceptions. This is why we as teachers have to be careful about what we say.

17.2. Immediate Applications of the Product Formula

The most striking application of the product formula (17.2) may be the explanation of the multiplication algorithm for decimals. This algorithm states that to multiply two decimals x and y,

(i) multiply x and y as if they are whole numbers by ignoring the decimal points, and

(ii) count the total number of decimal digits of x and y, say p, and put the decimal point back in the whole number obtained in (i) so that it has p decimal digits.

We now justify the algorithm using the example of 1.25×0.0067, noting at the same time that the reasoning in the general case is the same.

$$\begin{aligned} 1.25 \times 0.0067 &= \frac{125}{10^2} \times \frac{67}{10^4} \\ &= \frac{125 \times 67}{10^2 \times 10^4} \quad \text{(product formula)}. \end{aligned}$$

Now we multiply 125×67 in the numerator instead of 1.25×0.0067; this corresponds to (i). Therefore,

$$1.25 \times 0.0067 = \frac{8375}{10^2 \times 10^4}.$$

By equation (1.4) on page 30, $10^2 \times 10^4 = 10^{2+4}$, so that

$$1.25 \times 0.0067 = \frac{8375}{10^{2+4}}.$$

But by the definition of a decimal, the right side is 0.008375, and this corresponds exactly to (ii). In particular, we see that the basic reason that the number of decimal digits in the product equals the sum of the number of decimal digits in each factor is equation (1.4).

Remark. There has been an increased emphasis in the recent educational literature on *connections* in mathematics. The fact that the product formula is responsible for the multiplication algorithm is certainly a remarkable connection that deserves recognition.

We noted on page 223 the validity of the commutative law and the associative law for the addition of fractions. With the availability of the product formula, the same laws for multiplication are straightforward consequences of formula (17.2) and the corresponding laws for whole numbers. Thus, *fraction multiplication is also associative and commutative.*

Now that we can both add and multiply fractions, we can also verify the **distributive laws**. Introduce the notation $\pm$: the equality $a \pm b = c \pm d$ is an abbreviation for two equalities: $a + b = c + d$, and $a - b = c - d$. The distributive law for addition and subtraction then states that for fractions A, B, C, being nonzero,

(17.4) $$A \times (B \pm C) = (A \times B) \pm (A \times C).$$

This can be proved in two ways: algebraically and geometrically. The algebraic proof consists of a straightforward computation using the addition formula (14.3) on page 222 of Chapter 14 and the product formula (17.2), but we will leave that as an exercise.

We will give the geometric proof of the case of "+" in (17.4); the "−" case can be handled similarly. Consider a rectangle with one side equal to A, and such that the other side consists of two concatenated segments of lengths B and C. This gives rise to a partition of the rectangle into two smaller rectangles as shown.

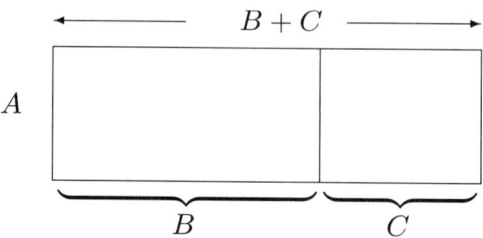

Then the distributive law (17.4) is merely the statement that the area of the big rectangle (left side of (17.4)) is equal to the sum of the areas of the two smaller rectangles (right side of (17.4)).

As an application of the distributive law, consider $181\frac{1}{6} \times \frac{3}{7}$. It is equal to
$$\left(181 + \frac{1}{6}\right) \times \frac{3}{7} = \frac{543}{7} + \frac{1}{14} = \frac{1086 + 1}{14} = 77\frac{9}{14}.$$
Of course one may also do this calculation without appealing to the distributive law for fractions:
$$181\frac{1}{6} \times \frac{3}{7} = \frac{1087}{6} \times \frac{3}{7} = \frac{1087}{14} = 77\frac{9}{14}.$$

This is the right place to revisit equivalent fractions. In section 13.3, **Proof of Theorem 13.1**, on page 209, we quoted a passage from a textbook, which purports to "prove" the theorem on equivalent fractions by multiplying a fraction by $\frac{k}{k}$ for any whole number k before multiplication of fractions is even defined. However, now that we have the product formula (17.2) at our disposal, this reasoning becomes valid:
$$\frac{cm}{cn} = \frac{c}{c} \times \frac{m}{n} = 1 \times \frac{m}{n} = \frac{m}{n}.$$
What this says is that having the product formula gives a different perspective on why the theorem on equivalent fractions is true, or if you like, the product formula makes it easier to remember this theorem.

17.2. Immediate Applications of the Product Formula

If you still insist on using this argument to *prove* the theorem on equivalent fractions, then be sure that you are aware of the need to first discuss multiplication of fractions before taking on the concept of equivalent fractions. (However, this is not a good move pedagogically, because the product formula is more difficult to prove than the theorem on equivalent fractions.)

As one more illustration of the power of the product formula (17.2), we now give the explanation of why the following **cancellation phenomenon** is valid (we cancel the ℓ's and the k's in the numerator and denominator):

(17.5) $$\frac{\ell}{k} \times \frac{kn}{\ell m} = \frac{n}{m}$$

for all whole numbers k, ℓ, m, n with $k \neq 0$, $\ell \neq 0$, and $m \neq 0$. Such cancellation is the direct consequence of the product formula (17.2) and the cancellation law (13.1):

$$\frac{\ell}{k} \times \frac{kn}{\ell m} = \frac{\ell kn}{\ell km} = \frac{n}{m}.$$

As special cases, we have the following two well-known facts:

$$\ell \times \frac{n}{\ell m} = \frac{n}{m}$$

and if both $k, \ell \neq 0$,

$$\frac{\ell}{k} \times \frac{k}{\ell} = 1.$$

The latter equality says that *every* nonzero fraction $\frac{k}{\ell}$ multiplied by its reciprocal is equal to 1. This simple fact says more than meets the eye: if we multiply both sides of this equality by an arbitrary fraction $\frac{m}{n}$, then we obtain the equality[2]

$$\frac{m}{n} \times \left(\frac{\ell}{k} \times \frac{k}{\ell} \right) = \frac{m}{n} \quad \text{for all fractions } \frac{m}{n} \text{ and } \frac{k}{\ell}, \frac{k}{\ell} \neq 0.$$

Now we rewrite this equality using the associative law, and we get

(17.6) $$\left(\frac{m}{n} \times \frac{\ell}{k} \right) \times \frac{k}{\ell} = \frac{m}{n} \quad \text{for } all \text{ fractions } \frac{m}{n} \text{ and } \frac{k}{\ell}, \frac{k}{\ell} \neq 0.$$

The reason this is interesting is that it provides the answer to the question: If a nonzero fraction $\frac{k}{\ell}$ is given, and another fraction $\frac{m}{n}$ is also given, is there a fraction A so that

$$A \times \frac{k}{\ell} = \frac{m}{n} \, ?$$

Of course, looking at the equality in (17.6), we immediately recognize that the answer is affirmative by letting $A = \frac{m}{n} \times \frac{\ell}{k}$.

We can leave this interesting information as is and go on to other aspects of multiplication, but it is instructive to push this line of reasoning to its

---

[2]See the comment on page 263 immediately below the definition of multiplication.

logical conclusion by making one more observation. As we said, by looking at (17.6), we saw that there was an easy answer for the preceding question, namely, letting $A = \frac{m}{n} \times \frac{\ell}{k}$ would work. More is true, however, in that *this is the only possible answer to the question and there is no other possibility.* What we are saying is that, if B is *any* fraction so that $B \times \frac{k}{\ell} = \frac{m}{n}$, then B must also be equal to $\frac{m}{n} \times \frac{\ell}{k}$. To see this, we multiply both sides of $B \times \frac{k}{\ell} = \frac{m}{n}$ by $\frac{\ell}{k}$ to get

$$\left(B \times \frac{k}{\ell}\right) \times \frac{\ell}{k} = \frac{m}{n} \times \frac{\ell}{k}.$$

But the left side is equal to $B \times (\frac{k}{\ell} \times \frac{\ell}{k})$, by the associative law, which is of course equal to $B \times 1 = B$. Hence we have proved that *if* $B \times \frac{k}{\ell} = \frac{m}{n}$, then necessarily

$$B = \frac{m}{n} \times \frac{\ell}{k}.$$

We have therefore proved the following lemma:[3]

Lemma 17.1. *We are given a nonzero fraction $\frac{k}{\ell}$. Then for every fraction $\frac{m}{n}$, there is a* unique *fraction A, i.e., one and only one A, so that*

$$A \times \frac{k}{\ell} = \frac{m}{n}.$$

In fact, $A = \frac{m}{n} \times \frac{\ell}{k}$.

In other words, A is obtained by *inverting* $\frac{k}{\ell}$ (in the sense of interchanging its numerator and denominator) and multiplying $\frac{m}{n}$. The full import of this lemma emerges only when we consider the division of fractions (see Chapter 18).

17.3. A Second Interpretation of Fraction Multiplication

The goal of this section is to give an interpretation of the multiplication of fractions that connects the geometric definition of fraction multiplication with the everyday usage of fractions. This interpretation could serve equally well as the definition of fraction multiplication.

As usual, we fix a number line. By the preceding discussion, the unit 1 on the line is not only the unit of length, but will also serve as the area of the unit square. Given fractions $\frac{m}{n}$ and $\frac{k}{\ell}$. Then the product $\frac{m}{n} \times \frac{k}{\ell}$ (in addition to being the area of a certain rectangle) is also the length of a segment as it is a point on the number line. So far, we have located this point only via a circuitous route: fix $\frac{m}{n}$ and $\frac{k}{\ell}$ on the given number line and

---

[3] In mathematics, a *lemma* is a theorem that is deemed to be of lesser interest, in the sense that it is usually somewhat technical in nature and probably not something you want to write home about.

17.3. A Second Interpretation of Fraction Multiplication

compute the area of the rectangle with sides $\frac{m}{n}$ and $\frac{k}{\ell}$ to obtain a point P on a second number line whose unit is the area of the unit square (i.e., the square whose side has length equal to the unit on the original number line), and then identify P as a point on the original number line after the unit of area has been identified with the unit of length. The position of P on the original number line is then $\frac{m}{n} \times \frac{k}{\ell}$.

We now show how to *directly* locate $\frac{m}{n} \times \frac{k}{\ell}$ on the number line:

$$(17.7) \qquad \frac{m}{n} \times \frac{k}{\ell} = \frac{m}{n} \text{ of the fraction } \frac{k}{\ell}.$$

This interpretation makes use of the terminology "$\frac{m}{n}$ of $\frac{k}{\ell}$" introduced in section 15.4, **The Concept of $\frac{m}{n}$ of $\frac{k}{\ell}$**, on page 245; you may wish to review that section before proceeding further. Let us first restate (17.7) by unraveling the definition of the word "of":

$$\frac{m}{n} \times \frac{k}{\ell} = \left\{ \begin{array}{l} \text{the length of } m \text{ concatenated parts when a segment} \\ \text{of length } \frac{k}{\ell} \text{ is divided into } n \text{ parts of equal length} \end{array} \right\}.$$

At the end of section 15.4 (Theorem 15.3), we already proved that the right side is equal to $\frac{mk}{n\ell}$, and the product formula shows that the left side is also. So (17.7) is true, and that should be the end of the story.

However, in view of the importance of (17.7), we will expand upon the preceding proof by explaining in more intuitive language why the right side of (17.7) is equal to $\frac{mk}{n\ell}$, which in turn is equal to $\frac{m}{n} \times \frac{k}{\ell}$.

First we look at a special case and prove that

$$\frac{4}{3} \text{ of } \frac{5}{7} = \frac{20}{21}.$$

The left side is, by definition, the length of the concatenation of four parts when the segment $[0, \frac{5}{7}]$ is divided into three equal parts. Since the length of $[0, \frac{5}{7}]$ is five copies of $\frac{1}{7}$, it is awkward to divide such a length into three equal parts. However, the theorem on equivalent fractions (page 204) says

$$\frac{5}{7} = \frac{3 \times 5}{3 \times 7} = \frac{5}{21} + \frac{5}{21} + \frac{5}{21}.$$

Thus $[0, \frac{5}{7}]$ is the concatenation of three segments each of length $\frac{5}{21}$. The length of one part when $[0, \frac{5}{7}]$ is divided into three equal parts is therefore $\frac{5}{21}$. The total length of four such parts is then $\frac{4 \times 5}{21}$, i.e., it is $\frac{20}{21}$, as claimed. Since $\frac{20}{21} = \frac{4}{3} \times \frac{5}{7}$ by the product formula, (17.7) is proved in this special case.

The proof of (17.7) in general is no different. We want to divide $[0, \frac{k}{\ell}]$ into n equal parts. By the theorem on equivalent fractions, the length $\frac{k}{\ell}$ of

the segment $[0, \frac{k}{\ell}]$ is equal to
$$\frac{k}{\ell} = \frac{nk}{n\ell} = \underbrace{\frac{k}{n\ell} + \cdots + \frac{k}{n\ell}}_{n}.$$
This sum then clearly exhibits $[0, \frac{k}{\ell}]$ as the concatenation of n segments each of length $\frac{k}{n\ell}$. Therefore the length of one part when $[0, \frac{k}{\ell}]$ is divided into n equal parts is $\frac{k}{n\ell}$. If we concatenate m of these parts, then the total length is
$$\underbrace{\frac{k}{n\ell} + \cdots + \frac{k}{n\ell}}_{m} = \frac{mk}{n\ell} = \frac{m}{n} \times \frac{k}{\ell},$$
where the last step is by the product formula. The proof of (17.7) is now complete.

Next, we rephrase the preceding proof of (17.7) slightly differently by breaking it down into simpler components. Pedagogically, this approach may be more suitable for classroom use.

Case I. This is the special case of (17.7) when $n = 1$, which says,
$$m \times \frac{k}{\ell} = m \text{ copies of } \frac{k}{\ell}.$$
To prove this, we use equality (17.3) on page 268, which says $m \times \frac{k}{\ell}$ is exactly m copies of $\frac{k}{\ell}$.

Case II. This is the special case of (17.7) when $m = 1$, which says,
$$\frac{1}{n} \times \frac{k}{\ell} = \text{ the length of one part when } \frac{k}{\ell} \text{ is divided into } n \text{ equal parts.}$$
To prove this, observe that $\frac{1}{n} \times \frac{k}{\ell} = \frac{k}{n\ell}$ (product formula). What we have to show is therefore that $\frac{k}{n\ell}$ is the length of one part when $\frac{k}{\ell}$ is divided into n equal parts. This is equivalent to showing that the length of the concatenation of n segments of length $\frac{k}{n\ell}$ is equal to $\frac{k}{\ell}$, i.e., n copies of $\frac{k}{n\ell}$ is equal to $\frac{k}{\ell}$. By Case I,
$$n \text{ copies of } \frac{k}{n\ell} = n \times \frac{k}{n\ell} = \frac{nk}{n\ell} = \frac{k}{\ell}.$$
This completes the proof of Case II.

Case III. This is the general case. To prove this, we apply the product formula and the associative law of multiplication to get
$$\frac{m}{n} \times \frac{k}{\ell} = \left(m \times \frac{1}{n}\right) \times \frac{k}{\ell} = m \times \left(\frac{1}{n} \times \frac{k}{\ell}\right).$$
Now by Case II,
$$\frac{1}{n} \times \frac{k}{\ell} = \text{ the length of one part when } \frac{k}{\ell} \text{ is divided into } n \text{ equal parts.}$$

17.3. A Second Interpretation of Fraction Multiplication

By Case I, therefore $m \times (\frac{1}{n} \times \frac{k}{\ell})$ is the length of m concatenated parts when $\frac{k}{\ell}$ is divided into n equal parts. Hence, $\frac{m}{n} \times \frac{k}{\ell}$ being equal to $m \times (\frac{1}{n} \times \frac{k}{\ell})$, we have

$$\frac{m}{n} \times \frac{k}{\ell} = \text{the length of } m \text{ parts when } \frac{k}{\ell} \text{ is divided into } n \text{ equal parts.}$$

This proves (17.7) completely.

The interpretation (17.7) of fraction multiplication justifies our instinctive practice, when confronted with the calculation of the weight of two-thirds of a $2\frac{1}{4}$-pound piece of beef, to simply multiply: $\frac{2}{3} \times 2\frac{1}{4} = \frac{3}{2}$ pounds. It also justifies, if the capacity of a cup is 8 fluid ounces, why "$3\frac{1}{4}$ cups" of orange juice has $3\frac{1}{4} \times 8$ fluid ounces of orange juice, as follows. The accepted meaning of "$3\frac{1}{4}$ cups" is of course

$$3 \text{ cups plus } \tfrac{1}{4} \text{ of a cup,}$$

which, according to (17.7), is equal to

$$(3 \times 8) + \left(\frac{1}{4} \times 8\right) = \left(3 + \frac{1}{4}\right) \times 8 = 3\frac{1}{4} \times 8 \text{ fluid ounces,}$$

which is exactly as claimed.

We can look at the preceding discussion as an interpretation of $3\frac{1}{4} \times 8$ as 3 *and* $\frac{1}{4}$ *copies of* 8, in the sense that it is equal to $8 + 8 + 8$ plus 1 part of 8 when it is divided into four equal parts. For the purpose of doing word problems, we will see that it would be to our advantage to expand upon this interpretation. We will use the informal language of

$$\frac{m}{n} \text{ copies of } \frac{k}{\ell}$$

to mean the following. If $\frac{m}{n}$ is a proper fraction, then $\frac{m}{n}$ copies of $\frac{k}{\ell}$ would mean exactly what is asserted in (17.7). If $\frac{m}{n}$ is an improper fraction, then we rewrite it as a mixed number $N\frac{b}{c}$, where N is a positive whole number and $\frac{b}{c}$ is a proper fraction. In that case, $\frac{m}{n}$ copies of $\frac{k}{\ell}$ would mean $\frac{k}{\ell} + \cdots + \frac{k}{\ell}$ (N times) plus the totality of b parts when $\frac{k}{\ell}$ is divided into c equal parts. We assert that

$$\left(\frac{m}{n} \times \frac{k}{\ell}\right) \text{ is } \frac{m}{n} \text{ copies of } \frac{k}{\ell}.$$

This is because, when $\frac{m}{n}$ is a proper fraction, this statement is just (17.7), and when $\frac{m}{n}$ is an improper fraction, then (with $\frac{m}{n} = N\frac{b}{c}$ as above),

$$\frac{m}{n} \times \frac{k}{\ell} = \left(N + \frac{b}{c}\right) \times \frac{k}{\ell} = \left(N \times \frac{k}{\ell}\right) + \left(\frac{b}{c} \times \frac{k}{\ell}\right).$$

Our assertion then follows from the fact that multiplication by a whole number is repeated addition as well as from (17.7).

For the illustration of this particular interface between everyday language and fraction multiplication, here are three examples.

EXAMPLE 1. A bucket has a capacity of 4.5 gallons. If $5\frac{2}{3}$ buckets of water fill a tank, what is the capacity of the tank?

The capacity of the tank is, by the given data, $5\frac{2}{3}$ copies of 4.5 gallons, and therefore equals

$$5\frac{2}{3} \times 4.5 = \frac{17}{3} \times \frac{45}{10} = \frac{17}{1} \times \frac{15}{10} = 25.5 \text{ gallons.}$$

EXAMPLE 2. How many buckets of water would fill a container if the capacity of the bucket is $2\frac{1}{2}$ gallons and that of the container is $6\frac{1}{3}$ gallons?

Suppose $\frac{m}{n}$ is a fraction so that $\frac{m}{n}$ buckets of water fill the container, then the capacity of the container—$6\frac{1}{3}$ gallons—is equal to $\frac{m}{n} \times 2\frac{1}{2}$ gallons. That is,

$$6\frac{1}{3} = \frac{m}{n} \times 2\frac{1}{2} = \frac{m}{n} \times \frac{5}{2},$$

which is equivalent to

$$\frac{19}{3} = \frac{m}{n} \times \frac{5}{2}.$$

It is easy to just *guess* what $\frac{m}{n}$ must be: $\frac{38}{15}$, but that would miss the point. A lot more can be said.

First and foremost, guessing is not needed. The fraction $\frac{m}{n}$ is one that, when multiplied by $\frac{5}{2}$, is equal to $\frac{19}{3}$. By Lemma 17.1 on page 272, this fraction $\frac{m}{n}$ must be equal to

$$\frac{19}{3} \times \frac{2}{5} \quad \left(= \frac{38}{15}\right).$$

This is our first application of this lemma, but hardly the last.

Next, we will directly deduce what $\frac{m}{n}$ has to be without using this lemma, as follows. Because multiplication is commutative, we have $\frac{19}{3} = \frac{5}{2} \times \frac{m}{n}$, so that by (17.7), we have

$$\frac{19}{3} = \frac{5}{2} \text{ of } \frac{m}{n}.$$

By the definition on page 246,[4] the preceding equality means: if we divide $[0, \frac{m}{n}]$ into two equal parts, then the concatenation of five of these parts has length $\frac{19}{3}$, as shown:

---

[4]It often comes down to precise definitions in mathematical reasoning.

17.3. A Second Interpretation of Fraction Multiplication

Let the length of one part when $[0, \frac{m}{n}]$ is divided into two equal parts be K. Then K is also the length of one part when $[0, \frac{19}{3}]$ is divided into five equal parts. Since
$$\frac{19}{3} = \frac{5 \times 19}{5 \times 3},$$
K is clearly equal to $\frac{19}{5 \times 3}$. From the picture, we see that $\frac{m}{n}$ is two copies of K, so
$$\frac{m}{n} = 2 \times \frac{19}{5 \times 3} = \frac{2}{5} \times \frac{19}{3}.$$
Thus $\frac{m}{n}$ is obtained by inverting $\frac{5}{2}$ and then multiplying by $\frac{19}{3}$. In particular, $\frac{m}{n} = \frac{39}{15}$ as before.

EXAMPLE 3. If a ribbon can be cut into $11\frac{1}{3}$ bows, each of length $\frac{7}{8}$ foot, how long is the ribbon?

The length of the ribbon is thus $11\frac{1}{3}$ copies of $\frac{7}{8}$ feet, and is therefore equal to
$$11\frac{1}{3} \times \frac{7}{8} = \frac{34}{3} \times \frac{7}{8} = \frac{119}{12} = 9\frac{11}{12} \text{ feet}$$
In other words, 9 feet and 11 inches.

It remains to observe that it is not at all obvious that the definition of fraction multiplication in terms of areas of rectangles would have the property indicated in (17.7). One way to achieve a more intuitive understanding of (17.7) is by looking at rectangles directly. As we have seen in the second proof of (17.7), writing $\frac{m}{n} \times A$ for a given fraction A as $m \times (\frac{1}{n} \times A)$ makes it quite clear that the key idea is to understand the special case $\frac{1}{n} \times A$ of (17.7). To make the drawing of pictures easier, let us say $n = 4$, but it will be abundantly clear that the number 4 plays no role in what follows. Now the number A, in addition to being a length, is also equal to $1 \times A$ and is therefore the area of the following rectangle:

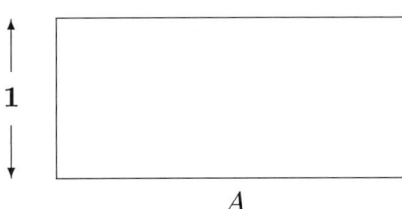

Thus $\frac{1}{4} \times A$ is $\frac{1}{4}$ of the area A, which according to section 15.4 is one part when A is divided into four parts of equal area. We can achieve this division by dividing the vertical sides of the rectangle into four segments of equal length and connecting corresponding division points, as shown below.

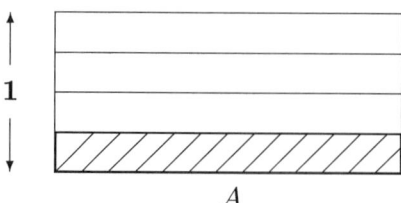

Therefore, by the definition of fraction multiplication, $\frac{1}{4} \times A =$ the area of the shaded rectangle, which is $\frac{1}{4}$ of A (area of the big rectangle).

17.4. Inequalities

We finish this discussion with the fraction analogues of one of the inequalities (2.7) on page 50. Let A, B, C, D be fractions. Then the desired inequality is

If $A > 0$, $AB < AC$ is equivalent to $B < C$.

We have to prove that if $B < C$, then $AB < AC$, and also if $AB < AC$, then $B < C$. Using the definition of multiplication as the area of rectangles (page 263), this is a simple exercise.

17.5. Linguistic vs. Mathematical Issues

The second interpretation of fraction multiplication, (17.7), brings up an educational issue that merits some comments. Recall that the way we approached (17.7) was by

(a) giving a precise definition of fraction multiplication,
(b) giving a precise definition of what it means to say "$\frac{m}{n}$ of a quantity",
(c) giving a precise proof of (17.7) on the basis of (a) and (b).

Each step is completely precise and none is open to doubt. When (17.7) is understood in this manner, we know why "three-fifths of the length of a rod" is equal to "$\frac{3}{5} \times$ (the length of the rod)".

It should be noted that this attitude towards verbal information is not universally accepted in mathematics education. Some believe that for school mathematics, a better way to teach it is to reverse the roles of mathematics and everyday language. In this view, everyday language is the primary source, and mathematics—especially fractions—is no more than a symbolic reflection of the language. To the extent that everyday language is often vague when it comes to quantitative information, doing mathematics *this way* then often becomes a guessing game: how well we can do mathematics depends on how well we understand the hidden meanings of everyday

17.5. Linguistic vs. Mathematical Issues

language. A good illustration of this point of view is given by the article [**Moy96**], in which the author describes a discussion in a sixth-grade classroom about what students did to "solve the following problems":

(1) The Davis family attended a picnic. Their family made up $\frac{1}{3}$ of the 15 people at the picnic. How many Davises were at the picnic?

(2) John ate $\frac{1}{8}$ of the 16 hot dogs. How many hot dogs did John eat?

(3) One-fourth of the hot dogs were served without relish. How many were served without relish?

Apparently, the students solved the problems in groups and then were challenged by the teacher to explain what mathematical operation should take the place of the preposition "of": is it + or ×? According to the teacher, knowing that "of" means *multiply* was important, because "If the algorithm for multiplying fractions was to make sense, they need to understand that *of* means multiply" and "understanding had to come first". In this view, "understanding" mathematics becomes synonymous with *understanding everyday language*.

This is a case of an attempt, by no means uncommon in mathematics education, to improve the teaching of fractions (in this case the multiplication of fractions) solely by appealing to students' understanding of the vagaries of everyday language (in this case, the meaning of "of"). In particular, such an education philosophy assumes that all children are hard-wired with the "understanding" that the word *of* signifies *multiplication*. Moreover, it further assumes that once children see that multiplication is the correct operation to use, they would see inevitably that $\frac{a}{b} \times \frac{c}{d} = \frac{ac}{bd}$.

It should be a matter of concern in mathematics education when a *sixth-grade* class does not show any interest in a mathematical definition of the multiplication of two fractions or why fraction multiplication should be interpreted as in (17.7). When vagueness replaces precision in a sixth-grade class, are these students adequately prepared to tackle algebra and geometry two or three years down the road? (See [**Wu09a**].)

Exercises

1. (a) Verify the distributive law (17.4) directly by expanding both sides using formulas (14.3) and (17.2). (b) Prove the inequality in section 17.4, **Inequalities**, on page 278. (*Hint*: This is a very simple proof provided you remember what the definition of fraction multiplication is.)

2. Use a calculator to do the whole number computations if necessary (and *only* for that purpose), compute: (a) $4\frac{2}{9} \times 6\frac{11}{13} = ?$ (b) $(15\frac{4}{17} \times 23\frac{9}{25}) - (16\frac{8}{19} \times 15\frac{4}{17}) = ?$ (c) $2\frac{7}{8} \times 14\frac{4}{5} \times 3\frac{1}{6} = ?$

3. (a) $3.15 \times \frac{19}{35} = ?$ (b) $0.00026 \times 4.7 = ?$ (c) $(0.516 \times \frac{14}{25}) - (0.016 \times \frac{14}{25}) = ?$ (d) $(3.05 \times 117) + (\frac{117}{3} \times 20.85) = ?$

4. (a) $\frac{3}{10} + \frac{5}{12} = ?$ (b) Kate worked on a math problem for 18 minutes without success. She came back, refocused, and got it done in 25 minutes. How much time did she spend on this problem altogether, and what does this have to do with part (4a)?

5. Without using (17.2), explain directly to a sixth grader why $\frac{3}{7} \times \frac{6}{5} = \frac{18}{35}$.

6. Write a word problem for the multiplication $38\frac{2}{5} \times 2\frac{1}{2}$ *without* using a rectangle.

7. (a) A rectangle has area 6 and a side of length $\frac{1}{3}$. What is the length of the other side? (b) A rectangle has area 0.375 and one side of length 5. What is the length of the other side? (c) A rectangle has area $\frac{7}{8}$ and one side of length $1\frac{1}{3}$. What is the length of the other side? (d) A small rectangle with sides $1\frac{2}{3}$ and $2\frac{1}{7}$ is contained in a larger rectangle with sides $12\frac{1}{3}$ and $6\frac{2}{5}$. Find the area of the region between these rectangles. (Use a four-function calculator.)

8. $8\frac{2}{50} \times 1250\frac{1}{2} = ?$ Do it without a calculator. (*Hint*: Use the distributive law.)

9. (*This is Exercise 9 on page 250. Now do it again using the concept of fraction multiplication.*) James gave a riddle to his friends: "I was on a hiking trail, and after walking $\frac{7}{12}$ of a mile, I was $\frac{5}{9}$ of the way to the end. How long is the trail?" Help his friends solve the riddle.

10. Which is heavier: $\frac{3}{4}$ of a ham that weighs $10\frac{3}{8}$ lbs. or $\frac{11}{25}$ of another ham that weighs 17.5 lbs.?

11. Given two fractions. Their difference is $\frac{4}{5}$ of the smaller one, while their sum is equal to $\frac{28}{15}$. What are the fractions? (*Hint*: Use the number line.)

12. Imitate the reasoning in Example 2 on page 276 to find a fraction $\frac{a}{b}$ so that $\frac{a}{b} \times \frac{5}{4} = \frac{2}{7}$. Do the same for $5\frac{1}{7} \times \frac{a}{b} = 2\frac{1}{2}$, and also $2\frac{4}{5} \times \frac{a}{b} = \frac{1}{2}$.

13. (a) $(1\frac{1}{4} \times 1\frac{1}{4} \times 1\frac{1}{4} \times 1\frac{1}{4} \times 1\frac{1}{4}) \times (1\frac{3}{5} \times 1\frac{3}{5} \times 1\frac{3}{5} \times 1\frac{3}{5}) = ?$ (b) $(2\frac{2}{3} \times 2\frac{2}{3} \times 2\frac{2}{3}) \times (4\frac{1}{8} \times 4\frac{1}{8} \times 4\frac{1}{8}) = ?$ No calculators. (c) $(\frac{7}{18} \times 3\frac{2}{3}) + (2\frac{1}{6} \times \frac{7}{18}) + (\frac{7}{18} \times 3\frac{1}{6}) = ?$

14. (a) Which is closer to $\frac{2}{7}$, $\frac{1}{3}$ or $\frac{5}{21}$? (b) Which is closer to $\frac{2}{3}$, $\frac{12}{19}$ or $\frac{9}{13}$?

15. Let A and B be two fractions such that $A < B$. Show that there is always a fraction C so that $A < C < B$. (*Caution*: After finding what you think is a good candidate for C, do not forget to actually prove that $A < C < B$. Also, compare Exercise 10 on page 250.)

16. (a) $16\frac{1}{2}$ cups of liquid would fill a punch bowl. If the capacity of the cup is $9\frac{1}{3}$ fluid ounces, what is the capacity of the punch bowl? Explain carefully why your answer is correct. (b) A rod can be cut into $18\frac{5}{8}$ short pieces that are $3\frac{1}{4}$ inches long. How long is the rod?

17. Exercise 18 on page 125 asks for a proof of the distributive law for division, namely, let k, m, n, be whole numbers. Let $n > 0$ and let k, m be multiples of n. Then
$$(m \div n) + (k \div n) = (m + k) \div n.$$
Now give a new proof of this result *without* the restriction that k, m be multiples of n. (Consult section 15.1.)

Chapter 18

Division of Fractions

Division of fractions and decimals are among the most elusive topics in elementary school mathematics. When properly presented, however, division of fractions is no different from the division of whole numbers; see the discussion surrounding definition (7.1) on page 97. We will stress this analogy while noting a critical difference between the two, namely, whereas the division of whole numbers requires the dividend to be a multiple of the divisor, there is no such requirement for the division of fractions. For fractions A and B ($B \neq 0$), the division of A by B *always* makes sense.

The sections are as follows:

> Informal Overview
>
> The Definition and Invert-and-Multiply
>
> Applications
>
> Comments on the Division of Decimals
>
> Inequalities
>
> False Doctrines

18.1. Informal Overview

In this section, we give an *intuitive* discussion of the meaning of the division of fractions (the *formal* discussion will be given in the next section). We wish to bring out the fact that the definition of fraction division is conceptually the same as the definition of whole number division and the fact that division is to multiplication as subtraction is to addition. We will suppress, at the initial stage of the discussion, the auxiliary restrictions that normally come with the precise definition of each of these concepts.

We begin with the concepts of subtraction and division between *whole numbers*. Let m and n be whole numbers.

Definition 1. $\boldsymbol{m - n}$ is the whole number k so that $m = k + n$ (see Chapter 5).

Definition 2. $\frac{m}{n}$ is the whole number k so that $m = kn$ (see Chapter 7).

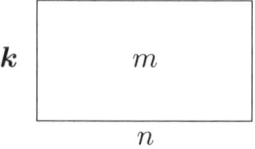

Notice that by changing addition to multiplication and subtraction to division in Definition 1, we get Definition 2 verbatim. We now follow this pattern in approaching fractions. Definition 3 is what we already know about fraction subtraction, and Definition 4 is what we "should" get if, indeed, whole numbers and fractions are not dissimilar. Thus, let A, B be fractions.

Definition 3. $\boldsymbol{A - B}$ is the fraction C so that $A = C + B$ (see Chapter 16).

Definition 4. $\frac{A}{B}$ is the fraction C so that $A = CB$.

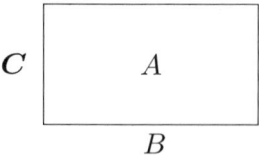

Definition 4 will turn out to be the correct definition of fraction division. Thus the concept of the division of fractions is a direct extension of the concept of the division of whole numbers. At the same time, the relation between division and multiplication is entirely analogous to that between subtraction and addition, as one can see by inspecting Definitions 1 and 2, and then Definitions 3 and 4.

Now we pay attention to the details. First of all, in Definition 1, clearly we need $m > n$; likewise, in Definition 3, we need $A > B$. In a similar vein, Definition 2 requires that $n > 0$, and Definition 4 requires that $B > 0$. These conditions are necessary, of course, but it is fair to say that they are not central to the understanding of subtraction or division. There is

18.1. Informal Overview

something more substantial, however, that is missing in Definitions 2 and 4. We have seen (Chapter 7) that $\frac{m}{n}$ does not make sense for whole numbers unless m is a multiple of n; indeed, the meaning of $\frac{m}{n}$ as $m = kn$ requires that m be a multiple of n. Since in Definition 4, the meaning of $\frac{A}{B}$ is likewise $A = CB$, should we not also require that A be a multiple of B (in the sense of $A = CB$ for some fraction C) before we can define the division of A by B? The surprise is that such a requirement is not necessary: so long as B is not zero, *every* fraction A can be written as CB for a unique fraction C. Obviously, this is something that needs a careful proof, and it will turn out that Lemma 17.1 on page 272 (in section 17.2) has already taken care of that.

Let us also make a preliminary exploration of the computation of $\frac{A}{B}$. What Definition 4 says is that $\frac{A}{B} = C$ means exactly $A = CB$. The proper way to understand $A = CB$ is via the discussion on page 275: the equality $A = CB$ means that A is C **copies of** B in the sense that 12 is 3 copies of 4, or 15 is $1\frac{1}{2}$ copies of 10. Thus intuitively, the division $\frac{A}{B}$ gives *the number of copies of B that are needed to make up A.*

We use this intuition to *make guesses* in some simple cases of what C must be in order to satisfy $A = CB$. Eventually, however, we will have to make use of Lemma 17.1.

EXAMPLE 1. $\dfrac{1}{\frac{1}{n}}$.

If we want a C so that $1 = C \times \frac{1}{n}$, $C = n$ would work. This confirms the intuition that n copies of $\frac{1}{n}$ is 1.

EXAMPLE 2. $\dfrac{5}{\frac{1}{3}}$.

We know 3 copies of $\frac{1}{3}$ equals 1, so 5×3 copies of $\frac{1}{3}$ would make up 5. We therefore guess that $\frac{5}{1/3} = 15$. Sure enough, $5 = 15 \times \frac{1}{3}$.

EXAMPLE 3. $\dfrac{\frac{15}{4}}{\frac{3}{4}}$.

If we use $\frac{1}{4}$ as a unit, then $\frac{15}{4}$ is just 15 and $\frac{3}{4}$ is just 3. The above division then asks for the division of 15 by 3, which is 5. We check: $\frac{15}{4} = 5 \times \frac{3}{4}$.

EXAMPLE 4. $\dfrac{\frac{2}{7}}{\frac{5}{3}}$.

How many copies of $\frac{5}{3}$ would make up $\frac{2}{7}$? Guessing is harder in this case, so we have to make use of Lemma 17.1 on page 272. We want a fraction C so that $\frac{2}{7} = C \times \frac{5}{3}$. The lemma says $C = \frac{2}{7} \times \frac{3}{5}$. Witness the emergence of "invert-and-multiply" (see also Example 2 on page 276).

We are now ready to turn to the formal discussion.

18.2. The Definition and Invert-and-Multiply

Division is an alternate, but equivalent, way of expressing multiplication, and this is true regardless of context. Although we have already stressed this point of view in whole numbers (Chapter 7), we will nevertheless begin with a brief review.

We teach children that $\frac{36}{9} = 4$ because 4 is the whole number so that $4 \times 9 = 36$. This then is the statement that 36 *divided by* 9 is the whole number which, when multiplied by 9, gives 36. In symbols, we may express the foregoing as follows: $\frac{36}{9}$ is by definition the number which satisfies $\frac{36}{9} \times 9 = 36$. Similarly, 72 *divided by* 24 is the whole number which, when multiplied by 24, gives 72, i.e., $\frac{72}{24}$ is the whole number which satisfies $\frac{72}{24} \times 24 = 72$. Likewise, $\frac{84}{7}$ is the whole number which satisfies $\frac{84}{7} \times 7 = 84$, etc. In general, given any two whole numbers a and b with $b \neq 0$, we always want the division $\frac{a}{b}$ to be the *whole number* so that $(\frac{a}{b})b = a$. Since this cannot happen if a is not a multiple of b, the precise formulation of the concept of division among whole numbers is this:

Definition. *Given whole numbers a and b, with $b \neq 0$ and a being a multiple of b, then **the division of** a **by** b, in symbols $\frac{a}{b}$, is the whole number so that the equality $a = (\frac{a}{b})b$ holds.*

As a consequence of this definition, we see that if we denote $\frac{a}{b}$ by c, then $a = cb$. Conversely, if we have whole numbers a, b, c so that $a = cb$, then c is what we mean by $\frac{a}{b}$. These two facts together lead to the following statement about division:

> Given whole numbers a, b, and c, with $b \neq 0$, the equality $\frac{a}{b} = c$ is equivalent to $a = cb$.

Recall the meaning of equivalence: This statement means that for three whole numbers a, b, and c with $b \neq 0$, if $\frac{a}{b} = c$, then $a = bc$, and if $a = bc$, then $\frac{a}{b} = c$. We have more to say about this equivalence on page 290; see especially footnote 1 on that page.

Before embarking on a discussion of the division of fractions, we first rephrase, in an equivalent form, Lemma 17.1 on page 272. (In this context, the discussion of Example 2 on page 276 becomes relevant.)

Lemma. *Given any two fractions A and B ($B \neq 0$), there is one and only one fraction C, so that $A = CB$.*

This lemma has already been proved in Chapter 17. However, we will now present a second proof that directly deduces what this C must be. For the purpose of understanding invert-and-multiply, this proof may be preferable.

18.2. The Definition and Invert-and-Multiply

The formal proof will be quite short. In order to make sure that the intuitive idea of the simple proof is not lost, we write out this proof in two specific cases. First, let $A = \frac{3}{2}$ and $B = \frac{4}{5}$. We look for a fraction C so that

$$\frac{3}{2} = C \times \frac{4}{5}.$$

By the commutativity of multiplication, we may rewrite it as $\frac{3}{2} = \frac{4}{5} \times C$. Thus we look for a C so that four-fifths of it is equal to $\frac{3}{2}$. By (17.7) on page 273, $\frac{4}{5} \times C$ is the length of four concatenated parts when C is divided into five equal parts. Since this product is equal to $\frac{3}{2}$, $\frac{3}{2}$ is the point on the number line indicated in the following picture:

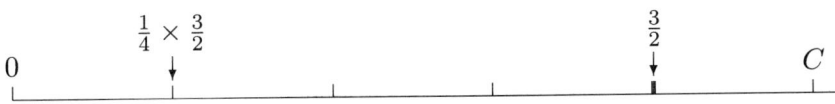

The first division point to the right of 0 is therefore one-fourth of $\frac{3}{2}$, which is $\frac{1}{4} \times \frac{3}{2}$ (by (17.7)), but by construction, this is also one-fifth of C. Therefore C is five copies of $\frac{1}{4} \times \frac{3}{2}$, so that by (17.7),

$$C = 5 \times \left(\frac{1}{4} \times \frac{3}{2}\right) = \frac{5}{4} \times \frac{3}{2}.$$

We have therefore proved that *any* C that satisfies $\frac{3}{2} = C \times \frac{4}{5}$ must be equal to $\frac{5}{4} \times \frac{3}{2}$. This shows the uniqueness of such a C. Of course it is routine to check that if $C = \frac{5}{4} \times \frac{3}{2}$, then $\frac{3}{2} = C \times \frac{4}{5}$. The proof is complete.

The preceding example might give the impression that because $B = \frac{4}{5} < 1$, the reasoning is easier. In order to dispel this illusion, we take up a similar example. Let $A = \frac{7}{3}$ and $B = \frac{5}{4}$. Again we look for a fraction C so that

$$\frac{7}{3} = \frac{5}{4} \times C.$$

As before, we may rephrase this as a search for a fraction C so that $\frac{7}{3} = \frac{5}{4} \times C$. Thus if we divided C into four equal parts, then the length of the concatenation of five of these parts is equal to $\frac{7}{3}$, as shown:

The first division point to the right of 0 is then one-fifth of $\frac{7}{3}$, which by virtue of (17.7) is equal to $\frac{1}{5} \times \frac{7}{3}$. Since C is four copies of $\frac{1}{5} \times \frac{7}{3}$, we have

$$C = 4 \times \left(\frac{1}{5} \times \frac{7}{3}\right) = \frac{4}{5} \times \frac{7}{3}.$$

Hence any fraction C satisfying $\frac{7}{3} = \frac{5}{4} \times C$ must be equal to $\frac{4}{5} \times \frac{7}{3}$. This shows that this C is unique. It is again trivial to prove that if $C = \frac{4}{5} \times \frac{7}{3}$, then C satisfies $\frac{7}{3} = \frac{5}{4} \times C$. The proof is complete.

Proof of Lemma. Let $A = \frac{m}{n}$ and $B = \frac{k}{\ell}$. Thus we are looking for a fraction C, so that
$$\frac{m}{n} = C \times \frac{k}{\ell}.$$
Let us assume that there is such a C, and we will find out what it is. Rewrite this as $\frac{m}{n} = \frac{k}{\ell} \times C$. By the second interpretation of fraction multiplication in (17.7), $\frac{k}{\ell} \times C$ is the length of the concatenation of k parts if we partition $[0, C]$ into ℓ equal parts. Therefore, $\frac{m}{n}$ is the length of the concatenation of these k parts. So we have the following picture:

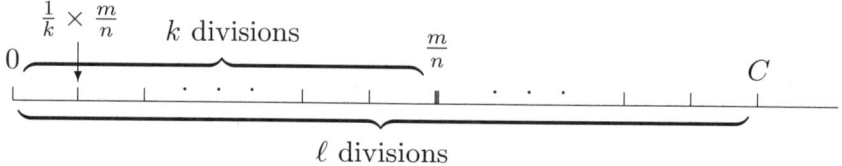

(This picture implicitly assumes $\frac{k}{\ell} < 1$, but as we have seen above, the validity of the reasoning is independent of this assumption.) In the following argument, we will refer to one of the parts in which $[0, C]$ has been divided into ℓ equal parts simply as a *part*. Observe that such a part is also one of the parts when $[0, \frac{m}{n}]$ is divided into k equal parts; see preceding picture. Therefore by (17.7), the length of a part is $\frac{1}{k} \times \frac{m}{n}$. Now recall that $[0, C]$ is ℓ copies of these parts. Therefore,
$$C = \ell \times \left(\frac{1}{k} \times \frac{m}{n}\right) = \frac{\ell}{k} \times \frac{m}{n}.$$
This shows that if a fraction C satisfies $\frac{m}{n} = C \times \frac{k}{\ell}$, then necessarily $C = \frac{\ell}{k} \times \frac{m}{n}$. Hence such a C is unique. It remains to observe simply that this value of C does work: $\frac{m}{n} = (\frac{\ell}{k} \times \frac{m}{n}) \times \frac{k}{\ell}$, i.e., $A = (\frac{\ell}{k} \times \frac{m}{n}) \times B$. The proof is complete. □

The proof of the lemma, either the preceding one or the one in Chapter 17, gives the exact value of the fraction C so that $A = CB$, namely, if $A = \frac{m}{n}$ and $B = \frac{k}{\ell}$, then $C = \frac{\ell}{k} \times \frac{m}{n}$, or, by the commutativity of multiplication,

(18.1) $$C = \frac{m}{n} \times \frac{\ell}{k}.$$

We now expand upon the content of the lemma. It says that if a fraction B is nonzero, then *every* fraction A is a **fractional multiple of B**, in the

18.2. The Definition and Invert-and-Multiply

sense that $A = CB$ for some *fraction* C. Taking $A = 1$, the lemma implies that there is exactly one fraction, which we will denote by $\boldsymbol{B^{-1}}$, so that $B^{-1}B = 1$. We already saw in Chapter 17 that if $B = \frac{k}{\ell}$, then B^{-1} is just the reciprocal of B, i.e., $B^{-1} = \frac{\ell}{k}$. We call this B^{-1} the **inverse** (or **multiplicative inverse**, to be precise) of B. In this notation, the above expression of C may be rewritten as
$$C = AB^{-1}.$$
For example, if $A = \frac{11}{5}$ and $B = \frac{23}{8}$, then the C that satisfies $A = CB$ is
$$AB^{-1} = \frac{11}{5} \times \frac{8}{23} = \frac{88}{115}.$$
We can easily verify that, indeed, $\frac{11}{5} = \frac{88}{115} \times \frac{23}{8}$.

We now give the definition of fraction division. It is, word for word, the same as the preceding definition of whole number division, with the exception that the lemma guarantees that it is not necessary to first assume that A is a fractional multiple of B.

Definition. *If A, B are fractions ($B \neq 0$), the **division of A by B**, or the **quotient of A by B**, denoted by $\frac{A}{B}$, is by definition the fraction so that $A = (\frac{A}{B})B$.*

By the preceding lemma, the fact that there is always a unique such $\frac{A}{B}$ is not in doubt. So the definition is meaningful. Moreover, with $\frac{k}{\ell}$ and $\frac{m}{n}$ given, the equality (18.1) implies that

(18.2) $$\frac{\frac{m}{n}}{\frac{k}{\ell}} = \frac{m}{n} \times \frac{\ell}{k}.$$

This is the famous **invert-and-multiply rule** for the division of fractions. We see that there is nothing mysterious about this rule: it is a simple consequence of the correct definition of division.

Equation (18.2) shows that the definition of $\frac{m}{n} / \frac{k}{\ell}$ is independent of equivalent fractions, in the following sense. If we have equivalent fractions $\frac{m}{n} = \frac{M}{N}$ and $\frac{k}{\ell} = \frac{K}{L}$, then we claim
$$\frac{\frac{m}{n}}{\frac{k}{\ell}} = \frac{\frac{M}{N}}{\frac{K}{L}}.$$
This is because, by (18.2), the left side is the area of a rectangle with sides $\frac{m}{n}$ and $\frac{\ell}{k}$ while the right side is the area of a rectangle with sides $\frac{M}{N}$ and $\frac{L}{K}$. But we know the sides $\frac{m}{n}$ and $\frac{M}{N}$ have the same length (by the definition of the equality of fractions), and since $\frac{\ell}{k} = \frac{L}{K}$ (because of $\frac{k}{\ell} = \frac{K}{L}$ and the cross-multiplication algorithm), the sides $\frac{\ell}{k}$ and $\frac{L}{K}$ also have the same length. Since the areas of rectangles with sides of the same lengths must be equal, the claim is proved.

As before, we note that, for fractions A, B, C, with $B \neq 0$,

the statement "$\frac{A}{B} = C$" is, by definition, equivalent to the statement "$A = CB$".

This is the precise meaning of the statement at the beginning of this section that *division is an alternate, but equivalent, way of writing multiplication.* To be explicit, for the three fractions A, B, C as given ($B \neq 0$), we have both of the following:

$$\text{if } \frac{A}{B} = C, \text{ then } A = CB,$$

$$\text{if } A = CB, \text{ then } \frac{A}{B} = C.$$

Notice that in order for this statement to make sense, we have to first give a precise meaning to the product CB and a precise meaning to the division $\frac{A}{B}$.[1]

In view of the interpretation of fraction multiplication in (17.7) on page 273, we can interpret $\frac{A}{B} = C$ as follows. Because $A = CB$, the discussion below (17.7) implies that $\frac{A}{B}$ is the fraction C so that C copies of B equals A. For example, the fact that

$$\frac{\frac{11}{5}}{\frac{23}{8}} = \frac{88}{115}$$

means that $\frac{88}{115}$ copies of $\frac{23}{8}$ is equal to $\frac{11}{5}$.

Activity. (a) Rewrite each of the following as a division:

$$\frac{14}{27} = \frac{2}{3} \times \frac{7}{9}, \quad 4\frac{8}{15} = 5\frac{2}{3} \times \frac{4}{5}, \quad \frac{a}{b} \times \frac{c}{d} = \frac{ac}{bd},$$

where $a, \ldots, d$ are nonzero whole numbers.

(b) Rewrite each of the following as a multiplication:

$$\frac{\frac{45}{7}}{3} = \frac{15}{7}, \quad \frac{3\frac{5}{8}}{\frac{6}{5}} = \frac{x}{y}, \quad \frac{87}{\frac{x}{y}} = \frac{3}{22},$$

where x, y are nonzero whole numbers.

(c) What is the fraction A in each of the following?

$$\frac{A}{\frac{6}{7}} = \frac{5}{14}, \quad \frac{1\frac{7}{8}}{A} = \frac{5}{2}, \quad \frac{A}{2\frac{4}{5}} = 2\frac{7}{9}.$$

---

[1] In your own classroom, avoid making the statement often found in textbooks and education materials, to the effect that "division is the inverse operation of multiplication". This statement cannot be understood unless you have already precisely defined the multiplication of fractions, the division of fractions, *and* have pointed out that if $\frac{A}{B} = C$, then $A = CB$, and if $A = CB$, then $\frac{A}{B} = C$.

It is time to clear up a subtle point about fraction division. Consider a fraction $\frac{7}{5}$. It now possesses two meanings *as a division*. On the one hand, it is the length of a part when a segment of length 7 is divided into five parts of equal length (see section 15.1, A Different View of a Fraction, on page 236). On the other hand, we look at $\frac{7}{5}$ as the division of the *fraction* 7 by the *fraction* 5 in the sense of the preceding definition (remember that each whole number is also a fraction). Question: Is there any inconsistency between these two meanings? Answer: None. This is because if $\frac{7}{5}$ has the first meaning, then the length of the concatenation of five segments of length $\frac{7}{5}$ is 7. Thus,[2] $5 \times \frac{7}{5} = 7$, or,

$$\frac{7}{1} = \frac{7}{5} \times \frac{5}{1}.$$

This says $\frac{7}{5}$ is the fraction C so that

$$\frac{7}{1} = C \times \frac{5}{1}.$$

In terms of the preceding definition of fraction division, the fraction C is the division of the fraction 7 by the fraction 5. In other words, this C also has the second meaning.

We have the following geometric interpretation of the division of fractions. Because the area of a rectangle is the product of its sides (definition of fraction multiplication), if two fractions A and B are given, with $B \neq 0$, the quotient $\frac{A}{B}$ is the length of one side of a rectangle whose area is A and the other side has length B.

18.3. Applications

We first give an alternate, computational approach to Example 1 in section 18.1. The special case of (18.2) where $m = n = k = 1$ yields, once again, that for any nonzero whole number n,

$$\frac{1}{\frac{1}{n}} = n.$$

The following is a typical application of the concept of fraction division in school mathematics and in the mathematics education literature. Notice the difference between the usual presentation and the one given here: *we give the explicit reason why division has to be used*.

---

[2]See equality (17.3) on page 268.

EXAMPLE. A rod $43\frac{3}{8}$ meters long is cut into pieces which are $\frac{5}{3}$ meters long. How many such pieces can we get out of the rod?

Let A be a fraction so that the rod can be cut into A such pieces, each being $\frac{5}{3}$ meters long. Then the discussion in section 17.3, especially Examples 1–3, implies that the length of the rod is A copies of $\frac{5}{3}$. Hence,

$$43\frac{3}{8} = A \times \frac{5}{3}.$$

By the definition of division, we get

$$A = \frac{43\frac{3}{8}}{\frac{5}{3}} = \frac{\frac{347}{8}}{\frac{5}{3}} = \frac{347}{8} \times \frac{3}{5} = 26\frac{1}{40}.$$

At the risk of harping on the obvious, recall the discussion in section 17.3, which explains why the equality $43\frac{3}{8} = A \times \frac{5}{3}$ means that, with $A = 26\frac{1}{40}$, 26 complete pieces of $\frac{5}{3}$ meters *plus* $\frac{1}{40}$ *of* $\frac{5}{3}$ *meters* add up to $43\frac{3}{8}$ meters. Since each piece is $\frac{5}{3}$ meters long, the leftover piece is $\frac{1}{40} \times \frac{5}{3} = \frac{1}{24}$ meters long.

Activity. A rod $15\frac{5}{7}$ meters long is cut into short pieces which are $2\frac{1}{8}$ meters long. How many short pieces are there?

We next revisit the motion problem first discussed in Example 1 of section 7.1 (see page 100).

We begin with a general definition of an object (or person) moving at *constant speed*. In section 7.1, we defined the motion to be of constant speed v if, once the unit of time is fixed, say an hour, then in any time interval of length equal to one hour, the distance traveled by the object or person is v miles. Now that we have fractions, this definition is no longer adequate because such a description gives no information about, for example, whether in any time interval of $\frac{1}{n}$ hour for any whole number $n > 0$, the distance traveled by the object or person is $\frac{v}{n}$ miles. We need a more sophisticated definition of constant speed.

Fixing a unit of time (seconds, minutes, hours, etc.), let a time interval of length t, be given. Suppose the object travels a total distance of d (cm, meters, km, etc.) during this time interval. Then the object's **average speed in the given time interval** is by definition $\frac{d}{t}$ (cm per second, etc.)

18.3. Applications

Here is an example to reinforce this concept. Suppose one tracks the movement of a car in the first hour; the data are summarized in the following table:

Time Interval	Distance traveled
0 – 10 minutes	11 miles
10 – 20 minutes	5 miles
20 – 30 minutes	9 miles
30 – 40 minutes	7 miles
40 – 50 minutes	6 miles
50 – 60 minutes	10 miles

This example brings out the fact that, for a general motion, it makes no sense to talk about its "average speed" without specifying *in which time interval* we are measuring this average speed. For example, the average speed from 0 to 10 minutes (which is $\frac{1}{6}$ of an hour) is $\frac{11}{1/6} = 66$ mph, but the average speed from 10 to 20 minutes is, $\frac{5}{1/6} = 30$ mph. We can go on. The average speed from 0 to 20 minutes is, $\frac{11+5}{2/6} = 48$ mph. The average speed from 10 to 40 minutes is $\frac{5+9+7}{3/6} = 42$ mph, and the average speed from 20 to 50 minutes is $\frac{9+7+6}{3/6} = 44$ mph. The average speed during the first hour (from 0 to 60 minutes) is 48 mph because

$$\frac{11+5+9+7+6+10}{1} = 48 \text{ mph.}$$

On the other hand, if we consider only the average speeds during the first three consecutive 20-minute intervals, then they are all 48 mph, because they are, respectively,

$$\frac{11+5}{\frac{20}{60}}, \quad \frac{9+7}{\frac{20}{60}}, \quad \frac{6+10}{\frac{20}{60}}.$$

Therefore, if we only look at the first three consecutive 20-minute intervals, we may be misled into believing that this is a motion "of constant speed".

In general, we say the object travels at **constant speed** v if the average speed $\frac{d}{t}$ in *any* time interval of length t, for *any* t, is always equal to the fixed constant v. (We assume that d, t, v are all fractions; see the discussion of FASM in Chapter 21 below.) This number v is then called **the speed of the motion**.

We should point out that the main weight of the definition of constant speed lies in the presence of the word "any". For example, the motion described in the preceding table is not one of constant speed because, although the average speeds in the time intervals from 0 to 20 minutes, 20 to 40 minutes, 40 to 60 minutes, and from 0 to 60 minutes are all equal to 48 mph, the fact that the average speed in the single time interval from 0 to 10 minutes being 66 mph is sufficient to show that 48 mph is *not* the average speed of the motion in any time interval. On the other hand, if we know that a motion of an object traveling for 2 hours is of constant speed, say 48 mph, then we have the assurance that its average speed in any time interval, be it from 1 to 12 minutes, from 0 to 2 hours, from 63 to 68 minutes, or from 77 to 111 minutes, it will *all* be 48 mph. In solving word problems, we will consistently make use of the assumption of constant speed in exactly this fashion.

In particular, a motion of constant speed v ft/sec travels exactly v ft in any time interval of 1 sec because, suppose in a given 1-sec time interval it travels s miles. Then its average speed in this time interval is $\frac{s}{1} = s$ ft/sec. Since we are assuming a constant speed of v ft/sec, the average speed s must equal v. Therefore, in the given time interval of 1 sec, the object travels v ft.

> *We call attention to the fact that the terminology of "average speed" unfortunately coincides with one from daily life, and it is easy, upon seeing this phrase to forget the mathematical definition and simply fall back on the common understanding of the word "average", i.e., add two speeds and divide by 2. Please make an effort to remember that you are now dealing with a precise mathematical concept and must therefore use it precisely according to its definition. In addition, we emphasize once more that the phrase "average speed", by itself, has no meaning because it only makes sense to talk about the average speed in a fixed time interval. The complete concept is "the average speed in a given time interval". It is only when the speed is constant that we can talk about "average speed" without mentioning the time interval in question, because in that case the average speed is the same regardless what time interval it is.*

We proceed to reconcile this definition of constant speed with the preliminary one given in section 7.1 by showing that a motion of constant speed in the present sense is also a motion of constant speed in the sense of section 7.1. This is easy: if we have a motion of constant speed v mph in the present sense, then in any one hour duration, the above reasoning shows that the distance traveled is v miles, as required by the definition in section 7.1.

More can be said, however. If we have a motion of constant speed in the sense just defined, then given a time interval of length T, let D be the distance traveled in this time interval. By definition of constant speed, we have $\frac{D}{T} = v$, where v is the speed. By the definition of division, we know that this is equivalent to $D = vT$, which is the same as $D = Tv$. This says

> *A motion of constant speed v travels the same distance Tv in any time interval of a given length T.*

18.3. Applications

If we think of v as the number of miles the object travels in one hour, then in T hours, then equality $D = Tv$ says that the total distance traveled is T copies of v. (See the discussion of fraction multiplication in section 17.3.)

The definition of average speed requires the division of a distance d by a length of time t. Since we only divide numbers on the same number line, d and t have to be put on the same number line before the division can be performed. As in the case of multiplication (see the beginning of section 17.1), *we identify the unit of distance with the unit of time to bring d and t to the same number line*, so that 1 can be one unit of time or one unit of distance. Such identifications will become increasingly common in Chapters 22 and 23.

In the school mathematics literature, one hardly ever finds a precise definition of the *constancy* of speed, so that all problems related to motion are done either by rote, by so-called common sense, by so-called unit rate, or by so-called proportional reasoning, but not by mathematical reasoning. *It is not good education when students are asked to do problems of motion without being taught what they need to know in the first place.* In this book, we will illustrate how to use precise definitions and mathematical reasoning to solve not only motion problems but all rate problems as well; see Chapter 22 and Chapter 23.

EXAMPLE. Suppose we embark on a full-day hike to the beach starting from park headquarters. The Bear Valley Trail is $12\frac{1}{3}$ miles. When we start off in the morning, we count on our ability to maintain a brisk pace of a constant speed of $3\frac{1}{2}$ miles an hour. On this basis, how long would it take us to get to the beach?

We will proceed in two ways.

First the naive approach. Each $3\frac{1}{2}$ miles takes an hour, so 7 miles will take two hours and $10\frac{1}{2}$ miles will take 3 hours. There are still $1\frac{5}{6}$ miles left in the hike ($1\frac{5}{6} = 12\frac{1}{3} - 10\frac{1}{2}$). Now $1\frac{5}{6}$ is roughly half of $3\frac{1}{2}$ because

$$2 \times \left(1\frac{5}{6}\right) = 2 \times \frac{11}{6} = \frac{11}{3} = 3\frac{2}{3},$$

which is roughly $3\frac{1}{2}$. Therefore if $3\frac{1}{2}$ miles takes 1 hour to walk, the constancy of the speed suggests that $1\frac{5}{6}$ miles will take roughly half an hour to walk. The time it takes to walk the $12\frac{1}{3}$ miles to get to the beach is then roughly $3 + \frac{1}{2} = 3\frac{1}{2}$ hours. This is as far as the naive approach can go.

Next, the precise approach. We will obtain the same result by more fully exploiting the given assumption of constant speed. Suppose it takes T hours to get to the beach. Then from $D = Tv$, we see that

$$12\frac{1}{3} = T \times 3\frac{1}{2}.$$

By the definition of division,
$$T = \frac{12\frac{1}{3}}{3\frac{1}{2}}.$$
Then we can invert-and-multiply to obtain $m = 3\frac{11}{21}$ so that it takes $3\frac{11}{21}$ hours to get to the beach. (Incidentally, $3\frac{11}{21}$ is close to $3\frac{1}{2}$.)

This reasoning is perfectly general[3] so that the second approach explains why,

> *if a certain motion has constant speed v, then the* total time *it takes to cover a distance of d (miles, km, etc.) is* $\frac{d}{v}$ *(hours, minutes, etc.)*

This is the standard formula for computing the time to travel a given distance.

Activity. A train running at constant speed takes $2\frac{1}{3}$ hours to go 125 miles. At the same speed, how long does it take to go 180 miles? (You do not need to "set up a proportion" to do this problem. If you insist on setting up a proportion, then you would be setting yourself up for the impossible task of explaining what it means.)

18.4. Comments on the Division of Decimals

We will prove that *all divisions of decimals are reduced to the division of whole numbers*. This is the easy part. The much more subtle part is to show how the division of whole numbers, which, by Chapter 15 and the discussion at the end of section 18.2, is just a fraction, leads to a decimal. Thus, we will have to deal with the conversion of a fraction to a decimal.

The explanation of the first part is best done by an example. We will show how the division
$$\frac{2.0498}{14.3}$$
is equal to the division of two whole numbers. By the FFFP for decimals (see section 13.4), we may rewrite it as $\frac{2.0498}{14.3000}$, and a simple application of invert-and-multiply and the cancellation phenomenon (17.5) then leads to a division of whole numbers:

$$\frac{2.0498}{14.3000} = \frac{\frac{20498}{10000}}{\frac{143000}{10000}} = \frac{20498}{10000} \times \frac{10000}{143000} = \frac{20498}{143000}.$$

The reasoning is perfectly general, so that a division of decimals is always equal to a fraction.

---

[3]Remember that although we are assuming all the numbers are fractions, we are invoking FASM of Chapter 21.

18.4. Comments on the Division of Decimals

We will now show, albeit imperfectly, how to get the decimal equal to a given fraction. For this purpose, it suffices to deal with *proper* fractions because, by use of division-with-remainder, an improper fraction is always the sum of a whole number with a proper fraction, e.g.,

$$\frac{51}{8} = \frac{(6 \times 8) + 3}{8} = 6 + \frac{3}{8}.$$

So if we can convert the proper fraction $\frac{3}{8}$ to a decimal (which we can, by section 13.2, and it is 0.375), then we can deal with $\frac{51}{8}$, namely,

$$\frac{51}{8} = 6 + 0.375 = 6.000 + 0.375 = 6.375.$$

An arbitrary improper fraction can be handled the same way.

From now on, we only deal with proper fractions. *Let this be understood in the remainder of this section.* We will give a partial explanation of why the *traditional method* of converting a fraction to a decimal is correct. We recall from Chapter 13 the description of this method in three steps:

(i) Add any number of zeros to the right of the numerator and then divide the resulting number by the denominator;

(ii) Obtain a decimal by the judicious placement of a decimal point in the quotient;

(iii) Equate this decimal with the original fraction.

The explanation we give below will only be a *partial* one because we will only touch lightly on the case of an infinite decimal. The explanation of what an infinite decimal means will be reserved for Chapter 41 in Part 5. We will be satisfied instead with "getting as many decimal digits as we want" in the general case.

As in Chapter 13, we first only consider *those fractions whose denominators are a product of 2's and 5's.*[4] We will show how the process of long division yields a finite decimal equal to a given fraction $\frac{m}{n}$ satisfying the above condition. First we recall from Chapter 12 what a finite decimal is:[5] it is a fraction with denominator equal to a power of 10. In order to get a fraction with a denominator equal to a power of 10 in this context, we use the direct approach: the cancellation phenomenon illustrated in equation (17.5) on page 177 implies that for *any* whole number k,

$$\frac{m}{n} = \left(\frac{m \cdot 10^k}{n}\right) \times \frac{1}{10^k}.$$

All we have to do now is prove that

---

[4] It is understood that not both 2 and 5 need be present at the same time; see Theorem 36.2 on page 479.

[5] Yes, it is always necessary to know your definitions.

(1) the fraction $\frac{m \cdot 10^k}{n}$ is equal to a whole number q for a sufficiently large k

because once it is done, we have

$$\frac{m}{n} = \left(\frac{m \cdot 10^k}{n}\right) \times \frac{1}{10^k} = q \times \frac{1}{10^k} = \frac{q}{10^k},$$

and $\frac{q}{10^k}$ is, by definition, a finite decimal.

Postponing for the moment the proof of the assertion (1), we illustrate this method of conversion with a simple fraction $\frac{3}{8}$. With $m = 3$ and $n = 8$, we see that $8 = 2^3$, and this power 3 suggests that we take $k = 3$ (or greater). Then the long division algorithm gives,

$$\frac{3 \times 10^3}{8} = \frac{3000}{8} = 375$$

```
         3 7 5
     8 ) 3 0 0 0
         2 4
         ─────
           6 0
           5 6
           ─────
             4 0
             4 0
             ─────
               0
```

Therefore,

$$\frac{3}{8} = \left(\frac{3 \times 10^3}{8}\right) \times \frac{1}{10^3} = 375 \times \frac{1}{10^3} = \frac{375}{10^3},$$

which, *by definition*, is 0.375. Perhaps some would consider this proof of $\frac{3}{8} = 0.375$ to be more comforting than the one we gave back in section 13.2.

Activity. (a) Why could we *not* give this explanation back in Chapter 13? (b) Discuss what would happen to the preceding conversion of $\frac{3}{8} = 0.375$ if the number k (the power of 10) had been chosen to be 2, or 4, or 5.

Another example: let $\frac{m}{n} = \frac{15}{32}$. Then we see that $32 = 2^5$, and the power 5 of 2^5 suggests taking $k = 5$ (or greater). Long division yields:

18.4. Comments on the Division of Decimals

$$\frac{15 \times 10^5}{32} = \frac{1500000}{32} = 46875$$

```
              4 6 8 7 5
        32 ) 1 5 0 0 0 0 0
              1 2 8
                2 2 0
                1 9 2
                  2 8 0
                  2 5 6
                    2 4 0
                    2 2 4
                      1 6 0
                      1 6 0
                            0
```

Therefore,
$$\frac{15}{32} = \left(\frac{15 \times 10^5}{32}\right) \times \frac{1}{10^5} = \frac{46875}{10^5} = 0.46875.$$

It remains to prove assertion (1), i.e., explain why, if n is a whole number equal to a product of 2's or 5's or both, and m is any whole number, then for a sufficiently large whole number k, $\frac{m \cdot 10^k}{n}$ is a whole number. In fact the whole number m is irrelevant, and we will prove something stronger:

If n is a whole number equal to a product of 2's or 5's or both, then for a sufficiently large whole number k, $\frac{10^k}{n}$ is a whole number.

A clue to the proof of this assertion can be obtained by looking at the two preceding examples to see directly why

$$\frac{10^3}{8} \quad \text{and} \quad \frac{10^5}{32}$$

are whole numbers. Observe that $10 = 2 \cdot 5$, so that for any whole number k, we have
$$10^k = 2^k \cdot 5^k.$$

Therefore, by virtue of the cancellation phenomenon (see Chapter 17), we have:
$$\frac{10^3}{8} = \frac{2^3 \cdot 5^3}{2^3} = 5^3, \quad \text{and} \quad \frac{10^5}{32} = \frac{2^5 \cdot 5^5}{2^5} = 5^5,$$

and of course 5^3 and 5^5 are both whole numbers. We see that we had to choose k to be 3 (or bigger) in the case of 8 because 8 is the *third* power of 2, and we had to choose $k = 5$ (or bigger) because 32 is the *fifth* power of 2.

It is now clear how to prove the assertion in general: Let $n = 2^a \cdot 5^b$, where a and b are whole numbers. If $a \leq b$, we choose $k = b$ so that

$$\frac{10^b}{n} = \frac{2^b \cdot 5^b}{2^a \cdot 5^b} = \frac{2^b}{2^a}.$$

Since $a \le b$, there are at least as many 2's in the numerator as in the denominator so that, after cancellation, $\frac{10^b}{n}$ is either equal to 1 or a product of 2's, which is a whole number. On the other hand, if $a > b$, we choose $k = a$ so that
$$\frac{10^a}{n} = \frac{2^a \cdot 5^a}{2^a \cdot 5^b} = \frac{5^a}{5^b}.$$
Again, since $a > b$, there are more 5's in the numerator than in the denominator, and $\frac{10^b}{n}$ is a product of 5's (there are $a - b$ of them). In either case, we have proved the assertion and, therewith, also the following theorem.

Theorem 18.1. *If the denominator n of $\frac{m}{n}$ is a product of only 2's or 5's, or both, then for a sufficiently large whole number k, the division of $m \cdot 10^k$ by n is a whole number q, and $\frac{m}{n}$ is equal to the finite decimal $\frac{q}{10^k}$.*

As we saw from the preceding reasoning, the number k in Theorem 18.1 only needs to be as big as the total numbers of factors of 2 in n, and also as big as the total number of factors of 5 in n.

To bring closure to this discussion, is worthwhile to point out that if we further assume the fraction $\frac{m}{n}$ to be reduced, then the condition that n be a product of only 2's and 5's is not only sufficient to guarantee that $\frac{m}{n}$ be equal to a finite decimal, but also *necessary*. For this, see Theorem 36.2 on page 479.

It remains to give a brief discussion of the conversion of a fraction $\frac{m}{n}$ to a decimal when its denominator n has a prime factor[6] which is not 2 or 5. In view of the theorem on equivalent fractions (Theorem 13.1 on page 204), we note right away that, for a general fraction which is not necessarily reduced, the statement that "its denominator has a prime factor which is not 2 or 5" carries almost no information. This can be seen from a simple example. Certainly the denominator of the fraction $\frac{3}{30}$ contains the prime factor 3, which is neither 2 nor 5, but since $\frac{3}{30} = \frac{1}{10}$, we may also consider it as a fraction whose denominator is a product of 2's and 5's. In fact, we already know that $\frac{3}{30} = \frac{1}{10} = 0.1$. For this reason, *we shall require all fractions in the remainder of this section to be* reduced.

The preceding paragraph implies that if, in a reduced fraction $\frac{m}{n}$, the denominator contains a prime factor other than 2 and 5, then $\frac{m}{n}$ is not equal to a finite decimal, but is equal to an *infinite* decimal; see Chapter 41. We will not pause to explain what an infinite decimal is, but see Chapter 41 for an informal discussion. Taking for granted the intuitive meaning of the *decimal expansion* of a fraction, however, we will get the first seven decimal digits of the decimal expansion of $\frac{2}{7}$. It will be seen that this example

---

[6]We assume a general familiarity with the concept of prime numbers at this point for the purpose of the discussion, but we will not use it in any essential way; see Chapter 33 for a fuller discussion.

18.4. Comments on the Division of Decimals

exhibits all facets of the general case, no matter how many decimal digits we care to get.

As before, we always have
$$\frac{2}{7} = \left(\frac{2 \times 10^7}{7}\right) \times \frac{1}{10^7} = \frac{20000000}{7} \times \frac{1}{10^7}.$$

We perform the long division of 20000000 by 7 and obtain the quotient and remainder:

$$20000000 = (2857142 \times 7) + 6$$

```
            2 8 5 7 1 4 2
        _____
    7 ) 2 0 0 0 0 0 0 0
        1 4
        ___
          6 0
          5 6
          ___
            4 0
            3 5
            ___
              5 0
              4 9
              ___
                1 0
                  7
                ___
                  3 0
                  2 8
                  ___
                    2 0
                    1 4
                    ___
                      6
```

Therefore,
$$\frac{2}{7} = \left(\frac{(2857142 \times 7) + 6}{7}\right) \times \frac{1}{10^7}$$
$$= (2857142 + \frac{6}{7}) \times \frac{1}{10^7}$$
$$= \frac{2857142}{10^7} + \left(\frac{6}{7} \times \frac{1}{10^7}\right) \quad \text{(distributive law)}$$

so that using the definition of the decimal notation, we get
$$\frac{2}{7} = 0.2857142 + \left(\frac{6}{7} \times \frac{1}{10^7}\right).$$

In this case, the finite decimal 0.2857142 is called **the 7-digit decimal expansion of** $\frac{2}{7}$.

Three observations are relevant here. The first is that we chose 10^7 to get the division-with-remainder of 2×10^7 by 7 because we decided ahead of time to get 7 decimal digits of the decimal expansion of $\frac{2}{7}$. However, if we are interested in getting, say, 13 decimal digits of the expansion, then we would choose 10^{13} to get the division-with-remainder of 2×10^{13} by 7. Second, the six digits 285714 are going to repeat ad infinitum if we continue with the long division because, in the above schematic long division on the right, the two bold-faced 2-by-2 blocks in the upper left and lower right corners are identical (see Chapter 42 for a more thorough discussion). The repetition of the digits in the quotient is therefore inevitable. Finally, the decimal 0.2857142 is a very good approximation to $\frac{2}{7}$ because their difference can be easily estimated:

$$\frac{2}{7} - 0.2857142 \;=\; \frac{6}{7} \times \frac{1}{10^7} \;<\; 1 \times \frac{1}{10^7},$$

and therefore

(18.3) $$0 \;<\; \frac{2}{7} - 0.2857142 \;<\; \frac{1}{10^7}.$$

This means if we take only the decimal 0.2857142 to represent $\frac{2}{7}$, the error is at most only as big as $1/10^7$. If we want the 13-digit decimal expansion of $\frac{2}{7}$, we would get, likewise, that the difference between $\frac{2}{7}$ and this 13-digit decimal is less than $\frac{1}{10^{13}}$. In any case, *the more digits you get in the decimal expansion of a fraction, the smaller the difference between the fraction $\frac{2}{7}$ and the resulting expansion.* In the school classroom, this knowledge may be even more important than any superficial explanation of what an infinite decimal is.

Note that we have just given a decimal estimation of $\frac{2}{7}$, namely, 0.2857142 together with the error of this estimation in the inequalities in (18.3) (see Chapter 10).

18.5. Inequalities

We have seen that if ℓ and n are nonzero whole numbers and $\ell < n$, then $\frac{1}{\ell} > \frac{1}{n}$, (see, for example, Example 4 on page 243). It turns out that this also holds if ℓ and n are fractions. Precisely,

If A and B are nonzero fractions and $A < B$, then $\frac{1}{A} > \frac{1}{B}$.

The proof is simple. Let $A = \frac{k}{\ell}$ and $B = \frac{m}{n}$. By the cross-multiplication algorithm (Theorem 15.1 on page 240), $A < B$ if and only if $kn < \ell m$. On the only hand, $\frac{1}{A} = \frac{\ell}{k}$ and $\frac{1}{B} = \frac{n}{m}$. Therefore, using the cross-multiplication algorithm once more, we have

$$A < B \text{ implies } kn < \ell m, \text{ implies } \tfrac{n}{m} < \tfrac{\ell}{k}, \text{ implies } \tfrac{1}{B} < \tfrac{1}{A}.$$

The assertion is therefore proved.

18.6. False Doctrines

Through the years, the division of fractions has perhaps received more defective treatments than any other topic in the mathematics of elementary school. It is vitally important that you as a teacher recognize not only what is right, but also what is wrong, in order to give correct guidance to your students. With this in mind, we proceed to discuss the three most common mistreatments of fraction division.

The first advocates that one way to achieve conceptual understanding of the division of fractions is by regarding it as *repeated subtraction*. This kind of thinking is likely the result of confusing the meaning of division among whole numbers with division-with-remainder (see the introduction of Chapter 7 on page 95). A popular item used to illustrate the *repeated-subtraction* approach to fraction division is the following:

> If 5 yards of ribbon are cut into pieces that are $\frac{3}{4}$ yards long to make bows, how many bows can be made?

The proposed solution subtracts $\frac{3}{4}$ from 5 repeatedly, until after the sixth time, what remains is

$$5 - \underbrace{\frac{3}{4} - \cdots - \frac{3}{4}}_{6} = \frac{1}{2}.$$

Since $\frac{1}{2} < \frac{3}{4}$, no further subtraction is possible. The conclusion is that six bows can be made from the ribbon, with $\frac{1}{2}$ yard left over. Observe that $\frac{1}{2}$ is $\frac{2}{3}$ of $\frac{3}{4}$ (which, one presumes, can be figured out by guess-and-check), so the leftover piece is equal to $\frac{2}{3}$ of a bow.

There are at least two criticisms of this approach to the division of fractions. First, it ignores the fact that division, as an arithmetic operation, must associate with two numbers another number (e.g., it associates with 36 and 9 the number 4). But the preceding approach to the division of 5 by $\frac{3}{4}$ associates 5 and $\frac{3}{4}$ *a pair* of numbers, namely, the quotient 6 and the remainder $\frac{1}{2}$, because

$$5 = \left(6 \times \frac{3}{4}\right) + \frac{1}{2}.$$

So conceptually this is not the right way to do division. A second criticism of the above repeated-subtraction method is that it says nothing about the division of fractions being an alternate way of expressing multiplication. The latter view of division is the one that makes sense in mathematics in general. The repeated-subtraction interpretation of division is used only for division-with-remainder among whole numbers.[7]

---

[7]Or in a Euclidean domain, where division has to be interpreted as a generalized form of division-with-remainder.

For a simple problem such as the preceding one about cutting a ribbon, the deficiency of the method of treating division as repeated-subtraction is likely to be hidden from the general view. Consider, however, the following problem:

How many $\frac{5}{3}$'s are there in $2543\frac{3}{8}$?

Even if you only want an approximate answer by using repeated subtraction, the number of repeated subtractions needed for this purpose would make the problem entirely inappropriate for elementary students. Worse, the repeated-subtraction mechanism totally breaks down in case of problems of the following type:

How many $\frac{5}{3}$'s are there in $\frac{1}{8}$?

Needless to say, both of these problems yield immediately to invert-and-multiply, and we have seen how invert-and-multiply is a simple consequence of the correct definition of fraction division. For a fuller discussion, see [**Wu05**].

A second approach to division of fractions that we shall discuss goes as follows. We try to find out what $\frac{m/n}{k/\ell}$ could mean. Using the theorem on equivalent fractions (see Theorem 13.1 on page 204), we get

$$\frac{\frac{m}{n}}{\frac{k}{\ell}} = \frac{\frac{m}{n} \times \ell n}{\frac{k}{\ell} \times \ell n} = \frac{m\ell}{nk},$$

and therefore (so the argument goes) it is a valid mathematical fact that

$$\frac{\frac{m}{n}}{\frac{k}{\ell}} = \frac{m\ell}{nk}.$$

This is then the invert-and-multiply rule.

Now the conclusion is superficially consistent with (18.2) as both seem to prove that the invert-and-multiply rule is correct, but the flaws in this approach are subtle. First of all, it does not explain what *division* means but only gives an answer, namely, $\frac{m\ell}{nk}$. This is a prime example of teaching by rote. Next, if we look carefully at the first step,

(18.4) $$\frac{\frac{m}{n}}{\frac{k}{\ell}} = \frac{\frac{m}{n} \times \ell n}{\frac{k}{\ell} \times \ell n},$$

then we realize that it does not make any sense because we do not as yet know what it means to divide fractions and, therefore, we do not know the meaning of either side of (18.4). To assert the equality of two things that we know nothing about is thus pure fantasy and not mathematics. This is the mathematical equivalent of the statement that "we saw two unicorns, one taller than the other, but after we measured their weights we found them to be the same". How do you measure the weights of two unicorns when there aren't any unicorns around? How can you decide that the division of one

18.6. False Doctrines

pair of fractions equals the division of another pair when we do not know how to divide any fractions?

There is another issue that is commonly overlooked: the reason for the equality in (18.4) is usually given in the literature as "using equivalent fractions", but all we know about equivalent fractions (Theorem 13.1 on page 204) is that $\frac{K}{L} = \frac{KM}{LM}$ where K, L, M are *whole numbers*, not fractions. Therefore the first step of such an argument, e.g., the equality (18.4), is seen to depend on something which is not yet known to be true. Such a presentation of the division of fractions is consequently misleading in multiple ways.

Note that the preceding argument can be rephrased as a heuristic one that shows that, under reasonable assumptions on the behavior of fractions, the fraction $\frac{m\ell}{nk}$ is the only reasonable answer to the given division of fractions. To complete the picture, it remains to point out that the application of so-called "equivalent fractions" in (18.4) will in fact be proved to be valid; see rule (b) on page 310. However, the proof of rule (b) requires a correct definition of the division of fractions as well as the validity of invert-and-multiply, whereas the goal of (18.4) is to prove invert-and-multiply!

A third approach is a variant of the second one, and it goes as follows. Again we try to find out what a division such as $\frac{m/n}{k/\ell}$ means, so we use equivalent fractions:

$$\frac{\frac{m}{n}}{\frac{k}{\ell}} = \frac{\frac{m}{n} \times 1}{\frac{k}{\ell} \times 1} = \frac{\frac{m}{n} \times \frac{\ell}{\ell}}{\frac{k}{\ell} \times \frac{n}{n}} = \frac{\frac{m\ell}{n\ell}}{\frac{nk}{n\ell}} = \frac{m\ell \times \frac{1}{\ell n}}{nk \times \frac{1}{\ell n}},$$

and thus the division $\frac{m/n}{k/\ell}$ becomes the division of $m\ell$ of the *new unit* $\frac{1}{\ell n}$ divided by nk of the same new unit. Naturally, the result is $\frac{m\ell}{nk}$. Therefore,

$$\frac{\frac{m}{n}}{\frac{k}{\ell}} = \frac{m\ell}{nk}.$$

The multiple logical defects of such an explanation of division are similar to those in the second approach.

Exercises

1. Recall that whole numbers are fractions. Now express the following as ordinary fractions for nonzero whole numbers a, b, and c: $a/\frac{b}{c}$, $\frac{a}{b}/c$, $12/\frac{4}{5}$, $\frac{4}{5}/8$, $5/\frac{1}{7}$, $\frac{1}{7}/5$.

2. Analyze the third approach to the definition of division of fractions in section 18.6 on **False Doctrines**, describe what the difficulties are, and make suggestions on how to make it into a heuristic argument for the correct definition of fraction division.

3. (a) How many $\frac{5}{13}$'s are there in $83\frac{2}{7}$? (b) How many blocks of 11 minutes are there in $7\frac{1}{2}$ hours? Calculate it in terms of minutes, and then calculate it in terms of hours. Compare.

4. Rosa walked $15\frac{2}{3}$ miles in $5\frac{1}{2}$ hours. What was her average speed? Assuming that she walked at constant speed, how many miles did she walk in each hour?

5. (a) A pizza parlor has a Learning Fractions Special.[8] Normally, it charges $\frac{m}{n} \times 8$ dollars for $\frac{m}{n}$ of a small pizza. During this special sale, it charges for each $\frac{1}{5}$ of a pizza the normal price of $\frac{1}{6}$ of a pizza. At the sales price, how much would $2\frac{2}{3}$ small pizzas cost? (b) If $\frac{4}{7}$ of a sack of rice is $4\frac{2}{3}$ the weight of 5 books, and if each book weighs $1\frac{1}{2}$ lbs., how much (in lbs.) does a sack of rice weigh?

6. (Sixth-grade Japanese exam question.) A train 132 meters long travels at 87 kilometers per hour and another train 118 meters long travels at 93 kilometers per hour. Both trains are traveling in the same direction on parallel tracks. How many seconds does it take from the time the front of the locomotive of the faster train reaches the end of the slower train to the time that the end of the faster train reaches the front of the locomotive on the slower one?

7. (a) Find a fraction q so that $28\frac{1}{2} = q \times 5\frac{3}{4}$. (b) Do the same for $218\frac{1}{7} = q \times 20\frac{1}{2}$. (c) Make up a word problem for each of (7a) and (7b).

8. It takes 2 tablespoons of a chemical to dechlorinate 120 gallons of water. Given that 3 *tea*spoons make up a tablespoon, how many *tea*spoons of this chemical are needed to dechlorinate 43 gallons? (Assume that if it takes n tablespoons of this chemical to dechlorinate w gallons of water, then $\frac{w}{n}$ is always a fixed constant no matter what n or w may be.)

9. Shawna used to spend $\frac{2}{3}$ of an hour driving to work. Now that her firm has moved further away, she spends $\frac{5}{6}$ of an hour driving to work at the

---

[8]I got this idea from my friend David Collins. We believe that if all pizza parlors buy into this idea, the national fractions achievement will improve.

same speed. If she drives 12 miles more nowadays, how far is her firm from her home?

10. Sam drove from Town A to Town B at 60 mph and drove back along the same freeway at 70 mph. If the distance between the towns is 73 miles, what was Sam's average speed for the round trip?

11. Convert the following fractions to decimals: (a) $\frac{7}{16}$, (b) $\frac{42}{125}$, (c) $\frac{9}{64}$, (d) $\frac{17}{625}$.

12. (a) Get the 7-digit decimal expansion of $\frac{3}{11}$. Can you estimate by how much this finite decimal differs from $\frac{3}{11}$? (b) Get the 6-digit decimal expansion of $\frac{2}{13}$, and estimate by how much this finite decimal differs from $\frac{2}{13}$.

13. The **perimeter** of a rectangle is by definition the sum of the lengths of its four sides. (a) Exhibit a rectangle with perimeter 20 cm and area $< \frac{1}{10^3}$ sq cm. (b) Can any rectangle with perimeter 20 cm have area > 25 sq cm? Why? (c) Show that given a number A and a number L, there is a rectangle with area equal to A sq cm, but with a perimeter that is bigger than L cm.

14. (a) Show that there is a rectangle with area < 1 sq cm and *perimeter* (see Exercise 13 for a definition) equal to 100 cm. (b) Given a number A and a number L, show that there is a rectangle with perimeter equal to L cm but with an area smaller than A sq cm.

15. Let a, d be whole numbers, and let q and r be the quotient and remainder of a divided by d. Also let Q be the fraction so that $a = Qd$. Determine the relationship among Q, q, and r. (*Caution*: Be very careful.)

16. (a) $\frac{1}{\frac{1}{2}(\frac{1}{3}+\frac{1}{4})} = ?$ $\frac{1}{\frac{1}{2}(\frac{1}{2/3}+\frac{1}{5/4})} = ?$ (b) If x, y are nonzero fractions, what is $\frac{1}{\frac{1}{2}(\frac{1}{x}+\frac{1}{y})}$? This expression for x and y turns up often enough to merit a name—the **harmonic mean** of x and y. (c) If x, y, u, v are nonzero fractions so that $x < u$ and $y < v$, prove that
$$\frac{xy}{x+y} < \frac{uv}{u+v}.$$

17. If $\frac{8}{11}$ of a number N exceeds half of N by 70, what is N? (a) Solve it by using the number line. (b) Can you do it a second way?

Chapter 19

Complex Fractions

We now come to a central topic in the study of fractions: the arithmetic operations of **complex fractions**,[1] which are by definition quotients of fractions, $\frac{a}{b}/\frac{c}{d}$, where $\frac{a}{b}$ and $\frac{c}{d}$ are fractions with $\frac{c}{d} \neq 0$. Since the preceding chapter already tells us that such a quotient $\frac{a}{b}/\frac{c}{d}$ is nothing other than the fraction $\frac{ad}{bc}$, in one sense, there is nothing more to be said about complex fractions beyond what we have already proved about fractions. Indeed, this topic is generally not even mentioned in most school textbooks or professional development materials on fractions. On the other hand, it is common in pre-algebra mathematics, and absolutely essential in algebra, to compute with complex fractions $\frac{A}{B}$, where A, B now stand for fractions, *as if A and B were whole numbers,* i.e., as if $\frac{A}{B}$ were an ordinary fraction with a whole-number numerator A and a whole-number denominator B. Insofar as such computations cannot be undertaken without justification, the purpose of this section is to explicitly call attention to these computational procedures and (of course) to provide proofs for them. Such procedures are central to any computation that arises from word problems, and you will verify to your own satisfaction that formulas (a)–(f) in section 19.1 are absolutely indispensable in the remainder of Part 2.

The sections are as follows:

The Basic Skills

Why Are Complex Fractions Important?

---

[1] This is strictly the terminology of school mathematics, and has nothing to do with *complex numbers*.

19.1. The Basic Skills

Writing A for $\frac{a}{b}$ and C for $\frac{c}{d}$, we can abbreviate the complex fraction $\frac{a/b}{c/d}$ to $\frac{A}{C}$. If A and C were whole numbers, then $\frac{A}{C}$ would be a fraction and we could apply our knowledge about the arithmetic operations of fractions to $\frac{A}{C}$. Naturally, we wish to know how much of this knowledge of $\frac{A}{C}$ when A and C are whole numbers carries over to the case when A and C are fractions. The answer: *virtually everything*. This will therefore be an impressive display of the power of symbolic notation, in that it makes possible a tremendous saving of mental energy by encoding two parallel developments using only one set of formulas.

In anticipation of the fact that a complex fraction $\frac{A}{C}$ will behave very much like an ordinary fraction, it is common to continue to call A the **numerator** and C the **denominator** of $\frac{A}{C}$.

We proceed to list the basic rules concerning complex fractions. Let $A, B, \ldots, F$ be fractions (which will be assumed to be nonzero in the event any of them appears in the denominator). We will omit the multiplication symbol "$\times$" between letters unless the circumstance calls for it. Thus $A \times B$ will be simply written as AB. With this understood, then the following set of rules are valid.

(a) $A \times \dfrac{B}{C} = \dfrac{AB}{C}$.

    Example: $\dfrac{22}{27} \times \dfrac{\frac{3}{5}}{\frac{13}{8}} = \dfrac{\left(\frac{22}{27}\right) \times \left(\frac{3}{5}\right)}{\frac{13}{8}}$.

(b) Cancellation law: If $C \neq 0$, then $\dfrac{AC}{BC} = \dfrac{A}{B}$.

    Example: $\dfrac{\frac{16}{5} \times \frac{7}{17}}{\frac{2}{3} \times \frac{7}{17}} = \dfrac{\frac{16}{5}}{\frac{2}{3}}$.

(c) $\dfrac{A}{B} = \dfrac{C}{D}$ if and only if $AD = BC$,

    $\dfrac{A}{B} < \dfrac{C}{D}$ if and only if $AD < BC$.

    Example: $\dfrac{\frac{4}{5}}{\frac{2}{3}} < \dfrac{\frac{13}{2}}{\frac{16}{3}}$ because $\dfrac{4}{5} \times \dfrac{16}{3} < \dfrac{2}{3} \times \dfrac{13}{2}$, and conversely.

(d) $\dfrac{A}{B} \pm \dfrac{C}{D} = \dfrac{(AD) \pm (BC)}{BD}$.

    Example: $\dfrac{0.3}{2.5} + \dfrac{5}{1.7} = \dfrac{(0.3 \times 1.7) + (2.5 \times 5)}{2.5 \times 1.7}$.

(e) $\frac{A}{B} \times \frac{C}{D} = \frac{AC}{BD}$.

Example: $\frac{12.5}{\frac{7}{8}} \times \frac{\frac{4}{5}}{6.2} = \frac{12.5 \times \frac{4}{5}}{\frac{7}{8} \times 6.2}$.

(f) Distributive law: $\frac{A}{B} \times \left(\frac{C}{D} \pm \frac{E}{F}\right) = \left(\frac{A}{B} \times \frac{C}{D}\right) \pm \left(\frac{A}{B} \times \frac{E}{F}\right)$.

Example: $\frac{0.5}{1.7} \times \left(\frac{\frac{2}{3}}{\frac{4}{5}} + \frac{\frac{6}{7}}{\frac{8}{9}}\right) = \left(\frac{0.5}{1.7} \times \frac{\frac{2}{3}}{\frac{4}{5}}\right) + \left(\frac{0.5}{1.7} \times \frac{\frac{6}{7}}{\frac{8}{9}}\right)$.

Here is the correspondence between the items on this list and their cognate facts in ordinary fractions:

(a) $\leftrightarrow$ (17.4),   (b) $\leftrightarrow$ (13.1),   (c) $\leftrightarrow$ (13.3) and (15.2)
(d) $\leftrightarrow$ (14.3) and (16.1),   (e) $\leftrightarrow$ (17.2),   (f) $\leftrightarrow$ (17.4).

We take immediate note of two interesting consequences of these rules. First, for fractions $A, B, \ldots, E$, we also have a **cancellation phenomenon** for complex fractions:

$$\frac{A}{BC} \times \frac{CD}{E} = \frac{AD}{BE}.$$

When $A, B, \ldots, E$ are whole numbers, we have already proved this in (17.5) of Chapter 17. When they are fractions, we know by (e) that the left side is $\frac{ACD}{BCE}$, which, by (b), is equal to the right side.

If $B = D = E = 1$, then this says $\frac{A}{C} \times C = A$. This is of course the definition of the fraction division $\frac{A}{C}$ discussed in section 18.2.

A second consequence is a different way to look at the definition of fraction division. Up to this point, we know as a matter of definition that, for *fractions A, B, and C*, $\frac{A}{C} = B$ means $A = BC$. What (a) and (b) can do is to allow us, in the case of *complex fractions A, B, and C*, to "transition" from $\frac{A}{C} = B$ to $A = BC$ by a computation, namely, we multiply both sides of $\frac{A}{C} = B$ by C to get

$$\frac{A}{C} \times C = BC.$$

But the left side is equal to A on account of the just-mentioned fact that $\frac{A}{C} \times C = A$. Thus we arrive at $A = BC$. This at least makes it easier for some people to make the link between $\frac{A}{C} = B$ and $A = BC$. However, this argument is not a proof that $\frac{A}{C} = B$ implies $A = BC$. Do you know why not? (See the end of this section for an answer.)

The algebraic proofs of (a)–(f) are entirely mechanical and somewhat tedious, and they are based on (13.1), (14.3), (15.2), (16.1), (17.2) and (17.4) (which will be used without comment in the following). For this reason, only the proofs of (a), (d), and (f) will be given for the purpose of illustration, and the proofs of the rest will be left as exercises.

In a sixth grade classroom, one or two such proofs ought to be presented, but perhaps no more than that. There is no doubt, however, that as a teacher you should point out, explicitly, the fact that *the usual formulas for fractions such as (13.1), (14.3), (15.2), (16.1), (17.2), (17.4) can be expanded to include complex fractions.*

Proof of (a). Let $A = \frac{k}{\ell}$, $B = \frac{m}{n}$, and $C = \frac{p}{q}$. Then $\frac{B}{C} = \frac{mq}{np}$, so that

$$A \times \frac{B}{C} = \frac{kmq}{\ell np}.$$

But

$$\frac{AB}{C} = \frac{\frac{km}{\ell n}}{\frac{p}{q}} = \frac{kmq}{\ell np} = A \times \frac{B}{C},$$

so (a) is proved. □

Proof of (d). With A, B, C as above and $D = \frac{r}{s}$, we have

$$\frac{A}{B} \pm \frac{C}{D} = \frac{\frac{k}{\ell}}{\frac{m}{n}} \pm \frac{\frac{p}{q}}{\frac{r}{s}} = \frac{kn}{\ell m} \pm \frac{ps}{qr} = \frac{knqr \pm \ell mps}{\ell mqr},$$

and

$$\frac{(AD) \pm (BC)}{BD} = \frac{\frac{kr}{\ell s} \pm \frac{mp}{nq}}{\frac{mr}{ns}} = \frac{\frac{(krnq) \pm (\ell smp)}{\ell snq}}{\frac{mr}{ns}}$$

$$= \frac{ns\big((krnq) \pm (\ell smp)\big)}{mr(\ell snq)} = \frac{krnq \pm \ell smp}{mr\ell q} = \frac{A}{B} \pm \frac{C}{D},$$

so (d) is proved. □

Proof of (f). Since each of $\frac{A}{B}$, $\frac{C}{D}$, etc. is a fraction, (f) is already implied by the distributive law (17.4). However, let us make use of this opportunity to prove the "−" case of (f) as we only proved the "+" case of (17.4). In the figure below, let $\mathcal{R}$ be the rectangle with one side $\frac{A}{B}$ and the other side $\frac{C}{D}$ (the big rectangle), and let the side of length $\frac{C}{D}$ be divided into two segments, one of length $\frac{C}{D} - \frac{E}{F}$ and the other of length $\frac{E}{F}$, and we form the corresponding rectangles, as shown. Let $\mathcal{R}_1$ be a rectangle with one side equal to $\frac{A}{B}$ and the other side $\frac{C}{D} - \frac{E}{F}$ (the smaller rectangle on the left), and let $\mathcal{R}_2$ be the rectangle with one side $\frac{A}{B}$ and the other side $\frac{E}{F}$ (the smaller rectangle on the right).

19.1. The Basic Skills

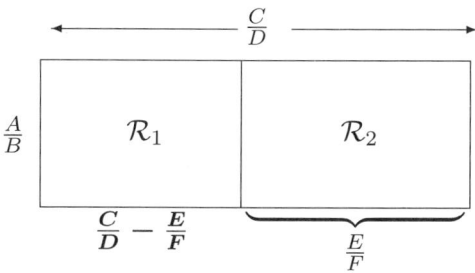

Then from the picture, we see that

$$\frac{A}{B} \times \left(\frac{C}{D} - \frac{E}{F}\right) = \text{area of } \mathcal{R}_1 = \text{area of } \mathcal{R} - \text{area of } \mathcal{R}_2$$
$$= \left(\frac{A}{B} \times \frac{C}{D}\right) - \left(\frac{A}{B} \times \frac{E}{F}\right).$$

Therefore (f) is proved. □

We give two applications that will be useful later.

EXAMPLE 1. If A, B, C are fractions and $B \neq 0$, then

$$\frac{A}{B} + \frac{C}{B} = \frac{A+C}{B}.$$

(Compare equation (14.2) on page 222.)

We can see this by a direct computation if we replace A by $\frac{k}{l}$, B by $\frac{m}{n}$, etc. More enlightening are the following two alternative arguments.

We can make use of (a) and (f):

$$\frac{A}{B} + \frac{C}{B} \overset{(a)}{=} \left(\frac{1}{B} \times A\right) + \left(\frac{1}{B} \times C\right)$$
$$\overset{(f)}{=} \frac{1}{B} \times (A + C) \overset{(a)}{=} \frac{A+C}{B}.$$

Or, we can use (d) and (b):

$$\frac{A}{B} + \frac{C}{B} \overset{(d)}{=} \frac{AB + CB}{BB} = \frac{(A+C)B}{BB} \overset{(b)}{=} \frac{A+C}{B}.$$

EXAMPLE 2. Let A and B be fractions and $B \neq 0$. Then

$$\underbrace{\frac{A}{B} + \cdots + \frac{A}{B}}_{j} = \frac{jA}{B}.$$

We can prove this directly, of course, but we choose to use Example 1 repeatedly:

$$\underbrace{\frac{A}{B} + \cdots + \frac{A}{B}}_{j} = \frac{2A}{B} + \underbrace{\frac{A}{B} + \cdots + \frac{A}{B}}_{j-2}$$
$$= \frac{3A}{B} + \underbrace{\frac{A}{B} + \cdots + \frac{A}{B}}_{j-3} = \cdots = \frac{jA}{B}.$$

EXAMPLE 3.   Locate approximately the following numbers on the number line:

$$\frac{119\frac{2}{5}}{57}, \quad \frac{54}{26\frac{1}{2}}.$$

We can compare these numbers by applying rule (c) to check which of the two,

$$119\frac{2}{5} \times 26\frac{1}{2} \quad \text{or} \quad 57 \times 54,$$

is larger. Observe that

$$119\frac{2}{5} > 119 \quad \text{and} \quad 26\frac{1}{2} > 26,$$

while $26 \times 119 = 3094 > 3078 = 54 \times 57$. Therefore, we know by rule (c) that

$$\frac{119\frac{2}{5}}{57} > \frac{119}{57} > \frac{54}{26} > \frac{54}{26\frac{1}{2}}.$$

To place these numbers on the number line, note that

$$\frac{54}{26\frac{1}{2}} \stackrel{(b)}{=} \frac{54 \times 2}{26\frac{1}{2} \times 2} = \frac{108}{53} = 2\frac{2}{53} > 2.$$

Hence $54/26\frac{1}{2}$ is to the right of 2, but not by much because $\frac{2}{53} = \frac{4}{106}$ is roughly $\frac{4}{100}$. On the other hand,

$$\frac{119\frac{2}{5}}{57} \stackrel{(b)}{<} \frac{120}{57} = 2\frac{6}{57} < 2\frac{6}{54} = 2\frac{1}{9}.$$

Therefore $119\frac{2}{5}/57$ is less than $\frac{1}{9}$ beyond 2. The picture will therefore look approximately like this:

We now explain why the argument we gave was not a proof that if $\frac{A}{C} = B$, then $A = BC$. The argument depends on rules (a) and (b) which depend on the invert-and-multiply rule, which in turn depends on the definition of fraction division in Chapter 18. In other words, to make this argument work, we had to implicitly assume from the beginning that $\frac{A}{C} = B$ is equivalent to $A = BC$. Therefore we cannot claim to prove something that we have assumed in the first place.

19.2. Why Are Complex Fractions Important?

The importance of the algebraic operations on complex fractions has not been properly recognized in the K–12 mathematics curriculum. On the most fundamental level, their importance will be seen in the next few sections where, without the basic rules (a)–(f) about complex fractions, most of the discussions concerning ratios, rates, and percents would stop dead in their tracks. For example, the myth that a percent[2] such as 4.125% *is a fraction with numerator* 4.125 *and denominator* 100 is still being perpetuated in the mathematics education literature. Unfortunately, 4.125% *is not a fraction with denominator equal to* 100 because, by definition (see section 12.2), the numerator of a fraction is a whole number and 4.125 is not a whole number but is a fraction. It is sloppiness of this nature that contributes to the overall neglect of precise definitions in school mathematics.

Complex fractions appear routinely in all kinds of situations, either in mathematics or in everyday life, and one needs to be able to compute with them with total ease. Consider the following routine computation with decimals:
$$\frac{0.06}{1.3} + \frac{0.5}{3.4} = ?$$
The addition formula (14.3) on page 222 cannot be applied directly because it is only good for fractions, and the above numbers are *not* fractions in the sense that the numerator 0.06 is a fraction ($= \frac{6}{100}$) and not a whole number, for example. Nevertheless, students are usually taught to simply carry out the addition using formula (14.3) by treating each of 0.06, 1.3, 0.5, and 3.4 as if it were a whole number:
$$\frac{0.06}{1.3} + \frac{0.5}{3.4} = \frac{3.4 \times 0.06 + 1.3 \times 0.5}{1.3 \times 3.4} = \frac{0.854}{4.42} = \frac{854}{4420}.$$
Fortunately, equality (d) of the preceding section justifies this computation.

As another example, again anticipating the discussion of percent in the next section, we see immediately that a common statement such as, "The Federal Reserve has announced that it will raise the prime interest rate from

---

[2] Here we anticipate the discussion in the next chapter on percent; however, beyond recognizing the symbol %, no knowledge of percent will be needed in the following.

7.2% to 7.7%" already makes use of complex fractions. This is because the Federal Reserve wants us to concentrate on the "numerator" 7.2 in 7.2% and realize that 7.2% is less than 7.7%, whereas we know that 7.2 is the numerator of *no* fraction but *is* the numerator of the *complex* fraction 7.2%. Moreover, the quickest way to see 7.2% < 7.7% is to make use of (c) of the preceding section.

Incidentally, the preceding statement, "...raise the prime interest rate from 7.2% to 7.7%", takes for granted that a percent is a number. This observation will be relevant in the next chapter.

Finally, if we look ahead to the middle school curriculum and recognize that fractions are an indispensable part of the foundation of algebra, then the importance of complex fractions becomes even more apparent. Consider, for example, the simplest kind of operation on rational expressions in a number x:

$$\frac{1}{1-x} + \frac{1}{1+x} = \frac{1 \cdot (1+x) + 1 \cdot (1-x)}{(1-x)(1+x)} = \frac{2}{1-x^2}.$$

This is an *identity* in x for all $x \neq \pm 1$. In particular, it asserts for the special case of $x = \frac{2}{3}$ that

$$\frac{1}{1-\frac{2}{3}} + \frac{1}{1-\frac{2}{3}} = \frac{1 \cdot (1+\frac{2}{3}) + 1 \cdot (1-\frac{2}{3})}{(1-\frac{2}{3})(1+\frac{2}{3})} = \frac{2}{1-(\frac{2}{3})^2}.$$

In other words, if we let B be the fraction $(1 - \frac{2}{3})$ and D be the fraction $(1 + \frac{2}{3})$, then we are claiming that

$$\frac{1}{B} + \frac{1}{D} = \frac{1 \cdot D + 1 \cdot B}{BD}.$$

The validity of this particular computation, and therefore the validity of the identity $\frac{1}{1-x} + \frac{1}{1+x} = \frac{2}{1-x^2}$ in general, depend on rule (d) of section 19.1. Naturally, the need for other items in rules (a)–(f) of section 19.1 arises in many other computations with rational expressions as well.

One may ask how to justify this identity for an x which is not a fraction, e.g., $\sqrt{2}$ or π. This issue will be taken up in Chapter 21.

Exercises

1. Give algebraic proofs of (b), (c), (e), and (f) of section 19.1.
2. Compute: (a) $12.53 \times \frac{3}{7}$. (b) $\frac{2.8}{51} / \frac{7}{1.7}$. (c) $\frac{11}{2.1} + \frac{1}{7}$ and simplify.
3. If $A = \frac{11}{5}$, $B = \frac{2}{7}$, $C = \frac{22}{21}$, $D = 2\frac{4}{5}$, $E = \frac{11}{7}$, and $F = \frac{5}{2}$, directly verify (a)–(e) of section 19.1.
4. If $A, B, \ldots, E$ are fractions, show that $\frac{A}{BC} + \frac{E}{BD} = \frac{AD+CE}{BCD}$. (This is the analogue of equation (15.2) on page 240.)
5. Consider the following statement made by a student: "I can find a fraction between $\frac{2}{7}$ and $\frac{3}{7}$: it is $\frac{2.5}{7}$." (a) Suppose the student tells you this right after you have just taught him the materials of Chapters 12–13. Would this be a correct statement? Explain. (b) On the other hand, if a student who knows all the materials in Chapters 12–19 makes the same statement, would he be justified in saying this? Also explain. (c) Can you find five distinct fractions between $\frac{1}{103}$ and $\frac{2}{103}$?
6. Put the following complex fractions approximately on the number line (compare Example 3 on page 314).
$$\frac{3\frac{89}{90}}{10\frac{1}{2}}, \quad \frac{\frac{12}{59}}{\frac{23}{47}}, \quad \frac{40\frac{1}{3}}{100}, \quad \frac{118}{297}.$$
7. Divide 88 into two parts A and B so that $\frac{A}{B} = \frac{2/3}{4/5}$. (This problem was in the 1875 California Exam for Teachers; see [**Shu86**]).
8. (a) Divide 88 into two parts A and B so that $\frac{A}{B} = \frac{2/3}{3/4}$. (b) Divide $\frac{2}{7}$ into two parts A and B so that $\frac{A}{B} = \frac{4}{5}$.
9. (a) Explain why
$$\frac{8\frac{1}{3}}{A} < \frac{8\frac{1}{2}}{A}$$
for any nonzero fraction A, and why
$$\frac{B}{5\frac{1}{2}} < \frac{B}{5\frac{1}{6}}$$
for any nonzero fraction B. (b) In general, suppose A, B, C are nonzero fractions. Show that if $A < B$, then $\frac{A}{C} < \frac{B}{C}$. On the other hand, if $B > C$, then $\frac{A}{B} < \frac{A}{C}$.

Chapter 20

Percent

This chapter introduces the language of "percent". The main point is that a *percent* is a number. Along the way, we will bring closure to the concept of *relative error* that first surfaced in the discussion of estimation in Chapter 10.

The sections are as follows:

Percent

Relative Error

20.1. Percent

A **percent** is, by definition, a complex fraction whose denominator is 100. The number $\frac{N}{100}$, where N is a fraction, is usually called $\boldsymbol{N}$ **percent**. Note that $\frac{N}{100}$ is usually written as $\boldsymbol{N}\%$. Thus "2 percent" means the fraction $\frac{2}{100}$, and "seven-and-a-half percent" means the complex fraction $7\frac{1}{2}/100$.

We emphasize that, in general, percents are *complex* fractions, rather than fractions, with a denominator of 100.

The preceding precise definition of **percent** *must* be given before we begin the mathematical discussion because mathematics is never about reasoning based on undefined concepts. In addition, the need of a precise definition in this case is particularly acute because the immense amount of nonlearning associated with percent can be directly traced to the absence of a definition, as we proceed to demonstrate. In standard texts, *percent* is introduced as "out of 100". Armed with this vague description, how are students supposed to work out precise mathematical problems? The following statement by a teacher in a remedial ninth grade class, quoted from

[**BGJ94**, pages 106–107], gives an eloquent testimony to the handicap under which students labor in this kind of instruction:

> *I have had a great deal of difficulty teaching the three types of percent problems:*
>
> *What is 15% of 20?*
> *What % of 20 is 3?*
> *3 is 15% of what number?*
>
> *Early in my teaching career, I used to teach my students to solve these problems the same way I had learned them. In the first type, students were taught to rename the percent as a decimal by moving the decimal point two places to the left and multiplying the two numbers together. In the second and third types, students were taught to divide the two numbers. ...*
>
> *This year I decided to teach percent through proportions. ... What percent of 20 is 3? Once again I had my students use grids to help them write and solve proportions. ...I taught students how to solve the related proportion by using cross-multiplication: $\frac{x}{100} = \frac{3}{20}$, where x represents 15% of 20. ...*

Two aspects of this quote are noteworthy. First, the absence of a definition of *percent* forces the teacher (and her students too) to guess the meaning of percent as it appears in each of the three types of problems she listed. No reason was given as to why each problem should be solved in the way described. To be blunt, we are bearing witness to a striking case of *teaching by rote*. By comparison, we will see that once our definition is accepted, the solutions of all percent problems are logical consequences of the definition and the facts we have already proved about fractions. Second, the teacher said explicitly that early in her career, she taught her students in the same way she herself was taught, i.e., because she was taught by rote, she too will teach by rote. If nothing else, you see why you need to to teach your own students correctly when it is your turn to teach.

We now put the definition to work. Students are told without explanation that "$N\%$ of something" can be interpreted as the totality of N parts when that something is divided into 100 equal parts. We now *prove* that this so-called interpretation is in fact a logical consequence of the definition. For definiteness, consider the claim:

> "$7\frac{1}{2}\%$ *of 512 dollars*" *means the totality of* $7\frac{1}{2}$ *parts when we divide* 512 *dollars into* 100 *equal parts.*

We now *provide the precise reasoning to back up this claim*. Because $7\frac{1}{2}\%$ is a fraction (because a complex fraction *is* a fraction), the second interpretation of fraction multiplication in (17.7) on page 273 implies that

$$7\frac{1}{2}\% \text{ of } 512 \;=\; 7\frac{1}{2}\% \times 512.$$

20.1. Percent

Now,
$$7\frac{1}{2}\% \times 512 \;=\; \frac{7\frac{1}{2}}{100} \times 512 \;=\; 7\frac{1}{2} \times \left(\frac{1}{100} \times 512\right).$$
By (17.7), $\frac{1}{100} \times 512$ is one part when 512 is divided into 100 equal parts. Therefore by the discussion below (17.7), we can interpret the product $7\frac{1}{2} \times (\frac{1}{100} \times 512)$ as follows:

$$\begin{aligned}7\frac{1}{2}\% \times 512 &= 7\frac{1}{2} \text{ copies of } (\tfrac{1}{100} \times 512) \\ &= 7\frac{1}{2} \text{ parts when 512 is divided into 100 equal parts.}\end{aligned}$$

This is then the reason why one can make that claim.

We want to give a perhaps oversimplified account of the origin of the concept of percent. Suppose you go to a supermarket to buy beef and you want as little fat as possible. One package says "$\frac{3}{28}$ of this package is fat"[1] and the other says "Fat content: $\frac{2}{15}$". Of course in real life, such oddball labeling does not occur, but the point of this story is to explain why they do not occur. So let us go on. Your concern is about which of the two kinds of beef has less fat per unit of weight. In other words, if we take a pound of beef from the first package, then the amount of fat is $\frac{3}{28} \times 1 = \frac{3}{28}$ pounds (notice that we have made use of the mathematical meaning of "of"; see Chapter 15). Similarly, a pound of beef from the second package has $\frac{2}{15}$ pounds of fat. Which is bigger: $\frac{3}{28}$ or $\frac{2}{15}$? You know your fractions by now, so mentally you compute by using the cross-multiplication algorithm: $3 \times 15 = 45 < 56 = 2 \times 28$, therefore $\frac{3}{28} < \frac{2}{15}$ and you pick the first package.

This kind of labeling is unacceptable, of course, because important information should be clear at a glance without requiring mental computations. One way to resolve this difficulty would be for the first meat packing company to also measure fat by dividing each unit weight into 15 instead of 28 equal parts. Had it done that, it would have found that $1\frac{17}{28}$ parts out of 15 consist of fat and would have therefore labeled the package as "$1\frac{17}{28}/15$ of this package is fat". (Notice how complex fractions appear naturally.) Then it would indeed be clear at a glance that the second package contains more fat inasmuch as $2 > 1\frac{17}{28}$. This then gives rise to the idea that if all meat packing companies agree to use the same denominator for their fat-content declarations, all shoppers would be able to make comparisons at a glance. In fact, the same goes for all kinds of quantitative comparisons: why not use the same denominator across the board?

Activity. Verify that the preceding statement about $\frac{3}{28} = \frac{1\frac{17}{28}}{15}$ is correct.

Quite miraculously, such a general agreement *was* reached at some point in the past, and people seemed happy to use 100 as the standard denominator. They even devised a new notation (%) and created a new word

---
[1] Such measurements are understood to be by weight.

(percent) in the process (Latin: *per centum*, meaning "by the hundred"). Thus instead of $\frac{3}{28}$, we want to express it as $C\%$, where C is some fraction, i.e., we want $\frac{3}{28} = C\%$, or, making use of (a) on page 310,

$$\frac{3}{28} = C \times \frac{1}{100}.$$

By the Lemma on page 286, we are guaranteed that there is such a C, but there is no need to memorize the formula for C in (18.1) on page 288. Indeed, if we multiply both sides of the preceding equation by 100, we get

$$C = 100 \times \frac{3}{28} = 10\frac{5}{7}.$$

The first package of beef would then be labeled, "$10\frac{5}{7}\%$ of this package is fat". Similarly, the second package would be labeled, "Fat content: $13\frac{1}{3}\%$", because if $\frac{2}{15} = \frac{y}{100}$, we get as before that

$$y = \frac{100 \times 2}{15} = 13\frac{1}{3}.$$

Now every shopper can tell without a moment of hesitation that the second package has more fat per unit weight because $13\frac{1}{3} > 13 > 11 > 10\frac{5}{7}$.

It remains to point out the element of unreality in this story: percents in ordinary labeling in a commercial context will usually be rounded off to the nearest 10^0 (see Chapter 10). Instead of $13\frac{1}{3}$ and $10\frac{5}{7}$, it would be, respectively, 13% and 11%. Therefore you should think of this beef story as a fairy tale that we are using merely to make a point. However, the method of computation of percents illustrated above is essential knowledge and, in the process, it also underscores the importance of formulas (a)–(f) in Chapter 19. All in all, you'd agree that this is not a bad fairy tale as such things go.

We have so far interpreted the notation of "%" as the denominator of a complex fraction or fraction, e.g., 13% means $\frac{13}{100}$. In ordinary language, however, "percent" is often used as an adjective, as in "I am getting 5 percent interest from my savings bank". There are educators and mathematicians alike who would argue that "5%" is not a number but an "action" or an "operator" (whatever they are), and should therefore be treated differently from ordinary fractions or complex fractions. Let us put such a statement in perspective. *Every* number, be it whole number or fraction, is often used as an "action" or an "operator", e.g., "three apples", or "a quarter pound of beef". To say that a percent is a number therefore in no way contradicts its common usage as an "operator". This book promotes the primacy of the concept of a number, because the belief is that if students have a firm grasp of what a number is, can explain all the arithmetic operations with numbers, and can compute with them fluently, then how we use numbers (as a noun or an adjective) will hardly be an issue. In particular, *there is no*

20.1. Percent

cause for singling out "percent" to make believe that it is anything different from a complex fraction. It is a *number*, end of discussion. (Compare the discussion on page 316.)

We now give seven examples to illustrate how percent is used. Observe that all the calculations are based strictly on our definition of percent, and that in all the calculations below, rules (a)–(f) of complex fractions in Chapter 19 are used extensively without comment.

EXAMPLE 1. Express the percent $33\frac{1}{3}\%$ as a fraction.

Note that $33\frac{1}{3} = \frac{3 \times 33 + 1}{3} = \frac{100}{3}$, so by invert-and-multiply,

$$33\frac{1}{3}\% = \frac{100}{3} \times \frac{1}{100} = \frac{1}{3}.$$

EXAMPLE 2. Express $\frac{5}{16}$ as a percent.

If $\frac{5}{16} = C\%$, then $\frac{5}{16} = C \times \frac{1}{100}$. Multiplying both sides by 100 gives

$$C = 100 \times \frac{5}{16} = 31\frac{1}{4}.$$

So $\frac{5}{16} = 31\frac{1}{4}\%$.

EXAMPLE 3. What is 45% of 70?

By our mathematical interpretation of the word "of" (see (17.7) on page 273), the answer is $45\% \times 70 = 31.5$.

EXAMPLE 4. 70 is 45% of what number?

Let that number be N. Then we are given that $70 = 45\% \times N = N \times 45\%$, and therefore by the definition of division in Chapter 18,

$$N = \frac{70}{45\%} = 70 \times \frac{100}{45} = \frac{7000}{45} = 153\frac{1}{3}.$$

EXAMPLE 5. What percent of 70 is 45?

Let us say 45 is $N\%$ of 70. So $45 = N\% \times 70$, and $45 = \frac{N \times 7}{10}$. Thus, $N \times 7 = 450$, and $N = \frac{450}{7} = 64\frac{2}{7}$. The answer is therefore $64\frac{2}{7}\%$.

EXAMPLE 6. 7 is 26 percent of a number L. What is L? How would you get an approximate answer *without* using pencil and paper?

We are given that $7 = 26\% \times L$. (Notice once again that we are using the mathematical interpretation of the word "of" as discussed on page 273.) We can write it as $7 = L \times \frac{26}{100}$ so that

$$L = \frac{7}{\frac{26}{100}} = 7 \times \frac{100}{26} = 26\frac{12}{13}.$$

If writing is not allowed, we would recognize that 26% is roughly 25%, which is $\frac{1}{4}$. So 7 is roughly $\frac{1}{4}$ of this L, and L is therefore roughly 28.

Activity. (a) Express $\frac{1}{85}$ as a percent. (b) What is 28% of 45? (c) 17 is 35 percent of which number? (d) 48 is what percent of 35?

The next example needs a little preparation in the form of an activity.

Activity. A bed costs $200. Because of unprecedented demand, the price went up 15% overnight. Then came a recession and the price tumbled down by the same 15% from its higher price. Does it get back to $200?

> *A comment about the potential pitfall in the teaching of word problems like this activity is in order. Textbooks generally take for granted that students know the meaning of phrases such as "the price went up 15%", "a discount of 20%", "stocks are down by 4%", or "a sales tax of $8\frac{1}{2}$%". It is possible that students have a vague idea about such phrases, but* having a vague idea is not enough for doing mathematics, *because they need to know the precise meaning of what they have to deal with. Therefore, it is incumbent on you as a teacher to give these phrases a precise definition. For example, the meaning of "the price went up 15%" is the following. Let the price be P dollars. Then the new price is $P + (P \times 15\%)$ dollars. In like manner, the meaning of the price tumbles down by 15% is that the new price is $P - (P \times 15\%)$ dollars. The related phrase of the price is discounted by 20% means the new price is $P - (P \times 20\%)$ dollars. And so on. Please keep firmly in mind the need to define each of these phrases when you teach!*

EXAMPLE 7. The price of the stock of a diaper company went up by 12%, and then went down by 12% the next day. Did it get back to its original price?

We are not given the price of the stock, and so must represent it by some symbol, say D dollars. If it went up by 12%, then the new price is

$$D + (12\% \times D) = (1 \times D) + (12\% \times D) = (1 + 12\%) \times D = 112\% \, D$$

dollars. If the low price went down by 12%, then the new price is

$$(112\% \times D) - 12\% \times (112\% \times D) = 1 \times \underbrace{(112\% \times D)} - 12\% \times \underbrace{(112\% \times D)}.$$

By the distributive law,

$$1 \times \underbrace{(112\% \times D)} - 12\% \times \underbrace{(112\% \times D)} = (1 - 12\%)\underbrace{(112\% \times D)}.$$

20.2. Relative Error

Therefore,
$$(112\% \times D) - 12\% \times (112\% \times D) = (1 - 12\%)\underbrace{(112\% \times D)}$$
$$= 88\% \times 112\% \times D$$
$$= \frac{9856}{10000} \times D.$$

Because $\frac{9856}{10000} < 1$, the new price is less than D dollars and therefore not as high as the original price.

We can go a step further by computing the percent the stock of Example 7 must go down the next day in order to drop back to the original price. Suppose it went down by $x\%$ the next day for a certain *number* x; then the new price would be computed exactly as above, except that 12 would be replaced by x everywhere:

$$(112\% \times D) - x\% \times (112\% \times D) = 1 \times \underbrace{(112\% \times D)} - x\% \times \underbrace{(112\% \times D)}$$
$$= (1 - x\%)(112\% \times D)$$
$$= (100 - x)\% \times 112\% \times D$$
$$= \frac{(100 - x) \times 112}{10000} \times D.$$

Thus the new price will be D dollars exactly when
$$(100 - x) \times 112 = 10000.$$

Remember that x is just a number, and we can compute with it the usual way. By the distributive law, we have $11200 - 112x = 10000$. Adding the same number $112x - 10000$ to both sides gives $112x = 1200$. Therefore,
$$x = \frac{1200}{112} = \frac{150}{14} = 10\frac{5}{7}.$$

In other words, the price of the stock would have to go down by $10\frac{5}{7}\%$ in order to equal its original price.

20.2. Relative Error

An excellent illustration of how percent is used to advantage is the concept of the *relative error* of an approximation which was first brought up in section 10.2.

We make approximations all the time in daily life, and such approximations must be understood in context. For example, a statement such as "three-sevenths of the people" does not make strict sense unless the number of people happens to be divisible by 7. For example, if there are 72 people, then "three-sevenths of the people" would mean, literally, $30\frac{6}{7}$ people, which is absurd. In this case, you probably round off in your own mind the number 72 down to 70, and then three-sevenths of 70 people become

$\frac{3}{7} \times 70 = 30$ people. This is an approximation that most people would consider acceptable. What we want to address, if only in a superficial way, is the question of what is meant by "acceptable", i.e., how to measure whether an approximation is reasonable.

To be more precise, suppose you tell someone "I'll meet you in 15 minutes". You probably already know that your walk to the meeting place would not be exactly 15 minutes. In this case, consider the scenario where you meet your friend five minutes late. So you arrive in 20 minutes. Thus the time it actually takes to get there is 20 minutes but your estimate was that it would be only 15. The error of your estimation of "fifteen minutes" is therefore $20 - 15 = 5$ minutes. The **absolute error** of your estimation is the difference

$$\textit{true value} - \textit{the estimated or approximate value}$$

in case the true value is greater than the estimated value, and is otherwise

$$\textit{estimated or approximate value} - \textit{the true value}.$$

The person waiting for you may not take offense at your tardiness, but he would probably take note of the absolute error in your estimation, i.e., the five-minute waiting time that he had to endure with grace.

The absolute error, five minutes in the case at hand, does not by itself show how seriously you erred in your estimation, however. To explain what this means, imagine that you had to drive from San Francisco to Los Angeles to visit a friend and you promised to be there at 6 PM. You arrived at 6:05 PM instead. The absolute error of your estimated time of arrival is therefore also five minutes, but such an absolute error was in all likelihood completely ignored by your friend as she praised you for your punctuality. The difference in the perception between this "five minutes" and the earlier "five minutes" lies in the context. In the first situation, the relative error,

$$\frac{\textit{absolute error}}{\textit{true value}},$$

is $\frac{5}{20}$, which is 25%. In the second situation, let us assume for simplicity that you left San Francisco at 11 AM because you thought it would only take 7 hours (= 420 minutes) of driving time. But the actual driving time turned out to be 7 hours and 5 minutes (= 425 minutes). The relative error therefore becomes $\frac{5}{425}$, which is $\frac{1}{85} = 1\frac{3}{17}\%$, and is less than 2% (see rule (c) on page 310). Thus, the five-minute error is considerably more significant in the first situation.

The quotient

$$\frac{\textit{absolute error}}{\textit{true value}}$$

for an approximation or estimation *when expressed as a percent* is called the **relative error** of the approximation or estimation. This is the number

20.2. Relative Error

(rather than the absolute error) that conveys how good or how bad the approximation is. The point of expressing the relative error as a percent is to make for an easy comparison between differing approximations to see which is better. The above simple example about "five minutes" is a good illustration of such a comparison.

Suppose we make an approximation of $67\frac{1}{2}\%$ by $\frac{2}{3}$. Let us compute the relative error of that approximation. The absolute error is

$$67\frac{1}{2}\% - \frac{2}{3} = 67\frac{1}{2}\% - 66\frac{2}{3}\% = \frac{5}{6}\%.$$

The relative error is therefore

$$\frac{\frac{5}{6}\%}{67\frac{1}{2}\%} = 1\frac{19}{81}\%.$$

Since $1\frac{19}{81} < 2$, the relative error is less than 2%.

Activity. Verify that the preceding computation resulting in $1\frac{19}{81}\%$ is correct.

Let us go back to our initial question of whether an approximation is good enough. While there is no absolute standard of "good enough", it is fair to say that if the relative error of an approximation is less than 5%, then for most everyday purposes, the approximation is generally good enough. If the relative error is 20% or more, it may not be so good. Even with these qualified statements in place, we still leave a large gray area between 5% and 20%. Subjective judgment in each case will have the final say.

Exercises

1. (a) What percent is 18 of 84? 72 of 120? (b) What percent of 125 is 36? (c) 24 is what percent of 33? (d) What is 15 percent of 75? (e) 16 percent of what number is 24? (f) 25 is 15 percent of which number?

2. Express the following as percents: (a) $\frac{1}{4}$, $\frac{7}{5}$, $\frac{3}{16}$, $\frac{17}{32}$, $\frac{34}{25}$, $\frac{24}{125}$, $\frac{18}{125}$. (b) $\frac{5}{12}$, $\frac{24}{7}$, $\frac{8}{15}$, $\frac{7}{3}$, $\frac{5}{6}$, $\frac{7}{48}$. Do you notice a difference between the answers to the two groups? Can you guess an explanation?

3. A shop plans to have a sale. One suggestion is to give all customers a 15% *discount after sales tax has been computed* (this means, if an article costs y after sales tax, then customers only pay $y - (15\% \times y)$ dollars). Another suggestion is to give a 20% discount before sales tax. If the sales tax is 5%, which suggestion would give the customer greater savings?

4. A bike is priced at $469.80 including an 8% sales tax. How much is the price of the bike before sales tax and how much is the sales tax?

5. A high-tech stock dropped 55% of its value in April to its present value of N. A stock broker tells his clients that if the stock goes up by 55% of its present value, then it will be back to where it was in April. Is he correct? If so, why? If not, by what percent must the stock in its present value of N rise in order to regain its former value?

6. After a 40% increase in value, a stock is now worth M. By what percent must its present value drop in order to get back to its former value?

7. A train goes between two towns at constant speed. If we want to shorten the travel time by 25%, by how many percent should we increase the speed of the train?

8. How much money would be in an account after two years if the initial deposit was $100 and the bank pays an interest of 5% at the end of each year? And at the end of three years? (No calculator.)

9. What is the relative error if we approximate $\frac{3}{8}$ by 37%? $\frac{30}{72}$ by $\frac{3}{7}$?

10. Five pounds of rice costs $4.50. The owner of a grocery store wants to increase sales by giving customers 15% more rice for the $4.50. Then he decides that it may be simpler to sell the five pounds for less to achieve the same goal of "more rice for the dollar". What should the sales price of five pounds of rice be so that the customers achieve the same amount of savings per pound of rice?

11. At one school, $\frac{3}{7}$ of the boys went to a football game. Among these, $\frac{5}{6}$ went the game by bus. What is the percent of the boys in this school who went to the game by bus?

Exercises

12. Given a positive number c (e.g., $c = 65$), one can approximate its **square root** $\sqrt{c}$ (i.e., the positive number so that $(\sqrt{c})^2 = c$) by the following method. Let s be the whole number whose square s^2 is closest to c (e.g., with $c = 65$, $s = 8$). Then $\frac{1}{2}(s+\frac{c}{s})$ is an approximation to $\sqrt{c}$. Use this fact to compute such an approximate value of $\sqrt{c}$ when $c = 65$, and again when $c = 83$. Then use a four-function calculator to compute the relative error of this approximation. (This is the first step of what is called an *iterative procedure* that leads to a computation of as many digits of the square root as one wishes. There is some evidence that the Babylonians knew about this procedure some four thousand years ago. From the point of view of calculus, this is a simple special case of a general method of approximation called *Newton's Iteration*.)

Chapter 21

Fundamental Assumption of School Mathematics (FASM)

Everything we have done so far has been restricted to fractions, including our discussion of percent. In the next section, however, we will have to deal with problems related to rates, and numbers which are not fractions—the so-called **irrational numbers**—cannot help but intrude. Because a discussion of irrational numbers requires a level of mathematical sophistication that is more appropriate for university mathematics courses, school mathematics limits itself in the main to fractions or, more generally, to rational numbers, which are positive and negative fractions (see Part 3). The fact that *school mathematics intentionally skirts a discussion of irrational numbers and yet tries to maintain the appearance of dealing with all numbers, rational or irrational,* seems not to have been made explicit in the literature until 2001 (see [**Wu02**, page 101]). The purpose of this chapter is to bring out certain aspects of school mathematics related to this fact.

The following discussion will on occasion invoke concepts and results not yet developed in this book. On the whole, we will keep it on an intuitive level so that even if you encounter a phrase or two that seems unfamiliar, you should just ignore it and forge ahead.

In real life, fractions (and, in fact, the even more special class of fractions, namely, finite decimals; see Chapter 38) are the only numbers that matter. For example, although we know the length of the diagonal of the unit square is $\sqrt{2}$, which is irrational, this length in an everyday context would usually be taken to be a decimal approximation to the true value of

$$\sqrt{2} = 1.4142135623730950488\cdots$$

such as 1.414. Therefore, fractions occupy a position of singular importance among numbers. We must acknowledge, however, that many **real numbers** (i.e., the points on the number line; see Chapter 8) are irrational and school mathematics has to deal with irrational numbers. After all, how can we avoid π, which is irrational,[1] if we have to compute the area or circumference of circles? Therefore, students sooner or later will have to perform computations with irrational numbers. Such being the case, how does the school mathematics curriculum cope with this situation? One would like to say that it at least makes an honest effort to confront it, but such is not the case. What one can *infer* from the common textbooks is that it *implicitly* asks students to make what we propose to call **The Fundamental Assumption in School Mathematics (FASM)**:

> *All the information about the arithmetic operations on fractions can be extrapolated to all real numbers.*

This is a profound assumption. It allows students to manipulate irrational numbers as if they were integers, even if the students have no idea what an irrational number is. Thus a typical student would write down the following without a moment's thought:

$$\frac{\pi}{7} + \frac{\sqrt{2}}{3\sqrt{5}} = \frac{3\pi\sqrt{5} + 7\sqrt{2}}{21\sqrt{5}},$$

or

$$\sqrt{2}\,(\sqrt{3} + \pi) = \sqrt{2}\,\sqrt{3} + \sqrt{2}\,\pi,$$

or

$$37 \times \pi = \pi \times 37.$$

They would also justify the first equality by rule (d) on page 310, the second by the distributive law, and the last by the commutative law of multiplication, in spite of the fact that every single one of these laws or identities has been proved only when the numbers are fractions. In other words, *FASM is implicitly at work.*

We note in passing that the first equality reveals why it is important to have a general formula for the addition of two fractions as in formula (14.3) on page 222, and *why the common way to define the addition of fractions by seeking the lcm of the denominators grievously distorts what fraction addition means.*

We hasten to add that FASM is in fact correct, but the explanation is more advanced than the mathematics of the school curriculum.

---

[1] A fact that is by no means obvious.

21. Fundamental Assumption of School Mathematics (FASM)

As a result of FASM, *we can now extend the definition of the division of fractions in section 18.2 to include the division of any two real numbers.* More precisely, FASM yields a version of the lemma on page 286 for real numbers A and B, so that given two real numbers A and B (rational or irrational), the *division of A by B*, denoted by $\frac{A}{B}$ (assuming $B \neq 0$), is now by definition the real number so that $(\frac{A}{B})B = A$. What is important for school mathematics is the fact that, *on a formal level*, FASM together with rules (a)–(f) on page 310 allow us to treat the division of real numbers *operationally* as the division of two fractions. Therefore, the division of real numbers can hardly be simpler from a computational point of view. With this understood, we are now in a position to take up the concepts of ratio and rate in the next section.

Chapter 22

Ratio and Rate

The concepts of "ratio" and "rate" are used extensively in school mathematics as well as in everyday life. Some textbooks define a ratio as a "quotient", which would seem to be in complete accord with what we are going to do until one notices that the concept of a "quotient" is nowhere to be found in these books. Here are samples of some other attempts to address what a "ratio" is:

> A ratio is a comparative index; it always makes a statement about one measurement in relation to another.

> A ratio is a comparison of any two quantities. A ratio may be used to convey an idea that cannot be expressed as a single number.

> A ratio is a comparison of two quantities that tells the scale between them. Ratios may be expressed as quotients, fractions, decimals, percent, or given in the form of $a : b$.

> A ratio is a way to describe a relationship between numbers. If there are 13 boys and 15 girls in a classroom, then the ratio of boys to girls is 13 to 15.

There is more verbiage of the same kind for "rate". However, to judge by the amount of nonlearning associated with these concepts, very little of what has been said seems to make sense to students (see the reference [**CRE96**]). In this section, we choose to pick up this terminology because (a) so long as the terminology is in use you have to get used to it, right or wrong, and (b) we will be discussing word problems that are usually classified as problems on *ratio* or *rate* in the education literature, and there is no avoiding this terminology. The truth is that *the terminology is completely irrelevant.* The only things that matter are the reasoning and the methods of solution.

It is the purpose of this chapter to address both ratio and rate as mathematical concepts by giving precise definitions and making statements on the basis of mathematical reasoning.

The sections are as follows:

Ratio

Why Ratio?

Rate

Units

Cooperative Work

22.1. Ratio

We are given two fractions M and N, both nonzero. Suppose we want to compare them *multiplicatively* rather than *additively*, in the sense that we want to consider the division $\frac{M}{N}$ rather than $M - N$.[1] M and N may be the weights of two hams measured in pounds, the volume of two pitchers of orange juice measured in liters, the lengths of two curves measured in cm, etc. We will generically refer to M and N in such situations as **quantities**. In each situation, it is understood that a specific unit has been chosen for the measurement and both M and N are measured in terms of the same unit (thus we do not want to compare the weight of two hams, one measured in pounds and the other in kilograms). In other words, M and N are points on the same number line. We say **the ratio of M to N is $a : b$**, where a and b are nonzero fractions, if

$$\frac{M}{N} = \frac{a}{b}.$$

Thus a ratio is simply a division, $\frac{a}{b}$, no less and no more.

Observe that a ratio, so defined, is a complex fraction. Observe also that, *in view of FASM, we may consider a ratio to be defined also for any two (positive) real numbers*. For example, the number π is the ratio of the circumference of a circle to its diameter so that, in this case, numbers that are not fractions will have to be used.

The strange notation $a : b$ was first introduced by Gottfried Leibniz (1646–1716).[2] Thus in this notation, equivalent fractions (Chapter 13) would take the form $a : b = ca : cb$ for all whole numbers a, b, and c, where $b, c \neq 0$. It is quite possible that this strange notation is part of the reason for the mystification of the concept of ratio. You will of course look past the notation and always remember that $a : b$ is just the division $\frac{a}{b}$. When the

---

[1] See the next section, 22.2, for comments about the difference between an additive comparison and a multiplicative comparison.

[2] Co-discoverer of calculus with Newton. In addition, he was a visionary in symbolic logic and computer technology, and one of the most important philosophers.

notation of $a:b$ is used, both a and b are *usually* (but not always) whole numbers.

Our first task is to win your confidence by demonstrating that the definition of ratio as a division leads to a clearer and more precise understanding of a ratio than any of the descriptions mentioned in the preamble of this chapter.

Consider the statement, "In an auditorium, the ratio of boys to girls is $4:5$". Before offering an analysis of this statement, we point out a common (and accepted) abuse of language in this context: the meaning of the phrase "the ratio of boys to girls is $4:5$" is that

the ratio of the number of *boys to* the number of *girls is* $4:5$.

Thus let the number of boys and girls be denoted by B and G, respectively. We are given that $\frac{B}{G} = \frac{4}{5}$. Keeping in mind that B and G are just whole numbers, we apply the cross-multiplication algorithm to conclude that the equality is equivalent to
$$\frac{B}{4} = \frac{G}{5}.$$
Let the common value of $\frac{B}{4}$ and $\frac{G}{5}$, which is a fraction, be denoted by U, then
$$B = 4U \quad \text{and} \quad G = 5U.$$
Now we put B and G on the number line where the unit 1 is "one person". Then we get the following picture.

```
   0    U                    B    G
   |----|----|----|----|----|----|
```

But this means that if we let U be the new unit on this number line, then $B = 4$ and $G = 5$, as shown:

```
   0    1                    4    5
   |----|----|----|----|----|----|
        U                    B    G
```

Usually when B and G are given as specific whole numbers (e.g., 36 boys and 45 girls), it will always be the case that B is a multiple of 4 and G is a multiple of 5, so that U is in fact a whole number. In general, the fact that $\frac{4}{5}$ is a reduced fraction will also imply that U is a whole number.[3] We may therefore conclude:

(1) Suppose the ratio of the number of boys B to the number of girls G is $4:5$. Then there is a whole number U so that
$$B = 4U \quad \text{and} \quad G = 5U.$$

---

[3]The proof of this fact requires something like the Euclidean Algorithm, which will be taken up on page 469 in Chapter 35. In particular, see problem 1 on page 472.

Thus the boys can be divided into four equal groups of U boys, and the girls can be divided into five equal groups of U girls.

We can go a little further. The above analysis makes use of the partitive interpretation of division for whole numbers on $\frac{B}{4}$ and $\frac{G}{5}$ (see Chapter 7). We can also use the measurement interpretation for the analysis. So again let $\frac{B}{4} = \frac{G}{5} = U$. Then
$$B = U \cdot 4 \quad \text{and} \quad G = U \cdot 5.$$
This means that if we divide the boys into equal groups of 4, then there are U such groups. Similarly, if we divide the girls into equal groups of 5, the number of such groups is also U. Thus:

(2) Suppose the ratio of the number of boys B to the number of girls G is $4:5$. Then if we divide the boys into equal groups of 4 and divide the girls into equal groups of 5, the number of groups is the same among boys or girls.

The preceding example is special in that the two quantities, "number of boys" and "number of girls", are whole numbers and, because A and B in the ratio $A:B$ also happen to be whole numbers, the number U has a chance to be a whole number as well (as indeed it was). Statement (2) above will not make sense if U $(= \frac{B}{4})$ is not a whole number. However, the reasoning leading up to statement (1) is perfectly general, and we state the general case as a theorem. Recall that, because of FASM, every number may be tacitly assumed to be a nonzero fraction in this discussion.

Theorem 22.1. *Suppose the ratio of two quantities M and N measured in terms of a given unit is $a:b$, where a and b are nonzero whole numbers. Then there is a fraction u so that*
$$M = au \quad \text{and} \quad N = bu.$$
In particular, if M is divided into $\boldsymbol{a}$ equal parts and N is divided into $\boldsymbol{b}$ equal parts, then the size of a part is equal to u in both cases.

The requirement of the theorem that the ratio be in terms of whole numbers rather than fractions is hardly restrictive, because if a, b are nonzero fractions, then $ab^{-1} = \frac{m}{n}$ for some nonzero whole numbers m and n, and
$$a:b = \frac{a}{b} = ab^{-1} = \frac{m}{n} = m:n.$$
Therefore we may replace $a:b$ by $m:n$, and now m and n are nonzero whole numbers.

Proof. By hypothesis, $\frac{M}{N} = \frac{a}{b}$. The cross-multiplication algorithm being valid for complex fractions (see rule (c) on page 310), we have $\frac{M}{a} = \frac{N}{b}$. Let the fraction u be defined by $u = \frac{M}{a} = \frac{N}{b}$. Then we have immediately

$M = au$ and $N = bu$. The last assertion of the theorem is a consequence of the definition of multiplication by a whole number: $au = u + u + \cdots + u$ (a times), and $bu = u + u + \cdots + u$ (b times); see (17.3) on page 268. The proof is complete. $\square$

If in the ratio $a : b$, the numbers a and b are no longer whole numbers but are fractions, then a weaker version of Theorem 22.1 is still true. In this case, recall from section 17.3 that $\frac{m}{n} \times \frac{k}{\ell}$ is $\frac{m}{n}$ copies of $\frac{k}{\ell}$ (see especially the discussion following (17.7) above Example 1).

Theorem 22.2. *Suppose the ratio of two quantities M and N measured in terms of a given unit is $a : b$, where a and b are nonzero fractions. Then there is a fraction u so that*

$$M = au \quad \text{and} \quad N = bu.$$

In particular, M is equal to $\boldsymbol{a}$ copies of u and N is equal to $\boldsymbol{b}$ copies of the same u.

Proof. It is essentially the same proof as for Theorem 22.1. By hypothesis, $\frac{M}{N} = \frac{a}{b}$. The cross-multiplication algorithm being valid for complex fractions, we have $\frac{M}{a} = \frac{N}{b}$. Let the fraction u be defined by $u = \frac{M}{a} = \frac{N}{b}$. Then we have immediately $M = au$ and $N = bu$. The last assertion of the theorem is a consequence of the second interpretation of fraction multiplication in (17.7) on page 273. The proof is complete. $\square$

In discussions of ratios in school mathematics, the case where a and b are whole numbers in $a : b$ is without a doubt the most important and the example of boys and girls above is typical. If students can achieve at least a basic understanding of this example, they would be in good shape. It may be speculated that if the phrase "the ratio of boys to girls is 4 : 5" were defined in textbooks as in either statement (1) or (2) on page 337, students would not be as confused about the meaning of *ratio*, but that hasn't happened yet. Furthermore, we now see the advantage of our definition of ratio simply as a division:

(i) It is simple and entirely unambiguous because it is a number (rather than something ineffable) and the concept of division has been carefully defined.

(ii) It entails as *logical consequences* the meanings of ratio given in statements (1) and (2).

(iii) It does not make use of any new concepts, only what we know about division.

(iv) Its formulation as a division lends itself to direct symbolic computations.

We emphasize the last point about the ease with which our definition of a ratio as a division lends itself to computations. Problem solving requires computation, and knowing a ratio $4:5$ as a fraction $\frac{4}{5}$ is far superior to knowing that it means "for every 4 boys there are 5 girls" as far as computation goes. What can one do with "for every 4 boys there are 5 girls"? In fact, what does this mean?

Here are some illustrative examples on ratios. One thing to emphasize about their solutions is that, while it is good to remember Theorems 22.1 and 22.2 and use them freely, one should be able to also obtain the solutions directly from the definition of a ratio.

PROBLEM 1. IN A CLASS OF 27 STUDENTS, THE RATIO OF BOYS TO GIRLS IS $4:5$. HOW MANY ARE BOYS AND HOW MANY ARE GIRLS?

We are going to solve the problem in four different ways, and each is worth knowing. Let the number of boys and girls be abbreviated by B and G, respectively. We are given that $\frac{B}{G} = \frac{4}{5}$ and $B + G = 27$.

First solution. By multiplying both sides of $\frac{B}{G} = \frac{4}{5}$ by the number G, we obtain $B = \frac{4}{5}G$. Substitute this value of B into $B + G = 27$ to get $\frac{4}{5}G + G = 27$, so that by the distributive law, $(\frac{4}{5} + 1)G = 27$. Thus $\frac{9}{5}G = 27$, and

$$G = \frac{27}{\frac{9}{5}} = 27 \times \frac{5}{9} = 15.$$

So $B = 27 - G = 12$.

One should always double-check: $\frac{12}{15} = \frac{3 \times 4}{3 \times 5} = \frac{4}{5}$, which is consistent with the ratio of boys to girls being $4:5$.

Second solution. This solution relies on a technique that may be new to some: we will show how to make the quantity $B + G$ appear in the given equation $\frac{B}{G} = \frac{4}{5}$ (and the reason we want $B + G$ to show up is of course the fact that we know $B + G = 27$). We add 1 to both sides of $\frac{B}{G} = \frac{4}{5}$,

$$1 + \frac{B}{G} = 1 + \frac{4}{5}$$

and use formula (14.4) on page 223 to compute the left side: $1 + \frac{B}{G} = \frac{G}{G} + \frac{B}{G} = \frac{G+B}{G}$. The right side is of course $1 + \frac{4}{5} = \frac{9}{5}$. Thus we have

$$\frac{G+B}{G} = \frac{9}{5}.$$

Now we see the appearance of $G + B$ on the left, and we know it is equal to 27. Therefore,

$$\frac{27}{G} = \frac{9}{5}$$

To find out what G is, we use the cross-multiplication algorithm again: $27 \times 5 = G \times 9$, or $9G = 135$. Multiplying both sides by $\frac{1}{9}$ gives $\frac{1}{9} \times 9G =$

22.1. Ratio

$\frac{1}{9} \times 135$, and $G = 15$. We now know there are 15 girls, and therefore 12 ($= 27 - 15$) boys, as before.

Third solution. From Theorem 22.1, we know that $B = 4u$ and $G = 5u$ for some number u. Since $B + G = 27$, we get $4u + 5u = 27$ so that $9u = 27$. So $u = 3$ and $B = 12$ and $G = 15$.

Fourth solution. This method makes use of picture drawing on the number line, and may be the most attractive of all four solutions to a beginner, but before we describe it we have to give it some perspective. While it is good to know about solutions that are accessible to picture drawing, please keep in mind that the kind of symbolic manipulation in any of the above solutions is so basic that you just have to learn them all no matter what. So learn about picture drawing, but learn the other three methods also.

Again, we start by multiplying both sides of $\frac{B}{G} = \frac{4}{5}$ with G to get $B = \frac{4}{5} \times G$. Here is the critical step: we interpret this equality by making use of the second interpretation of fraction multiplication in (17.7) on page 273. Therefore, B is the number of students in four of the parts when the girls are divided into five equal parts. Here is a pictorial representation: if the unit of the number line is one student, then G is a whole number on this number line and we can divide the segment $[0, G]$ into five equal parts.

Because the length of $[0, B]$ is the concatenation of four of the parts above, $G + B$, being the length of the concatenation of $[0, G]$ and $[0, B]$, is the concatenation of $5 + 4$ of these parts, as shown.

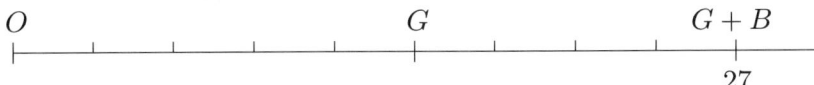

Now $G + B = 27$ and $[0, 27]$ is therefore divided into nine equal parts. Each part then has length 3, and therefore $G = 5 \times 3 = 15$ and $B = 4 \times 3 = 12$, the same as before.

Because the above solutions are vaguely reminiscent of the kind of solutions of ratio problems you have been exposed to, there is the danger that you would take the above solutions for granted as more of the same. Let it be pointed out, therefore, that every *step of the preceding reasoning is based on precise definitions or theorems that we have proved. For example, we used the definition of ratio as a division and the proven interpretation of fraction multiplication in (17.7). Never did we appeal to your a priori understanding of what a ratio is or your implicit understanding that fraction multiplication means "of ", in sharp contrast with the usual practice in the teaching of these ratio problems.*

There is a related concept that we should introduce. If A, B, C are three quantities which satisfy
$$A:B = \ell:m \quad \text{and} \quad B:C = m:n$$
for some fractions ℓ, m, n, then we write $\boldsymbol{A:B:C = \ell:m:n}$ in place of the two pairs of ratios. We also say that $\boldsymbol{A}$, $\boldsymbol{B}$, $\boldsymbol{C}$, **are in the ratio** $\boldsymbol{\ell:m:n}$. With this understood, we claim that

(22.1) $\qquad A:B:C = \ell:m:n \quad \text{implies} \quad \dfrac{A}{\ell} = \dfrac{B}{m} = \dfrac{C}{n}.$

To see this, we use a method already used in the solution of Problem 1. By repeated application of the cross-multiplication algorithm, $\frac{A}{B} = \frac{\ell}{m}$ is equivalent to $Am = B\ell$, which is in turn equivalent to $\frac{A}{\ell} = \frac{B}{m}$. The proof of $\frac{B}{m} = \frac{C}{n}$ is entirely similar. This then proves (22.1).

Activity. If A, B, C, are in the ratio $3:8:5$ and $B = 20$, what are A and C?

It is not difficult to see that the converse of (22.1) is also true (see Exercise 1 on page 355). In fact, (22.1) shows us how to extend this definition to define more than three quantities in a certain ratio.

PROBLEM 2. BENOIT, CARL, AND DAVIDA CHIP IN TO BUY A HI-FI SYSTEM. THE COST IS $440, AND THEIR CONTRIBUTIONS (IN THE ORDER OF BENOIT, CARL, AND DAVIDA) ARE IN THE RATIO OF $2:5:4$. HOW MUCH DID EACH CONTRIBUTE?

Let the contributions of Benoit, Carl, and Davida be $\$B$, $\$C$, and $\$D$, respectively. Then we are given: $B + C + D = 440$ and, on account of (22.1), $\frac{B}{2} = \frac{C}{5} = \frac{D}{4}$. Let k be the common value of $\frac{B}{2}$, $\frac{C}{5}$, and $\frac{D}{4}$. Thus
$$\frac{B}{2} = k, \quad \frac{C}{5} = k, \quad \text{and} \quad \frac{D}{4} = k,$$
so that
$$B = 2k, \quad C = 5k, \quad D = 4k.$$
Therefore $B + C + D = 2k + 5k + 4k = 11k$. But $B + C + D = 440$, so $11k = 440$, which implies $k = 40$. The answer is then: Benoit contributed $2k = 80$ dollars, Carl $5k = 200$ dollars, and Davida $4k = 160$ dollars. To double-check: $80 + 200 + 160 = 440$ and $\frac{B}{C} = \frac{80}{200} = \frac{2}{5}$, $\frac{C}{D} = \frac{200}{160} = \frac{5}{4}$.

Activity. Solve Problem 2 by imitating the fourth solution of Problem 1.

Before discussing the next problem, it would be advisable to first do the following activity.

Activity. A school district has a teacher-student ratio of $1:30$. If the number of students is 450, how many more teachers does the district need to hire in order to improve the ratio to $1:25$?

22.1. Ratio

PROBLEM 3. A SCHOOL DISTRICT HAS A TEACHER-STUDENT RATIO OF 1 : 24. IF THE NUMBER OF STUDENTS IS S, IN TERMS OF S, HOW MANY MORE TEACHERS DOES THE DISTRICT NEED TO HIRE IN ORDER TO IMPROVE THE RATIO TO AT LEAST 1 : 20?

Having done the preceding activity, you already know how to handle the situation if the number of students is known to you (e.g., 450). Although it is no longer known, the underlying reasoning does not change. Let us therefore continue the same discussion, letting T denote the number of teachers before the hiring. It is given that

$$\frac{T}{S} = \frac{1}{24}.$$

Now suppose the district hires N new teachers. The total number of teachers will be $N + T$, and we want the new teacher-student ratio $\frac{N+T}{S}$ to be at least $\frac{1}{20}$. In other words, we want N so big that

$$\frac{N+T}{S} \geq \frac{1}{20}.$$

We can work directly with this inequality, but if you are not yet used to inequalities, we can just work with the equality to solve for T, and then remember that our answer should be any number that is at least as big as the solution. Thus we try to solve for N so that $\frac{N+T}{S} = \frac{1}{20}$. The left side can be simplified:

$$\frac{N+T}{S} = \frac{N}{S} + \frac{T}{S} = \frac{N}{S} + \frac{1}{24}.$$

Therefore, to solve $\frac{N+T}{S} = \frac{1}{20}$, it is equivalent to solving

$$\frac{N}{S} + \frac{1}{24} = \frac{1}{20}.$$

But this is equivalent to

$$\frac{N}{S} = \frac{1}{20} - \frac{1}{24} = \frac{1}{120}.$$

Hence if $\frac{N}{S} = \frac{1}{120}$, the new teacher-student ratio would be equal to $\frac{1}{20}$.

Of course the equality $\frac{N}{S} = \frac{1}{120}$ is equivalent to the equality $N = \frac{S}{120}$. This means that if the number of new teachers N is equal to $\frac{S}{120}$, then the new teacher-student ratio would be equal to $\frac{1}{20}$. Therefore to achieve a teacher-student ratio of *at least* $\frac{1}{20}$, the district must hire at least $\frac{S}{120}$ new teachers. The fraction $\frac{S}{120}$ will not be a whole number for most S's (for example, if $S = 250$), so the number of new teachers to be hired is any whole number equal to or bigger than $\frac{S}{120}$.

22.2. Why Ratio?

It was pointed out that the ratio of A to B is a comparison between the two quantities A and B in the multiplicative context. For example, if A is the number of legs of all the chairs in the room and B is the number of those same chairs, then $A = 4B$, so that $\frac{A}{B} = 4$, or, $A : B = 4$. This then gives out the clear information that to each chair there are four legs.

The question that readily comes to mind is: Why compare multiplicatively if comparing additively using subtraction, $A - B$, is available? At least for the case of chairs and legs, one can see why the additive comparison is less desirable. Indeed, what would $A - B$ be? The answer is that it all depends on what B is. If there are 10 chairs (so $B = 10$), then $A - B = 30$, but if there are 17 chairs (so $B = 17$), $A - B = 51$. In general, $A - B = 4B - B = 3B$. So giving the difference of $A - B$ fails to convey, succinctly, the vital information that there are four legs to each chair.

Another well-known example where the multiplicative comparison by use of the ratio is superior to the additive comparison is the case of comparing the circumference C of a circle to its diameter D. Since $C - D = (\pi - 1)D$, but $\frac{C}{D} = \pi$, the preference for the use of ratio is clear.

A deeper understanding of ratio will have to await the study of linear functions in algebra. In the meantime, we want you to be convinced, at least, that it is worth your effort to understand what ratio is about.

22.3. Rate

Having defined ratio, we now approach the topic of "rate". Let two quantities M and N be given, so that they are now **of different types**, which means they are points on different number lines (such as M being the total distance traveled in a given time duration so that it is a point on a number line whose unit is one foot, one mile, etc., and N is the length of the time duration in question so that N is a point on another number line whose unit is one minute, one hour, etc.) Then the division $\frac{M}{N}$ will not make any sense unless M and N can be put on the same number line. This we can do by identifying the units on these number lines. That done, the division $\frac{M}{N}$ is called a **rate**. Note that we have done this kind of identification before, e.g., in the definition of the multiplication of fractions (see the introduction of Chapter 17) and in the definition of *average speed in a given time interval* in section 18.3, page 292.

Beyond the fact that a "rate" is a division, there is nothing more that needs to be said. For example, if we ask you to go back to Chapter 18 to look over the discussion of constant speed in section 18.3, **Applications**, you will see with hindsight that every statement we made about motions of constant speed is a statement about "rate", and yet *we never once discussed what a*

22.3. Rate

"*rate*" *was.* That should give you the confidence to ignore all the verbiage that goes with "rate" beyond the fact that it is a division.

The prototypical examples of a "rate" problem are those about *constant speed.* Recall that a motion is of *constant speed* if there is a constant v so that, for *any* time interval of length T, the distance D traveled during this time interval satisfies

$$\frac{D}{T} = \text{ the constant } v.$$

As we have already devoted long passages to a discussion of speed problems in Chapter 18, we will turn to some other variations on this theme here.

Next to constant speed, the most common example of a rate problem is one about water flow *at constant rate.* As is always the case in mathematics, we have to make sense of the terminology before we can embark on solving problems. The concept of water flowing at a constant rate is so similar to that of a motion of constant speed that we will dispense with all preliminary discussions and come straight to the point. Suppose during a time interval of t minutes, a total of w gallons of water comes out of the faucet. We say the **average rate of water flow in the time interval of t minutes** is[4]

$$\frac{w \text{ gallons}}{t \text{ minutes}}.$$

The **unit of the rate** is gallons per minute, or in abbreviation, **gal/min**. By definition, **the rate of the water flow is constant** if, using notation as above, there is a fixed number r so that the quotient $\frac{w}{t}$ is always equal to r for *any* time interval t. In this case, the meaning of the **rate of the water flow** is unambiguous: it is r gal/min. Furthermore, the amount of water coming out of the faucet in *any* 1-minute interval is exactly r gallons, because if the amount of water that comes out of the faucet in any given 1-minute interval is s gallons, then the average rate in the 1-minute interval, $\frac{s}{1}$ is equal to r. So $r = \frac{s}{1} = s$.

Let it be said emphatically that our interest in solving rate problems lies not only in the problems themselves but also in the *precise reasoning* used to arrive at the solutions. We want to demonstrate that, if the concept of the *constancy* of the rate is clearly enunciated and understood, no guesswork or mystery would be involved in the solutions.

A main purpose of the following discussion is therefore to call attention to the need in school mathematics for clearly explaining the concept of a constant rate *(of motion, water flow, work, etc.) before plunging into problem-solving. This need is very great indeed.*

---

[4] We will not repeat what has already been said, to the effect that to discuss the rate of water flow, we have to identify two number lines, the one where 1 stands for the unit of volume, and the other where 1 stands for the unit of time.

PROBLEM 4. A FULLY OPEN FAUCET (WITH A CONSTANT RATE OF WATER FLOW) TAKES 25 SECONDS TO FILL A CONTAINER OF $3\frac{1}{2}$ CUBIC FEET. AT THE SAME RATE, HOW LONG DOES IT TAKE TO FILL A TANK OF $11\frac{1}{2}$ CUBIC FEET?

Most people probably know how to do this problem, and most people would probably do it by "setting up a proportion".[5] The reason Problem 4 is taken up for discussion here is precisely to analyze this rote procedure and also to show how such problems can be solved by mathematical reasoning.

Let the time it takes to fill the tank be t seconds, and let the (constant) rate of water flow be r cubic feet per second. Then we know

$$\frac{11\frac{1}{2}}{t} = r.$$

We are also given that

$$\frac{3\frac{1}{2}}{25} = r.$$

Comparing these two equations, we conclude

(22.2) $$\frac{11\frac{1}{2}}{t} = \frac{3\frac{1}{2}}{25}.$$

Cross-multiply to get $25 \times 11\frac{1}{2} = 3\frac{1}{2}t$, i.e., $\frac{7}{2}t = \frac{575}{2}$. It would simplify matters if we multiply through by 2 to get $7t = 575$, and therefore $t = 82\frac{1}{7}$ seconds. So it takes $82\frac{1}{7}$ seconds to fill the tank.

Now in the standard approach to problems of this type, the emphasis is on "setting up the proportion" in (22.2), but the reason for such an equality is rarely, if ever, given. Generations of students have been mystified by this procedure of setting up proportions. But we see clearly that there is no mystery. *If* we assume the constancy of the water flow, then equation (22.2) is a very natural consequence of this assumption. So as a teacher, *you must explain clearly not only the meaning of the constancy of the water flow, but also how this assumption is used each time one solves such problems.* In particular, please do not ever again ask them to set up a proportion with no explanation. Life is too short for that.

Activity. Paul rides a bicycle at constant speed. It takes him 25 minutes to go $3\frac{1}{2}$ miles. At the same speed, how long would it take him to go $11\frac{1}{2}$ miles?

PROBLEM 5. A FAUCET WITH A CONSTANT RATE OF WATER FLOW FILLS A TUB IN 15 MINUTES. IF THE RATE OF WATER FLOW IS REDUCED BY 15%, HOW LONG WOULD IT TAKE TO FILL THE SAME TUB?

For the purpose of solving this problem, we should bring out more clearly what is given and also how to rephrase the desired conclusion in symbolic

---

[5]Or, equivalently, by using the concept of "unit rate".

22.3. Rate

language. First, if we denote by V the total volume of water (in gallons) coming out of this faucet in a 15-minute interval, then V is exactly the capacity of the tub. In this 15-minute interval, the average rate of water flow is thus $\frac{V}{15}$ gal/min. The rate of the water flow from this faucet is assumed to be constant, therefore if this constant rate is denoted by r gallons per minute, then $r = \frac{V}{15}$. Now suppose at the reduced rate of $(85\%)r$ gallons per minute, it takes t minutes to get V gallons out of the faucet (i.e., fill the same tub). Then we also have

$$\frac{V}{t} = (85\%)r,$$

from which we get immediately

$$r = \frac{100}{85} \times \frac{V}{t}.$$

But we had $r = \frac{V}{15}$, so

$$\frac{V}{15} = \frac{100}{85} \times \frac{V}{t}.$$

Multiplying both sides by $\frac{1}{V}$ gives

$$\frac{100}{85t} = \frac{1}{15}.$$

Cross-multiply to get $85t = 1500$, and we obtain

$$t = \frac{1500}{85} = 17\frac{11}{17}.$$

Hence it would take $17\frac{11}{17}$ minutes to fill the tub at the reduced water flow.

Clearly the same argument would show that, under the assumption of constant water flow, if it takes 31 minutes to fill a tub, and if we reduce the water flow by 25%, then it would take $\frac{3100}{75}$ minutes to fill the tub. In like manner, we arrive at an algorithm to solve all such problems, as follows. Assuming always constancy of the rate of the water flow, if it takes k minutes to fill a tub, and if we reduce the rate of the water flow by $N\%$, then it would take

$$\frac{k \times 100}{100 - N} \quad \text{minutes}$$

to fill the same tub. It is this algorithm that has been passed around in school classrooms, with all kinds of heuristic arguments to justify why it is true. However, our reasoning makes it clear that, if the meaning of the constancy of the rate of the water flow is understood, the algorithm is a straightforward consequence of this assumption of constancy and *there is no need to memorize the algorithm*. This is the message we want to emphasize, again and again.

Activity. Sunil mows lawns at a constant rate of r sq ft per minute (this means, if in a time interval of t minutes he mows A sq ft of lawn, then the quotient $\frac{A}{t}$ equals r for any t). He mows a certain lawn in 15 minutes. If he reduces his rate to 85% r sq ft per minute, how long would it take him to mow the same lawn?

22.4. Units

Many teachers are concerned about getting students to use the correct unit in such so-called rate problems. As a result of this concern, something called "dimension analysis" has sprung up to help students learn about changing one unit to another. Dimension analysis *is* used extensively in science and engineering as a quick check on the correct use of units because one can imagine that in physics, for example, all kinds of units have to be used to fit the occasion. Thus for the study of a motion within the lab in a time interval of three seconds, one might have second thoughts about using *miles per hour* as the unit of speed; perhaps *feet per second* or *meters per second* would be more appropriate. But even in physics, dimension analysis cannot replace the knowledge of why a unit of acceleration is m/sec^2 or a unit of momentum is kg-m/sec. There is need for a basic understanding of the processes involved. Why is this relevant? Because in school mathematics, dimension analysis is taught as a rote skill. While it is possible to explain the procedures used in dimension analysis, it is ultimately not worth the trouble to do so because the units used in school mathematics are so simple that it is far easier to forget the rote skill and just do what is needed directly. Once again, the key issue here is whether the concepts of average rate and constant rate have been presented with sufficient clarity. If students know the definitions and can follow the definitions faithfully, then they will see that there is no mystery to changing units. In this short section, we illustrate this point of view with two examples.

Suppose water comes out of a faucet at a constant rate of 5 gallons per minute. We show how to express this rate in terms of quarts and seconds. In other words, how many quarts of water come out of the faucet each second? Recall that the meaning of *constant* rate is that the average rate over any time interval is the same number, so we look at the average rate of the water flow in a time interval of one minute. We use a one-minute interval because it is given that in this time interval, 5 gallons come out of the faucet. Now one minute is 60 seconds, and each gallon is 4 quarts. Since 5 gallons is $5 \times 4 = 20$ quarts, it is therefore given that 20 quarts come out of the faucet every 60 seconds. The *average rate* in a 60 second time interval is, *by definition*, the quotient

$$\frac{20 \text{ (quarts)}}{60 \text{ (seconds)}} = \frac{20}{60} \text{ quarts per second} = \frac{1}{3} \text{ quarts/second}.$$

Since we are assuming a constant rate, we see that this average rate is in fact the constant rate, i.e., $\frac{1}{3}$ qt/sec.

Once we are more used to this reasoning, we would do the conversion directly without further ado, as follows:
$$\frac{5 \text{ gallons}}{1 \text{ minute}} = \frac{5 \times 4 \text{ qt}}{60 \text{ sec}} = \frac{20}{60} \text{ qt/sec} = \frac{1}{3} \text{ qt/sec}$$

As another example, suppose an object travels at a constant speed of $85\frac{1}{5}$ ft/sec. What is its speed in terms of mph (miles per hour)? We know 1 mile = 5280 ft, so $85\frac{1}{15}$ ft = $85\frac{1}{15} \times \frac{1}{5280}$ mi = $\frac{29}{1800}$ mi. On the other hand, 1 hour is 3600 seconds, so 1 second is $\frac{1}{3600}$ hour. The object therefore travels $\frac{29}{1800}$ mi in a time interval of $\frac{1}{3600}$ hr. By definition of average rate, the average rate of the motion in a time interval of $\frac{1}{3600}$ hr is the quotient
$$\frac{\frac{29}{1800}}{\frac{1}{3600}} \text{ mph } = 58 \text{ mph}.$$

Again, having gone through this process once, we can now compute more simply:
$$85\frac{1}{15} \text{ ft/sec} = \frac{85\frac{1}{15} \times \frac{1}{5280}}{1 \times \frac{1}{3600}} \text{ mph } = 58 \text{ mph}.$$

22.5. Cooperative Work

Another kind of popular rate problem is related to two or more people doing work together, each working at a constant rate: mowing a lawn, painting a house, etc. Students tend to be befuddled by how such work problems are solved. Textbooks do provide templates for solving these problems, but the templates are usually as mysterious as the problems themselves. As a teacher, the important thing is therefore not just to teach the correct solution but also to explain the thinking behind the solution *in a way that makes mathematical sense*. Of course, all it takes is to understand the *constancy* of the rate of work, as we shall see.

Again, constant rate is assumed throughout. *In addition, in all the problems of cooperative work, it is always assumed that when two or more people work together on a job, they work on separate parts of the job in question without interfering with each other.* Thus in the following problem, we tacitly assume that Eric and Regina can mow different parts of the lawn without interfering with each other throughout.

PROBLEM 6. IT TAKES REGINA 10 HOURS TO MOW A LAWN (IT IS A *very big* LAWN), BUT IT TAKES ERIC 12 HOURS. IF THEY MOW THE SAME LAWN TOGETHER, HOW LONG WOULD IT TAKE THEM TO FINISH MOWING THE LAWN?

We approach the problem as follows. To be able to use the definition of constant rate of mowing, we have to be able to express the fact that, for example, Regina mows the total area of the lawn in 10 hours. Not knowing what the area of the lawn is, *we give it a name, say A sq ft, in order to be able to express the given data precisely.* So if R (sq ft per hr) denotes the (constant) rate at which Regina mows the lawn, the given assumption is that

$$R = \frac{A}{10}.$$

Similarly, if E (sq ft per hr) denotes the rate at which Eric mows the lawn, then

$$E = \frac{A}{12}.$$

We are going to show that, when Regina and Eric work together, they also mow the lawn at a constant rate.[6] So fix a time interval t. Then in this interval, Regina's average rate of mowing the lawn is by definition $\frac{B}{t}$ sq ft/hr, where B is the total area of the lawn she mows in these t hours. But her rate being constant, $\frac{B}{t} = R$, which implies that $B = Rt$ sq ft. Thus Regina mows Rt sq ft. Likewise, in the same time interval, Eric mows Et sq ft. Together, then, they mow $Rt + Et = (R+E)t$ sq ft in this time interval of t hours, so that the average rate at which they mow the lawn together in this time interval is

$$\frac{(R+E)t}{t} = R + E \text{ sq ft/hr.}$$

Observe that this average rate *in the time interval of t hours turns out to be independent of t*; by definition of constant rate, we see that, working together, they mow the lawn at the constant rate of $R + E$ sq ft/hr. (This must have been what you expected in the first place. Having gone through it once, you will be able to assume it next time, but not before.)

It follows that if it takes them T hours of mowing together to finish mowing A sq ft (i.e., the whole lawn),

$$R + E = \frac{A}{T}.$$

Multiplying both sides by $\frac{T}{R+E}$ and using rules (e) and (b) of Chapter 19, we get

$$T = \frac{A}{R+E}.$$

---

[6]In elementary or middle school, this fact is always taken for granted. As a teacher, however, you must find out for yourself *why* this is true. You have to convince yourself that mathematics does not depend on personal beliefs or special private convictions. It something is true, it can be explained logically.

22.5. Cooperative Work

Since we know $R = \frac{A}{10}$ and $E = \frac{A}{12}$, we have
$$T = \frac{A}{\frac{A}{10} + \frac{A}{12}}.$$

Using rule (b) of Chapter 19 again, we can cancel the A from the numerator and denominator of the last complex fraction by multiplying both by $\frac{1}{A}$. Then
$$T = \frac{1}{\frac{1}{10} + \frac{1}{12}} = \frac{1}{\frac{11}{60}} = 5\frac{5}{11}.$$

So the time it takes them both to do it together is $5\frac{5}{11}$ hours.

Note that there were no tricks and no subtle turns in the above reasoning. Everything is just a straightforward computation using the *constancy* of the rate at which Regina or Eric mows the lawn.

The traditional method of solution is a bit different, and it goes like this: If Regina mows the whole lawn in 10 hours, then she gets $\frac{1}{10}$ of the job done in 1 hour. Likewise, Eric gets $\frac{1}{12}$ of the job done in 1 hour. Together they get $\frac{1}{10} + \frac{1}{12} = \frac{11}{60}$ of the job done in one hour. So the time it takes them to do the job completely is $\frac{1}{11/60} = 5\frac{5}{11}$ hours.

Quite apart from the issue that there are gaps in the above reasoning (which will be taken up next), what most likely strikes a beginner about the preceding solution is that the strategy seems entirely different from the strategy used in other rate problems about motion or water flow. This solution does not mention the constancy of Regina's or Eric's rate of work. Moreover, the reasoning that if they can do $\frac{11}{60}$ of the job in one hour, then they would finish the job in $\frac{1}{11/60}$ hours seems plausible but also tantalizingly elusive. A beginner would find this reasoning difficult to follow. Let us spell out where the subtlety lies. Suppose in each hour, Regina and Eric mow $\frac{1}{6}$ (instead of $\frac{11}{60}$) of the lawn; then it is clear that in 6 hours they would finish mowing the lawn, and $6 = \frac{1}{1/6}$. But when they mow $\frac{k}{\ell}$ of the lawn in one hour and $\frac{k}{\ell}$ is not something simple like $\frac{1}{6}$ (e.g., for the case at hand, $\frac{k}{\ell} = \frac{11}{60}$), then it becomes much less transparent as to why the division
$$\frac{1}{\frac{k}{\ell}}$$
would yield the *precise* answer to how long it takes them to finish mowing the lawn. In this case, only rigorous reasoning can satisfactorily explain this procedure, but no such reasoning was given.

We will now restore sanity by making sense of this solution. *Where is constant rate used implicitly?* Answer: The assertion that "if Regina mows the whole lawn in 10 hours, then she gets $\frac{1}{10}$ of the job done in 1 hour" depends on the assumption of constancy, in the following way. Let the total area of the lawn be A sq ft and let her constant rate be R sq ft/hr, then we saw above that $R = \frac{A}{10}$ sq ft/hr. In other words, she mows $\frac{1}{10} \times A$ sq ft in each hour. Of course $\frac{1}{10} \times A$ is $\frac{1}{10}$ of the lawn (see the second interpretation of fraction multiplication in (17.7) in Chapter 17), and this is the explanation for the quote. Similarly, Eric gets $\frac{1}{12}$ of the job done in 1 hour. If they work together, then as we have seen, they mow at the constant rate of

$$\frac{A}{10} + \frac{A}{12} = \left(\frac{11}{60} \times A\right) \text{ sq ft/hr,}$$

which is $\frac{11}{60}$ of the lawn in each hour. Now we come to the final step. If it takes the two of them together T hours to mow the lawn (which is A sq ft), then their average rate in this time interval of T hours is $\frac{A}{T}$ sq ft/hr, by definition. But their rate being constant, it is also $(\frac{11}{60} \times A)$ sq ft/hr. Thus

$$\frac{A}{T} = \left(\frac{11}{60} \times A\right).$$

Therefore,

$$T = \frac{A}{\left(\frac{11}{60} \times A\right)} = \frac{1}{\frac{11}{60}},$$

as claimed. Thus the traditional method of solution is correct, even if the correctness is a little difficult to discern.

When students are already thoroughly familiar with the correct solution, then they would notice that A, which is the total number of sq ft in the lawn, plays no role in the final answer. It is at this point that they can attempt to present their solution in the traditional way which talks about "the job" but makes no mention of A: If Regina mows the whole lawn in 10 hours, then she gets $\frac{1}{10}$ of the job done in 1 hour. Likewise, Eric gets . . . . But it makes no sense to present such an abbreviated solution with no explanation.

Along the line of the preceding paragraph, we can present a pictorial solution of the problem if we change the data of Problem 6 a bit. (For a beginner, a pictorial solution is helpful, but all students eventually have to learn how to make use of the constancy of the rate of work to solve the problem.) So suppose Eric can finish mowing the lawn in 15 hours (instead of 12). Then in one hour, Regina can mow $\frac{1}{10}$ of the lawn and Eric, $\frac{1}{15}$ of the lawn. Together, they can mow $\frac{1}{10} + \frac{1}{15} = \frac{1}{6}$ of the lawn in an hour. Here is a pictorial representation of the situation. If we think of the lawn as a square, then the areas mowed by Regina and Eric in one hour are darkened in the squares below:

22.5. Cooperative Work

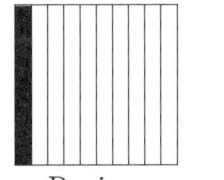
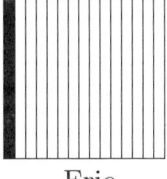

Regina   Eric

Therefore, the total area mowed by Regina and Eric together in one hour is obtained by combining the two darkened areas, and the combined area is also darkened below. The latter now represents the area that Regina and Eric working together can mow in one hour. Pictorially, one sees that 6 of these areas make up the whole square and therefore it takes 6 hours for Regina and Eric to finish mowing the lawn:

Regina and Eric

The following two activities should be compared with this problem about Regina and Eric mowing the lawn. See also Exercise 2 at the end of the chapter. We strongly recommend that these Activities be discussed in class.

Activity. Two water pipes drain into a tank. It takes the first pipe 10 hours to fill the tank, but it takes the second pipe 12 hours. If both water pipes drain into the tank at the same time, how long does it take them to fill the tank? (Remember, the rate of water flow is always assumed to be constant.)

Activity. Two shuttle trains go between cities A and B. It takes the first train 10 hours to make the trip, but it takes the second train 12 hours. Suppose now the first train is at city A and the second train is at city B and they take off at the same time on parallel tracks. How long will it be before they meet?

We conclude with an example of three people working together:

PROBLEM 7. JOSHUA, LI, AND MANFRED WORK AT A CONSTANT RATE, AND THEY ARE GOING TO PAINT A HOUSE TOGETHER. IT IS ESTIMATED THAT, INDIVIDUALLY, IT WOULD TAKE THEM 18 HOURS, 15 HOURS, AND 16 HOURS, RESPECTIVELY, TO PAINT THE WHOLE HOUSE. HOW LONG WILL IT TAKE THEM TO DO IT TOGETHER?

Let the rates at which Joshua, Li, and Manfred paint the house be J, L, and M sq ft/hr, respectively, and let there be H sq ft in the whole house. Since they paint at a constant rate, we have from the definition of constancy

(and the data of the problem) that
$$J = \frac{H}{18}, \quad L = \frac{H}{15}, \quad M = \frac{H}{16} \quad \text{(all in sq ft/hr)}.$$
Now if Joshua, Li, and Manfred paint the house together, then arguing exactly as in the preceding Problem 6, we can conclude that they paint at a constant rate of $J + L + M$ sq ft/hr, which is incidentally $\frac{37}{180} \times H$ sq ft/hr. So if it takes them t hours to paint the whole house, the average rate of painting in this time interval of t hours is $\frac{H}{t}$ sq ft/hr. By the assumption of constant rate, we have
$$\frac{H}{t} = J + L + M = \frac{37}{180} \times H.$$
We are interested in what t is. Multiplying both sides of the preceding equation by $\frac{t}{(37/180)\,H}$, we obtain, by rules (b) and (e) of Chapter 19, that
$$t = \frac{1}{\frac{37}{180}} = 4\frac{32}{37}.$$
The answer is therefore 4 and $\frac{32}{37}$ of an hour, or approximately 4 hours and 52 minutes.

Exercises

1. (a) We defined $A : B : C = \ell : m : n$ to mean $\frac{A}{B} = \frac{\ell}{m}$ and $\frac{B}{C} = \frac{m}{n}$, and proved that if $A : B : C = \ell : m : n$, then equality (22.1) holds. Now prove the converse: if equality (22.1) holds, then $A : B : C = \ell : m : n$.
   (b) Prove that every ratio $A : B$, where A and B are fractions, is equal to a ratio $c : d$, where c and d are whole numbers.

2. Two shuttle trains go between cities A and B. Both run at constant speed. It takes the first train 3 hours to make the trip, but it takes the second train 5 hours. Suppose now the first train is at city A and the second train is at city B and they take off at the same time on parallel tracks. How long will it be before they meet?

3. I drove from Town A to Town B at an average speed of 50 mph, and I drove back from Town B to Town A at an average speed of 60 mph. The round trip took $14\frac{2}{3}$ hours. How far apart are the towns?

4. I walked a trail of 12 miles and came back. My average speed for the round trip was $3\frac{1}{2}$ mph. On the way out, I maintained a constant speed of 3 mph. On the way back, I also walked with a constant speed. What was my speed on the way back?

5. The ratio of girls to boys in an auditorium is 9 : 7. If there are 336 students in the auditorium, how many are girls and how many are boys?

6. The ratio of liquor to fruit juice in a punch is 3 : 7 by volume. In the liquor the ratio of alcohol to non-alcoholic liquid is 43 : 57, also by volume. If there are 200 fluid ounces of punch, how many fluid ounces of alcohol are there?

7. Two people A and B walk straight towards each other at constant speed. A walks $1\frac{1}{2}$ times as fast as B. If they are 2000 feet apart initially, and if they meet after $2\frac{1}{2}$ minutes, how fast does each walk?

8. Peter, Gary, and Josh share \$165. The ratio of the amount belonging to Peter to the amount belonging to Gary and Josh combined is 3 : 2. Gary has \$14 more than Josh. How much does each one have?

9. We want to make a nonalcoholic fruit punch so that the amount of grape juice, the amount of lemon juice, and the amount of orange juice are in the ratio of 3 : 1 : 2. If we want 25 cups, how much grape juice, lemon juice, and orange juice should we use?

10. Because of drought, each faucet was fitted with a water-saving device to reduce the rate of its water flow by 40%. Now that the drought is over, the device is removed. Assuming as always that the rate of water flow is constant, how long does it take to fill a tank if it used to take 20 minutes?

11. In a town, two-thirds of the men are married to five-sevenths of the women. (a) Which is bigger: the number of men or the number of women? Answer this in two different ways. (b) What is the ratio of men to women? (c) If there are K married couples in this town, what is the difference between the number of men and the number of women in terms of K?

12. Recall Exercise 7 on page 317. Divide 88 into two parts A and B so that $\frac{A}{B} = \frac{2/3}{4/5}$. Do this problem again by imitating the method of the Fourth solution of Problem 1 on page 341.

13. Given that 0.69 of a pound of date sugar costs $6.20, how much date sugar could $16 buy?

14. At 8 AM two bikers, 40 miles apart, start biking towards each other. They meet at 11 AM. One biker pedals at a constant speed of 7.5 mph. What is the average speed of the other biker?[7]

15. Two faucets pour into a tub. The first faucet alone can fill the tub in 18 minutes, and the second faucet alone can fill the tub in 22 minutes. Assume the constancy of the rates of the water flow as usual. Let the first faucet be turned on for 4 minutes before the second faucet is turned on, and t minutes later the tub is filled. What is t?

16. Yesterday Yvonne drove at a constant speed of 42 mph from Town A to Town B and finished the trip in 2 hours and 24 minutes. On the way back via the same route, she again drove at constant speed and cut her time by 12.5%. How fast did she drive on the way back?

17. Joshua, Li, and Manfred mow lawns at a constant rate. How long would it take the three of them to mow a lawn if, for the same lawn, it takes Joshua and Li 2 hours to mow it together, Li and Manfred 3 hours to mow it together, and Joshua and Manfred 4 hours to mow it together?

18. Tom and May want to go from Town A to Town B. May leaves Town A 30 minutes ahead of Tom and she drives at a steady 45 mph. If Tom drives at a steady 50 mph, how many hours after May leaves will Tom catch up with her? (You may assume that Town B is very far from Town A.)

19. May drives from Town A to Town B at a constant speed and gets there in $2\frac{1}{4}$ hours. Tom drives from Town A to Town B at a constant speed of 48 mph; he leaves Town A half an hour after May and gets to Town B a third of an hour after she does. If the distance between Town A and Town B is 100 miles, what is May's speed?

---

[7]Problem from Tony Gardiner.

Chapter 23

Some Interesting Word Problems

In this section, we discuss four word problems to illustrate the diverse applications of fractions. These four are chosen for discussion because each contains an element of surprise. Notice that in each of the problems with the exception of the third, FASM is implicitly invoked.

PROBLEM 1. PAUL RODE HIS MOTORBIKE TO LANTERNTOWN AT A CONSTANT SPEED OF 15 MILES PER HOUR. FOR THE RETURN TRIP, HE DECIDED TO INCREASE HIS (STILL CONSTANT) SPEED TO 18 MILES AN HOUR. WHAT WAS THE AVERAGE SPEED OF HIS ROUND TRIP?

Recall that the concept of *average speed* was introduced in section 18.3. *According to this definition*, we need to resist the temptation of computing the "average speed" as the average of the two speeds, $\frac{15+18}{2}$, but rather must compute the division of the total distance traveled in the round trip by the travel time of the round trip. We emphasize that this is a matter of acclimating students to the use of precise definitions in mathematics, and it *has nothing to do with students' so-called conceptual understanding*.

Because the distance from Paul's hometown to Lanterntown is not known, the total distance of the round trip is therefore also not known. We seem to be stuck. In a situation like this, it is always a good policy to first try solving a simpler problem. We will assume that Paul rides the motorbike 40 miles each way and see if we can solve this simpler problem first. Thus the total distance of the round trip is $40 + 40 = 80$ miles. We have to find the total travel time. On the way out to Lanterntown, Paul covered 15 miles each hour. The number of hours needed to cover 40 miles is therefore $\frac{40}{15}$ hours (see the discussion of motion in section 18.3). On the way back,

Paul's speed is 18 miles per hour, and so the travel time by similar reasoning is $\frac{40}{18}$ hours. The round trip therefore took $\frac{40}{15} + \frac{40}{18}$ hours altogether. Hence,

$$\text{(23.1)} \qquad \text{average speed} = \frac{80}{\frac{40}{15} + \frac{40}{18}}.$$

A computation then gives the average speed as $16\frac{4}{11}$.

Activity. Verify that the answer is correct.

It happens sometimes in mathematics that, while doing something is good, *not* doing it could be even better. In this case, suppose you do not worry about getting an explicit number for an answer but just stare at equation (23.1). You might notice that the two 40's in the denominator on the right side cry out for the application of the distributive law:

$$\left(\frac{40}{15} + \frac{40}{18} \right) = \left(\frac{1}{15} + \frac{1}{18} \right) \times 40.$$

When this is done, then the number 40 appears in both the numerator (because $80 = 2 \times 40$) and the denominator of the right side of (23.1) and can therefore be cancelled, using the cancellation law (b) on page 310. Thus we get

$$\text{average speed} = \frac{2}{\frac{1}{15} + \frac{1}{18}}.$$

If we rewrite it as

$$\text{average speed} = \frac{1}{\frac{1}{2}\left(\frac{1}{15} + \frac{1}{18}\right)}.$$

then we recognize the average speed in this case as the harmonic mean[1] of the two numbers 15 and 18, the speeds of the two legs of the round trip. The distance between Paul's hometown and Lanterntown doesn't even appear! Is this an accident? Let us try again:

Activity. Do Problem 1 by assuming in addition that the distance between Paul's hometown and Lanterntown is 65 miles.

Assuming that you have done the activity, let us now do the problem in full generality. Let the distance between Paul's hometown and Lanterntown be D miles. On the way out, the travel time is $\frac{D}{15}$ hours, and on the way back, the travel time is $\frac{D}{18}$ hours. Since the total distance of the round trip is $2D$ miles, the average speed is then

$$\frac{2D}{\frac{D}{15} + \frac{D}{18}} \text{ mph.}$$

---

[1] See Exercise 16 on page 307 for the definition of this term.

23. Some Interesting Word Problems 359

Apply the distributive law to the denominator to get
$$\left(\frac{D}{15} + \frac{D}{18}\right) = \left(\frac{1}{15} + \frac{1}{18}\right) D.$$

The average speed is therefore (using rule (b) on page 310 again)
$$\frac{2D}{\left(\frac{1}{15} + \frac{1}{18}\right) D} = \frac{2}{\left(\frac{1}{15} + \frac{1}{18}\right)} = \frac{1}{\frac{1}{2}\left(\frac{1}{15} + \frac{1}{18}\right)}.$$

This shows that the average speed is the harmonic mean of the two speeds on the way and on the way back. As we noted, if we do the calculations, the answer is $16\frac{4}{11}$. But the fact that the harmonic mean appears out of nowhere reveals an unsuspected structure in the problem. In Exercise 2 on page 364, this line of reasoning is pushed to its logical conclusion.

PROBLEM 2. A TRAIN GOES BETWEEN TWO TOWNS AT CONSTANT SPEED. BY INCREASING THE SPEED BY A THIRD, BY HOW MANY PERCENT IS THE TRAVEL TIME SHORTENED?

It is always a good habit to state clearly at the outset what our goal is. To this end, we have to systematically work through the given data with the help of precise definitions. There are two sets of travel times: the *initial* travel time (**I**) and the subsequent *faster* travel time (**F**). The amount of time shortened as a result of the increase in speed is **I** − **F**. Therefore, our goal is to express *in percent* the quotient of
$$\frac{\mathbf{I} - \mathbf{F}}{\mathbf{I}}.$$

Let us proceed to compute each of **I** and **F** in terms of speed as the problem is about the change of speed. We do not know the distance between the two towns, so for the sake of discussion, let us call it D. For the same reason, let us call the initial speed s. Then
$$\mathbf{I} = \frac{D}{s}.$$

Just as in Problem 1, we adopt the strategy that, in order to get some intuitive feeling for the problem, we first do a simpler problem. Let us assume that the distance D between the two towns is 120 miles and the initial speed s is 60 mph. The original travel time **I** is then $\frac{120}{60} = 2$ hours. If the speed is increased by a third, then the new speed is $60 + (\frac{1}{3} \times 60) = 80$ mph, so that the new travel time **F** would be $\frac{120}{80} = 1\frac{1}{2}$ hours. The amount of time shortened is
$$\mathbf{I} - \mathbf{F} = 2 - 1\frac{1}{2} = \frac{1}{2} \text{ hour},$$
which (by inspection) is 25% of the original travel time of 2 hours.

Next we try $D = 200$ miles and $s = 30$ mph. Then the two travel times are $6\frac{2}{3}$ hours and $\frac{200}{40} = 5$ hours. The time saved is then $6\frac{2}{3} - 5 = 1\frac{2}{3}$ hours.

Because
$$\frac{1\frac{2}{3}}{6\frac{2}{3}} = \frac{\frac{5}{3}}{\frac{20}{3}} = \frac{1}{4},$$
again the time saved is 25% of the original travel time. One should do many such special cases until one detects a certain pattern. In this case, it is that the answer is 25% over and over again. It remains to verify this answer in general.

We now have the confidence to do the general case. Since s is the initial speed, the initial travel time **I** is $\frac{D}{s}$. The increased speed is a third greater than s, hence it is
$$s + \frac{1}{3}s = \left(1 + \frac{1}{3}\right)s = \frac{4}{3}s.$$
The shorter travel time is then
$$\mathbf{F} = \frac{D}{\frac{4}{3}s} = \frac{3}{4}\frac{D}{s} = \frac{3}{4}\mathbf{I}.$$
It follows that
$$\mathbf{I} - \mathbf{F} = \mathbf{I} - \frac{3}{4}\mathbf{I} = \left(1 - \frac{3}{4}\right)\mathbf{I} = \frac{1}{4}\mathbf{I},$$
where we have used the distributive law. Therefore, we have
$$\frac{\mathbf{I} - \mathbf{F}}{\mathbf{I}} = \frac{\frac{1}{4}\mathbf{I}}{\mathbf{I}} = \frac{1}{4},$$
by rule (b) on page 310. In terms of percent, it is easily seen to be 25%. So the travel time is shortened by 25 percent.

This is the right place to point out a special feature in the preceding application of the distributive law: $\mathbf{I} - \frac{3}{4}\mathbf{I} = (1 - \frac{3}{4})\mathbf{I}$. *We do not know what* **I** *is*, but we can confidently apply the law anyway because the law says $(ac - bc) = (a - b)c$ *no matter what a, b, c may be*. This is an example of the power of generality.

The next two problems come from Russia.

PROBLEM 3. FRESH CUCUMBERS CONTAIN 99% WATER BY WEIGHT. 300 LBS OF CUCUMBERS ARE PLACED IN STORAGE, BUT BY THE TIME THEY ARE BROUGHT TO MARKET, IT IS FOUND THAT THEY CONTAIN ONLY 98% OF WATER BY WEIGHT. HOW MUCH DO THESE DRIED-UP CUCUMBERS WEIGH?

Since 99% of 300 lbs is just water, there are $\frac{99}{100} \times 300 = 297$ lbs of water, and hence only $300 - 297 = 3$ lbs of solids. By the time the cucumbers are brought to market, some water has evaporated but the 3 lbs of solids remain unchanged, of course. Since 98% is water, the solids are now 2% of the total

23. *Some Interesting Word Problems*   361

weight. Hence if the total weight at market time is w lbs, we see that $3 = \frac{2}{100} \times w$. Using rules (b) and (e) of Chapter 19, we see that

$$w = \frac{100 \times 3}{2} = 150 \text{ lbs.}$$

We can also present a solution by the use of pictures. If the cucumbers are 99% water (by weight), then 1% is solids. We may symbolically represent the total weight of the cucumbers as a square. After dividing the square into 100 small squares of the same size (i.e., same weight), we darken one of the small squares to indicate that it represents solids:

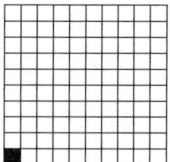

Note that since the total weight is 300 lbs, each small square is 3 lbs. Next we look a the dried-up version of the cucumbers. The solids portion remains unchanged (in weight), so the shaded small square remains the same size. But the solids are now 2% of the total weight because it is given that 98% of the weight is water. Since 2% is $\frac{1}{50}$, the shaded square is one among 50 small squares. The picture now becomes:

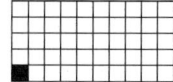

Since each small square is 3 lbs, the total weight, being 50 of the small squares, is now 150 lbs.

Discussion. A mindless application of the method of "setting up a correct proportion" would have produced the following. Let w be the weight of the cucumbers when they are brought to market. Then,

$$\frac{99/100}{300} = \frac{98/100}{w}.$$

Of course this gives $w = \frac{98}{99} \times 300 = 296.97\cdots$. This is one reason why it is so dangerous to try to teach "setting up a proportion".

PROBLEM 4. THERE IS A BOTTLE OF WINE AND A KETTLE OF TEA. A SPOON OF TEA IS TAKEN FROM THE KETTLE AND POURED INTO THE BOTTLE OF WINE. THE MIXTURE IS THOROUGHLY STIRRED AND A SPOONFUL OF THE MIXTURE IS TAKEN FROM THE BOTTLE AND POURED INTO THE KETTLE. IS THERE MORE TEA IN THE BOTTLE OR MORE WINE IN THE KETTLE? ALSO DO THE PROBLEM *without* ASSUMING THAT THE MIXTURE HAS BEEN STIRRED.

We do the stirred version first. Let the amount of wine in the bottle, the amount of tea in the kettle, and the capacity of the spoon be b cc, k cc, and s cc, respectively ("cc" means "cubic centimeter"). Using b, k, and s, we can compute the amount of tea in the bottle and the amount of wine in the kettle. In order to make you feel more comfortable with the use of symbols, we start off by assigning some specific values to b, k, and s.

So let us say, $b = 1000$ cc, $k = 2500$ cc, and $s = 5$ cc. After a spoonful of tea has been added to the bottle of wine, the amount of liquid in the bottle is $1000 + 5 = 1005$ cc. The fraction of tea in the mixture is therefore $\frac{5}{1005}$, and the fraction of wine in the mixture is $\frac{1000}{1005}$. A spoonful of the thoroughly stirred mixture would therefore contain

$$\tfrac{5}{1005} \times 5 = \tfrac{25}{1005} \text{ cc of tea} \quad \text{and} \quad \tfrac{1000}{1005} \times 5 \text{ cc of wine}$$

(see (17.7) in section 17.3). When this spoonful is poured into the kettle of tea, there would be

$$\frac{1000}{1005} \times 5 = \frac{5000}{1005} \text{ cc of wine in the kettle.}$$

On the other hand, the mixture in the bottle originally had 5 cc of tea, but since $(\frac{5}{1005} \times 5) = \frac{25}{1005}$ cc have been taken away, the amount of tea left in the bottle is

$$5 - \frac{25}{1005} = \frac{5000}{1005} \text{ cc,}$$

which is the same as the amount of wine in the kettle.

One can repeat this with different choices of values for b, k, and s.

By the way, you may have noticed that the number 2500 never appeared in the above discussion.

Now we can begin the general argument. After a spoonful of tea has been added to the bottle of wine, the amount of liquid in the bottle is $(b+s)$ cc. The fraction of tea in the mixture is $\frac{s}{b+s}$, and the fraction of wine in the mixture is $\frac{b}{b+s}$. A spoonful of the thoroughly stirred mixture would therefore contain

$$\left(\frac{s}{b+s}\right) s \text{ cc of tea} \quad \text{and} \quad \left(\frac{b}{b+s}\right) s = \frac{bs}{b+s} \text{ cc of wine}$$

(again, see (17.7) in section 17.3). When this spoonful is poured into the kettle of tea, there would be

$$\frac{bs}{b+s} \text{ cc of wine in the kettle.}$$

On the other hand, the mixture in the bottle originally had s cc (one spoonful) of tea, but since $\left((\frac{s}{b+s})s\right)$ cc has been taken away, the amount of tea

left in the bottle is
$$s - \left(\frac{s}{b+s}\right)s = s - \frac{s^2}{b+s} = \frac{s(b+s)}{b+s} - \frac{s^2}{b+s} = \frac{sb+s^2-s^2}{b+s} = \frac{sb}{b+s} \text{ cc},$$
which is the same as the amount of wine in the kettle.

We again take note that k did not figure at all in the discussion of the problem.

Now the "unstirred" case. Suppose the spoonful of mixture—to be taken from the bottle to the kettle—contains T cc of tea and W cc of wine, where T and W are some positive numbers that satisfy $T + W = s$ (recall that, as before, s denotes the capacity of the spoon). Therefore when the spoonful of mixture is poured into the kettle, the amount of wine in the kettle is W cc. On the other hand, the bottle of mixture originally had s cc of tea. But with T cc of the tea taken away by the spoon, only $(s-T)$ cc of tea is left in the bottle. Since $(s-T) = W$, the amount of tea in the bottle is thus equal to the amount of wine in the kettle, as before.

The surprising aspect of the second solution is that, since it does not depend on any assumption about whether or not the mixture has been stirred, it supersedes the first solution. Thus the precise calculations of the first solution were completely unnecessary! Nevertheless, the first solution is a valuable exercise in thinking about fractions and should not be thought of as a waste of time.

Exercises

1. Paul rode his motorbike to Lanterntown at an *average speed* of 15 miles per hour. On the way back, he decided to go faster and his *average speed* was 18 miles an hour. What is the average speed of his round trip?

2. Paul rode his motorbike to Lanterntown at an average speed of x miles per hour. On the way back, his average speed was y miles an hour. What is the average speed of his round trip?

3. A law firm has a men-to-women ratio of 5 : 1. The firm wants to reduce it to 4 : 1. What percentage increase in women would make this increase possible?

4. Because of drought, each faucet is now fitted with a water-saving device to reduce the rate of water flow by 35%. If it takes 10 minutes to fill a tub now, how long did it use take to fill it before the drought? (Assume the rate of water flow stays constant throughout.)

5. For alcoholic beverages, "200 proof" means "100%" by volume, so that "120 proof" means "60% by volume". Suppose 150 bottles of 120-proof vodka were left in the vault without the lid on, and by the time the mistake was discovered, 90% of the alcohol had evaporated. Assuming for simplicity that there was no evaporation of the remaining fluid, what is the proof of the vodka now, and how many bottles' worth is there?

6. Given two bottles of liquor (of different sizes), one is 50 proof and the other 140 proof (see the preceding problem for the meaning of *proof*). Suppose the two bottles contain the same amount of nonalcoholic fluid, say 2 cups. What is the amount of alcohol in each bottle in terms of cups?

7. Explain to a sixth grader directly why $\frac{2}{5}$ of a sack of rice weighing 12 pounds would weigh $\frac{2}{5} \times 12$ pounds. You may assume that he knows the product formula (17.2) on page 267, but you should begin by reminding him the definition of "$\frac{2}{5}$ of 12".

8. A water tank contains 271 gallons of water when it is $\frac{19}{23}$ of its full capacity. What is its full capacity?

9. Mr. Dennis took his students to a play and he was disconcerted by the fact that only $\frac{2}{5}$ of the students showed up. If 52 students showed up, how many students did Mr. Dennis have?

10. The women in a policy think tank used to comprise $\frac{2}{3}$ of the workforce. After getting three more women, the ratio of women to workforce is now 11 : 16. How many women were there before?

11. A bottle of 70% rubbing alcohol (i.e., alcohol is 70% of the liquid) was left open and three days later, the liquid had become a 50% alcohol. Percentage-wise, how much alcohol had evaporated if the water had negligible evaporation?

12. Colin and Lynn saw a CD set that they wanted to buy, but neither had enough money. Lynn could pay for 70% of the cost, and if Colin would contribute $\frac{2}{3}$ of what he had, they could take home the set and Colin would have $9 left. How much is the CD set, and how much money did Colin and Lynn have individually?

13. There are two recipes for making banana bread. One calls for $\frac{5}{8}$ cups of sugar for 4 cups of flour, and another calls for $\frac{9}{10}$ cups of sugar for 6 cups of flour. All other things being equal, which banana bread would taste sweeter?

>If you ever write a recipe like this, you should not get into the culinary business!! Learn to use easier fractions, such as $\frac{3}{4}$ or $\frac{2}{3}$, but certainly not $\frac{9}{10}$. On the other hand, if you decide that you do not want to do this problem because the numbers are not "real-world", then (1) you do not know much about fractions, and (2) unless you are willing to learn more about fractions, you should think twice before deciding to pursue a career as an elementary teacher. By the way, there are two ways to think about which banana bread is sweeter: which has more sugar in each cup of the sugar-flour mixture, or which sugar-flour ratio is bigger. The two ways are of course equivalent. See Theorem 15.2 on page 244.

Chapter 24

On the Teaching of Fractions in Elementary School

The following is a slight revision of an unsuccessful grant proposal written in October of 1999 to NSF-EHR, asking for $50,000 to teach fractions to teachers in the manner of Part 2 of this book.

It is widely recognized that there are at least two major bottlenecks in the mathematics education of grades K–8: the teaching of fractions and the introduction of algebra. Both are in need of an overhaul. I hope to make a contribution to the former problem by devising a new approach to elevate teachers' understanding of fractions.

The need for a better knowledge of fractions among teachers has no better illustration than the following anecdote related by the mathematician Herb Clemens ([**Cle95**]):

> Last August, I began a week of fractions classes at a workshop for elementary teachers with a graph paper explanation of why $\frac{2}{7} \div \frac{1}{9} = 2\frac{4}{7}$. The reaction of my audience astounded me. Several of the teachers present were simply terrified. None of my protestations about this being a preview, none of my "Don't worry" statements had any effect.

This situation cries out for improvement. Through the years, there has been no want of attempts from the mathematics education community to improve the teaching of fractions ([**Lam99**], [**BC89**], [**LB98**], to name just a few), but real success has proven to be elusive. By analyzing these attempts and the existing texts on fractions, for both schools and professional development, one detects certain persistent problematic areas in both the theory and practice, and they can be briefly described as follows:

(1) Most concepts related to fractions are never clearly defined. Fraction, mixed number, decimal, percent, ratio, etc.—each is usually presented by way of a metaphor, e.g., $\frac{1}{3}$ is *like* what you get when you cut a pizza into three "equal" pieces. But mathematics is *precise* and cannot be taught by metaphors alone, or if it is taught by metaphors, then we should not ask students to do precise computations or require them to reason precisely.[1]

(2) The linguistic complexities associated with the common usage of fractions are emphasized from the beginning at the expense of the underlying mathematical simplicity of the concept.

(3) The meaning of each of the four arithmetic operations is almost never given, so that the rules of the operations have to be made up on an ad hoc basis, unrelated to the usual four operations on whole numbers with which students are familiar.

(4) *Mathematical* explanations of essentially all aspects of fractions are lacking.

These four problems are interrelated and are all fundamentally mathematical in nature. For example, if one never gives a clear-cut definition of a fraction, one is forced to "talk around" every possible interpretation of the many guises of fractions in daily life in an effort to overcompensate. A good example is the over-stretching of a common expression such as "a third of a group of fifteen people" into a main theme in the teaching of fractions ([**Moy96**]). Or, instead of offering *mathematical* explanations to children of what the usual algorithms mean and why they are reasonable—a simple task *if* one starts from a precise definition of a fraction—algorithms are justified through "connections among real-world experiences, concrete models and diagrams, oral language, and symbols" ([**Hui98**, page 181]; see also [**LB98**] and [**Sha98**]). It would be so much simpler and so much more to the point if an honest mathematical explanation were given. It is almost as if one makes the concession from the start: "We will offer everything but the real thing".

Let us look more closely at the way fractions are introduced in the classroom. Children are told that a fraction $\frac{c}{d}$, with positive integers c and d, is simultaneously at least five different objects (cf. [**Lam99**] and [**RSLS98**]):

(a) parts of a whole: when an object is equally divided into d parts, then $\frac{c}{d}$ denotes c of those d parts.

---

[1] See [**Wu10b**].

(b) the size of a portion when an object of size c is divided into d equal portions.

(c) the quotient of the integer c divided by d.

(d) the ratio of c to d.

(e) an operator: an instruction that carries out a process, such as "$\frac{2}{3}$ of".

It is clear that even those children blessed with an overabundance of faith would balk at accepting a concept so magical as to fit all these descriptions all at once. How could this glaring "crisis of confidence" in fractions be consistently overlooked? More importantly, such an introduction to a new topic in mathematics is contrary to every mode of mathematical exposition that is deemed acceptable by modern standards. Yet, even Hans Freudenthal, a good mathematician at the time he switched over to mathematics education, made no mention of this central credibility problem in his Olympian pronouncements on fractions ([**Fre83**]).

Of the existence of such a crisis of confidence, there is no doubt. In 1996, a newsletter for teachers from the mathematics department of the University of Rhode Island devoted five pages of its January issue to "Ratios and Rational Numbers" ([**CRE96**]). The editor wrote:

> This is a collection of reactions and responses to the following note from a newly appointed teacher who wishes to remain anonymous:
>
> "On the first day of my teaching career, I defined a rational number to my eighth grade class as a number that can be expressed as a ratio of integers. A student asked me: What exactly are ratios? How do ratios differ from fractions? I gave some answers that I was not satisfied with. So I consulted some other teachers and texts. The result was confusion ...."

This is followed by three pages worth of input from teachers as well as the editor on this topic, each detailing his or her inconclusive findings after consulting existing texts and dictionaries (!).[2]

In a similar vein, Lamon wrote: "As one moves from whole number into fraction, the variety and complexity of the situation that give meaning to the symbols increases dramatically. Understanding of rational numbers involves the coordination of many different but interconnected ideas and interpretations. There are many different meanings that end up looking alike when they are written in fraction symbol" ([**Lam99**, pages 30–31]). All the while, students are told that no one single idea or interpretation is sufficiently clear to explain the "meaning" of a fraction. *This is a pedagogical disaster,* and to explain why this is so, let us draw an analogy. Suppose you

---

[2] Allow me to make a trite observation: on technical matters related to science or mathematics, it is usually a fatal mistake to appeal to dictionaries as the ultimate arbiters of truth.

are trying to get driving directions to a small town. Which would you prefer: getting fifty written suggestions on what landmarks to watch out for, what to do at each fork of the road, and how to interpret each road sign along the way, or simply *getting a clearly drawn road map*? So long as nobody believes in giving clear-cut definitions in the teaching of fractions, students will forever "do" fractions without any idea of what they are doing ([**LB98**], [**Lam99**]). After all, what else can they do when they are consistently fed information that is mathematically garbled and incorrect? More pertinent is the following question:

> *Why do we blame students for not learning fractions when it certainly is our fault that we do not teach the subject in a way that can be learned?*

For example, it is certainly difficult for children to be told that a fraction is a piece of pie (according to interpretation (a) above), and then be taught that fractions can be multiplied. *How can you multiply two pieces of pie?*[3]

Sometimes one could "get by" a mathematical concept without a precise definition if its rules of operation are clearly explained. Conjecturally, that was how Europeans in the 16th and 17th centuries dealt with negative numbers. In the case of fractions, however, this is not true even when interpretation (b) of fractions is used. In book after book, fractions are added, multiplied, and divided with no explanation of what the operations mean and with no effort to relate these operations to the same operations on whole numbers. Even when such an attempt is made, the good intention is sabotaged by mathematical flaws, such as the attempt to explain the division of fractions as repeated subtraction (cf. [**PSS00**, page 219]).

That fractions induce math anxiety and mathphobia is no longer news (cf. [**Ash02**]). Informal surveys among teachers also consistently reveal that many of their students simply give up learning fractions at the point of the introduction of addition. It is probably not just a matter of being confused by gcd and lcm, which are central to the traditional way of adding fractions, but more likely a feeling of bewilderment and disgust at being forced to learn a new way of doing addition that seems to bear no relationship to what they already know about addition, namely the addition of *whole numbers*. This then brings us to the problem area (3) at the beginning of this article. We see, for example, that Bezuk and Cramer ([**BC89**, page 156]) willingly concede that

> Children must adopt new rules for fractions that often conflict with well-established ideas about whole numbers.

New rules? In mathematics, one of the ultimate goals is to achieve simplicity. In the context of learning, it is highly desirable, perhaps even mandatory,

---

[3]I have appropriated this question from a delightful article by Kathleen Hart in [**Har00**].

that we convey this message of simplicity to students. However, when we tell students that a concept as simple as *addition* must be different for whole numbers and fractions, we are certainly teaching them incorrect mathematics (see the introduction of Chapter 14). Even when students are willing to suspend disbelief to go along on such a weird journey, they pay a dear price. Indeed, there are recurrent reports of students at the University of California at Berkeley and at Stanford University claiming in their homework and exam papers that $\frac{a}{b} + \frac{a}{c} = \frac{a}{b+c}$ and $\frac{a}{b} + \frac{c}{d} = \frac{a+c}{b+d}$.

All in all, a mathematician approaching the subject of fractions in school mathematics cannot help but be struck by the total absence of the characteristic features of mathematics: precise definitions as starting point, logical progression from topic to topic, and most importantly, explanations that accompany each step. This is not to say that the teaching of fractions in elementary school should be rigidly formal from the beginning. Fractions should be informally introduced no later than the second grade (because even second graders need to worry about drinking "half a glass" of orange juice!), and there is no harm done in allowing children to get acquainted with fractions in an intuitive manner up to about the fourth grade. An analogy may be helpful here. The initial exploration of fractions may be taken as the "data-collecting phase" of a working scientist: just take it all in and worry about the "how and why" later. In time, however, the point will be reached when said scientist must sit down to organize and theorize about his or her data. So it is that when students reach the fifth grade ([**CAF99**]) or the sixth grade ([**PSS00**]), their mathematical development cannot go forward unless "miracles", such as having one object $\frac{c}{d}$ enjoying the five different properties of (a)–(e) above, are fully explained and rules such as $\frac{a}{b} / \frac{c}{d} = \frac{ad}{bc}$ are justified. And it is at this critical juncture of students' mathematical education that I hope to make a contribution.

The work done on the teaching of fractions thus far has come mainly from the education community. Perhaps because of the recent emphasis on situated learning, discussions of the teaching of fractions in the education research literature tend to stay *at the source*, in the sense that attention is invariably focused on the interpretation of fractions in a "real world" setting. Such an emphasis seems to ignore the fact that when fractions are presented to beginners in a variety of contexts with the attendant myriad interpretations, they tend to get confused. Because students are not given the mathematical underpinnings of these interpretations, they end up being exposed only to the raw data but not the theory-building that makes sense of fractions (see [**Wu08**]). They are denied access to learning about an essential component of doing mathematics, namely, *when confronted with complications, try to abstract in order to achieve understanding.* Even children should be exposed to correct mathematical thinking as early as possible, and in a manner as simple as possible. Students' first serious encounters with the computation of fractions—generally in the fifth and sixth grades—would be the right moment in the school curriculum to begin emphasizing the abstract component of mathematics and make the abstraction a key point of classroom instruction (again, see [**Wu08**]). By so doing, one would also be giving students a head start in their quest to learn algebra. The ability to abstract, so essential in algebra, should be taught gently and early in the school curriculum, which would mean during the teaching of fractions. By giving abstraction its due in teaching fractions, we would be easing students' passage to algebra as well (cf. [**Wu01**], [**NMP08a**], [**NMP08b**], [**Wu09a**]).

It takes no insight to conclude that two things have to happen if mathematics education in K–8 is to improve: there must be textbooks that treat fractions logically, and teachers must have the requisite mathematical knowledge to guide their students through this rather sophisticated subject. I propose to take up the former problem by writing a book to improve teachers' understanding of fractions.

The first and main objective of this book is to give a treatment of fractions and decimals for teachers of grades 5–8 which is mathematically correct in the sense that everything is explained and the explanations are sufficiently elementary to be understood by elementary school teachers. In view of what has already been said above, an analogy may further explain what this book hopes to accomplish. Imagine that we are mounting an exhibit of Rembrandt's paintings, and a vigorous discussion is taking place about the proper lighting to use and the kind of frames that would show off the paintings to best advantage. Good ideas are also being offered on the printing of a handsome catalogue for the exhibit and the proper way to publicize the exhibit in order to attract a wider audience. Then someone takes a closer look at the paintings and realizes that all these good ideas might come to naught because some of the paintings are fakes. So finally people see the need to focus on the most basic part of the exhibit—the paintings—before allowing the exhibit to go public. In like manner, what this book would try to do is to call attention to the need of putting the *mathematics* of fractions in proper order before considering pedagogical strategies and classroom activities in the actual teaching. See [**Wu08**] for further discussions along this line.

Looking ahead, this book—and similar volumes for middle school and high school teachers—which tries to pull the teaching of fractions and rational numbers into the mathematical mainstream, can only be a beginning. Much remains to be done in terms of bringing mathematical integrity back to school textbooks, pre-service professional development, education research, and the school mathematics curriculum. All this requires long-term collaborations between mathematicians and educators. A general discussion of the framework for such a collaboration is given in [**Wu06**].

Part 3

Rational Numbers

Chapter 25

The (Two-Sided) Number Line

We are going to revisit the number line which, up to now, has been scrutinized only on the right side of the point we designate as 0. It is time that we make full use of the entire number line, both to the left and to the right. Because we already have the fractions to the right of 0, we now look at the collection of numbers (i.e., points on the number line) to the left of 0 that are obtained by reflecting the fractions across 0. The fractions together with their reflected images will be seen to form a number system in the sense that we can perform the four arithmetic operations on them all, in a way that is consistent with the operations already defined on the fractions. This number system, called the **rational numbers**, is the subject of this chapter.

Recall that a **number** is by definition a point on the number line (Chapter 8, page 128). We now look at all the numbers as a whole. Take any point p on the number line—p could be on either side of 0 and, in particular, it does not have to be a fraction—and we will define its **mirror reflection** p^* as follows. If $p = 0$, we define 0^* to be 0. If $p \neq 0$, then p^* is the point of the number line on the opposite side of 0 so that p and p^* are equidistant from 0. Thus we have

(25.1) $\qquad 0^* = 0 \qquad$ and $\qquad p^{**} = p$

for any point p. The first is just a definition, and the second is nothing but a succinct way of expressing the fact that reflecting a point across 0 twice in succession brings the point back to itself. Here are two examples of reflecting two points p and q in the manner described:

Because the fractions are to the right of 0, the numbers such as 1^*, 2^*, or $(\frac{9}{5})^*$ are to the left of 0. Here are some examples of the reflections of fractions (remember that fractions include whole numbers):

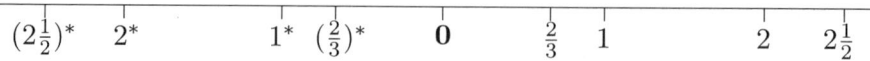

The set of all the fractions and their mirror reflections with respect to 0, i.e., the numbers $\frac{m}{n}$ and $(\frac{k}{l})^*$ for all whole numbers k, l, m, n ($l \neq 0$, $n \neq 0$), is called the **rational numbers**. Recall that the whole numbers are contained in the fractions. The set of whole numbers and their mirror reflections,

$$\ldots, 3^*, 2^*, 1^*, 0, 1, 2, 3, \ldots$$

is called the **integers**. Using "⊂" to denote "is contained in", we therefore have

whole numbers ⊂ integers ⊂ rational numbers.

Next, we extend the concept of **order** among numbers from fractions to all numbers: for any x, y on the number line, $\boldsymbol{x < y}$ means that x is to the left of y. An equivalent notation is $\boldsymbol{y > x}$.

In particular, the symbols $x < y$ have the same meaning as before when x and y are fractions.

Numbers which are to the right of 0 (thus those x satisfying $x > 0$) are called **positive**, and those which are to the left of 0 (thus those that satisfy $x < 0$) are **negative**. So 2^* and $(\frac{1}{3})^*$ are negative, while all nonzero fractions are positive. *The number 0 is, by definition, neither positive nor negative.*

You are undoubtedly accustomed to writing, for example, 2^* as -2 and $(\frac{1}{3})^*$ as $-\frac{1}{3}$. You also know that the "−" sign in front of -2 is called the **negative sign**. So you may wonder why we employ this $*$ notation and avoid mentioning the negative sign up to this point. The reason is that the negative sign, having to do with the operation of subtraction, simply will not figure in our considerations until we begin to subtract rational numbers. Moreover, the terminology of "negative sign" carries certain psychological baggage that may interfere with the learning of the basic facts about rational numbers the proper way. For example, if $a = -3$, then there is nothing "negative" about $-a$, which is 3. It is therefore best to withhold introducing the negative sign until its natural appearance in the context of subtraction in Chapter 27.

Chapter 26

A Different View of Rational Numbers

In the same way that we added more numbers to whole numbers to get the fractions, we have now added more numbers to fractions to obtain the rational numbers. Before proceeding any further, we should ask why we bother with rational numbers. To answer this question, we take up the problem of solving equations.

If we ask which whole number x has the property that when multiplied by 7 it equals 5, the answer is obviously "none". By allowing x to be a fraction instead of just being a whole number, the situation changes radically: the problem now has an answer and it is $\frac{5}{7}$. We express this fact by saying that $x = \frac{5}{7}$ is **a solution of the equation** $7x = 5$. Similarly, the equation $3x = 14$ has a solution $x = \frac{14}{3}$, and in general if m, n are an arbitrary pair of *whole numbers* ($m \neq 0$), then the equation $mx = n$ has solution $x = \frac{n}{m}$. Recall that in order for $\frac{n}{m}$ to make sense as a solution, the concept of multiplication must be first extended from the whole numbers to fractions.

It may be mentioned that fractions are sometimes defined in advanced mathematics as the set of all solutions of the equation $mx = n$, where m and n are whole numbers. For school mathematics, however, this is not at all a good definition of a fraction. The logical development that such a definition entails requires a level of mathematical sophistication that is not appropriate for the school environment. In fact, the acquisition of this kind of abstract thinking is one of the major hurdles that stands in the way of college math majors. For this reason, the professional development of teachers should steer clear of such seductive definitions unless absolutely necessary.

In any case, once fractions are introduced, they exceed our expectations by solving a larger class of equations, namely, all equations of the type $Mx = N$, where M and N are themselves *fractions* (rather than just whole numbers). For example, from section 18.2, we see that the solution of $\frac{2}{7}x = \frac{3}{5}$ is $\frac{7}{2} \times \frac{3}{5}$, which is $\frac{21}{10}$. Similarly, the equation $\frac{11}{3}x = \frac{2}{5}$ has solution $\frac{3}{11} \times \frac{2}{5}$, which is $\frac{6}{55}$. More generally, if A, B are an arbitrary pair of *fractions* ($B \neq 0$), then the solution of the equation $Bx = A$ is the fraction $B^{-1}A$.

Let us now consider the same problem for the two numbers 7 and 5, but with multiplication replaced by addition, i.e., we ask which whole number x has the property that when added to 7 it equals 5. The answer is again that there is no such whole number, so that the equation $7 + x = 5$ *has no solution among whole numbers*. Unfortunately, allowing x to be a fraction does not improve the situation this time around.

Now if we add the negative fractions—the mirror reflections of the fractions—to fractions to obtain the rational numbers, and if the concept of addition is correctly extended from fractions to the rational numbers, then there will be solutions to every equation $a + x = b$ for any fractions a and b. For example, we shall see that $x = 2^*$ is the solution to $7 + x = 5$, and $y = (\frac{1}{4})^*$ is the solution to $\frac{1}{2} + y = \frac{1}{4}$. Then, as before, once the rational numbers are introduced this way, they provide solutions to all equations of the type: $A + x = B$, where A and B are now *rational numbers* (rather than just fractions).

It is entirely legitimate to formally introduce rational numbers as the set of all solutions of the equation $a + x = b$ where a and b are any fractions. This is a fairly standard procedure in advanced mathematics, but again it is not a good strategy to define rational numbers this way for the professional development of teachers.

But we are ahead of ourselves. Before we can talk about solving $a + x = b$ for arbitrary fractions a and b, we must first deal with the problem of how to add and multiply rational numbers. This will be the task of the next three chapters. It will be seen that, once the arithmetic of the rational numbers has been firmly established, every linear equation $Ax + B = Cx + D$ for *rational numbers A, B, C, and D* (i.e., not just fractions), with $A - C \neq 0$ will have a solution x among rational numbers. See Exercise 5 on page 420.

Our experience with fractions in Part 2 has conditioned us to expect that when addition and multiplication are well understood, subtraction and division will be straightforward because the latter are nothing more than alternate ways of writing the former. In fact, subtraction and division will be formally replaced by addition and multiplication by the end of Chapter 30.

The remainder of Part 3 will be devoted mainly to a discussion of these four basic operations on rational numbers.

Chapter 27

Adding and Subtracting Rational Numbers

To approach the addition and subtraction of rational numbers, it would be best if we step back and reflect a little on our journey from whole numbers to fractions. At the beginning of Part 2, we had just added many more points of the number line to the whole numbers. Our task then was to make sense of adding and subtracting these new numbers, in a way that would not contradict what we already knew about whole numbers. In other words, we wanted to *define* the addition and subtraction of fractions so that, if the fractions happen to be whole numbers, the new addition and subtraction would coincide with the existing operations on whole numbers. We found that we could do it by appealing to the concept of concatenating segments.

Our situation now is that we can do addition and subtraction on fractions, i.e., the rational numbers on the right of 0. We are going to slightly extend the concept of concatenating segments and, in so doing, we bring the rational numbers to the left of 0 into the fold at least as far as addition and subtraction are concerned. The way this is done is to make every rational number into a "vector", and then define the addition of vectors. This addition of rational numbers will turn out to be the same as the addition of fractions when the rational numbers are positive. The surprising twist here is that the *subtraction of rational numbers will end up being defined in terms of addition.*

This chapter may be called the "down-to-earth" approach to the addition of rational numbers. However, a full understanding of rational numbers, even at the level of school mathematics, requires a higher level of abstraction. In terms of students' learning trajectory, the intellectual jump from fractions to rational numbers is roughly comparable to the intellectual jump from whole numbers to fractions. The next chapter will take up the addition and subtraction of rational numbers again, but will reformulate it in more abstract language. Far from abstraction for abstraction's sake, this reformulation will smooth the passage into the realm of multiplication in Chapter 29.

The sections are as follows:

> Definition of Vectors
>
> Vector Addition for Special Vectors
>
> Addition of Rational Numbers
>
> Explicit Computations
>
> Subtraction as Addition

27.1. Definition of Vectors

While the addition of fractions involves only the concatenation of segments, the addition of rational numbers will need something slightly more complicated than segments. Specifically, it needs the concept of a *vector*, which is, naively, a segment with a "direction". In greater detail, a **vector** is a segment on the number line together with a designation of one of its two endpoints as a **starting point** and the other as an **endpoint**. We shall refer to the length of the segment as the **length** of the vector, and say the vector is **left-pointing** if the endpoint is to the left of the starting point, and **right-pointing** if the endpoint is to the right of the starting point. The **direction** of a vector refers to whether it is left-pointing or right-pointing. We denote vectors by placing an arrow above the letter, e.g., $\vec{A}$, $\vec{x}$, etc., (our convention here will be a right-pointing arrow above the letter; actual vectors may be right- or left-pointing); in pictures we put an arrowhead at the endpoint of a vector to indicate its direction. For example, the vector $\vec{K}$ below is left-pointing and has length 1, with a starting point at 1^* and an endpoint at 2^*, while the vector $\vec{L}$ is right-pointing and has length 2, with a starting point at 0 and an endpoint at 2.

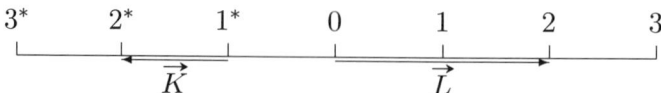

For the purpose of studying rational numbers, we do not need to consider arbitrary vectors, only very special ones that are intrinsically tied to a number, i.e., a point on the number line. Let x be a number (not necessarily rational), then we define the vector $\vec{x}$ to be the one with its starting point at 0 and endpoint at x. It follows from the definition that, *if x is a nonzero*

fraction, then the segment of the vector $\vec{x}$ is exactly $[0, x]$. Here are two examples: $\overrightarrow{1.5}$ and $\overrightarrow{3^*}$.

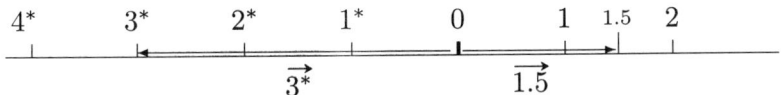

A **special vector** is a vector whose starting point is 0. Thus if x is a number, then $\vec{x}$ is a special vector. Conversely, if a special vector has y as its endpoint, then it is equal to $\vec{y}$.

Activity. Check the last assertion.

The advantage of dealing with special vectors is that their starting point is always 0. To determine the endpoint of a special vector, it suffices to know its length and its direction. Hence:

> *A special vector is completely determined by its length and its direction.*

For example, if $\vec{x}$ has length 3.7 and is left-pointing, then $x = (3.7)^*$. If $\vec{y}$ has length $\frac{23}{7}$ and is right-pointing, then $y = \frac{23}{7}$.

27.2. Vector Addition for Special Vectors

As a prelude to defining the addition of rational numbers, we describe how to add special vectors. Let x and y be numbers. Then the **sum** $\vec{x} + \vec{y}$ is the special vector whose endpoint is the new endpoint of $\vec{y}$ *after* we slide $\vec{y}$ along the number line until its original starting point (which is 0) rests at x.

In intuitive language, the sum $\vec{x} + \vec{y}$ is the special vector whose endpoint is obtained by "*going from the point x along a vector with the same length and same direction as $\vec{y}$.*" For example, here is the pictorial representation of $\vec{3} + \overrightarrow{1^*}$: it is the special vector $\vec{2}$ because its endpoint is indicated by the vertical up-arrow:

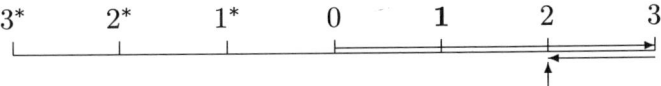

As another example, the sum $\overrightarrow{3^*} + \overrightarrow{1.2}$ is the special vector $\overrightarrow{(1.8)^*}$ whose endpoint is indicated by the down-arrow:

Because $\vec{x}$ and $\vec{x^*}$ have the same length but opposite directions, we see that
$$\vec{x} + \vec{x^*} = \vec{0}$$

Activity. Check this!

The following observation is critical in dealing with the addition of vectors:

Key Observation. *Let x and y be rational numbers, then*

(i) *the direction of the sum $\vec{x} + \vec{y}$ is the direction of the vector whose length is greater, and*

(ii) *the length of the sum $\vec{x} + \vec{y}$ is the sum of the lengths of the vectors if they have the same direction, and is the difference[1] of the lengths of the two vectors if they have different directions.*

Proof. The verification of (i) follows directly from the definition of adding special vectors for the case that $\vec{x}$ and $\vec{y}$ have the same direction. If they have different directions, then the following pictures take care of all possibilities. First, suppose $\vec{x}$ is left-pointing and is longer than $\vec{y}$. Then $\vec{x} + \vec{y}$ is left-pointing and the tip of the vertical arrow is the endpoint of $\vec{x} + \vec{y}$.

If $\vec{x}$ is left-pointing but is shorter than y, then $\vec{x} + \vec{y}$ would be right-pointing and the tip of the vertical arrow is the endpoint of $\vec{x} + \vec{y}$.

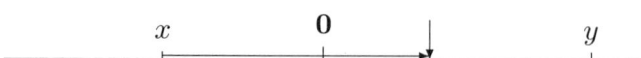

Next, suppose $\vec{x}$ is right-pointing but is longer than $\vec{y}$. The tip of the vertical arrow is the endpoint of $\vec{x} + \vec{y}$, which is right-pointing.

Finally, $\vec{x}$ is right-pointing and is shorter than $\vec{y}$. Then $\vec{x} + \vec{y}$ is left-pointing.

---

[1] We understand "difference" to be the longer length minus the shorter length. In particular, it is always positive.

The validity of (ii) then also follows from the same pictures because the distance of the endpoint of $\vec{x} + \vec{y}$ from 0 is always the difference of the longer length and the shorter length. The proof of the Key Observation is complete. □

There are some easy consequences of the Key Observation. The most obvious is the following.

(iii) *If the two rational numbers x and y are both positive (respectively, both negative), then $\vec{x} + \vec{y}$ is right-pointing (respectively, left-pointing), and the length of $\vec{x} + \vec{y}$ is the sum of their lengths.*

Moreover, insofar as the sum of any two special vectors is determined by the direction and the length of the sum vector and insofar as the direction and length of the sum vector are independent of the order of the addition (see (i) and (ii) in the Key Observation), we see the following.

The addition of special vectors is commutative, i.e., $\vec{x} + \vec{y} = \vec{y} + \vec{x}$ for all rational numbers x and y.

The Key Observation also implies that

(iv) *If x, y, z are rational numbers, the length of $(\vec{x} + \vec{y}) + \vec{z}$ is the difference of $\mathcal{L}$ and $\mathcal{R}$, where $\mathcal{L}$ is the sum of the lengths of the left-pointing vectors among $\vec{x}$, $\vec{y}$, and $\vec{z}$, and $\mathcal{R}$ is the sum of the lengths of the right-pointing vectors.*

(v) *If x, y, z are rational numbers, then $(\vec{x} + \vec{y}) + \vec{z}$ is left-pointing if, in the notation of (iv), $\mathcal{L} > \mathcal{R}$, and it is right-pointing if $\mathcal{R} > \mathcal{L}$.*

The verifications of (iv) and (v) also follow from a case-by-case argument using picture drawing. Here are two examples where $\mathcal{L} > \mathcal{R}$. In the first picture, $\vec{x}$ is left-pointing, $\vec{y}$ and $\vec{z}$ are right-pointing, and the length of $\vec{x}$ exceeds the sum of the lengths of $\vec{y}$ and $\vec{z}$, so that $\mathcal{L}$ is the length of $\vec{x}$ and $\mathcal{R}$ is the sum of the lengths of $\vec{y}$ and $\vec{z}$. The picture shows $(\vec{x} + \vec{y}) + \vec{z}$ as the special vector whose endpoint is the tip of the vertical down-arrow in the following picture.

The next picture shows $(\vec{x} + \vec{y}) + \vec{z}$, where $\vec{x}$ and $\vec{y}$ are right-pointing, $\vec{z}$ is left-pointing, the length of $\vec{z}$ exceeds the sum of the lengths of $\vec{x}$ and $\vec{y}$, $\mathcal{L}$ is the length of $\vec{z}$, while $\mathcal{R}$ is the sum of the lengths of $\vec{x}$ and $\vec{y}$. Again:

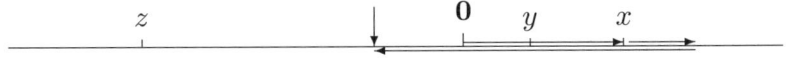

The rest of the proof of (iv) and (v) will be left as an exercise.

To verify the associative law of addition, one must prove $(\vec{x} + \vec{y}) + \vec{z} = \vec{x} + (\vec{y} + \vec{z})$. Using the commutativity of addition, this is equivalent to proving $(\vec{x} + \vec{y}) + \vec{z} = (\vec{y} + \vec{z}) + \vec{x}$. Observations (iv) and (v) imply that the two sides of the latter have the same length and same direction. Thus,

The addition of special vectors is associative. (See Exercise 2 on page 394.)

We note that, up to this point, we have not discussed whether z is a rational number when x and y are rational numbers and $\vec{z}$ is the special vector so that $\vec{z} = \vec{x} + \vec{y}$. However, if x and y are rational numbers, then the length of $\vec{x} + \vec{y}$ is a positive rational number z so that either $\vec{z}$ or $\vec{z^*}$ has the same length and same direction as $\vec{x} + \vec{y}$. Hence, $\vec{x} + \vec{y}$ is equal to either $\vec{z}$ or $\vec{z^*}$. Since both z and z^* are rational numbers, we have proved that

If x and y are rational numbers, then there is a rational number z so that $\vec{z} = \vec{x} + \vec{y}$.

27.3. Addition of Rational Numbers

The relevance of vector addition to the addition of rational numbers can be seen by letting x and y be *fractions* and considering the vector $\vec{x} + \vec{y}$. By consequence (iii) of the Key Observation in the preceding section, the endpoint of $\vec{x} + \vec{y}$ is exactly $x + y$. This suggests that we could have defined the addition of fractions using vectors. It also suggests that when x and y are no longer fractions but arbitrary *rational numbers*, the number $x + y$ "should be" the endpoint of $\vec{x} + \vec{y}$. We now follow up on this suggestion.

Formally, we define **the sum $x + y$** for any two numbers x and y to be the endpoint of the vector $\vec{x} + \vec{y}$. In other words,

$$x + y = \text{ the endpoint of } \vec{x} + \vec{y}.$$

Put another way, $x + y$ is defined to be the point on the number line so that the special vector $\overrightarrow{(x+y)}$ satisfies

$$\overrightarrow{(x+y)} = \vec{x} + \vec{y}.$$

It follows immediately from the definitions of the addition of special vectors and the last observation of the preceding section that

(vi) *If x and y are rational numbers, then $x + y$ is a rational number whose distance from x is equal to the length of y, and is to the left of x if y is negative and to the right of x if y is positive.*

By (ii) of the Key Observation, if x and y are rational numbers, so is $x + y$. (Compare Exercise 1 on page 394.) The definition of addition also implies that
$$0 + x = x + 0 = x \quad \text{for any rational number } x$$
and that
$$x + x^* = 0 \quad \text{for any rational number } x.$$
The second property is of course a consequence of the observation made in the last section that $\vec{x} + \vec{x^*} = \vec{0}$. In a sense, the fact that, for any x, we can find an x^* so that $x + x^* = 0$ is the first inkling we have that rational numbers are good for something. Indeed, if instead of rational numbers we only have fractions, then it never happens that $A + B = 0$ so long as one of A, B is nonzero.

We showed in the preceding section that the addition of special vectors satisfies both the associative law and the commutative law. Therefore

(vii) *The addition of rational numbers satisfies both the associative law and the commutative law.*

From (iii) of the preceding section and (vii) above, we obtain that

(viii) *If x and y are both positive, then $x + y$ is equal to the usual addition of fractions, but if they are negative, then $x + y$ is negative and is the mirror reflection of the point $x^* + y^*$.*

As we mentioned at the beginning of this chapter, we want a definition of the addition of rational numbers which coincides with the addition of fractions when both rational numbers are positive. The statement (viii) says that we are on the right track.

27.4. Explicit Computations

We are now in a position to explicitly compute the sum of any two rational numbers. Since we know that $0 + x = x$ for any rational number x, we may as well consider the sum of two *nonzero* rational numbers only. Let us consider then two arbitrary *positive* rational numbers s and t, so that s^* and t^* will be negative. The four sums,
$$s + t, \quad s^* + t^*, \quad s + t^*, \quad s^* + t,$$
then exhaust all possibilities of the sum of two nonzero rational numbers. Because addition is commutative, we can also forget the last one, $s^* + t$, since we already have $s + t^*$ (don't forget: s and t stand for arbitrary positive rational numbers). In addition, we have just seen that the sum $s + t$ is nothing more than ordinary fraction addition, according to (viii). We will therefore concentrate on the middle two sums, $s^* + t^*$ and $s + t^*$.

The computation of $s^* + t^*$ is settled by (vi), which implies that
$$s^* + t^* = (s+t)^*$$
for arbitrary fractions s and t. One should verify through explicit picture drawing with vectors that, for example,
$$\left(\frac{4}{3}\right)^* + \left(\frac{2}{5}\right)^* = \left(\frac{4}{3} + \frac{2}{5}\right)^* \quad \text{and} \quad 2^* + 5 = (2 + 5^*)^*.$$
However, for a later need, it would be convenient to record something more general. We claim that for any *rational numbers* x and y (thus x and y are not necessarily positive),

(27.1) $\qquad\qquad\qquad x^* + y^* = (x+y)^*.$

This also follows from (vi). Indeed, since * does not change length, the lengths of $\vec{x^*} + \vec{y^*}$ and $\vec{x} + \vec{y}$ are equal, by (ii) of the Key Observation. But since $\vec{x^*} + \vec{y^*}$ and $\vec{x} + \vec{y}$ clearly have opposite directions, $\vec{x^*} + \vec{y^*}$ and $\vec{x} + \vec{y}$ are special vectors with the same length and opposite directions. Thus if $\vec{x} + \vec{y} = \vec{z}$ for some rational number z, then $\vec{x^*} + \vec{y^*} = \vec{z^*}$. Therefore, according to the definition of the sum of rational numbers, $x + y = z$ and $x^* + y^* = z^*$. This says the same thing as (27.1).

It remains to consider
$$s + t^*.$$
There are two cases to consider: $s \geq t$ and $s < t$; for example, $7 + 3^*$ and $2 + 5^*$, respectively. We have in these cases:
$$7 + 3^* = 4 = (7-3) \quad \text{and} \quad 2 + 5^* = 3^* = (5-2)^*,$$
as the following pictures show.

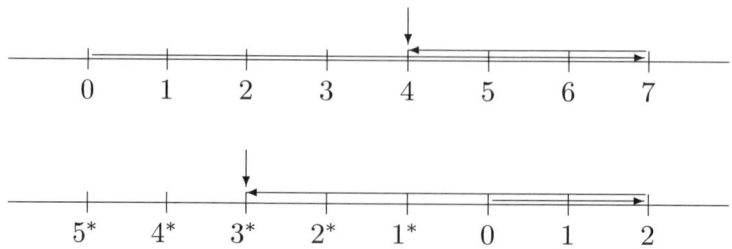

Similarly, one can draw pictures and use (iv) and (v) on page 385 to get
$$\frac{15}{4} + \left(\frac{5}{3}\right)^* = \frac{15}{4} - \frac{5}{3}$$
and
$$\frac{7}{6} + 2^* = \left(2 - \frac{7}{6}\right)^*$$
because:

27.4. Explicit Computations

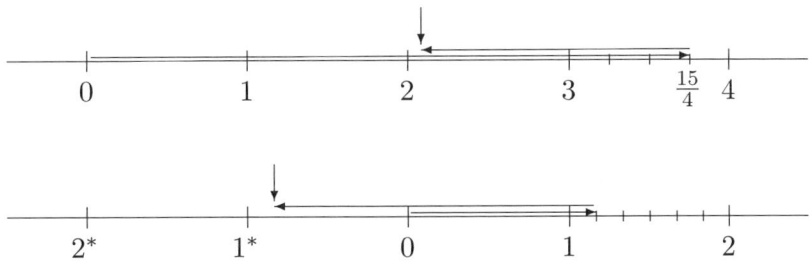

It is also important to gain conviction about these computations, without looking at pictures, by invoking precise definitions. Let us do this for at least one case, e.g.,

$$2 + 5^* = (5-2)^*.$$

By definition, $2 + 5^*$ is the endpoint of the sum of the vectors $\vec{2}$ and $\vec{5^*}$. We now use the Key Observation on page 384 to conclude that the length of the sum vector is the difference of the lengths 5 and 2 (thus $5 - 2$) and its direction is left-pointing because $\vec{5^*}$ is longer than $\vec{2}$. Since the endpoint of a left-pointing vector of length $5-2$ is exactly $(5-2)^*$, the assertion is proved.

Activity. Compute $3 + (4\frac{5}{6})^*$ by drawing the vectors and by direct computation using the Key Observation.

These examples lend credence to the following assertions:

(27.2) $\qquad s + t^* = (s - t) \quad \text{if } s \geq t,$

(27.3) $\qquad s + t^* = (t - s)^* \quad \text{if } s < t.$

Both are immediate consequences of the Key Observation. Consider equation (27.2). By the definition of rational number addition (page 386), it suffices to prove that $\vec{s} + \vec{t^*} = \overrightarrow{s - t}$. Since the length of $\vec{t^*}$ is t and, by hypothesis, $s \geq t$, $\vec{s} + \vec{t^*}$ is right-pointing or 0, by (i) of the Key Observation. Now (ii) of the Key Observation implies that the length of $\vec{s} + \vec{t^*}$ is $s - t$. Thus $\vec{s} + \vec{t^*}$ and $\overrightarrow{s-t}$ have the same length and the same direction and are therefore equal.

As to (27.3), again it suffices to prove $\vec{s} + \vec{t^*} = \overrightarrow{(t-s)^*}$. By (i) of the Key Observation, $\vec{s} + \vec{t^*}$ is left-pointing because the length t of $\vec{t^*}$ exceeds the length s of $\vec{s}$. Moreover, (ii) of the Key Observation implies that the length of $\vec{s} + \vec{t^*}$ is $(t - s)$ because $s < t$ by hypothesis. Since $\overrightarrow{(t-s)}$ is also a left-pointing special vector $(t - s > 0$ by hypothesis) with length $(t - s)$, we see that $\vec{s} + \vec{t^*} = \overrightarrow{(t-s)^*}$, as claimed.

Using (27.2), we have $5.1 + 2.8^* = (5.1 - 2.8) = 2.3$. Using (27.3), we have $(\frac{8}{7})^* + \frac{1}{3} = (\frac{8}{7} - \frac{1}{3})^* = (\frac{17}{21})^*$.

Activity. Compute $\frac{7}{8} + (\frac{1}{3})^*$, $(7\frac{2}{3})^* + 9\frac{1}{4}$, $401 + 193.7^*$.

Activity. Compute $12\frac{1}{3} + 57^*$, $\frac{11}{12} + (2\frac{1}{2})^*$, $(71\frac{1}{3})^* + 68\frac{1}{5}$.

We can now compute the sum of any two rational numbers.

27.5. Subtraction as Addition

The explicit computation of the addition of rational numbers leads to the following insight: *the subtraction of fractions becomes addition in terms of rational numbers*, in the following sense. Consider one of the equalities in the preceding section,

$$\frac{15}{4} + \left(\frac{5}{3}\right)^* = \frac{15}{4} - \frac{5}{3}.$$

Now $\frac{15}{4} - \frac{5}{3}$ is just ordinary subtraction between fractions (because $\frac{15}{4} > \frac{5}{3}$). If we read the equality backwards, then we have

$$\frac{15}{4} - \frac{5}{3} = \frac{15}{4} + \left(\frac{5}{3}\right)^*,$$

so that the subtraction becomes the *addition* of the two rational numbers $\frac{15}{4}$ and $(\frac{5}{3})^*$. As we have seen in the last section, this phenomenon persists in a more general context. Let s, t be any two *fractions* so that $s \geq t$. Then we see from (27.2) that

$$s - t = s + t^* \quad \text{when } s \geq t.$$

A significant feature of this equality is that while the left side makes sense only when $s \geq t$, the right side $s + t^*$ *always makes sense for* any rational numbers s *and* t. This suggests a way to make sense of **subtraction** for any two rational numbers: we *define*, in general, for any two rational numbers x and y,

$$x - y \stackrel{\text{def}}{=} x + y^*.$$

When x, y are fractions and $x > y$, this meaning of $x - y$ coincides with the meaning of subtracting fractions as given in Part 2. In other words, we did not create a new meaning for subtraction on a whim, but, rather, put it in the proper perspective so that what we used to know about subtraction now holds in a broader context.

As a consequence of the definition of $x - y$, we have

$$0 - y = y^*.$$

because $0 + y^* = y^*$. Common sense dictates that we **abbreviate $0 - y$ to $-y$**. So we have

$$-y = y^*.$$

Of course, we now adopt the common terminology and call $-y$ the **negative** of y and "$-$" the **negative sign**. But keep in mind that if $y = 5^*$, then

27.5. Subtraction as Addition

$-y = 5$, which is *positive*. A better terminology would be to call $-y$ **minus** y, and "$-$" the **minus sign**; see [**Lan88**, page 8].

Activity. *Use the definition of rational number addition to calculate:*
(a) $2.1 - 7$, (b) $7^* - 3$, (c) $7.5^* - 12.6$, (d) $5^* - 2^*$, (e) $(\frac{2}{5})^* - (\frac{8}{7})^*$.

At this point, we abandon the notation of y^* and replace it by $-y$. Many of the preceding equalities will now assume a more familiar appearance. For example, (25.1) becomes

$$-0 = 0 \quad \text{and} \quad -(-x) = x \quad \text{for all rational numbers } x.$$

From the preceding definition of $-y$ for a rational number y, and from (25.1), we have

(27.4) $$-y = y^*, \quad -y^* = y.$$

Similarly, the definition of subtraction can now be restated as

(27.5) $$x - y \stackrel{\text{def}}{=} x + (-y).$$

This is the normal notation that one comes across in mathematics, but conceptually, (27.5) is less transparent than $x - y = x + y^*$ because the latter lays bare the fact that subtracting y is the same as adding the mirror reflection of y.

Equation (27.1) now becomes the well-known "rule for removing parentheses": if x and y are any rational numbers, then

(27.6) $$-(x + y) = -x - y \quad \text{for all rational numbers } x \text{ and } y.$$

Applying (27.6) to x and y^* for two arbitrary rational numbers x and y^*, we get $-(x + y^*) = -x - y^*$. By the definition of subtraction, the left side is $-(x - y)$, while the right side is $-x + (y^*)^* = -x + y$, using (25.1). Therefore, we have that

(27.7) $$-(x - y) = -x + y \quad \text{for all rational numbers } x \text{ and } y.$$

By now the message must be clear: to understand anything about subtraction, try to understand the addition of rational numbers better.

Remark. Students' usual approach to (27.6) and (27.7) is to "use the distributive law to multiply by -1", i.e., they believe (27.6) is true because

$$-(x + y) = (-1)(x + y) = (-1)x + (-1)y = -x + (-y) = -x - y,$$

and similarly (27.7) is true because

$$-(x - y) = (-1)(x - y) = (-1)x - (-1)y = -x - (-y) = -x + y.$$

Both computations are correct, but they are clumsy proofs of (27.6) and (27.7). These computations make use of the multiplicative fact that $(-1)x = -x$ for all rational numbers x, which will be proved in Theorem 29.2 on page 408. However, a conceptual understanding of (27.6) and (27.7) should include the realization that they only involve the addition and subtraction of rational numbers, as we have seen, but nothing about multiplication. For this reason, the simple proofs of (27.6) and (27.7) above are valuable.

Activity. Compute directly, and also compute by using (27.6) and (27.7): $5 - (\frac{1}{2} - 8)$, $1 - (-\frac{1}{3} + \frac{5}{4})$, $\frac{2}{3} - (5 - \frac{4}{7})$.

As another application of (27.6), we now tie up a loose end by revisiting identity (5.5) on page 75, which asserts that for whole numbers l, m, n, a, b, and c, we have

$$(l + m + n) - (a + b + c) = (l - a) + (m - b) + (n - c).$$

We are finally going to give a proof of this fact by proving something more general: for any *rational numbers* x, y, z, a, b, and c,

(27.8) $\qquad (x + y + z) - (a + b + c) = (x - a) + (y - b) + (z - c).$

The proof will show that the same identity is true when the number of rational numbers within the parentheses is changed from three to any positive integer n.

We preface the proof of (27.8) by the observation that *Theorem 2.1 on page 41 remains valid when whole numbers are replaced by rational numbers.* The reason is that the reasoning used in its proof is purely an exercise in the use of the associative law and the commutative law, and it has nothing to do with the fact that the symbols x, y, etc., are whole numbers. Since we now know that the addition of rational numbers satisfies the same two laws, Theorem 2.1 also extends to rational numbers.

With this understood, the proof of (27.8) goes as follows.

$$(x + y + z) - (a + b + c) = (x + y + z) + (a + b + c)^*,$$

by the definition of subtraction. Using (27.8) twice in succession, we get

$$(a + b + c)^* = (a + b)^* + c^* = a^* + b^* + c^*,$$

where we have made use of the analogue of Theorem 2.1 for rational numbers to do away with parentheses. We can likewise use the same theorem to write

$$\begin{aligned}(x + y + z) - (a + b + c) &= x + y + z + a^* + b^* + c^* \\ &= x + a^* + y + b^* + z + c^* \\ &= (x + a^*) + (y + b^*) + (z + c^*) \\ &= (x - a) + (y - b) + (z - c),\end{aligned}$$

where the last line is by the definition of subtraction. This proves (27.8). It is clear from the proof that once again the truth of (27.8) hinges on the fact that, by definition, subtraction among rational numbers is just addition.

This definition of subtraction not only enables us to talk about the subtraction of any two rational numbers (see Exercise 17 on page 84, for instance), but also reveals the fact that, in the context of rational numbers, subtraction is just addition. This revelation will be seen in Chapter 29 to extend the validity of the distributive law to include not only the addition of numbers but subtraction as well.

Exercises

1. Let x and y be rational numbers. Explain in detail, (a) why $x + y$ is a rational number, and (b) why $x + x^* = 0$ for any number x.

2. For special vectors $\vec{x}$, $\vec{y}$, and $\vec{z}$, prove the associative law $(\vec{x}+\vec{y})+\vec{z} = \vec{x}+(\vec{y}+\vec{z})$, (a) when the length of $\vec{x}$ exceeds the sum of the lengths of $\vec{y}$ and $\vec{z}$, and when x is negative and y and z are positive, and (b) when the length of $\vec{x}$ is less than the sum of the lengths of $\vec{y}$ and $\vec{z}$ but exceeds either of the lengths of $\vec{y}$ and $\vec{z}$, and when x is negative and y and z are positive.

3. Use vectors to directly compute the following: (a) $\frac{1}{2} + (1\frac{1}{4})^*$, (b) $\frac{2}{3} + \frac{5}{6}$, (c) $2\frac{1}{3}+4.1^*$, (d) $99.5+101^*$, (e) $(\frac{2}{7})^*+(\frac{6}{5})^*$, (f) $13.4^*+2\frac{3}{4}$, (g) $(\frac{12}{5})^*+\frac{2}{3}$, (h) $(\frac{11}{5})^* + 4\frac{3}{4}$.

4. Compute the following: (a) $-3\frac{1}{2} + \frac{7}{8}$, (b) $1\frac{2}{3} - 6.18$, (c) $-4.7 + 2\frac{1}{3}$, (d) $-11\frac{2}{5} - 2.6$.

5. For all rational numbers x and y, (a) if $x+y = x$, then $x = 0$, and (b) if $x+y = 0$, then $x = y^*$ and $y = x^*$.

6. When you try to explain to a seventh grader why equation (27.7) above is true, he protests that he doesn't need to know. He says if you want to show $-(4-6) = -4+6$, he can compute both sides and see that each is equal to 2. Likewise, the validity of $-(5-(-1)) = -5+(-1)$ can be checked by computing both sides and finding that each is equal to -6, etc. The way he sees it, he can always compute like this to check whether an equality is valid or not, so he wonders why he should bother with a proof of (27.7). What do you say to him? Moreover, after convincing him that he should learn it, can you explain to him why (27.7) is true?

7. Chapter 26 begins the discussion about the use of rational numbers for solving equations. (a) Along this line, show that if x is a solution of $7 + x = 5$, then $x = 2^*$. (b) Let A and B be rational numbers. Show that every equation of the form $A + x = B$ has a solution x which is a rational number.

Chapter 28

Adding and Subtracting Rational Numbers Redux

The approach to adding and subtracting rational numbers through the use of vectors is perfectly respectable, but the drawback is that there is no corresponding approach to the multiplication and division of rational numbers. For the latter, a more abstract viewpoint will be needed. In preparation, we now explain how one could treat the addition and subtraction of rational numbers from the same abstract viewpoint. The advantage of doing this is that, since you have already seen a complete development of addition and subtraction in Chapter 27, seeing a *familiar* topic treated a little more abstractly will be less of a shock. This will prepare you for your encounter with the unavoidably abstract treatment of multiplication.

The principal distinction between the abstract treatment in this and the next chapter and the mathematics of Parts 1 and 2 is that in the latter, one usually proves that two numbers A and B are equal by a computation that starts with A and ends with B. For example, the assertion about the sum of unit fractions, $\frac{1}{m} + \frac{1}{n} = \frac{m+n}{mn}$, can be proved as follows:

$$\frac{1}{m} + \frac{1}{n} = \frac{n}{mn} + \frac{m}{mn} = \frac{m+n}{mn}.$$

On the other hand, the proof of the equality of two numbers in an abstract setting, more likely than not, requires a conceptual repositioning of the two numbers so that, from the new perspective, the two numbers are seen to share a common characteristic property and are therefore equal. There are many examples of this kind of reasoning in this and the following sections, and we need only point to the proofs of Basic Facts 1, 2, 3 and equations (28.1) and (28.2) in this chapter as illustrations. It takes a reorientation of the mindset cultivated in Parts 1 and 2 in order to get used to this new way of thinking. In the school classroom, this is the kind of learning difficulty a teacher has to be sensitive about.

Surprisingly, the more abstract approach is actually closer to the historical development of the subject. What we do is to *take for granted* that the rational numbers behave in exactly the same way as the fractions, in the sense that they obey the associative, commutative, and distributive laws, and that 0 and $*$ behave the way they "should": $x + 0 = x$ and $x + x^* = 0$ for all rational numbers x. On this basis, we will make use of mathematical reasoning to *deduce* all the usual properties of the rational numbers. The first person in the West[1] to confront negative numbers extensively was Diophantus of Alexandria, who probably lived around AD 250. He arrived at a set of rules about how negative numbers should behave; one infers from his work that his rules were probably the result of his expectation that these new numbers "behave in exactly the same way as the fractions, in the sense that they obey the associative, commutative, and distributive laws". A complete understanding of negative numbers had to wait until the nineteenth century.

Please keep in mind that, at least as far as logical deduction is concerned, *you must approach this section as if you have never seen the material in Chapter 27*. We are starting afresh, assuming only the material in Chapters 25–26.

The sections are as follows:

The Assumptions on Addition

The Basic Facts

Explicit Computations

Basic Assumptions and Facts, Revisited

28.1. The Assumptions on Addition

The central idea of our approach to the addition and multiplication of rational numbers is that, since we already expect their addition and multiplication, no matter how they are defined, to satisfy the associative, commutative, and distributive laws, we may as well *assume all these properties at the outset*. There is also a good practical reason for taking the associative, commutative, and distributive laws for granted: we are so used to having them around that it would take a major rethinking of our usual computational routine if they turned out to be unavailable. Historically, that was exactly what happened when people before the eighteenth century had to

---

[1] The Chinese document *Nine Chapters on the Mathematical Art* [**KCL99**] treats negative numbers routinely. This document was a compilation of mathematical knowledge in China through the centuries, starting from about 10th century BC to about the 1st century AD.

deal with the "new" numbers such as negative numbers and complex numbers. The working motto was pretty much, "No matter what they are, let us treat them like any other number." Those people were lucky, because this attitude turned out to serve them well. "Lucky" refers to the fact that we now know it is not true that every "number system" necessarily obeys these same laws, but fortunately, the real and complex numbers do. In particular, so do the rational numbers, and this is where we are.

Thus restricting ourselves to addition for the moment, we make the **fundamental assumption** that

(A1) *Given any two rational numbers x and y, there is a way to add these to get a unique rational number $x + y$ so that, if x and y are fractions, $x + y$ is the same as the usual sum of fractions. Furthermore, this addition of rational numbers satisfies the associative and commutative laws.*

There are two more natural assumptions to make.

(A2) *If x is any rational number, $x + x^* = 0$.*

(A3) $x + 0 = x$ *for any rational number x.*

Assumption (A2) makes it official that, for example, $2 + 2^* = 0$. This puts all the mirror reflections of the fractions in their proper place. Thus 2^* is a solution of the equation $2 + x = 0$. In the next section we will show that it is the only solution. As for (A3), it is not as trivial as it appears: if x is a negative rational number, then $x + 0$ is an unknown quantity at the moment. Thus it takes an explicit assumption to get $x + 0 = x$ for *any* rational number x.

Because we are assuming that addition among rational numbers is commutative, (A2) and (A3) imply that

(A2′) *If x is any rational number, $x^* + x = 0$;*

(A3′) $0 + x = x$ *for any rational number x.*

28.2. The Basic Facts

It remains for us to get to know the basic properties of the addition of rational numbers. The first three basic facts in this section will appear to be too obvious. Perhaps, but it is in the nature of such abstract considerations that obvious facts sometimes lead to surprising conclusions, as you will see when you come to their applications in the succeeding sections.

Basic Fact 1. For all rational numbers x and y, if $x + y = x$, then $y = 0$.

Activity. Try proving the following special case before approaching the general case: if y satisfies $67 + y = 67$, then $y = 0$. Don't forget, you are supposed to make use of only (A1)–(A3).

This property is, first of all, nice to know, because what else could this y be? However, this property is also useful because if we want to prove that a rational number y is actually equal to 0, all we need to do is to find a rational number x so that $x + y = x$. Then Basic Fact 1 would conclude for us that indeed $y = 0$. You may *think* that you would never have to show any number is 0, but wait until you get to Chapter 29 on multiplication (e.g., see the proof of (M3) in section 29.1).

Here is the reason for Basic Fact 1. Given $x + y = x$, adding x^* to both sides gives the same number on account of the uniqueness assumption on the sum of two numbers in (A1). Hence $x^* + (x + y) = x^* + x$, which, by the associative law, is equivalent to $(x^* + x) + y = x^* + x$. But $x^* + x = 0$ by virtue of (A2'), so $0 + y = 0$, which means $y = 0$, by (A3').

Basic Fact 2. For all rational numbers x and y, if $x + y = 0$, then $y = x^*$ and $x = y^*$.

This statement, which is vaguely related to Basic Fact 1, turns out to be very powerful in applications. Very often we want to prove that for a given rational number x, another rational number y is in fact equal to x^*. (If you have any doubts, look at Basic Fact 3 immediately following, but there are more of the same in the next chapter, e.g., Theorem 29.2 on page 408.) Then Basic Fact 2 tells us that it suffices to prove $x + y = 0$. You will get to see this line of reasoning several more times in the remainder of Part 3.

We will just prove $y = x^*$, as the proof of $x = y^*$ would follow by taking * of both sides of $y = x^*$ and using (25.1). Now $x + y = 0$ implies that $x^* + (x + y) = x^* + 0$ (just add x^* to both sides). Using the associative law, we have $(x^* + x) + y = x^* + 0$, and hence $0 + y = x^*$, by (A2') and (A3). Therefore, $y = x^*$ by applying (A3') to the left side.

Basic Fact 3. For all rational numbers x and y, $(x + y)^* = x^* + y^*$.

We have seen in Chapter 27 that this is the rule for "removing parentheses". It is possible to give an elementary proof using a case-by-case argument, but the one we are going to give makes use of Basic Fact 2 and is far more enlightening. What Basic Fact 3 asserts is that $x^* + y^*$ is the mirror reflection of $(x + y)$. According to Basic Fact 2, it suffices to prove that

$$(x + y) + (x^* + y^*) = 0.$$

By repeated use of the associative and commutative laws, we have

$$(x + y) + (x^* + y^*) = (x + x^*) + (y + y^*) = 0 + 0 = 0$$

(by (A2) and (A3)). So we are done.

Activity. Compute: $(2\frac{6}{7})^* + (3\frac{2}{5})^*$, $(\frac{24}{11})^* + (156\frac{1}{2})^*$.

28.3. Explicit Computations

With these basic facts out of the way, we are in a position to explicitly compute the sum of any two rational numbers.[2] These computations have all been carried out in the preceding chapter by use of vectors, but the difference in the methods used is quite striking. Since we already know how to add 0 to any number, by (A3) and (A3$'$), we need only deal with positive rational numbers. Thus, let s and t be arbitrary positive rational numbers, i.e., they are nonzero fractions. Now the sum $s+t$ is already known (Part 2), so there remain only three possibilities:

$$s^* + t^*,$$
$$s + t^*,$$
$$s^* + t.$$

From Basic Fact 3, we have

$$s^* + t^* = (s+t)^*,$$

so this takes care of the first possibility. For example, $(\frac{2}{3})^* + (\frac{4}{5})^* = (\frac{2}{3} + \frac{4}{5})^* = (\frac{22}{15})^*$.

Because $s^* + t = t + s^*$ by the commutative law, the third possibility above follows from the second possibility. (Because this may be confusing, let us spell out explicitly what we are saying: knowing how to compute $s+t^*$ *for all positive rational numbers s and t* implies knowing how to compute $s^* + t$ *for all positive rational numbers s and t*.) Thus it suffices to take care of the second possibility, $s + t^*$. This breaks up into two cases: $s \geq t$ and $s < t$. Consider the first case. We claim that

(28.1) $\qquad\qquad s + t^* = s - t \quad \text{if } s \geq t.$

To get some feeling for (28.1), let us see why $9 + 5^* = 4$ ($= 9 - 5$) *from the point of view of (A1)–(A3)*. Since we do not know how to directly compute $9 + 5^*$, we do the next best thing by working with 4 on the right side. We have $4 + 5 = 9$, so that by adding 5^* to both sides, we get $(4+5)+5^* = 9+5^*$. But

$$(4+5) + 5^* = 4 + (5 + 5^*) = 4 + 0 = 4,$$

where we have used the associative law. So from $(4+5) + 5^* = 9 + 5^*$, we get $4 = 9 + 5^*$, as desired.

---

[2] Once more: we only assume we know (A1)–(A3) but cannot assume anything from Chapter 27.

The general case is no different. This is because, if $s \geq t$, then $s - t$ makes sense as a subtraction between fractions, and $(s - t) + t = s$. Add t^* to both sides and we get
$$((s - t) + t) + t^* = s + t^*.$$
The left side is equal to (by the associative law)
$$(s - t) + (t + t^*) = (s - t) + 0 = s - t.$$
Thus we have $s - t = s + t^*$, which is exactly what we want.

It remains to tackle the second possibility: $s < t$. We have to show that
(28.2) $$s + t^* = (t - s)^* \quad \text{if } s < t.$$
Again, let us first verify it for a special case: $3 + 8^* = 5^* \;(= (8-3)^*)$. If we follow the reasoning in the proof of (28.1), we would work with the right side. But the right side is a negative number about which not much is known at this point, so we have to do something different. What we observe is that we can make use of (25.1) on page 375 and Basic Fact 3 of the preceding section to rewrite the left side:
$$(3 + 8^*) = (3 + 8^*)^{**} = (3^* + 8^{**})^* = (3^* + 8)^* = (8 + 3^*)^*.$$
Now we can apply (28.1) to $8 + 3^*$ to get $8 + 3^* = 8 - 3 = 5$. Thus $(3 + 8^*) = (8 + 3^*)^* = 5^*$, as desired.

For the general case, we follow the same idea. Apply Basic Fact 3 and (25.1) to rewrite the left side of (28.2) as
$$s + t^* = (s + t^*)^{**} = (s^* + t^{**})^* = (s^* + t)^*.$$
Thus we have to prove
$$(s^* + t)^* = (t - s)^*.$$
It suffices to prove $(s^* + t) = (t - s)$, or what is the same thing,
$$t + s^* = t - s.$$
Since $t > s$, this is just (28.1) with s and t interchanged. The proof of (28.2) is complete.

In summary, we have shown how to compute the sum of any two rational numbers *strictly on the basis of assumptions* (A1)–(A3).

28.4. Basic Assumptions and Facts, Revisited

As before, we define, in general, for any two rational numbers x and y
$$x - y \stackrel{\text{def}}{=} x + y^*,$$
and we are led to write, for exactly the same reason as before, that
$$-y = y^*$$
and abandon the notation of y^* by replacing it with $-y$.

28.4. Basic Assumptions and Facts, Revisited

In this new notation, (A2) and Basic Fact 2 become, respectively,

for any rational number x, $x + (-x) = -x + x = 0$;

if $x + y = 0$, then $x = -y$ and $y = -x$.

When we use the new notation in Basic Fact 3, we again obtain the rules for "removing parentheses" for any rational number x and y:

$$-(x+y) = -x-y,$$
$$-(x-y) = -x+y.$$

This time around, we use Basic Fact 3 to conclude the first, and the second follows from the first because $-(x-y) = -(x+y^*) = -x-y^* = -x+(y^*)^* = -x+y$, where we have used (25.1).

By now the message must be clear: in order to understand anything about subtraction, try to understand the addition of rational numbers better.

Remark. We have now arrived at exactly the same place as the end of Chapter 27. Every statement about the addition of rational numbers in these two chapters has now received two different proofs. *From now on, we are free to avail ourselves of either collection of facts in these two chapters* to do the addition and subtraction of rational numbers.

Exercises

1. Compute on the basis of (A1)–(A3): (a) $-3.6 - 4\frac{7}{8}$. (b) $-3.6 + 4\frac{3}{8}$. (c) $2\frac{2}{5} - 7\frac{3}{4}$. (d) $2\frac{1}{7} - 6.5$. (e) $12 - ((-6.7) + (-4.4))$. (f) $7 - (2.5 - 3\frac{2}{3})$. (g) $1 - ((-\frac{2}{3}) + \frac{5}{6})$. (h) $(401.5 - 247.6) - (\frac{14}{5} - 8.2)$. (i) $(-7003.2 + 5277.4) - (\frac{1}{5} - 3\frac{2}{3})$.

2. Repeat Exercise 7 on page 394 by using (A1)–(A3) instead. In other words: (a) Show that if x is a solution of $7 + x = 5$, then $x = 2^*$. (b) Let A and B be rational numbers. Show that every equation of the form $A + x = B$ has a solution x which is a rational number.

3. (*This problem brings closure to the discussion of the subtraction algorithm in Chapter 5, specifically page 79.*) Explain why the following method of doing the subtraction $756 - 389$ is correct.

    Do the following column-by-column subtractions in any order, from left to right or right to left, or if you prefer, starting from the middle:

    $$\begin{array}{r} 756 \\ -389 \\ \hline 4[\![-3]\!][\![-3]\!] \end{array}$$

    where each $[\![-3]\!]$ indicates the result of *the column-wise subtraction in that column*. To get the final answer, the method says:

    > Treat $4[\![-3]\!][\![-3]\!]$ as if it were a whole number with digits 4, $[\![-3]\!]$, and $[\![-3]\!]$, and write it out in expanded form.

    Thus:

    $$4[\![-3]\!][\![-3]\!] = \{4 \times 10^2\} + \{(-3) \times 10^1\} + \{(-3) \times 10^0\}.$$

Chapter 29

Multiplying Rational Numbers

The Most Frequently Asked Question in school mathematics is arguably the one about *negative times negative is positive.* This is one indication that the teaching of the multiplication of rational numbers in schools is in disarray. It would seem that the teaching of this topic in schools mainly consists of a set of rules to be memorized without benefit of reasoning. Over the years, all kinds of analogies have been offered to explain why negative times negative is positive, but in the last analysis, none of them makes any sense. What is apparently not realized by many people is that the *mathematical* explanation is far simpler than any of these analogies.

We will present the multiplication of rational numbers the same way we do any topic in mathematics. In such a discussion, there will always be a set of clearly stated hypotheses and conclusions, and we will try to get from hypotheses to conclusions as simply as we know how by the use of mathematical reasoning. As we mentioned earlier in section 28.1 (page 396), our principal hypothesis is that the arithmetic of rational numbers is like that of whole numbers or fractions: they obey the associative and commutative laws for addition and multiplication as well as the distributive law that connects the two. This hypothesis is not extravagant, because it is consistent with our common perception of what a "number system" ought to be. As we pointed out before, this was also more or less the way mathematicians before the eighteenth century came to grips with negative numbers.

There is no doubt that the climax of any discussion of the multiplication of rational numbers in the school classroom is still the fact that negative times negative is positive. Because of its importance, we prove this fact in two stages, the first stage for integers (page 407) and then the general case (page 409). It is hoped that the extended exposure to the underlying reasoning will make this seemingly mystifying fact more accessible to students as a whole.

This sections are as follows:

> The Assumptions on Multiplication
> The Equality $(-m)(-n) = mn$ for Whole Numbers
> Explicit Computations
> Some Observations

29.1. The Assumptions on Multiplication

We approach multiplication the same way we did addition in the last chapter, namely, we make a similar **fundamental assumption**:[1]

(M1) *Given any two rational numbers x and y, there is a way to multiply them to get a unique rational number xy so that, if x and y are fractions, xy is the same as the usual product of fractions. Furthermore, this multiplication of rational numbers satisfies the associative, commutative, and distributive laws.*

We also assume the obvious fact that

(M2) *If x is any rational number, then $1 \cdot x = x$.*

Again we point out the need to make such an explicit assumption, because it includes the fact that, for example, $1 \cdot (-\frac{2}{3}) = -\frac{2}{3}$, which was unknown up to this point.

The analogue of (A3) on page 397 is

(M3) $0 \cdot x = x \cdot 0 = 0$ *for any rational number x.*

However, instead of being another assumption, (M3) turns out to be provable. For the proof, the idea is to make use of the existing *mathematical* characterization of the number 0 in terms of addition, namely, it is the only rational number W so that with x given, $x + W = x$ (this is a consequence of (vi) of section 27.3 and is also Basic Fact 1 of section 28.2). So we only need to prove that $x + 0 \cdot x = x$. The proof goes as follows:

$$\begin{aligned} x + 0 \cdot x &= 1 \cdot x + 0 \cdot x && \text{(by (M2))} \\ &= (1+0)x && \text{(distributive law (M1))} \\ &= 1 \cdot x \\ &= x && \text{(by (M2) again)}. \end{aligned}$$

---

[1] See (A1)–(A3) of section 28.1 on page 397.

So we have $x + 0 \cdot x = x$. As we said, Basic Fact 1 of section 28.2 or (vi) of section 27.3 now shows $0 \cdot x = 0$. By the commutativity of multiplication (M1), we also have $x \cdot 0 = 0$, thereby proving (M3).

We wish to make explicit a fruitful idea that is embedded in the preceding proof. What this proof does is to prove something about multiplication (i.e., $0 \cdot x = 0$) by making use of a fact already known about addition (i.e., if $x + \mathcal{W} = x$, then $\mathcal{W} = 0$). To make such a proof possible, we have to be able to transfer information from addition to multiplication. At this stage of our mathematical development, the only tool at our disposal for this transfer is the distributive law, which is a general statement about the relationship between addition and multiplication—in fact, the only one so far. In the rest of this chapter, the distributive law will be exploited, again and again, for this purpose, namely, to transfer the known information about the addition of rational numbers as recorded in Chapters 27 and 28 to the multiplication of rational numbers.

29.2. The Equality $(-m)(-n) = mn$ for Whole Numbers

Now that we have introduced multiplication among rational numbers, our first task is to find out how multiplication is related to the existing operations, in particular, addition and the mirror reflection *. The relationship between addition and multiplication is codified by the distributive law, which is part of the assumption in (M1). As to the operation *, it is natural to ask whether the order of applying multiplication and * is interchangeable. In other words, given two rational numbers x and y, if we get their mirror reflections first and then multiply (thus x^*y^*), how is it related to the number obtained by multiplying them first before applying the mirror reflection (thus $(xy)^*$)? If multiplication is replaced by addition, the order *is* interchangeable; $x^* + y^* = (x+y)^*$ (see equation (27.1) on page 388). *In the case of multiplication, however, the order matters* because $x^*y^* \neq (xy)^*$, or in the notation of the minus sign, $(-x)(-y) \neq -(xy)$. As is well known, the correct answer is

$$(-x)(-y) = xy \qquad \text{for all rational numbers } x \text{ and } y.$$

This fact, the bane of many middle school students, can be given a very short proof. We will present this proof at the end of the next section. For the middle school classroom, such a proof is too sophisticated to be given at the outset; maybe it should never be given in schools except for the gifted few. Instead we will give a more leisurely proof by first taking a detour through the more familiar terrain of the integers to see why $(-m)(-n) = mn$ for all *whole numbers* m and n. There is a reason for singling out the whole numbers. It is not only easier to learn (it is that for sure!), but it is also far easier to *teach*, as we shall see presently. If you can get all your elementary and middle school students to believe, for example, that $(-1234)(-5678) = 1234 \times 5678$, then you have done very well indeed.

We begin with the simplest special case of this assertion: the case of $x = y = 1$. This will turn out to be the critical case.

Theorem 29.1. $(-1)(-1) = 1$.

Proof. Let x denote $(-1)(-1)$. Our goal is to show $x = 1$. We should ask ourselves: if we have a number x, how can we tell whether it is 1 or not? One way is to try to see if $(-1) + x = 0$. If it is, then $\vec{x}$ is the right-pointing vector of length 1 because it must go from -1 back to 0 (see picture), and therefore $x = 1$.[2]

With this in mind, we now compute

$$\begin{aligned}
(-1) + x &= 1 \cdot (-1) + (-1)(-1) &&\text{(by (M2) and the definition of } x\text{)} \\
&= \bigl(1 + (-1)\bigr) \cdot (-1) &&\text{(distributive law)} \\
&= 0 \cdot (-1) &&\text{(by (M3))} \\
&= 0.
\end{aligned}$$

So we know $x = 1$, i.e., $(-1)(-1) = 1$. The proof is complete. □

We can rephrase this proof to give it a more algebraic flavor; this reformulation will turn out to be more in line with the proof in general that $(-x)(-y) = xy$ for all rational numbers x, y. It goes as follows:

We have $(-1) + 1 = 0$ (see (A2) on page 397). Multiply each side by (-1), and the uniqueness statement in (M1) says $(-1)\bigl((-1) + 1\bigr)$ is equal to $(-1) \cdot 0$. Apply the distributive law and we get

$$(-1)(-1) + 1 \cdot (-1) = 0 \cdot (-1).$$

By (M3), the right side is 0. As for the left side, by (M2), it is equal to $(-1)(-1) + (-1)$. Therefore,

$$(-1)(-1) + (-1) = 0.$$

Adding 1 to both sides and using the associative law of addition, we get

$$(-1)(-1) + ((-1) + 1) = 0 + 1.$$

Now the left side is $(-1)(-1)$ and the right side is 1. Theorem 29.1 is proved once more.

---

[2]This is entirely similar to what one does in chemistry: to see whether a solution is acidic or not, dip a piece of litmus paper (pH test strip) into the solution and see if it turns red.

29.2. The Equality $(-m)(-n) = mn$ for Whole Numbers

Activity. Practice explaining as if to a seventh grader, in two different ways, why $(-1)(-1) = 1$ by using your neighbor as a stand-in for the seventh grader. (Theorem 29.1 is so basic to understanding rational number multiplication that this Activity is strongly recommended.)

We can now give the proof that $(-m)(-n) = mn$ for all *whole numbers* m and n. Let us first do a special case: why is $(-2)(-3) = 6$? This is because, by the rule on removing parentheses ((27.6) on page 391), $-2 = -(1+1) = (-1) + (-1)$, and similarly, $-3 = (-1) + (-1) + (-1)$. Thus,

$$\begin{aligned}
(-2)(-3) &= \big((-1)+(-1)\big) \cdot \big((-1)+(-1)+(-1)\big) \\
&= \underbrace{(-1)(-1) + \cdots + (-1)(-1)}_{6} \quad \text{(distributive law)} \\
&= \underbrace{1 + \cdots + 1}_{6} \quad \text{(Theorem 29.1)} \\
&= 6.
\end{aligned}$$

In the same manner, we can show even fifth graders why $(-3)(-4) = 12$, $(-5)(-2) = 10$, etc. A teacher can probably win the psychological battle over students' disbelief of *(negative)* × *(negative)* = *positive* by these very concrete computations.

The general proof of $(-m)(-n) = mn$ is essentially the same. So let m, n be whole numbers. We first prove that

$$(-1)(-m) = m \quad \text{for any whole number } m.$$

This is because (see Basic Fact 3 on page 398),

$$-m = -\underbrace{(1 + \cdots + 1)}_{m} = \underbrace{(-1) + \cdots + (-1)}_{m},$$

so that

$$\begin{aligned}
(-1)(-m) &= (-1)\Big(\underbrace{(-1) + \cdots + (-1)}_{m}\Big) \\
&= \underbrace{(-1)(-1) + \cdots + (-1)(-1)}_{m} \quad \text{(distributive law)} \\
&= \underbrace{1 + \cdots + 1}_{m} \quad \text{(Theorem 29.1)} \\
&= m,
\end{aligned}$$

as desired. Hence, we have

$$\begin{aligned}
(-n)(-m) &= \big(\underbrace{(-1)+\cdots+(-1)}_{n}\big)(-m) & ((27.6)) \\
&= \underbrace{(-1)(-m)+\cdots+(-1)(-m)}_{n} & \text{(distributive law)} \\
&= \underbrace{m+\cdots+m}_{n} \\
&= nm = mn.
\end{aligned}$$

29.3. Explicit Computations

We next turn to explicit computations of multiplication for rational numbers in general. Thus let x, y be rational numbers. What is xy? If $x = 0$ or $y = 0$, we have just seen that $xy = 0$. We may therefore assume both x and y are nonzero. It is therefore sufficient to consider *positive* fractions s and t in general, and find out the products st, $(-s)t$, $s(-t)$, and $(-s)(-t)$. There is an immediate simplification, however. We do not need to consider the product st because we already know how to multiply s and t as fractions (see (M1) on page 404). Furthermore, since multiplication of rational numbers is assumed to be commutative, knowing $(-s)t$ for all positive s and t is equivalent to knowing $s(-t)$ for all positive s and t. Therefore it suffices to prove that for all positive fractions s and t,

(29.1) $\qquad\qquad\qquad (-s)\,t = -(s\,t),$

(29.2) $\qquad\qquad\qquad (-s)(-t) = s\,t.$

Let us now prove these assertions. Underlying both proofs is the following generalization of Theorem 29.1.

Theorem 29.2. *For any rational number x, the number $(-1)x$ is the mirror reflection of x. In symbols: $(-1)x = -x$.*

Proof. Here is the picture we want to be valid:

$$\underset{x}{\,\rule{1pt}{6pt}\,} \qquad\qquad \underset{O}{\,\rule{1pt}{6pt}\,} \qquad\qquad \underset{(-1)x}{\,\rule{1pt}{6pt}\,}$$

At this point, we recall a remark we made after Basic Fact 2 on page 398, to the effect that there will be occasions when one wants to know whether a number is the mirror reflection of a given number. This is exactly the situation we are in right now: we want to know if the number $(-1)x$ is the mirror reflection of x. According to Basic Fact 2, it suffices to prove that

$$x + (-1)x = 0.$$

29.3. Explicit Computations

For this purpose, we make use of the distributive law,

$$x + (-1)x \stackrel{(M2)}{=} 1 \cdot x + (-1)x = \bigl(1 + (-1)\bigr)x = 0 \cdot x \stackrel{(M3)}{=} 0.$$

We are done. □

If we let $x = (-1)$, then Theorem 29.2 yields Theorem 29.1 as a corollary: $(-1)(-1) = 1$.

We now give the proofs of (29.1) and (29.2). Let s and t be positive fractions. Then by Theorem 29.2, $(-s)\,t$ is equal to $\bigl((-1)s\bigr)t$ which, by associativity, is equal to $(-1)(st)$. By Theorem 29.2 again, $(-1)(st) = -(st)$. Altogether, we have $(-s)\,t = -(st)$ and this proves (29.1).

The proof of (29.2) is no different. With s and t as above, Theorem 29.2 shows that $(-s)(-t) = \bigl((-1)s\bigr)\bigl((-1)t\bigr) = (-1)(-1)(s\,t)$, where the last step makes use of the commutativity and associativity of multiplication. By Theorem 29.1, $(-1)(-1) = 1$, and therefore $(-s)(-t) = s\,t$. This is (29.2).

Remark. At this point, we can look back and see more clearly why (29.2) is easier to prove when x and y are positive integers (see the preceding section). The key to the proof of (29.2) is Theorem 29.2, $(-1)x = -x$, and the proof of Theorem 29.2 when x is a fraction requires a detour through Basic Fact 2. However, when x is a positive integer, e.g., $x = 3$, the proof of Theorem 29.2 is much more direct:

$$\begin{aligned}
(-1)(3) &= (-1)(1 + 1 + 1) \\
&= (-1) \cdot 1 + (-1) \cdot 1 + (-1) \cdot 1 \\
&= (-1) + (-1) + (-1) \\
&= -3.
\end{aligned}$$

The proof of $(-1)m = -m$ for an arbitrary positive integer m is the same.

We now know how to multiply all rational numbers, at least in principle, but we should know the limitations of (29.1) and (29.2). In order to make use of them, one must always distinguish positive numbers from negative numbers before we can carry out a computation because, as it stands, both s and t have to be positive. We should have analogues of (29.1) and (29.2) for *arbitrary* rational numbers s and t, and this is our next goal. Following the general CONVENTION of *letting the expression $-xy$ denote $-(xy)$*, we will prove the following.

Basic Fact 4.[3] $(-x)y = -xy$ for all *rational numbers x, y*.

Basic Fact 5. $(-x)(-y) = xy$ for all *rational numbers x, y*.

We will offer two proofs of these facts. The first one comes cheap: simply retrace the steps of the proofs of (29.1) and (29.2) and observe that, in fact, the proofs remain valid even when s and t are arbitrary rational numbers

---

[3] We continue the numbering of Basic Facts from page 397.

(exercise). The second proof is the sophisticated proof one normally finds in textbooks on abstract algebra; this is the short proof mentioned on page 405. We present this proof for the purpose of giving one demonstration of the power of abstraction. It goes straight from Basic Fact 2 (page 398) and the distributive law to the desired goals *without* benefit of Theorems 29.1 and equation (29.1).

To prove Basic Fact 4, let x, y be rational numbers, and we have to prove that $(-x)y$ is the mirror reflection of xy. By Basic Fact 2, it suffices to prove $(-x)y + xy = 0$. But

$$\begin{aligned} (-x)y + xy &= ((-x) + x)y && \text{(distributive law)} \\ &= 0 \cdot y \\ &= 0 && \text{(by (M3))}. \end{aligned}$$

This then proves Basic Fact 4. Next, to prove Basic Fact 5, we have to prove that $(-x)(-y)$ is the mirror reflection of $-xy$. Again, using Basic Fact 2, we only need to prove $(-x)(-y) + (-xy) = 0$. We compute

$$\begin{aligned} (-x)(-y) + (-xy) &= (-x)(-y) + (-x)y && \text{(Basic Fact 4)} \\ &= (-x)((-y) + y) && \text{(distributive law)} \\ &= (-x) \cdot 0 \\ &= 0 && \text{(by (M3))}. \end{aligned}$$

We have proved Basic Fact 5.

This proof of Basic Fact 5 gives the shortest proof of (negative) × (negative) = positive.

29.4. Some Observations

Theorem 29.2 gives us another way to think of equation (27.6) in Chapter 27, to the effect that $-(x+y) = -x - y$ for all rational numbers x and y. Here is the alternate reasoning:

$$\begin{aligned} -(x+y) &= (-1)(x+y) && \text{(Theorem 29.2)} \\ &= (-1)x + (-1)y && \text{(distributive law)} \\ &= -x - y && \text{(Theorem 29.2)}. \end{aligned}$$

As remarked earlier, conceptually this is not a good proof because the identity $-(x+y) = -x - y$ has nothing to do with multiplication and therefore the use of the distributive law in the proof is not germane to the purpose. However, most students would prefer this proof of the identity to the earlier one on page 391 because it is more computational.

29.4. Some Observations

We also tie up a loose end by proving the following form of the **distributive law for subtraction**, which is again commonly taken for granted:

$$x(y - z) = xy - xz \quad \text{for all rational } x, y, z.$$

Indeed, by using the ordinary distributive law, we have $x(y - z) = x(y + z^*) = xy + xz^* = xy + x(-z) = xy + (-xz)$, where the last step uses Basic Fact 4. By the definition of subtraction, $xy + (-xz) = xy - xz$, as desired.

An interesting consequence of the distributive law for subtraction is a second proof of (27.7) of Chapter 27 on "removing parentheses": $-(x-y) = -x + y$. We now prove it as follows:

$$\begin{aligned}
-(x - y) &= (-1)(x - y) &&\text{(Theorem 29.2)} \\
&= (-1)x - (-1)y &&\text{(dist. law for subtraction)} \\
&= -x - (-y) &&\text{(Theorem 29.2)} \\
&= -x + y.
\end{aligned}$$

For most students, the distributive law for subtraction becomes natural after a while, and this proof of $-(x - y) = -x + y$ may be the only one they will remember.

Exercises

1. Explain directly to a sixth grader why $(-3)(-4) = 12$. Do not assume that he knows anything about (29.2) or Basic Fact 5.

2. Compute: (i) $(-\frac{3}{2})(0.64 - \frac{4}{3})$. (ii) $(-4)(-1\frac{1}{2} + \frac{1}{4})$. (iii) for any rational numbers a, b, c, $(-a)(-b + c) - a(b - c)$ and $(a - b)(a + b) - a(-b)$. (iv) $165 - 560(\frac{3}{4} - \frac{8}{7})$.

3. Suppose m and n are positive integers. Find a proof of (29.1) in this special case, i.e., $(-m)n = -mn$, which is as elementary as possible.

4. Give detailed proofs of Basic Facts 4 and 5 by imitating the proofs of (29.1) and (29.2).

5. Consider each of the following two statements about any rational number x:
   (a) $\frac{1}{2}x < x$.
   (b) $(-5)x > x$.
   If it is always true or always false, prove. If it is sometimes true and sometimes false, give examples to explain why.

6. The following is a standard argument in textbooks to show, for example, that $(-2)(-3) = 6$:
   ```
   Consider the sequence of products
    ......    4 × (−3) = −12, 3 × (−3) = −9, 2 × (−3) = −6,
   1 × (−3) = −3,  0 × (−3) = 0,    (−1)(−3) = a,  (−2)(−3) = b,
   (−3)(−3) = c,  (−4)(−3) = d,          ......
   
   Observe the pattern that, for m×(−3) as m decreases
   to 0, each product increases by 3.  To continue this
   pattern beyond 0, one should assign 3 to a, 6 to b,
   9 to c, 12 to d, and so on, because (−1)(−3) = 0 + 3
   ```
   $= 3$, $(-2)(-3) = 3+3 = 6$, $(-3)(-3) = 6+3 = 9$, $(-4)(-3) = 9 + 3 = 12$.
   Is this a valid argument? What are the implicit assumptions used? Write a critique. (*Hint:* If you were to write down precisely what this so-called pattern says, it would be the statement that $(n - 1)(-3) = n(-3) + 3$ for any positive integer n.)

7. (a) I have a rational number x so that $5 - (3 - 2x) = (1 - \frac{8}{3}x)$. What is this x? (b) Same question for $(1 - 2x) - (x + 1) = \frac{2}{3}x + \frac{1}{2}$.

Chapter 30

Dividing Rational Numbers

It remains to discuss the division of rational numbers. With the concept of multiplication firmly in place and in view of the fact that division is nothing but a different way of expressing multiplication (cf. definition (7.1) on page 97 and the definition on page 289), we will see that there is no difference between the definition of division for rational numbers and the definition of division for fractions.

The sections are as follows:

> Definition of Division and Consequences
>
> Rational Quotients

30.1. Definition of Division and Consequences

First we prove the analogue of the lemma on page 286, namely:

Theorem. *Let a nonzero rational number y be given. Then for any rational number x, there is a unique (i.e., one and only one) rational number z such that $x = zy$.*

When $y > 0$, the theorem reduces to the lemma on page 286. Since this is the second time around, we will give a slightly more abstract proof of the theorem.

We first work out the proof for a special case. Is there a rational number z that satisfies $\frac{1}{3} = (-5)z$, and if so, is it unique? We will approach this problem by first ignoring whether there is such a z and asking, instead, *suppose* there is such a z, what must it be? If we find out as a result that this z can only be a specific number, then we will have proved the *uniqueness* part of the theorem.

So let us assume that a certain number z satisfies $\frac{1}{3} = (-5)z$. We notice that, by Basic Fact 4 on page 409, $(-\frac{1}{5})(-5) = 1$. Therefore to "get rid of the -5 in $(-5)z$" on the right, we multiply both sides by $-\frac{1}{5}$ to get

$$\left(-\frac{1}{5}\right) \times \frac{1}{3} = \left(-\frac{1}{5}\right) \times (-5)z.$$

Since the left side is equal to $-\frac{1}{15}$ by Basic Fact 4, we get $z = -\frac{1}{15}$. In other words, any z that satisfies $\frac{1}{3} = (-5)z$ must be equal to $-\frac{1}{15}$. We have proved the uniqueness of such a z. Knowing this, we verify that indeed $\frac{1}{3} = (-5)(-\frac{1}{15})$, so that we have also proved that $z = -\frac{1}{15}$ satisfies $\frac{1}{3} = (-5)z$.

You may have observed that the key point is to get $-\frac{1}{5}$ so that when it is multiplied by -5, we get 1. Of course, we get this more or less by inspection. In general, to a given number y, $y \neq 0$, a number y^{-1} so that $yy^{-1} = 1$ is called a **multiplicative inverse** of y. It will turn out that the multiplicative inverse of a rational number is always unique, as we now show.

Proof. *Part 1.* First we prove the existence of a unique multiplicative inverse of a given nonzero y.

If $y > 0$, then the existence of a multiplicative inverse of y has been proved in the lemma on page 286. Let us concentrate then on the case $y < 0$. We have $(-y) > 0$, and therefore there is a y' so that $(-y)y' = 1$. By Basic Fact 4, we have $(-y)y' = y(-y')$ as both are equal to $-yy'$. Hence we get $y(-y') = 1$ and y too has a multiplicative inverse $-y'$.

Now we prove the uniqueness of the multiplicative inverse of a rational number y, $y \neq 0$. Suppose there are rational numbers a and b so that both are multiplicative inverses of y, i.e., $ay = by = 1$. Multiplying both sides of $1 = ay$ by b, we get

$$b = ayb = a(by) = a \cdot 1 = a.$$

Thus $a = b$ and the multiplicative inverse of y is unique. We will simply denote it by y^{-1}.

Part 2. Next we prove the uniqueness part of the theorem. In general, with x and y as given, if z is any rational number such that $x = zy$, then

$$xy^{-1} = (zy)y^{-1} \quad \text{so that} \quad xy^{-1} = z(yy^{-1}) = z.$$

Thus all such z must be equal to xy^{-1}, and uniqueness is proved.

Part 3. To prove the existence part of the theorem, we directly verify that $z = (xy^{-1})$ satisfies $x = zy$. This is because of the associative law (see

30.1. Definition of Division and Consequences

(M1)):
$$zy = (xy^{-1})y = x(y^{-1}y) = x \cdot 1 = x.$$
The proof is complete. □

This theorem can be paraphrased in suggestive language: If y is a rational number, then the product zy will be referred to as a **multiple of** y, or a **rational multiple of** y to distinguish it from the case where z is a whole number if there is possible confusion. Then what the theorem says is that, if a nonzero rational number y is given, then every rational number x can be expressed as a multiple of y in one and only one way.

Remark. The uniqueness of z proves the useful fact, often asserted without any justification, that for numbers x and y,
$$xy = 0 \text{ and } y \neq 0 \quad \text{implies} \quad x = 0.$$
Indeed, from $0 \cdot y = 0$ and the given fact that $xy = 0$, we have expressed 0 as a multiple of y in two different ways. The uniqueness part of the theorem shows that $0 = x$. In view of the omnipresence of this fact in everyday mathematical arguments, however, it would be desirable to give a proof that is less subtle: Using "$\Longrightarrow$" to denote "implies", we have
$$y \neq 0, \; xy = 0 \Longrightarrow (xy)y^{-1} = 0 \cdot y^{-1}$$
$$\Longrightarrow x(yy^{-1}) = 0$$
$$\Longrightarrow x \cdot 1 = 0$$
$$\Longrightarrow x = 0.$$

We can now give the formal definition for the division of rational numbers: Let x, y, z be rational numbers with $y \neq 0$. Then by definition
$$\frac{x}{y} \quad \text{is the rational number which satisfies} \quad x = \left(\frac{x}{y}\right) y.$$

The number $\frac{x}{y}$ is called the **quotient** or the **division** x by y. We know from the theorem above that the quotient always exists, and for given x and y, there is only one such quotient, namely, xy^{-1}. Therefore there is no possibility of confusion in the notation $\frac{x}{y}$ for the quotient. Furthermore, we now see explicitly that the division of x by y is just the *multiplication* of x by the multiplicative inverse of y. In this precise sense, division is just a different way of expressing multiplication.

Using this definition of division, we wish to clear up a standard confusion in the study of rational numbers. The equalities
$$\frac{3}{-7} = \frac{-3}{7} = -\frac{3}{7}$$
are generally used in school classrooms without comment or explanation. In fact, the notation is itself a source of confusion. We know the meaning of

$\frac{3}{7}$, and also that of its mirror reflection $-\frac{3}{7}$. But $\frac{3}{-7}$? $\frac{-3}{-7}$? What could they mean?

Let us supply the necessary clarification and explanation. Because -3, -7, etc. are rational numbers, the symbols
$$\frac{-3}{7}, \quad \frac{3}{-7}, \quad \frac{-3}{-7}, \quad \text{etc.,}$$
make sense as divisions of rational numbers. Therefore we see from the proof of the theorem (with $z = xy^{-1}$) that
$$\frac{3}{-7} = 3 \times (-7)^{-1} = 3 \times \left(-\frac{1}{7}\right) = -\frac{3}{7},$$
where we have made use of Basic Fact 4 in the last step. In a similar fashion, we have $\frac{-3}{7} = -\frac{3}{7}$. More generally, the same reasoning supports the assertion that if k and ℓ are whole numbers and $\ell \neq 0$, then
$$\frac{-k}{\ell} = \frac{k}{-\ell} = -\frac{k}{\ell}$$
and
$$\frac{-k}{-\ell} = \frac{k}{\ell}.$$
We may also summarize these two formulas in the following statement:
$$\frac{-a}{b} = \frac{a}{-b} = -\frac{a}{b} \quad \text{for any two integers } a \text{ and } b, \text{ with } b \neq 0.$$

This formula will be seen to be a special case of basic facts about so-called rational quotients, to be introduced in the next section. We have singled it out for a separate discussion because of its frequent appearance in everyday computations with rational numbers. In particular, it implies that

> *every rational number can be written as the quotient of two integers.*

For example, $-\frac{17}{4} = \frac{-17}{4}$. We can further refine this to read

> *every rational number can be written as the quotient of two integers so that the denominator is a whole number.*

This is because if the b in $\frac{a}{b}$ is negative, we can rewrite it as $\frac{-a}{-b}$.

30.2. Rational Quotients

All the facts established for complex fractions in Chapter 19 will now be seen to be true for *quotients of rational numbers*. For lack of a better name, we shall refer to the latter as **rational quotients**. We now list the analogues of most of the basic properties (a)–(f) of complex fractions in Chapter 19. Let $X, Y, Z, \ldots$ be rational numbers so that they are nonzero where appropriate

30.2. Rational Quotients

in the following. Then $\frac{X}{Y}$ is an example of a rational quotient; we call X its **numerator** as usual, and Y its **denominator**.

(a) (*Generalized equivalent fractions*)
$$\frac{X}{Y} = \frac{ZX}{ZY} \text{ for any nonzero rational number } Z.$$

(b) $\frac{X}{Y} \pm \frac{Z}{W} = \frac{XW \pm ZY}{YW}.$

(c) $\frac{X}{Y} \times \frac{Z}{W} = \frac{XZ}{YW}.$

(d) (*Distributive law*)
$$\frac{U}{V}\left(\frac{X}{Y} \pm \frac{Z}{W}\right) = \left(\frac{U}{V} \times \frac{X}{Y}\right) \pm \left(\frac{U}{V} \times \frac{Z}{W}\right).$$

> *Compared with the corresponding items for complex fractions, it may be noticed that the analogue of the cross-multiplication algorithm for "<" is missing. Indeed, the presence of negative numbers adds complexity to the comparison of rational numbers. This issue will be dealt with in the next chapter (see especially page 424).*

For the proofs, one can grind it out by writing each of $X, Y, \ldots$ as $\pm\frac{k}{l}$, where k, l are whole numbers, and then check each of (a)–(d) as assertions about complex fractions. This can certainly be done, but it would be extremely tedious. Because we are more used to abstract arguments at this point, we will introduce a new way of proving things that is shorter and more enlightening, but also more sophisticated. We will make repeated use of the *uniqueness* part of the above theorem.

To prove (a), for example, let $A = \frac{X}{Y}$, $B = \frac{ZX}{ZY}$, and we will prove that $A = B$. By the definition of division of rational numbers, we have $X = AY$ and $ZX = B(ZY)$. But $X = AY$ implies $ZX = Z(AY)$, which implies $ZX = A(ZY)$. When compared with $ZX = B(ZY)$, we see that $A = B$ by the uniqueness part of the theorem in section 30.1 (with $x = ZX$ and $y = ZY$ in the theorem).

> *We explicitly caution against incorrect reasoning at this stage in the passage from*
> $$A = \frac{X}{Y} \quad \text{to} \quad X = AY,$$
> *and from*
> $$B = \frac{ZX}{ZY} \quad \text{to} \quad ZX = B(ZY).$$
> *It is tempting to think that either is the result of an appropriate cancellation, but that would be wrong. For example, it would*

appear that by multiplying both sides of $A = \frac{X}{Y}$ by Y, one gets $AY = \frac{X}{Y} \times Y$, so that by simplifying the right side, one arrives at $AY = X$. However, unless we already know that (a) and (c) are true, we do not get $\frac{X}{Y} \times Y = X$ because, at this stage, we are still trying to prove (a) and are therefore in no position to do any kind of cancellation. Rather, the equality $X = AY$ is the result of the *definition of the division of X by Y*. Similarly, one obtains $ZX = B(ZY)$ from $B = \frac{ZX}{ZY}$ by virtue of the *definition of dividing ZX by ZY*.

We repeat, there is no "cancellation" of any kind in the preceding proof of (a). But of course, once we have proved (a)–(d), we can cancel as much as we want.

To prove (c), let $A = \frac{X}{Y}$, $B = \frac{Z}{W}$, and $C = \frac{XZ}{YW}$. We want to show $AB = C$. Again, by the definition of division, we get, respectively,

$$AY = X,$$
$$BW = Z,$$
$$C(YW) = XZ.$$

Multiplying the first and second equalities together, we get

$$AB(YW) = XZ,$$

and comparing with the third equality, we have, by the uniqueness part of the theorem in section 30.1, that $AB = C$ (in the theorem, let $x = XZ$ and $y = YW$).

The proofs of the other two are similar and will be left as exercises.

These formulas may seem unnecessarily abstract, but they have interesting, practical consequences. For example, let $X, Y, \ldots$ be rational numbers as before. Then

(30.1) $$\left(\frac{X}{Y}\right)^{-1} = \frac{Y}{X}.$$

This is because, by (c) and (a),

$$\frac{Y}{X} \times \frac{X}{Y} = 1,$$

and therefore by the uniqueness of the theorem in section 30.1, equation (30.1) is proved. Also, we have the most general form of **invert and multiply**: for all rational quotients $\frac{X}{Y}$ and $\frac{Z}{W}$,

(30.2) $$\frac{\frac{X}{Y}}{\frac{Z}{W}} = \frac{X}{Y} \times \frac{W}{Z}.$$

30.2. Rational Quotients

This is because, by the definition of division, the left side is $\frac{X}{Y} \left(\frac{Z}{W}\right)^{-1}$, and we can now apply
$$\left(\frac{Z}{W}\right)^{-1} = \frac{W}{Z}$$
to finish the proof of (30.2).

For example, this form of invert and multiply fully justifies
$$\frac{\frac{-3}{5}}{\frac{2.4}{-7}} = \frac{(-3)(-7)}{5 \times 2.4}.$$

We emphasize that, contrary to common practice in textbooks, every step of this computation has been carefully explained and there is no hand-waving of any kind.

Exercises

1. Prove properties (b) and (d) of rational quotients.
2. Compute the following and simplify your answers: (a) $\frac{4}{-15} + \frac{-5}{6}$. (b) $\frac{-2}{5} - \frac{-4}{-3}$. (c) $(-15) \times \frac{4}{-35}$. (d) $\frac{27}{-16} \times \frac{-28}{45}$. (e) $(\frac{-39}{8} \times \frac{9}{11}) + (\frac{39}{-8} \times \frac{-5}{33})$.
3. Compute and simplify: (a) $\frac{3.6}{5} \times \frac{-2.5}{0.9}$. (b) $\frac{7}{1.2} + \frac{5}{-1.8}$. (c) $\frac{4}{27} - \frac{-1}{3.6}$.
4. Explain to a seventh grader why $-4/\frac{1}{-3} = 12$ and why $1/\frac{4}{-5} = -\frac{5}{4}$. Assuming only a knowledge of the multiplication of rational numbers, you have to explain the meaning of division in a way a seventh grader can understand.
5. (a) Find a rational number x so that $\frac{7}{4} - 3x = \frac{1}{2}x + 1\frac{3}{4}$. (b) Bring closure to the discussion of Chapter 26 by showing that if A, B, C, and D are rational numbers and $A - C \neq 0$, then every linear equation of the form $Ax + B = Cx + D$ has a solution.
6. (a) Two numbers satisfy the property that their sum is 10 and their *difference* (i.e., the larger subtracting the smaller) is 21. What are these numbers? (b) Two numbers satisfy the property that their sum is $3\frac{1}{2}$ and their difference is $21\frac{1}{4}$. What are these numbers?
7. Two numbers are given. Suppose twice the smaller number falls short of the bigger number by 3, but $3\frac{1}{3}$ of the smaller number exceeds the bigger number by 29. What are these numbers?

Chapter 31

Ordering Rational Numbers

In this section, we bring closure to the discussions of the ordering of numbers in Chapters 1, 2, and 16. We will take a serious look at the comparison of rational numbers and prove several basic inequalities that are useful in school mathematics.

The sections are as follows:

Basic Inequalities

Powers of Rational Numbers

Absolute Value

31.1. Basic Inequalities

Recall the definition of $x < y$ between two rational numbers x and y in Chapter 25: it means x is to the left of y on the number line.

$$\underset{x \quad\quad y}{\rule{5cm}{0.4pt}}$$

In the following, we will use "$\implies$" to denote "implies", and also "$\iff$" to denote "is equivalent to".

(A) For any rational numbers x, y, $x < y \iff -x > -y$.

The simplest illustration is the fact that $1 < 2 \iff -1 > -2$, which is obvious when you think about the number line.

In general, if $x < 0 < y$, then $-x > 0$ while $-y < 0$ and there is nothing to prove. Therefore we need only to attend to the cases where x and y **have**

421

the same sign, i.e., are both positive or both negative. If $0 < x < y$, then we have the following picture:

$$\begin{array}{ccccc} -y & -x & 0 & x & y \end{array}$$

On the other hand, if $x < y < 0$, then we have

$$\begin{array}{ccccc} x & y & 0 & -y & -x \end{array}$$

In both cases, the truth of $-x > -y$ is obvious.

(B) For any rational numbers x, y, z, $x < y \Longrightarrow x + z < y + z$.

It is easy to be convinced of the truth of (B) with a few simple experiments using specific numbers. For example, $2 < 4 \Longrightarrow (2 + 25) < (4 + 25)$ (because $27 < 29$) and $\big(2 + (-55)\big) < \big(4 + (-55)\big)$ (because $-53 < -51$).

In general, the proof is somewhat delicate. Suppose $x < y$. Because of the commutativity of addition, it suffices to prove $z + x < z + y$, or equivalently, the endpoint of the vector $\vec{z} + \vec{x}$ is to the left of the endpoint of the vector $\vec{z} + \vec{y}$. By the definition of vector addition on page 383, both vectors $\vec{z} + \vec{x}$ and $\vec{z} + \vec{y}$ are obtained by placing the starting points of $\vec{x}$ and $\vec{y}$, respectively, at the endpoint of $\vec{z}$, and the endpoints of the displaced $\vec{x}$ and $\vec{y}$, respectively, will be $z + x$ and $z + y$. Since by hypothesis, the endpoint of $\vec{x}$ is to the left of the endpoint of $\vec{y}$, the conclusion is immediate.

The following picture shows the case where $x > 0$, $y > 0$ and $z < 0$:

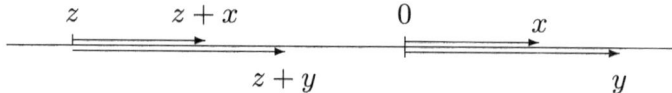

(C) For any rational numbers x, y, $x < y \Longleftrightarrow y - x > 0$.

First, we prove that $x < y \Longrightarrow y - x > 0$. By (B), $x < y$ implies $x + (-y) < y + (-y)$, which is equivalent to $x - y < 0$. By (A), we get $-(x - y) < 0$, which is equivalent to $y - x > 0$, by (27.7) on page 391. Conversely, we prove $y - x > 0 \Longrightarrow x < y$. Again we use (B): $y - x > 0$ implies that $(y - x) + x > 0 + x$, which is equivalent to $y > x$, as desired.

It should be remarked that sometimes (C) is taken as the definition of $x < y$.

(D) For any rational numbers x, y, z, if $z > 0$, then $x < y \Longrightarrow xz < yz$.

Intuitively, the fact that $z > 0$ means multiplication by z does not change the sign (positive or negative) of the number it multiplies. Thus, if $z = 5$, then $287 > 0$ implies $(5 \times 287) > 0$ and if $-23 < 0$, then $\big(5 \times (-23)\big) < 0$. Thus multiplication by a positive number such as 5 simply expands every positive number by a factor of 5 to the right of 0, and expands every negative number by a factor of 5 to the left of 0. It follows that if $0 < x < y$, $5x$ remains to the left of $5y$, and if $x < y < 0$, then $5x$ also remains to the left of $5y$. Of course if $x < 0 < y$, then $5x$ is negative and $5y$ is positive and, trivially, $5x < 5y$.

Formally, we give two proofs of (D). First, we make use of (C). If $z > 0$ and $x < y$, we have to prove $xz < yz$, which, by (C), is equivalent to proving $(yz - xz) > 0$. Now $(yz - xz) = (y - x)z$, and $y - x > 0$ by (C) and the hypothesis that $x < y$. Since $z > 0$, we have $(y - x)z > 0$, and therefore $(yz - xz) > 0$, as desired.

A second proof uses the definition of fraction multiplication. Given $z > 0$ and $x < y$, if $x < 0 < y$, then $xz < 0$ and $yz > 0$ and there would be nothing to prove. Therefore we need only consider the cases where x and y have the same sign. If $x, y > 0$, then x, y, and z are fractions and xz and yz are then areas of rectangles with sides of length x, z, and y, z, respectively. Since $x < y$, clearly the rectangle corresponding to yz has a greater area. Hence $yz > xz$. Next, suppose $x, y < 0$, then we get $(-x), (-y) > 0$. Moreover $x < y$ implies $(-y) < (-x)$, by (A). Thus we know from the preceding argument that $(-y)z < (-x)z$, which implies $-yz < -xz$ (Basic Fact 4 on page 409), and therefore $yz > xz$, by (A) again.

(E) For any rational numbers x, y, z, if $z < 0$, then $x < y \implies xz > yz$.

To students, this is the most fascinating inequality of the lot because the multiplication by a negative number *reverses* an inequality. To get an intuitive idea of why this is so, first look at some concrete examples involving the simple case of $z = -5$ in (E):

$2 < 3$, but $-10 > -15$.
$-5 < 1$, but $25 > -5$.
$\frac{1}{3} < \frac{1}{2}$, but $-\frac{5}{3} > -\frac{5}{2}$.

One can see why these are true by placing these numbers on the number line and simply observing their positions. Next we prove the special case of (E) where $z = -2$ and $0 < x < y$; it will be seen that this special case is the heart of the matter. Thus we want to understand how $x < y$ can lead to $(-2)x > (-2)y$. The inequality $x < y$ means x is to the left of y:

$$0 \quad x \quad y$$

Consider first $2x$ and $2y$: they are the result of pushing x and y further to the right of 0 by doubling their respective distances from 0:

$$0 \quad 2x \quad 2y$$

If we reflect them across 0, then we get the following picture:

$$-(2y) \quad\quad -(2x) \quad\quad 0 \quad\quad 2x \quad\quad 2y$$

Observe that the relative positions of $2x$ and $2y$ have been reversed, namely, whereas $2y$ is to the right of $2x$, $-(2y)$ is now to the *left* of $-(2x)$. But by Basic Fact 4, $-(2x)$ is equal to $(-2)x$, and likewise $-(2y) = (-2)y$. Therefore $(-2)y < (-2)x$. This is the essence of (E).

The proof of (E) in general is as simple as the preceding intuitive discussion. Let $z = -w$, where w is now *positive*. Since $x < y$, (D) implies that $wx < wy$. By (A), $-wx > -wy$. But Basic Fact 4 says $-wx = (-w)x = zx$, and $-wy = (-w)y = zy$. So $zx > zy$. The proof is complete.

We give a second proof making use of (C). Given $z < 0$ and $x < y$, we have to prove $xz > yz$. By (C), this is equivalent to proving $xz - yz > 0$, i.e., $(x - y)z > 0$. From the hypothesis $x < y$ and (C), $y - x > 0$, which implies $-(y - x) < 0$, by (A). Thus $x - y < 0$ (by (27.7) on page 391). Since z is also negative, the product of the two negative numbers z and $x - y$ is positive, i.e., $(x - y)z > 0$, as desired.

(F) For any rational number x, $x > 0 \iff \frac{1}{x} > 0$.

This is because $x(\frac{1}{x}) = 1$ and therefore $x(\frac{1}{x}) > 0$. This implies x and $\frac{1}{x}$ have to be both positive, or both negative.

(G) For any rational numbers x, y, z, let $x < y$. If $z > 0$, then $\frac{x}{z} < \frac{y}{z}$; if $z < 0$, then $\frac{x}{z} > \frac{y}{z}$.

This is an immediate consequence of (D)–(F).

Finally, we bring closure to the discussion of the cross-multiplication algorithm for rational numbers. This algorithm cannot be literally true because, for example, $\frac{1}{-2} < \frac{-1}{4}$ (because $-\frac{1}{2} < -\frac{1}{4}$), but $1 \times 4 > (-2)(-1)$. However, if we add a mild restriction, then the algorithm is restored. Recall that if x and y are rational numbers, then $\frac{x}{y} = \frac{-x}{-y}$ (see formula (a) on generalized equivalent fractions in the preceding section). Thus every rational quotient may be assumed to have a positive denominator. Then we have:

(H) *Cross-multiplication algorithm for rational numbers.* Let x, y, z, w be any rational numbers, and let $y, w > 0$. Then $\frac{x}{y} < \frac{z}{w} \iff xw < yz$.

Because $y, w > 0$, $\frac{x}{y} < \frac{z}{w}$ implies $(\frac{x}{y})yw < (\frac{z}{w})yw$ (according to (E)), which is equivalent to $xw < yz$ by formula (a) and (c) on rational quotients (see page 417). Conversely, suppose $xw < yz$. Now, $y, w > 0$ implies $\frac{1}{y}, \frac{1}{w} > 0$ (by (F)). Thus, $\frac{1}{yw} > 0$. By (E), $xw < yz$ implies $xw(\frac{1}{yw}) < yz(\frac{1}{yw})$, which is equivalent to $\frac{x}{y} < \frac{z}{w}$.

31.2. Powers of Rational Numbers

This section makes some remarks about how inequalities of rational numbers behave when both sides are squared or cubed, or in fact *raised to any integer power*, which we now define. If n is a positive integer, we define for any rational number x that $\boldsymbol{x^n} = xx \cdots xx$ (n times). If $x \neq 0$, we further define $x^0 = 1$ and
$$x^{-n} = \frac{1}{x^n}.$$
Still with a nonzero x, for any integer m, positive or negative, x^m is called the **m-th power of x**, and m is called the **exponent** of x^m. When $m = 2$ or 3, x^m is just the usual notion of **x-squared** or **x-cubed**, as the case may be. Note that

(31.1) $\qquad x^m x^n = x^{m+n} \qquad$ for any integers m and n.

We only need (31.1) for positive m and n in what follows, and for this special case, (31.1) is simplicity itself. For the special case $x = 10$, (31.1) was already proved for positive m and n; see (1.4) on page 30. The proof for a general x is the same. When m and n are arbitrary integers, the reasoning for (31.1) is not much different from the positive case. For example, let us see why $x^3 x^{-5} = x^{-2}$. By formulas (a) and (c) for rational quotients,
$$x^3 x^{-5} = xxx \frac{1}{xxxxx} = \frac{xxx}{xxxxx} = \frac{1}{xx} = x^{-2}.$$
For more details, see the discussion after equation (39.1) on page 500 for the case $x = 10$.

Let x, y be nonzero rational numbers for the rest of this chapter.

One usually thinks of squaring a number as making it bigger. Such is the case for 3 or 5: $3^2 = 9 > 3$ and $5^2 = 25 > 5$. However, $(\frac{1}{2})^2 = \frac{1}{4} < \frac{1}{2}$. Similarly, $(\frac{2}{7})^2 = \frac{4}{49} < \frac{2}{7}$. The critical difference is that $3 > 1$ and $5 > 1$, but $\frac{1}{2} < 1$ and $\frac{2}{7} < 1$. We now elucidate this phenomenon in general.

(I) (i) If $x > 1$, then $x^2 > x$, and more generally, $x^m > x^n$ for whole numbers $m > n$. (ii) If $0 < x < 1$, then $x^2 < x$, and more generally, $x^m < x^n$ for whole numbers $m > n$.

Indeed, because $x > 0$ and $x > 1$, (D) implies that $x \cdot x > 1 \cdot x$, i.e., $x^2 > x$. In general if $m > n$, then we shall prove $x^m > x^n$. For, from $x^2 > x$ and $x > 1$, we get $x^2 > 1$. Multiply both sides of the last inequality by x and we get $x^3 > x^2$, and because $x^2 > 1$, we also get $x^3 > 1$. Repeating, we get $x^4 > 1$, $x^5 > 1$, ..., $x^{m-n} > 1$. Now multiply both of sides of $x^{m-n} > 1$ by x^n, which is positive, and (D) implies that $x^n x^{m-n} > x^n \cdot 1$. By (31.1), $x^n x^{m-n} = x^m$, so $x^m > x^n$, as desired.

On the other hand, if $0 < x < 1$, then we have $x > 0$ and $1 > x$, so again by (D), $1 \cdot x > x \cdot x$, which is to say, $x > x^2$. The proof of the general case $x^m < x^n$ for whole numbers $m > n$ proceeds in exactly the same way as before. Thus we first show $1 > x^2$ (because $x > x^2$ and $1 > x$), and then $1 > x^3$, $1 > x^3$, ..., $1 > x^{m-n}$. Now multiply both sides of $1 > x^{m-n}$ by x^n and use (D) to conclude $x^n > x^n x^{m-n} = x^m$, by (31.1).

Remark. If $x < 1$ but x is not positive, it will not be true in general that $x^2 < x$. For example, $-4 < 1$, but $(-4)^2 = 16 > (-4)$.

(J) If $0 < x < y$, then $x^n < y^n$ for any whole number $n > 0$.

From $x > 0$ and $x < y$, we obtain $x^2 < xy$ by (D). From $y > 0$ and $x < y$, we obtain $xy < y^2$ also by (D). Together, we get $x^2 < xy < y^2$. So $x^2 < y^2$. Now multiply both sides of the last inequality by x and use (A) to conclude $x^3 < xy^2$. But we can also multiply both sides of $x < y$ by y^2 to get $xy^2 < y^3$, using (D). Putting $x^3 < xy^2$ and $xy^2 < y^3$ together, we obtain $x^3 < y^3$. Repeating the same pattern of deduction gets us the final result $x^n < y^n$ after n steps.

31.3. Absolute Value

Intrinsically tied to any discussion of inequalities is the notion of the **absolute value** $|x|$ of a number x, which is by definition the **distance** of x from 0 (i.e., the length of the segment $[x, 0]$ or $[0, x]$, depending on whether x is negative or positive). In particular, $|x| \geq 0$ no matter what x may be. If b is a positive number, then the set of *all* numbers x so that $|x| < b$ consists of all the points x of distance less than b from 0, indicated by the thickened segment below:

It follows that the inequality for a point x,
$$|x| < b,$$
is equivalent to the two inequalities $-b < x \leq 0$ and $0 \leq x < b$, which can be combined into a **double inequality**
$$-b < x < b.$$
In the usual notation for intervals on the number line, we use $(-b, b)$ to denote the usual segment $[a, b]$ *without* the two endpoints $-b$ and b. Then we have

(31.2) $\qquad |x| < b \quad \text{is equivalent to} \quad x \in (-b, b).$

Absolute value has a nice property: for all numbers x, y,
$$|x|\,|y| = |xy|.$$

31.3. Absolute Value

This can be proved by a case-by-case examination of the four cases where x and y take turns being positive and negative. The reasoning is routine and will be left as an exercise.

Having introduced absolute value, we must now face the question that is asked by most teachers (not to mention innumerable students), but hardly ever answered. Namely, why is this concept considered to be important? As one educator noted, in the school classroom, absolute value is usually taught as a topic disconnected from everything else in the curriculum; it is barely touched on, and is not understood except as a kind of rote procedure ("take off the minus sign if there is one"). Teachers feel handicapped by being made to teach something for which they don't see any relevance.

It is not possible in an elementary text to give a wholly satisfactory answer to this question. The importance of absolute value emerges mainly in the more advanced portion of mathematics, such as when we come face-to-face with the concept of limits and the inequalities that are intrinsically associated with limits. Nevertheless, we can point out that the use of the *idea* of absolute value is commonplace in science and everyday life. For example, in making measurements, errors occur. In order to quantify the maximum possible error in a measurement, one usually states the result of a measurement as a number *plus or minus a certain number*. The best example of this genre may be the measurement of the speed of light in a vacuum mentioned in section 10.3, page 148. Around 1975, the speed of light was found to be 299,792,458 ±1.2 meters per second. Precisely, this means if c denotes the speed of light in meters per second,[1] then

$$299{,}792{,}458 - 1.2 \ \leq \ c \ \leq \ 299{,}792{,}458 + 1.2.$$

In view of (B) above, this may be rewritten as

$$-1.2 \ \leq \ c - 299{,}792{,}458 \ \leq \ 1.2.$$

By observation (31.2) above, this double inequality is equivalent to the absolute value inequality

$$|c - 299{,}792{,}458| \ \leq \ 1.2.$$

In other words, the common scientific writing such as 299,792,458 ±1.2 is nothing but an alternate way of expressing a mathematical statement involving the use of absolute values. Thus, we can at least see the *need* for the concept of absolute value. Without going into details, one can also point out that the concept of *absolute error* of an estimate (see Chapter 10) is another concept that hides the use of absolute value.

Physicists, and almost all scientists, use the notation 299,792,458 ± 1.2 in place of $|c - 299{,}792{,}458| \leq 1.2$. However, the use of the absolute value notation is essential in advanced mathematics because, much more than a

---

[1] This is the universal symbol for the speed of light; "c" stands for *constant*.

language, absolute value becomes an indispensable tool in many basic computations. It would be at least awkward, and often impossible, to carry out extensive computations in many situations using only the alternate notation of "299,792,458 ±1.2". The so-called *triangle inequality* among numbers,

$$|x + y| \leq |x| + |y| \quad \text{for all numbers } x \text{ and } y,$$

or the *Cauchy–Schwarz inequality* for four numbers a, b, c, d,

$$|ac + bd| \leq \sqrt{a^2 + b^2} \sqrt{c^2 + d^2}$$

may serve to give a hint of the fundamental role played by the concept of absolute value in mathematics.

Exercises

1. Give direct proofs of the special case of (G) where x, y, z are integers.
2. If a is a positive rational number, then: (i) $a > 1$ implies $\frac{1}{a} < 1$, and (ii) $0 < a < 1$ implies $\frac{1}{a} > 1$ (cf. Example 4 on page 243). (Note that, for part (ii), it is important that a be positive. For example, $-3 < 1$ but $\frac{1}{-3} = -\frac{1}{3} < 1$.)
3. Which is greater? (a) $(1.4)(-3)$ or $-5 + \frac{3}{4}$. (b) $\frac{-4}{5}$ or $(-5)\frac{5.1}{7.5}$. (c) $\frac{-2}{3}/\frac{4}{7}$ or $(\frac{14}{-3})(\frac{2}{7.5})$.
4. Show that if $x > 1$, then $x^n > 1$ for any positive integer n. Also show that if $-1 < x < 1$, then $-1 < x^n < 1$ for any positive integer n.
5. Show that if $x < y < 0$, then $x^2 > y^2$, $x^3 < y^3$. What can you say about x^n and y^n for a positive integer n? Prove it.
6. Can you see why if x and y are any two rational numbers (in particular, they could be negative), then $x^2 - 2xy + y^2 \geq 0$?
7. If x is a rational number, is it true that $x < 1$ implies $\frac{1}{x} > 1$? If it is true, explain. If not, formulate a true statement, and prove it.
8. Prove that for any rational number x and y, $|x| \cdot |y| = |xy|$.

Part 4

Number Theory

Part Preview

So far we have expanded our domain of study from whole numbers to fractions and then to rational numbers. At each stage of this expansion, we had to face a conceptual jump from a known system to a more complicated system with many new numbers added to the known system. Now we take a step backward by returning to whole numbers, with only an occasional reference to the integers. This time we will not be facing any new numbers or trying to extend the arithmetic operations, yet it will not do to mislead you into believing that things will be simple and easy again. We will be looking into the relationship between addition and multiplication among whole numbers, which is part of *number theory*, and you may be surprised to learn that this is far from a simple matter. In fact, one of the main goals of mathematics is exactly to understand this relationship.

Among the highlights of this chapter will be an algorithm that reduces a fraction to one in lowest terms and a characterization of those fractions that are equal to finite decimals (Theorems 36.1 and 36.2 of Chapter 36). The proofs of these theorems require a brief detour through the most rudimentary aspects of number theory. More precisely, we will study in some depth the elementary concept of division-with-remainder. To this end, you will be asked to look at whole numbers in a completely different light: they will no longer be just tools for computations but, rather, objects of intrinsic interest. You will learn about the intricate pattern—*thought pattern*—in which the whole numbers are tied together by the two operations of addition and multiplication. You will learn how to work through many concrete examples (using a four-function calculator if necessary), not for numerical accuracy per se, but to observe how they exhibit this pattern. For example, whereas you normally use division-with-remainder just to get the quotient with little thought given to the remainder, now you will uncover the role played by the remainder in helping us come to grips with the greatest common divisor of two numbers. In fact, this part of the book—Part 4—may be regarded as nothing more than "A Meditation on the Importance of the Remainder in Division-with-Remainder". If you are mentally prepared for this shift in focus, you will find the task ahead a lot easier.

Chapter 32

Divisibility Rules

We will be looking at the divisibility properties of whole numbers ≤ 11. You have undoubtedly known for a long time that 3 divides 759 because 3 divides $7 + 5 + 9 = 21$, whereas 3 does not divide 277 because 3 does not divide $2 + 7 + 7 = 16$. The purpose of this chapter is to explain this and other simple divisibility tests.

The sections are as follows:

Review of Division-with-Remainder

Generalities about Divisibility

Divisibility Rules

32.1. Review of Division-with-Remainder

The one fact that runs through the whole chapter is the *theorem on division-with-remainder*, Theorem 7.1 on page 105, which we will now recall. Given whole numbers a and d, with $d > 0$, there is always a *unique* whole number q (in the sense that there is one and only one such q) with the property that

(32.1) $$qd \leq a < (q+1)d$$

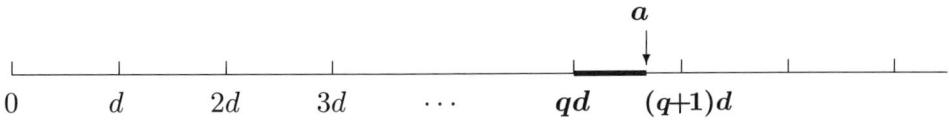

This q is characterized by the fact that qd is the largest multiple of q not to exceed a. This q is called the **quotient** of a **divided by** d. If qd happens to equal a, then $a = qd$ and q *is* the quotient of a divided by d in the sense of Chapter 7.

Denote $a - qd$ by r. We call r the **remainder** of a divided by d. We have $qd + r = qd + (a - qd) = a$. Therefore equation (32.1) can be rephrased as follows. Given whole numbers a and d with $d > 0$, there are unique whole numbers q and r so that

(32.2) $\qquad a = qd + r, \qquad$ where r satisfies $\quad 0 \leq r < d$.

The whole number r is the length of the thickened segment in the preceding picture.

Recall also the fact that, on the one hand, one can get the quotient q and remainder r just by patiently counting the whole number multiples of the divisor d, and on the other hand, the long division algorithm is an efficient way to get the quotient and the remainder.

In this chapter, you will witness the first piece of evidence that the remainder in division-with-remainder has an important role to play.

32.2. Generalities about Divisibility

The discussion below gains in simplicity if we extend the concepts of multiples and divisibility to integers (they are defined in Chapter 7 for whole numbers). We say an integer x is a **multiple** of another integer y ($y \neq 0$) if there is an integer z so that $x = zy$. In symbols: we write **$y|x$** for **y divides x** and $y \neq 0$. The symbol $y|x$ is not one that is commonly used in K–12, but it is universally accepted in mathematics. The reason we use it here is that, when we start to discuss divisibility in earnest, the incessant repetition of "this divides that" or "this doesn't divide that" in writing can become tedious. Other commonly used expressions include: y is a **divisor** of x, or y is a **factor** of x. If y does not divide x, we also use the symbol **$y \nmid x$**. If x and y are whole numbers and $y \neq 0$, then $y|x \iff$ the division-with-remainder of x by y has remainder 0. Recall that the symbol "$\iff$" means "is equivalent to", or "if and only if".

Note that *0 is a multiple of every nonzero integer*, and that *every nonzero integer divides 0*.

Having introduced the concept of one integer dividing another, we note nevertheless that our main focus remains the whole numbers. The extension of the concept of divisibility to integers is merely a technical contrivance to streamline the exposition. Some examples would help. We know $4|28$, but does -4 divide 28? Yes, because $28 = (-7) \times (-4)$. Does -4 divide -28? Yes again because $-28 = 7 \times (-4)$. What about 4 and -28? Indeed 4 divides -28 because $-28 = (-7) \times 4$. So we see that the divisibility of one integer by another is essentially the same as the divisibility of one whole number by another, and the only difference is the need to adjust the sign (i.e., $+$ or $-$) in each case.

32.2. Generalities about Divisibility

The intuitive idea of divisibility is so ingrained in us that some of you would be hard pressed to take the preceding definition seriously. Up to this point, we have had no occasion to make use of this concept in any substantial way, and so it would not have made any difference. But in this chapter, where the main theme is divisibility itself, it would be a good idea to develop a conditioned reflex to think, upon hearing "d divides a", that it means $a = qd$ for some whole number q. This will save you a lot of grief in trying to understand the following theorems and their proofs.

The next two lemmas will be used often in this chapter, and usually implicitly.

Lemma 32.1. *Suppose A, B, C are integers which satisfy $A = B + C$, and suppose a nonzero integer x divides any two of A, B, and C. Then x must also divide the third.*

We first look at a couple of examples. Let $A = 72$ and $C = 24$, and suppose $72 = B + 24$. Now we recognize that both 24 and 72 are multiples of 6 (i.e., *divisible by* 6), and therefore it is plausible that B would just be some multiple of 6 too. Indeed it is easy to see that B is equal to $72 - 24 = 48$, and sure enough, 48 is a multiple of 6 since $48 = 8 \times 6$. It is instructive to try to arrive at the same conclusion without the explicit computation. From the assumptions of divisibility, we know that $72 = m \times 6$ and $24 = n \times 6$ for some whole numbers m and n. (The fact that m and n can be explicitly determined to be 12 and 4, respectively, will play no role, as we shall see.) Therefore, by the distributive law,

$$B = 72 - 24 = (m \times 6) - (n \times 6) = (m - n) \times 6.$$

Since $m - n$ is an integer, the definition of divisibility says $6|B$, as desired. Another example is to let $B = -187$ and $C = 119$, and suppose $A = -187 + 119$. In this case, it is less obvious that there is any number dividing both 119 and -187, but there is: 17. So the lemma implies that $17|A$. Again, we can compute explicitly to get $A = -68$ and since $-68 = (-4) \times 17$, we see that $17|A$. But, again, we can do this without any explicit computations. Since 17 divides -187 and 119, there are integers m and n such that $-187 = m \times 17$ and $119 = n \times 17$. Then the distributive law implies that

$$A = -187 + 119 = (m \times 17) + (n \times 17) = (m + n) \times 17.$$

Because $m + n$ is an integer, the definition of divisibility implies that $17|A$.

Lemma 32.1 is often used in the following form:

> Suppose A, B, C are integers which satisfy $A = B + C$. Suppose a nonzero integer x divides one of them, say B. Then x divides $A \iff x$ divides C.

Indeed, let $x|B$. Then $x|A$ implies x divides both A and B and therefore $x|C$ by Lemma 32.1. Conversely, still assuming $x|B$, suppose $x|C$. Then x divides both B and C and therefore, by Lemma 32.1, $x|A$, as desired.

We are now in a position to prove Lemma 32.1 in general.

Proof. This proof is an exercise in the use of the definition of divisibility. Let $A = B + C$ and let the integer x divide B and C. We have to show $x|A$. The hypothesis means that $B = wx$ and $C = zx$ for integers w and z. Therefore,
$$A = B + C = wx + zx = (w+z)x,$$
where the last step uses the distributive law. Because $w + z$ is an integer, A is a multiple of x, and so x divides A, as claimed. Next, suppose x divides A and C, and we will show why x divides B. By hypothesis, we have $A = wx$ and $C = zx$ for some integers w and z. Therefore,
$$B = A - C = wx - zx = (w-z)x,$$
where the last step is by the distributive law for subtraction. Because $w - z$ is an integer, x divides B. The proof of the remaining case is the same. The proof is complete. □

Lemma 32.2. *Let A, B, C be integers. (a) If A divides B and B in turn divides C, then A also divides C. (b) If A does not divide C, then no multiple of A can divide C.*

Let us begin with some comments and examples before giving the proof. Part (a) is more intuitive if it is stated in an equivalent form in terms of *multiples*: if C is a multiple of B and B is a multiple of A, then C is a multiple of A. For example, 72 is a multiple of 12, and 12 is a multiple of 4, then it appears plausible that 72 is a multiple of 4. Symbolically: if $72 = a \times 12$ for some integer a (never mind that we know a is 6), and $12 = b \times 4$ for some integer b, then it is straightforward to check that $72 = a \times 12 = a \times (b \times 4) = (ab) \times 4$, which clearly exhibits 72 as a multiple of 4 (because ab is an integer). So why is part (a) not stated in terms of *multiples*? Because it is usually more natural, in context, to speak in terms of divisibility. For part (b), again it may be more intuitive when stated in terms of multiples: If C is not a multiple of A, then it cannot be a multiple of any of the following numbers either: $\pm 2A$, $\pm 3A$, $\pm 4A$, $\pm 5A$, etc. Thus, if 37 is not a multiple of 3, then it stands to reason that 37 cannot be a multiple of 15 or -15 (which are equal to 5×3 and $(-5) \times 3$).

Proof. We first prove (a). By hypothesis, $B = mA$ for some integer m (because A divides B) and $C = nB$ for some integer n (because B divides C). Therefore we may substitute the value of B given by the former equation into the latter equation to get
$$C = nB = n(mA) = (nm)A.$$

But nm is an integer, so C is a multiple of A, which means by definition that A divides C, as desired.

Next we prove (b). Suppose A does not divide C. Let B be a multiple of A and we will show that B cannot divide C. Now, between B and C, there are only two possibilities: either B divides C, or B does not divide C. If we can eliminate the first possibility, then we would be left with the second one, i.e., B does not divide C. So suppose B divides C. Then we would have that A divides B (because B is a multiple of A), and B divides C. By part (a), A must divide C. But we are given that A does not divide C, so the idea that B divides C leads to a conclusion that directly contradicts what is given. Hence B does not divide C, as desired. The proof is complete. □

We now pause to confer official status to a general method that we have just used to prove part (b), which is the concept of **proof by contradiction**. First we make explicit a fact that is integral to mathematics as a discipline:

> *Either a mathematical statement is true or its **negation** (i.e., logical opposite) is true, and there is no other possibility.*

This is a drastic statement, because it means that in order to prove something is true, all you have to show is that its negation is false. If a similar doctrine were implemented in real life in every situation, all of us would be in plenty of trouble most of the time. But in mathematics, this is *the* standard modus operandi. For example, we said above that either B divides C, or B does not divide C, and there is no other possibility. It may take you some time to get used to it, but it would be worth the effort because many proofs in number theory resort to this kind of reasoning.

32.3. Divisibility Rules

We can now turn to the simple rules about the divisibility of a whole number by a small number between 1 and 12. The overriding fact concerning these rules is the following. Let us call a number of the form 10^k for a whole number k a **power of 10**. Then:

> *All divisibility rules are consequences of the behavior of powers of 10 when they are divided by these small numbers.*

First, the divisibility by 2. Formally, we say a whole number is **even** if it is divisible by 2, and **odd** otherwise, i.e., *not* divisible by 2. A few comments about these two definitions are in order. We all have gut feelings about what an *even* number is and what an *odd* number is, so the following discussion can easily get entangled with these strong gut feelings. Part of learning mathematics is to put our own beliefs on hold in such situations and allow the logical argument to unfold. The advantage of the above definitions of evenness and oddness is that we know right away *every whole number* *is either even or odd*, because a whole number is either divisible or not divisible by 2. But we are used to saying things like "the number 21,374 is even because its last digit is 4, which is even, and the number 6,241 is odd because its last digit is 1, which is odd" and, in the face of these definitions, we are now left wondering, How do we know that we can check evenness (i.e., divisibility by 2) and oddness (i.e., nondivisibility by 2) just by inspecting the last digit? We answer this question with the following **divisibility rule for 2**:

A whole number is even $\iff$ *its ones digit is* 0, 2, 4, 6, *or* 8, *and is odd* $\iff$ *its ones digit is* 1, 3, 5, 7, *or* 9.

Proof. We will prove the assertion about even numbers and leave the one about odd numbers as an exercise. First suppose a whole number n is even and suppose its ones digit is k. By looking at the expanded form of n, we have $n = 10b + k$ for some whole number b. (For example, if the whole number is 867, then its ones digit 7 satisfies $867 = (10 \times 86) + 7$.) Then in the equation $n = 10b + k$, 2 divides n (by the hypothesis that n is even), and 2 divides $10b$ (because $10b = (5b) \times 2$). By Lemma 32.1, 2 must divide k. Since k is a single-digit number, $k = 0, 2, 4, 6$ or 8. Conversely, suppose the ones digit k of a given whole number n is $k = 0, 2, 4, 6$ or 8, then as before we have $n = 10b + k$ for some whole number b. Now 2 divides k, and also 2 divides $10b$ because $10b = (5b) \times 2$ for some integer $5b$. By Lemma 32.1 again, 2 must divide n and therefore n is even. The proof is complete. □

When is a whole number divisible by 3? The **divisibility rule for 3** is the following:

A whole number is divisible by 3 $\iff$ *the sum of its digits is divisible by* 3.

To see whether 84 or 674 is divisible by 3, for instance, we add up all the digits, $8+4$ and $6+7+4$, and the rule implies that the divisibility of the latter numbers by 3 will decide the divisibility of 84 and 674 by 3, respectively. Since $8 + 4 = 12$ and $3|12$ and $6 + 4 + 7 = 17$ and $3 \nmid 17$, we conclude that $3|84$ but $3 \nmid 674$. Sure enough, we have $84 = 26 \times 3$ and $674 = (224 \times 3) + 2$.

What is the reason behind this divisibility rule? It is of course nothing more than division-with-remainder at work: when a power of 10 is divided by 3, the remainder is always 1. Thus, $10 = (3 \times 3) + 1$, $100 = (33 \times 3) + 1$, $1000 = (333 \times 3) + 1, \ldots$, and in general,

$$10^n = (\underbrace{33 \cdots 3}_{n} \times 3) + 1.$$

32.3. Divisibility Rules

We put this simple fact to work by explaining why the fact that 3 divides 84 is equivalent to $3|(8+4)$. Recalling that since $84 = (8 \times 10) + 4$, we get

$$\begin{aligned} 84 &= (8 \times 10) + 4 \\ &= 8 \times \bigl((3 \times 3) + 1\bigr) + 4 \\ &= \bigl((8 \times 3) \times 3\bigr) + (8 + 4) \quad \text{(dist. law and assoc. law).} \end{aligned}$$

Because 3 already divides $(8 \times 3) \times 3$, if 3 divides $8+4$, it must also divide 84, and vice versa, in view of Lemma 32.1. Thus the fact that $3|84$ is equivalent to the fact that $3|(8+4)$.

Similarly, we can show why $3 \nmid (6+7+4)$ is equivalent to $3 \nmid 674$. First, observe that

$$\begin{aligned} 674 &= (6 \times 100) + (7 \times 10) + 4 \\ &= 6 \times \bigl((33 \times 3) + 1\bigr) + \Bigl(7 \times \bigl((3 \times 3) + 1\bigr)\Bigr) + 4 \\ &= \Bigl(\bigl((6 \times 33) + (7 \times 3)\bigr) \times 3\Bigr) + (6 + 7 + 4). \end{aligned}$$

Because 3 already divides $\bigl((6 \times 33) + (7 \times 3)\bigr) \times 3$, Lemma 32.1 implies as before that $3|674$ *is equivalent to* $3|(6+7+4)$.

The proof of the rule in general is entirely similar.

Activity. Do you have a quick way to see why 123,456,789 is divisible by 3, and why 67814235 is divisible by 3?

Next, we have the **divisibility rule for 4**:

A whole number n is divisible by 4 $\iff$ the two-digit number formed by n's last two digits (i.e., its tens digit and ones digit) is divisible by 4.

Before explaining why, let us put it to use by testing whether a number such as 572 or 93,386 is divisible by 4. This rule says we look only at the last two digits 72 of 572 and the last two digits 86 of 93,386, respectively, and use these to decide if 4 divides 572 and 93,386. Because $4|72$ and $4 \nmid 86$, we conclude that $4|572$ and $4 \nmid 93,386$. The verification of the correctness of these conclusions is immediate: $572 = 143 \times 4$ and $93,386 = (23,346 \times 4) + 2$.

The reason for this rule is that every multiple of 100 is divisible by 4. For example,
$$572 = 500 + 72.$$
So 4 divides 500. It follows from Lemma 32.1 that $4|572$ is equivalent to $4|72$. Similarly,
$$93,386 = 93,300 + 86.$$
Because 4 divides 93,300 $(= 933 \times 100)$, $4|86$ is equivalent to $4|93,386$, by virtue of Lemma 32.1. If 572 and 93,386 are replaced by a general number n, the rule is proved in exactly the same way.

The **divisibility rule for 5** is much simpler:

A whole number is divisible by 5 $\iff$ *its last digit (i.e., the ones digit) is either* 0 *or* 5.

The proof is left as an exercise.

The **divisibility rule for 9** is the same as that for 3:

A whole number is divisible by 9 $\iff$ *the sum of its digits is divisible by* 9.

The reason is the fact that when a power of 10 is divided by 9, the remainder is always 1. For example, $100{,}000 = (11{,}111 \times 9) + 1$, and in general

$$10^n = (\underbrace{11\cdots 1}_{n} \times 9) + 1.$$

The details of the proof of the validity of this rule are similar to the case of divisibility by 3 and are left as an exercise.

There is a divisibility rule for 7, but it is too complicated to be worth knowing. There is also a divisibility rule for 11. For its statement, define the **alternating sum of the digits of a whole number** to be the sum of its "odd" digits (i.e., the ones digit, the hundreds digit, the ten-thousands digit, etc.) minus its "even" digits (the tens digit, the thousands digit, the hundred-thousands digit). For example, the alternating sum of the digits of 517634 is $(1+6+4) - (5+7+3) = -4$.

Then the **divisibility rule for 11** states:

A whole number is divisible by 11 $\iff$ *the alternating sum of its digits is divisible by* 11.

Let us apply it to 517. The alternating sum of the digits of 517 is $(5+7)-1 = 11$. The rule then guarantees that 517 is divisible by 11, which is certainly true since $517 = 47 \times 11$. If the number is 517634, then the alternating sum of its digits is -4, as we have seen. Since $11 \nmid (-4)$, we conclude that $11 \nmid 517634$. Sure enough, $517634 = (47057 \times 11) + 7$.

The reason behind this rule, as usual, is due to the behavior of 10^n (n is a whole number) when it is divided by 11: if n is odd, then

$10^1 = (1 \times 11) + (-1), \qquad 10^3 = (91 \times 11) + (-1),$
$10^5 = (9091 \times 11) + (-1), \quad 10^7 = (909091 \times 11) + (-1),$

and in general if we write every odd number $n \geq 3$ as $n = 2k+1$ for a whole number $k \geq 1$ (see Exercise 1(b) below),

$$10^n = (\underbrace{9090\cdots 90}_{2(k-1)} 91 \times 11) + (-1).$$

32.3. Divisibility Rules

There is an analogous statement for 10^n when n is even:
$$10^2 = (9 \times 11) + 1, \qquad 10^4 = (909 \times 11) + 1,$$
$$10^6 = (90909 \times 11) + 1, \quad 10^8 = (9090909 \times 11) + 1,$$
and in general if $n = 2k$ for a whole number $k \geq 2$,
$$10^n = (\underbrace{9090 \cdots 90}_{2(k-1)} 9 \times 11) + 1.$$

These two facts together with the expanded form of a whole number then lead to the proof of the divisibility rule. Consider 517, for example:
$$517 = (5 \times 10^2) + (1 \times 10) + 7$$
$$= 5 \times \big((9 \times 11) + 1\big) + \Big(1 \times \big((1 \times 11) + (-1)\big)\Big) + 7$$
$$= \Big(\big((5 \times 9) + (1 \times 1)\big) \times 11\Big) + 5 - 1 + 7$$
$$= (46 \times 11) + \big((5 + 7) - 1\big).$$

Since 11 divides 46×11, Lemma 32.1 shows that $11|517 \iff 11|((5+7)-1)$.

The details of the proof in the general case are left to an exercise.

Exercises

1. (a) Prove that the product of two odd numbers is odd, the product of two even numbers is even, and the product of an odd number and an even number is even. (Caution: we are asking for a general proof, not the working out of concrete examples.) (b) *Prove* that every odd number is of the form $2n + 1$, where n is a whole number. (Caution: we are asking for a logical deduction from the definition of an odd number.) (c) Prove that no even number can divide an odd number. (d) Can an odd number divide an even number?

2. Prove the divisibility rule for 2 in the case of an odd number.

3. Prove the divisibility rule for 3 in the case of a 5-digit number, say, 59143.

4. Prove the divisibility rule for 5.

5. Prove the divisibility rule for 9.

6. Prove the divisibility rule for 11. To avoid notational complexity, write out a proof only for numbers with 7 digits.

7. Can you formulate a divisibility rule for 6 and prove it?

8. Can you formulate a divisibility rule for 8 and prove it? (Think about the divisibility rule for 4.)

9. Without using a calculator, check the following and explain your method: Is 164253 a multiple of 6? Is 57201 a multiple of 27? Is 111146 a multiple of 8? Which numbers between 1 and 10 divide 255780?

10. Does 12 divide 5,106,963,042? Does 15 divide 72,402,935?

11. The symbol 4! is a short way of writing $4 \cdot 3 \cdot 2 \cdot 1$, and for any positive integer n, $n!$ is a short way of writing $n(n-1)(n-2) \cdots 3 \cdot 2 \cdot 1$. So 4! hours is the same as 1 day, and 5! minutes is the same as 2 hours. How many weeks is the same as 10! seconds?[1]

12. Prove that if $k|\ell$ and $m|n$ for integers k, ℓ, m, n, then also $km|\ell n$.

13. Explain *clearly* to a sixth grader why, if m, n, and a are nonzero whole numbers and $am = an$, then $m = n$. Can you do it two different ways?

14. Explain why the square of a whole number must have a ones digit that is either 1, 4, 5, 6, or 9.

---

[1] Problem from Tony Gardiner.

Chapter 33

Primes and Divisors

One of the most charming features of number theory is that some of its deepest problems, problems that have resisted solution for centuries, can be understood by fifth graders. This chapter introduces the terminology of primes and divisors, and by the end of the chapter, we will be discussing unsolved problems about primes.

The sections are as follows:

>Definitions of Primes and Divisors
>
>The Sieve of Eratosthenes
>
>Some Theorems and Conjectures about Primes

33.1. Definitions of Primes and Divisors

Every whole number a has at least two divisors, namely 1 and a itself, but these divisors are nothing to write home about. If $b|a$ but $b \neq 1$ or a, then we call such a number b a **proper divisor** of a and such an a a **proper multiple** of b. Thus 12 is a proper multiple of 3, and 3 is a proper divisor of 12. On the other hand, 7 has no proper divisor, nor does 101, although it takes more effort to check this fact about 101. Put differently, a whole number b is a proper divisor of a whole number a if $b|a$ and $2 \leq b \leq (a-1)$; a is a proper multiple of b if a is a multiple of b and $a > b$. There is a simple but useful **Observation**:

> If b is a proper divisor of a whole number n and $n = bc$ for some whole number c, then c is also a proper divisor of n.

This is because if $c = 1$, then $b = n$, which is contrary to the properness of b, and similarly if $c = n$, then $b = 1$, which is impossible for the same reason. Thus $2 \leq c \leq n - 1$ and c is a proper divisor of n, as claimed. If $n = bc$

445

where b is a proper divisor (and hence so is c), we call bc a **factorization** of n.

We have already noted that some whole numbers have no proper divisors, e.g., 2, 3, 5, 7, 11, ... A whole number that is greater than 1 and has no proper divisor is called a **prime**, or a **prime number**. Note that 2 is the only even prime. A whole number that is greater than 1 and is not a prime is called a **composite**. *By definition, 1 is neither a composite nor a prime.*

There are many reasons for our enduring interest in primes—some of the simpler ones will surface presently—but a fundamental one is that they are the basic building blocks of the positive integers, in a sense that we shall make explicit in the next chapter.

Checking whether or not a whole number n is a prime is a formidable task: you have to check all the numbers from 2 to $n-1$ to see whether they divide n. For large n, it will always remain formidable. However, there are some simplifications that make the checking of primality less painful, at least when $n < 1000$. For example, checking the primality of 97 would seem to require the inspection of 95 divisions: 2|97? 3|97? $\cdots$ 95|97? and 96|97? But it is immediately clear that about half of these are unnecessary: *no whole number exceeding $\frac{97}{2}$ can divide 97.* To see this, suppose b is a whole number exceeding $\frac{97}{2}$ and $97 = bc$ is a factorization of 97; we are going to show that this supposition leads to something absurd. For then, c being a proper divisor (see the Observation above), $c \geq 2$. Therefore $97 = bc \geq b \times 2 > \frac{97}{2} \times 2 = 97$, and we have $97 > 97$. It is absurd, and therefore a whole number exceeding $\frac{97}{2}$ cannot be a divisor of 97.

In general, consider a whole number n. We claim that *no whole number k exceeding $\frac{n}{2}$ can be a divisor of n.* Suppose there is a factorization $n = kc$ for a k greater than $\frac{n}{2}$. Since c is a proper divisor, $c \geq 2$. Therefore we have $n = kc \geq k \times 2 > \frac{n}{2} \times 2 = n$ and we get $n > n$, which is absurd. So the claim is correct.

We conclude, therefore, that if there are no divisors of a whole number n among numbers at least 2 but at most $\frac{n}{2}$, then there would be no divisors of n at all, and therefore such an n must be a prime. Much more is true, however, but for the more general statement we have to formalize the concept of the square root of a positive number. If x is a positive number, then a positive number s so that $s^2 = x$ is called a **positive square root of x**. It is known that every positive number x has a unique positive square root, which is denoted by $\sqrt{x}$, so that we may legitimately call $\sqrt{x}$ *the* positive square root of x. We will not linger over these rather subtle concepts for now because they are merely incidental to our purpose.

Theorem 33.1. *If a whole number n has no divisor k which satisfies $2 \leq k \leq \sqrt{n}$, then n is a prime.*

33.1. Definitions of Primes and Divisors

Let us use Theorem 33.1 to check, for example, whether 193 is a prime. Noting that $\sqrt{193} \approx 13.9$,[1] we only need to check whether any of the whole numbers 2, 3, 4, ..., 11, 12, 13 is a divisor of 193.[2] There is a further reduction. It suffices to only check whether any of 3, 5, 7, 9, 11, 13 is a divisor of 193 because, 193 being odd, there is no need to bother with the even numbers; see Exercise 1(c) on page 444. Since $1 + 3 + 9 = 13$ and $3 \nmid 13$, $3 \nmid 193$; therefore 9 is also not a divisor of 193 (Lemma 32.2 (b) of Chapter 32, although one could derive the same conclusion from the fact that $9 \nmid 13$). Clearly 5 is not a divisor of 193 (see the divisibility rule for 5 in section 32.3), so it remains to check whether 7, 11, or 13 divide 193. This is painless, and the answer is no. So 193 has no divisor which is at least 2 and at most $\sqrt{193}$. By Theorem 33.1, 193 is a prime.

It is instructive, however, to directly prove that, if there is no proper divisor of $193 \leq \sqrt{193} \approx 13.9$, i.e., < 14, then 193 must be a prime. Indeed, suppose 193 is not a prime. Then we would have a factorization $193 = bc$. Since each of b and c is a divisor of 193, it must be at least 14. Hence their product bc is at least $14 \times 14 = 196$. But $bc = 193$, so we have the absurd conclusion that 193 is at least 196. Clearly we have made a mistake in assuming that 193 is not a prime, and so 193 must be a prime after all.

The preceding argument seems to depend on knowing explicitly that each of b and $c \geq 14$. A little reflection on the process reveals, however, that such explicit information is not necessary. The only thing we need is that, knowing each of b and c is a divisor, it cannot be $\leq \sqrt{193}$. Thus both b and $c > \sqrt{193}$, and their product exceeds $\sqrt{193} \times \sqrt{193}$ which, by the definition of a square root, is just 193. Therefore $bc > 193$. But $193 = bc$, so we arrive at the conclusion that $193 > 193$, which is impossible. This is then enough to show that we should not have assumed 193 not to be a prime in the first place. So again, 193 is a prime. Now we realize that our reasoning is in fact perfectly general and we have essentially achieved a proof of the theorem.

Next, we formalize an inequality that has already been used above. Let s, t, u, v be *positive* numbers. We claim that

(33.1) \qquad if $s < t$ and $u < v$, then $su < tv$.

For the proof, note that in Chapter 17 we proved in general that if A, B, C are fractions, then

(33.2) \qquad $A > 0$ and $B < C$ imply $AB < AC$.

By FASM (Chapter 21), (33.2) may be taken to be valid for all numbers A, B, C regardless of whether they are fractions or not. On this basis, we

---

[1] Feel free to use your four-function calculator for this kind of estimate.
[2] By comparison, the preceding method would require us to check all the whole numbers from 2 to 96 ($97 > (193/2)$).

prove (33.1) as follows: Using (33.2) twice, we have

$$u > 0 \quad \text{and} \quad s < t \quad \text{implies} \quad su < tu,$$
$$t > 0 \quad \text{and} \quad u < v \quad \text{implies} \quad tu < tv.$$

By the transitivity of $<$ (see Chapter 16), the two together imply that $su < tv$, which is (33.1), as desired.

We are now ready to prove Theorem 33.1 in general. It amounts to no more than replacing 193 in the previous argument by n.

Proof. Suppose no whole number k satisfying $2 \leq k \leq \sqrt{n}$ divides n, and we have to prove that n is a prime. To achieve a proof by contradiction, we will show that it is wrong to assume that n is not a prime. So let $n = km$ be a factorization of n. This implies that both k and m are proper divisors of n and therefore, by hypothesis, k and m must satisfy $\sqrt{n} < k$ and $\sqrt{n} < m$. By inequality (33.1) then,

$$n = \sqrt{n} \cdot \sqrt{n} < km = n,$$

so that $n < n$. A contradiction. The theorem is proved. □

Activity. Is 337 a prime? Is 373 a prime?

33.2. The Sieve of Eratosthenes

As an application of the preceding theorem, we give a systematic and simple method of getting *all* the primes in a fixed range. For illustration, we are going to write down all the primes ≤ 144. (We choose 144 because $\sqrt{144} = 12$.) For this and other purposes, we want to solidify our reasoning by proving the following useful lemma.

Lemma 33.2. *For positive numbers s and t, $s < t \iff \sqrt{s} < \sqrt{t}$.*

Proof. The proof of the lemma is very similar to that of the second statement in (2.7); see page 50. We first prove that

$$\sqrt{s} < \sqrt{t} \quad \text{implies} \quad s < t.$$

Indeed, because $\sqrt{s} < \sqrt{t}$, we use (33.1) to conclude that $\sqrt{s} \cdot \sqrt{s} < \sqrt{t} \cdot \sqrt{t}$. Since $\sqrt{s} \cdot \sqrt{s} = s$ and $\sqrt{t} \cdot \sqrt{t} = t$, we have $s < t$.

The converse is more interesting. We have to prove that

$$s < t \quad \text{implies} \quad \sqrt{s} < \sqrt{t}.$$

Now there are only three possibilities for the numbers $\sqrt{s}$ and $\sqrt{t}$: $\sqrt{s} < \sqrt{t}$, $\sqrt{s} = \sqrt{t}$, and $\sqrt{s} > \sqrt{t}$ (the trichotomy law for real numbers; see section 15.3 on page 239 and Chapter 21). Of course we want only the first possibility to hold and, for that purpose, it suffices to prove that the other two possibilities are not compatible with the given hypothesis that $s < t$. The second possibility is clearly hopeless, because $\sqrt{s} = \sqrt{t}$ immediately

33.2. The Sieve of Eratosthenes

implies $\sqrt{s} \cdot \sqrt{s} = \sqrt{t} \cdot \sqrt{t}$, so that we have $s = t$. This directly contradicts the hypothesis of $s < t$. Now look at the third possibility, which we rewrite as $\sqrt{t} < \sqrt{s}$. By what we have just proved, this leads to $t < s$, and this too contradicts the hypothesis of $s < t$. Therefore $\sqrt{s} < \sqrt{t}$ is the only option left, which is what we want. The lemma is proved. □

We can now proceed with the determination of all the primes up to 144. We claim:

> *For a whole number $k \leq 144$, it is a prime if it is not divisible by any of the eleven numbers* 2, 3, ..., 11, 12.

(Recall that 12 is the square root of 144, and that is the point.) Let us first make sure that this claim is correct. Suppose we have a whole number k so that $2 \leq k \leq 144$. By Theorem 33.1, to check whether k is a prime, it suffices to check whether any of the whole numbers ℓ satisfying

$$2 \geq \ell \leq \sqrt{k}$$

is a divisor of k. But how big is $\sqrt{k}$? By the lemma, the fact $k \leq 144$ implies that $\sqrt{k} \leq \sqrt{144}$. In other words, $\sqrt{k} \leq 12$. It follows that to check the primality of any whole number k which is at most 144, it suffices to check whether any whole number ℓ satisfying $2 \leq \ell \leq 12$ is a divisor of k. This proves the above claim.

What this claim provides is a general method to get around the boring, one by one, divisibility tests for each of the numbers ≤ 144, in the following way. A number that is divisible by one of the numbers 2, 3, ..., 11, 12 is also a multiple of one of them. Hence, if we simply

> *remove all the proper multiples of each of* 2, 3, 4, ..., 12 *from the list of whole numbers ≤ 144, what remains cannot be divisible by any of* 2, 3, ..., 11, 12 *and therefore must be primes.*

This method of producing primes is called the **Sieve of Eratosthenes**. Eratosthenes (ca. 276–194 BC) lived slightly after Euclid, but was a contemporary of Archimedes. He was a mathematician and an astronomer, and one of his notable achievements is that he gave an astonishingly accurate measurement—some 23 centuries ago—of the circumference of the earth, with an error of about 11%. Such accuracy is remarkable considering the primitive state of scientific measurements at the time.

Let us do the Sieve of Eratosthenes for the special case of 1, 2, 3, ..., 143, 144 in greater detail. We first remove all the proper multiples of 2, i.e., all the even numbers > 2. In the process, we have also removed all multiples of 4, 6, 8, 10, and 12. Therefore our task has been simplified considerably because we need only remove in our next step all the proper multiples of 3, 5, 7, 9, 11 among all the odd numbers from 3 to 143. This is a good exercise in skip-counting. For example, after removing all the proper multiples of 3, we get

$$\begin{array}{cccccccccccc}
2 & 3 & 5 & 7 & 11 & 13 & 17 & 19 & 23 & 25 & 29 & 31 & 35 \\
37 & 41 & 43 & 47 & 49 & 53 & 55 & 59 & 61 & 65 & 67 & 71 & 73 \\
77 & 79 & 83 & 85 & 89 & 91 & 95 & 97 & 101 & 103 & 107 & 109 & 113 \\
115 & 119 & 121 & 125 & 127 & 131 & 135 & 139 & 141 & 143 & & &
\end{array}$$

Since a multiple of 9 is automatically a multiple of 3, we only have to remove all the proper multiples of 5, 7, and 11 from this list to get all the primes. The proper multiples of 5 are easily recognized:

$$\begin{array}{cccccccccccc}
2 & 3 & 5 & 7 & 11 & 13 & 17 & 19 & 23 & \boxed{25} & 29 & 31 & \boxed{35} \\
37 & 41 & 43 & 47 & 49 & 53 & \boxed{55} & 59 & 61 & \boxed{65} & 67 & 71 & 73 \\
77 & 79 & 83 & \boxed{85} & 89 & 91 & \boxed{95} & 97 & 101 & 103 & 107 & 109 & 113 \\
\boxed{115} & 119 & 121 & \boxed{125} & 127 & 131 & \boxed{135} & 139 & 141 & 143 & & &
\end{array}$$

Then do the same to the multiples of 7 and 11. At the end, we get the complete list of the 34 primes up to 144:

$$\begin{array}{cccccccccccc}
2 & 3 & 5 & 7 & 11 & 13 & 17 & 19 & 23 & 29 & 31 & 37 \\
41 & 43 & 47 & 53 & 59 & 61 & 67 & 71 & 73 & 79 & 83 & 89 \\
97 & 101 & 103 & 107 & 109 & 113 & 127 & 131 & 137 & 139 & &
\end{array}$$

33.3. Some Theorems and Conjectures about Primes

We conclude this chapter with some general information about primes. While the statements of the theorems to follow are easy to understand, the proofs, as a rule, involve very advanced mathematics.

The attempt to understand prime numbers began in earnest about four hundred years ago, and has been a major preoccupation of mathematicians for the past two centuries. There is no sign that this interest will diminish any time soon. We will prove in section 36.4 that the totality of primes is infinite, a fact already known to Euclid 23 centuries ago. A theorem of P. G. L. Dirichlet (1805–1859) says that much more is true, and we describe it now. Two whole numbers a and b are said to be **relatively prime** if their only **common divisor**—a whole number that divides both a and b—is 1. Please note that saying two numbers are relatively prime says nothing about the possible primality of either individual number. For example, 12 and 175 are relatively prime, but neither 12 nor 175 comes close to being a prime. Of course, if p is a prime, and if n is not a multiple of p, then p and n are relatively prime. With this understood, the **arithmetic progression determined by any two whole numbers a and b** is the collection of numbers $a + b$, $2a + b$, $3a + b$, $4a + b$, ..., $an + b$, ... Dirichlet's theorem says that if a and b are relatively prime, then the arithmetic progression determined by a and b already contains an infinite number of primes. If $a = b = 1$, then a and b are certainly relatively prime. An inspection would reveal that the arithmetic progression determined by the numbers

$a = 1$ and $b = 1$ is in fact all the whole numbers greater than 1. In this sense, Dirichlet's theorem includes the simple statement of Euclid about the infinity of primes. If a and b are distinct primes, then it follows from the definition that they are relatively prime. Take $a = 3$ and $b = 7$, write out 30 terms of the arithmetic progression $3 + 7$, $6 + 7$, $9 + 7$, $12 + 7$, ... to convince yourself that indeed there are many primes there. You should try using other relatively prime whole numbers, e.g., 6 and 25.

Next, you can see from the list of primes up to 144 that there are many **twin primes** there, i.e., primes that are consecutive odd numbers. For example: 3 and 5, 5 and 7, 29 and 31, 71 and 73, 137 and 139. Inspection of all known tables of primes reveals that there seem to be arbitrarily large twin primes, e.g., 1091 and 1093, 2591 and 2593, 5741 and 5743, 9281 and 9283, etc. As of 2010, the largest known pair of twin primes has 58711 digits; see, for example, http://mathworld.wolfram.com/TwinPrimes.html.

Whether or not there is in fact an infinite number of twin primes has been an open problem for at least two centuries. Another seemingly innocuous question is whether every even number is the sum of two prime numbers. For example, $10 = 3 + 7$, $90 = 11 + 79$, $184 = 137 + 47$, $250 = 113 + 137$, etc. So far, no one has uncovered an even number which is not a sum of two primes. This question was first raised by the amateur mathematician C. Goldbach in 1742, and is therefore known as the **Goldbach Conjecture**. Although there are now theorems that are tantalizingly close to the conjecture,[3] the conjecture itself has remained unsolved to this day.

In connection with twin primes, you may ask whether there are **triplet primes**, meaning three primes which are consecutive odd numbers. Certainly 3, 5, 7 are triplet primes, but we shall now prove that these are the only triplet primes. Let us consider three consecutive odd numbers k, $k+2$, and $k+4$ where k is an odd number. We want to show that if $k > 3$, then one of them must be composite. What tool can we use to prove this? Yes, the division-with-remainder. Divide k by 3, and we get $k = 3n + r$, where $r = 0, 1$, or 2. First suppose $r = 0$, then $k = 3n$. Because $k > 3$, we have $n > 1$, so that k itself is already composite. Next suppose $r = 1$. Then $k + 2 = (3n + 1) + 2 = 3(n + 1)$ and $k + 2$ is composite. Finally, suppose $r = 2$, then $k + 4 = (3n + 2) + 4 = 3(n + 2)$ and this time it is $k + 4$ that is composite. In any case, there are no triplet primes other than 3, 5, 7. Now think about the twin prime problem again, and it becomes more intriguing.

It has long been conjectured in connection with questions about twin primes that there are arithmetic progressions of arbitrary (finite) lengths

---

[3]Two theorems proved by extremely difficult methods in the 20th century would serve to give you an idea: a theorem of I. M. Vinogradov says that any odd number bigger than a certain (very big) number is the sum of at most three primes, and a theorem of J. R. Chen says that every even number beyond 4 is the sum of a prime and another number which is either a prime or a product of two primes.

consisting entirely of primes. In other words, given any whole number k, the conjecture is that there are k equi-spaced primes (as points on the number line). For example, there are five equi-spaced primes "at our doorstep",

$$5, \quad 11, \quad 17, \quad 23, \quad 29,$$

with distance 6 between successive ones. The subtlety of such a progression is that these *are not consecutive primes* because, for example, the prime number 7 is between 5 and 11, the prime number 13 is between 11 and 17, etc. Another relatively simple example of this genre is the following six equi-spaced primes,

$$7, \quad 37, \quad 67, \quad 97, \quad 127, \quad 157,$$

with a distance of 30 between successive ones. Thus the conjecture is true for $k = 5$ and $k = 6$. For larger values of k, the conjecture is far from obvious. There are actually ten equi-spaced primes which are relatively small:

$$199, \quad 409, \quad 619, \quad 829, \quad 1039, \quad 1249, \quad 1459, \quad 1669, \quad 1879, \quad 2089,$$

with a distance of 210 between successive ones. So the conjecture is true for $k = 10$ as well. However, before 2004, the conjecture was not known even for $k = 23$. In 2004, Ben Green and Terence Tao surprised the whole world by proving this conjecture for all k. Their proof does not exhibit specific such progressions for each whole number k, and only guarantees their existence. (For example, one can be sure that the statement "there is a book in every library" is true even if no book is explicitly cited for each and every library.) The largest k for which such a progression has been found is $k = 25$, which was discovered in May of 2008. You should check http://hjem.get2net.dk/jka/math/aprecords.htm periodically for the ongoing saga.

At the other extreme, it is easy to show that there are arbitrarily long strings of consecutive whole numbers that are composites. First, recall from section 14.6 the notation $n!$ (read: n *factorial*) to be the product of all whole numbers from 1 to n. Thus $4! = 1 \times 2 \times 3 \times 4 = 24$, $6! = 1 \times 2 \times 3 \times 4 \times 5 \times 6 = 720$, etc. Then we claim that the string

$$n! + 2, \quad n! + 3, \quad \ldots, \quad n! + n$$

consists of composites. Indeed, 2 divides $n! + 2$, 3 divides $n! + 3$, ..., and n divides $n! + n$. Since there are $n - 1$ numbers in this string, this shows that as n gets arbitrarily large, we get arbitrarily long strings of composites.

The factorials get large very quickly. The number 10! may *seem* small, but it is 3628800. The number 50! is in fact bigger than any number that will ever come up in daily life, including the national budget or the national deficit.[4] It is roughly 3×10^{64}. To get an idea of how big this number is, the age of the universe is about 2×10^{10} years, which is about 5×10^{16}

---

[4]This is bigger than the estimate by Archimedes of the number of grains of sand needed to fill up the universe as he knew it, which is about 10^{63}.

33.3. Some Theorems and Conjectures about Primes

seconds. And 100! is positively unimaginable: about 9×10^{157}. So if we want a string of 100 consecutive composites, this method requires that we use numbers with about 157 digits. If we only want something modest, such as 6 consecutive composites, then 7! = 5040 and the string in question becomes

$$5042, 5043, 5044, 5045, 5046, 5047.$$

To give this string some perspective, we note that in fact 5039 and 5051 are both primes, but 5041, 5048 and 5049 are not (you may prove this as an exercise). So we get in fact a string of not 6 but 9 consecutive composites from 5041 to 5049 as a bonus. A more startling fact is that it is not at all necessary to go beyond 5000 in order to get 9 consecutive composites. Indeed, there is a string of 13 consecutive composites staring us in the face all this time: 114, 115, ..., 125, 126 are all composites! (Check our list of primes from 1 to 144 above to verify this assertion.) So the use of $n!$ to get $n-1$ consecutive composites is a correct general statement, but it is by no means the most efficient method in specific situations.

With the availability of the calculator, you can play around with reasonably large numbers to check effortlessly many of the assertions above and may even come up with better ones. Please do.

Exercises

1. Prove that if a whole number n is not divisible by any *prime* p satisfying $2 \le p \le \sqrt{n}$, then n is a prime.
2. Check the primality of the following four numbers: 247, 293, 461, 873.
3. (a) Find a whole number n so that both $6n+1$ and $6n-1$ are composites.
   (b) Prove that there are infinitely many whole numbers n so that both $6n+1$ and $6n-1$ are composites. (For part (b), you need to know some standard algebraic identities.)
4. Show that for a whole number n which is at most 20736, it is a prime if it has no divisor among the first 34 primes. (Look up the list of primes in the text.)
5. Among the first 40 terms of the arithmetic progression $3n+2$, where n is a whole number, which of them are primes? What about $4n+1$? Do you notice anything special about the primes of the latter? (*Hint*: Can you express each prime of the second case as a sum of two squares?)
6. Express each of the following as a sum of two primes: 88, 96, 162, 246, 254, 278.
7. Using a calculator to help with the calculations if necessary, show that 5041 and 5049 are composites. (The first number will try your patience a bit.)
8. Write down two distinct consecutive strings of five composites in addition to those given above.
9. (a) *Prove* that if p and q are primes and $p|q$, then $p = q$. (b) Prove that if m, n are integers and $m|n$, $n|m$, then $m = \pm n$.
10. Use the method of the Sieve of Eratosthenes to determine all the primes among whole numbers up to 289.

Chapter 34

The Fundamental Theorem of Arithmetic (FTA)

You are used to factoring small numbers into a product of primes, e.g., $12 = 2 \times 2 \times 3$, $18 = 2 \times 3 \times 3$, etc. You have also seen how easy it is to factor, for example, 193 into a product of primes, namely, 193 itself. (We follow the CONVENTION of also referring to the writing of a single prime by itself as "a product of primes".) This kind of prior experience may have conditioned you to take for granted that every number can be factored into a product of primes. Things change drastically, however, when small numbers are replaced by big numbers. Take for example the number 2,147,483,647. How do we factor this into a product of primes?

Let us first agree on a piece of terminology: by the **prime decomposition** of a whole number we mean its expression as a product of primes.

To deal with 2,147,483,647, we have to pursue—in this technological age of ours—two distinct routes of inquiry depending on how much technology we allow ourselves to use. Let us begin by limiting ourselves to the use of only hand-held calculators but no computers. To get the prime decomposition of 2,147,483,647, we begin by checking whether it has any divisors among numbers from 2 to about

$$\sqrt{2{,}147{,}483{,}647} \approx 46340.95.$$

As we have seen, by eliminating all the even numbers, we only look at roughly half of these 46340 numbers as potential divisors. With further simplifications down the road, e.g., eliminating multiples of 3 also eliminates a third of these 20000 divisions, let us make a *very* optimistic estimate and say that we only need to check 5000 divisions[1] to see if 2,147,483,647 has a divisor. Do you have the patience? And if not, how do you know that we will be able to express 2,147,483,647 as a product of primes?[2]

We can be far more modest and ask simply if 14933 is a product of primes. Even this much easier problem may tax your patience. Try it. The answer is that $14933 = 109 \times 137$, and both 109 and 137 are known to be primes. By now, you may have already gotten a better appreciation of the fact that it is *not* obvious that every whole number, no matter how large, is a product of primes.

Now it is time for us to pursue the second route of inquiry. You may say that the discussion about using calculators is silly because with the advent of powerful computers, such checking could be done easily and the preceding discussion has no relevance. Yes and no. For the computers of 2010, getting the prime decomposition of a ten-digit number such as 2,147,483,647 is almost instantaneous. This is the "yes" part. But computers have limitations too. We have been looking only at "small" numbers! Thus, even a twenty-digit number would pose no threat to a computer. However, if we try to find the prime decomposition of a 400-digit number, then in the foreseeable future, even all the computers in the world working together for a hundred years will not be able to get it done.

What is interesting is that this defect (if we can call it that) in technology has been turned into a big asset. Computer scientists succeeded in using the inherent limitation of computers to create the so-called **public key cryptosystems** that safeguard all our financial and confidential transactions on the Internet. Roughly what this means is that a secret message can be encoded in numbers, and although the procedure of the conversion from sentences to numbers is an open secret, the number-coded messages cannot be decoded, say within a hundred years, precisely because the decoding would require the factorization of a very large number into primes and these primes were used in the encoding. The security of the secret message therefore depends critically on the inability of the computers to factor very large numbers (at the moment, 400 digits is more than large enough). So even when access to the most powerful computers is allowed, the basic problem is still the same: how do you *know* that every number, such as one with 500 digits, has a prime decomposition?

You may get the idea by now that the brute force approach to an understanding of prime numbers is not advisable. We need mathematical reasoning, i.e., mathematical proofs, to help us decide if something is true.

---

[1]There is a good mathematical reason to settle for this number as the likely number of divisions that needs to be performed; it has to do with the so-called Prime Number Theorem.

[2]The number 2,147,483,647 is actually a prime.

34. The Fundamental Theorem of Arithmetic (FTA)

It is important to recall at this point that together with the prime decomposition of a number, we frequently make implicit use of the **uniqueness** of the decomposition. In other words, we not only use the fact that, say, $12 = 2 \times 2 \times 3$, but we also tend to take for granted that the three primes 2, 2, and 3 are the only three primes that ever come up in a prime decomposition. In saying this, we mean of course that these primes 2, 2, 3 are **unique up to rearrangement** in the sense that we could write the decomposition as $12 = 2 \times 2 \times 3$ or $12 = 2 \times 3 \times 2$ or $12 = 3 \times 2 \times 2$, but we regard these as the same decomposition. Let it be understood therefore that in the following,

the uniqueness of prime decomposition means uniqueness up to rearrangement.

Now the uniqueness of the prime decomposition of 12 is easy to check directly. But for a 500-digit number whose prime decomposition we cannot even lay our hands on explicitly, how can we be sure that there is not some other way of getting a decomposition with a different collection of primes?

The purpose of this discussion is to prepare you for the essential part of a proof of the next theorem, which affirms that, indeed, *every* number must have a prime decomposition, and that there is only one such. Obviously we will not be able to produce the explicit decomposition in every case, as was duly noted above, but we shall argue by logic that there must be one and only one. It turns out that the proof is important, because it introduces a useful and beautiful fact known as the Euclidean Algorithm.

Theorem 34.1 (The Fundamental Theorem of Arithmetic (FTA)).
Every whole number ≥ 2 has a prime decomposition, and the decomposition is unique.

One can appreciate this theorem, at the most superficial level to be sure, by observing that 3×8 and 4×6 are both decompositions of 24 into products of smaller numbers, and yet these two decompositions have no factors in common. Therefore if the factors of a *prime* decomposition of every number must stay the same, then this uniqueness must be due to something special about primes. The proof of the theorem then must invoke a fundamental property of primes that is not shared by a general number. The property in question turns out to be quite sophisticated (see Theorem 34.2 later in this chapter) and its proof is nothing short of astounding (see Chapter 35).

Let us first bring closure to an earlier remark, to the effect that the primes are the building blocks of the positive integers. What the Fundamental Theorem of Arithmetic says is that, *as far as multiplication is concerned*, one can obtain all whole numbers ≥ 2 solely by the operation of multiplication if one is given all the primes. The uniqueness part of the theorem in fact says more: if we ignore the order in which the primes are listed, then different collections of primes must yield distinct whole numbers, so that by systematically multiplying out the distinct collection of primes, we get every whole number ≥ 2 exactly once.

You may appreciate this last statement a little more by comparing it with the corresponding situation using addition instead of multiplication: one can obtain all the positive integers solely by the operation of addition if one is given the number 1. Thus the presence of the primes reveals that, *for whole numbers, the concept of multiplication is much more complicated than that of addition*. If you think back on the initial statement of this chapter, to the effect that we are going to study the interplay between the addition and multiplication of whole numbers, you see that there may be something to this interplay after all.

We begin with the **proof** of the easy part of the theorem, the *existence of the prime decomposition*. Let us first make an **Observation**:

> If n is composite, then it has a **prime divisor** (i.e., a divisor that is also a prime).

Indeed, n being a composite, it has at least one proper divisor d. Let d_0 be the smallest among all the proper divisors of n. We claim that d_0 must be a prime. If not, then d_0 has a proper divisor a, where $1 < a < d_0$. But because $a|d_0$ and $d_0|n$, we see that $a|n$ (Lemma 32.2). This would mean d_0 is not the smallest proper divisor of n, a contradiction. So we had better accept the fact that d_0 is a prime. But then d_0 is a prime divisor of n, proving the Observation.

We shall look at some concrete numbers before presenting the general proof of the existence of a prime decomposition. Let us begin with 193. We have seen that 193 is itself a prime decomposition of the number 193. Next we try 391. It turns out that 17 is a prime divisor, and $\frac{391}{17} = 23$. Thus $391 = 17 \times 23$. We recognize that 23 is a prime, so this is already a prime decomposition. Next we try, say, 1729. We get 7 as a prime divisor, and $\frac{1729}{7} = 247$. So $1729 = 7 \times 247$. If 247 is a prime, we are finished again, but 13 is a prime divisor of 247. Because $247 = 13 \times 19$, we have $1729 = (7 \times 13) \times 19$. Fortunately, 19 is a prime, so $1729 = 7 \times 13 \times 19$ is a prime decomposition of 1729. Let us try one more example: 16269. Because 3 divides $1 + 6 + 2 + 6 + 9 = 24$, 3 is a prime divisor of 16269. We have $16269 = 3 \times 5423$. It so happens that 5423 is not a prime because $11|(3 + 4 - (5 + 2))$, so that 11 divides 5423. (See the divisibility rule for 11 on page 442.) So $5423 = 11 \times 493$, and therefore $16269 = (3 \times 11) \times 493$. If 493 is a prime, we would be finished again, but $17|493$ and $493 = 17 \times 29$. Thus $16269 = 3 \times 11 \times 17 \times 29$, and this is a prime decomposition because every factor is now a prime.

In general, let n be a given whole number ≥ 2. If n is a prime, then n is its own prime decomposition. If not, by the Observation, it has a prime divisor p. Let $n = pA$ for some whole number A. If A is a prime, we are

34. The Fundamental Theorem of Arithmetic (FTA)

finished because then pA is already a prime decomposition of n. If not, then A has a prime divisor q, again by the Observation, and we have $A = qB$ for some whole number B. In this case, we have

$$n = pqB.$$

If B is a prime, then pqB is a prime decomposition for n. If not, there would be a prime divisor r for B, say $B = rC$ for some whole number C. Then

$$n = pqrC.$$

And so on. Note that since $A = qB$ and $q \geq 2$, we have $A \geq 2B$. Similarly $B \geq 2C$, etc. Thus we have a sequence of positive integers A, B, C, ..., so that B is at most half of A, C is at most half of B and therefore at most $1/2^2$ of A, the next number D after C will then be at most $1/2^3$ of A, etc. We see that A, B, C, ..., is a strictly decreasing sequence of *positive* integers and therefore after a finite number of steps the sequence must end on a prime. For example, suppose $n < 2^7$ and we get after five steps the following factorization of n into a product of primes p, q, r, s, t, and a number E:

$$\begin{aligned} n &= pA \\ &= pqB \\ &= pqrC \\ &= pqrsD \\ &= pqrstE. \end{aligned}$$

Then we claim that E has to be a prime. If not, the factorization continues and $E = uF$, where u is a prime and F is a positive integer ≥ 2. Then we have

$$n = pqrstE = pqrstuF \geq 2^7,$$

and this means $2^7 > n \geq 2^7$, which is absurd. Therefore $n = pqrstE$ is a prime decomposition of n. This completes the proof that every whole number has a prime decomposition.

Activity. Find the prime decompositions of 252 and 1119.

It remains to prove the uniqueness of the prime decomposition. Let us first look at a number such as 91 to see why its prime decomposition is unique. The prime decomposition of 91 is 7×13, and we want to know why this is the only possible prime decomposition of 91. Suppose 91 is a product of two or more other primes, then at least one of them is < 13. Indeed, if every one of them is ≥ 13, then their product is $\geq 13 \times 13 = 169 > 91$ so that it cannot be equal to 91. Thus we try all the primes < 13 to see which of them divides 91. Among 2, 3, 5, 7, and 11, only 7 does and the only possible prime decomposition of 91 is therefore 7×13.

Activity. Prove the uniqueness of the prime decomposition of 841 ($841 = 29 \times 29$).

The preceding proof of the uniqueness of the prime decomposition of 91 ($= 7 \times 13$) hinges on the fact that if p is a prime and $p|(7 \times 13)$, then $p = 7$ or $p = 13$. We managed to prove this by looking at all the primes from 2 to $\frac{91}{2}$ and eliminating them one by one. For a general n, we need a suitable replacement of this fact because we will not always get a list of primes from 2 to $\frac{n}{2}$; for example if n is very large, say having 500 digits, then (as we already noted above) no computer can churn out this list in our lifetime. What is unclear is what this replacement ought to be. It will have to be quite subtle because a number can easily divide a product of two numbers without being equal to either. For example, $6|(3 \times 8)$ but $6 \neq 3$ and $6 \neq 8$, or $25|(10 \times 10)$ but $25 \neq 10$. Of course one cannot fail to notice that in these two examples, none of the numbers is a prime. What we need is something like: if $p|rs$ and p, r, and s are primes, then $p = r$ or $p = s$. Having said that, we cannot offer an explanation of how the primality of these numbers could affect the outcome. But Euclid, and most likely his predecessors too, already understood this phenomenon 23 centuries ago. Here is what is already contained in Euclid's work:

Theorem 34.2. *If p is a prime and $p|ab$ for whole numbers a and b, then $p|a$ or $p|b$.*

Of course, if both a and b are primes, then $p|a$ means $p = a$ and $p|b$ means $p = b$ and we recover the desired conclusion above.

It is not obvious how to make use of the primality of p in this situation. The key idea of the proof, as presented by Euclid, turns out to come from a *completely different direction*: how to express the greatest common divisor of two whole numbers in terms of the numbers themselves. This is the *Euclidean Algorithm* for whole numbers, which is the concern of the next chapter.

Exercises

1. Give a second proof of the observation that the smallest divisor of any number must be a prime by making use of Exercise 1 on page 454.

2. Find the prime decomposition of each of the following with the help of a hand-held calculator: 1595, 1911, 7316, 15041, 30349.

3. Assume Theorem 34.2. (a) Explain why if p, q_1, q_2, ..., q_k are primes and $p|(q_1 q_2 \cdots q_k)$, then p must be one of the q_1, ..., q_k. (b) Prove that if p and q are primes and $p|q^n$ for a whole number n, then $p = q$ and $n > 0$. Is this true when p or q ceases being a prime?

4. Let m be the product of the first k primes, where k is a positive integer (thus if $k = 7$, $m = 2 \times 3 \times 5 \times 7 \times 11 \times 13 \times 17$), and let $n = m + 1$. Prove that either n is a prime bigger than any of the first k primes, or n has a prime factor bigger than any of the first k primes.

5. Can a prime number simultaneously divide n and $n + 1$ for a whole number n?

Chapter 35

The Euclidean Algorithm

The sections are as follows:

> Common Divisors and Gcd
>
> Gcd as an Integral Linear Combination

35.1. Common Divisors and Gcd

Given two numbers a and b with not both a and b equal to zero,[1] c is called the **greatest common divisor** of a and b if c is the biggest among all the **common divisors** of a and b, i.e., all the whole numbers that divide both a and b. In symbols, we write $\gcd(a, b)$ for c. Put another way, if we collect all the divisors of a and all the divisors of b, then the numbers common to both collections (e.g., the number 1) are the common divisors of a and b. The biggest number of the latter list is by definition $\gcd(a, b)$. For example, consider 24 and 54; the lists of their divisors are, respectively,

$$2, 4, 6, 8, 12,$$
$$2, 3, 6, 9, 18, 27.$$

The common divisors of 24 and 54 are therefore

$$2, 6,$$

and $\gcd(24, 54) = 6$.

For another example, consider 693 and 210. The lists of their divisors are, respectively:

---

[1] If we allow both a and b to be 0, then because every whole number divides 0, $\gcd(0, 0)$ would be undefined.

$$3, 7, 9, 11, 21, 33, 63, 77, 99, 231,$$
$$2, 3, 5, 6, 7, 10, 14, 15, 21, 30, 42, 70, 105.$$

The common divisors of 693 and 210 are therefore
$$3, 7, 21,$$
and $\gcd(693, 210) = 21$. Clearly, $\gcd(5, 12) = \gcd(4, 9) = 1$, and in general, *two whole numbers a and b are relatively prime exactly when* $\gcd(a, b) = 1$.

Note that $\gcd(a, 0) = a$ for any nonzero whole number a, and $\gcd(b, 1) = 1$ for every whole number b.

The purpose of this chapter is to produce a simple algorithm[2] that yields the gcd of any two numbers. It may seem paradoxical that, whereas producing the prime decomposition of a large number has been seen to be impractical, we should now assert that we can produce the gcd of any two numbers with relative ease. The fundamental reason is that by performing one division-with-remainder, such as dividing 693 by 210 (which is straightforward),
$$693 = (3 \times 210) + 63,$$
one can simplify the search for the gcd of these relatively large numbers to the search of the divisors of the relatively small number 63. This is because, by Lemma 32.1, $\gcd(693, 210)$ divides two of the three numbers in this division-with-remainder, namely 693 and 3×210, and therefore it divides also the third number 63. On the basis of this simple observation, we now make some refinements to arrive at a systematic method to get the gcd of two numbers. This is the Euclidean Algorithm.

Activity. Use only one division-with-remainder to determine the gcd of 665 and 7353.

The Euclidean Algorithm is best understood in terms of examples. Let us begin with the determination of the gcd of 3008 and 1344. The division-with-remainder gives:

(35.1) $$3008 = \left(2 \times \boxed{1344}\right) + \boxed{320},$$

where we have put the divisor 1344 and the remainder 320 in boxes for a reason that will be clear in the ensuing discussion. What is worthy of emphasis is that, whereas in school mathematics the remainder in a division is the least interesting object, here it plays a central role. Thus you will see that there is much more to the simple-looking equation (35.1) than meets the eye.

We are looking for the gcd of 3008 and 1344, and we have already seen that it is among the divisors of the remainder 320. At this point, we can

---

[2]Recall that an algorithm is a mechanical procedure consisting of a finite number of steps.

35.1. Common Divisors and Gcd

adopt a lazy attitude and tell ourselves that 320 is plenty small and we can easily get all its divisors. Because gcd(1344, 320) is among these divisors, we can just test all the divisors of 320 to see which is the largest one that also divides 1344, and that would give us gcd(1344, 320), and hence also gcd(3008, 1344). This would be correct, but it is not very good mathematics because we can do far better.

The key observation is that gcd(3008, 1344) is the same as the gcd of the two smaller numbers 1344 and 320. The is so because the set of *all* common divisors of 3008 and 1344 happens to coincide with the set of all common divisors of 1344 and 320 (and therefore the biggest common divisor is the same for either set). To see this, let c be a common divisor of 3008 and 1344. To show it is also a common divisor of 1344 and 320, it suffices to show that c divides 320. But we already know this. Conversely, if c is a common divisor of 1344 and 320, then it divides both numbers on the right-hand side of (35.1) and therefore also divides (by Lemma 32.1 again) the left-hand side, which is 3008. So c is also a common divisor of 3008 and 1344. This then proves that

$$\gcd(3008, 1344) = \gcd(1344, 320).$$

The whole point of the key observation is that, if applying the division-with-remainder once has the beneficial effect of reducing the computation of the gcd of two big numbers 3008 and 1344 to that of two (correspondingly) smaller numbers 1344 and 320, then by iterating the process in like manner will ultimately reduce the computation of the gcd of two given numbers, no matter how large, to that of two numbers so small that we can simply read off the answer. Let us put this to work: we have seen that gcd(3008, 1344) = gcd(1344, 320), and now we apply the division-with-remainder once more to 1344 and 320:

$$(35.2) \qquad 1344 = \left(4 \times \boxed{320}\right) + \boxed{64}.$$

By the previous reasoning, we know that gcd(1344, 320) = gcd(320, 64), so that

$$\gcd(3008, 1344) = \gcd(320, 64).$$

Once more, we apply the division-with-remainder to 320 and 64:

$$320 = \left(5 \times \boxed{64}\right) + \boxed{0},$$

and therefore

$$\gcd(3008, 1344) = \gcd(320, 64) = \gcd(64, 0) = 64.$$

We have therefore computed the desired gcd.

In summary, the reason for the success of this method is that at each stage, we change the computation of the gcd of two numbers to that of two

(correspondingly) smaller numbers:
$$\gcd(3008, 1344) = \gcd(1344, 320),$$
$$\gcd(1344, 320) = \gcd(320, 64),$$
$$\gcd(320, 64) = \gcd(64, 0) = 64,$$
so that at the end we can simply read off the gcd.

35.2. Gcd as an Integral Linear Combination

We mentioned at the end of Chapter 34 that our goal is not just to obtain the gcd of 3008 and 1344, but to express this gcd in terms of 3008 and 1344 themselves. We now show how to achieve this goal by essentially rewriting equations (35.1) and (35.2) in reverse order.

It would be appropriate to mention that it is in the following computations that negative numbers must be brought in for the purpose of computing the gcd of two *whole numbers*.

We first rewrite equation (35.2) as

(35.3) $\qquad 64 = 1344 - (4 \times 320).$

But by equation (35.1), $320 = 3008 - (2 \times 1344)$. Substituting this expression for 320 into equation (35.3) gives
$$64 = 1344 - \Big(4 \times \big(3008 - (2 \times 1344)\big)\Big).$$

One's conditioned reflex upon seeing all the arithmetic operations on the right side of the preceding equality is to immediately carry them out. That would be the wrong reaction, for two reasons. If we do the arithmetic on the right side, we are already guaranteed that it will end up with the answer 64, so there is no point. More importantly, our goal is to express 64 in terms of 3008 and 1344, so it will be to our advantage to simplify the right side *but keep the numbers* 3008 *and* 1344 *intact* in the process. With this in mind, we now make use of the identity (27.7) on page 391 to write,
$$64 = 1344 - (4 \times 3008) + (8 \times 1344)$$
$$= (9 \times 1344) - (4 \times 3008).$$

Or, in other words,

(35.4) $\qquad 64 = (9 \times 1344) + \big((-4) \times 3008\big).$

(Notice that we have just made use of Basic Fact 4 on page 409.)

Equation (35.4) expresses the gcd of 1344 and 3008 as the sum of their integral multiples—in the sense that 9×1344 is the 9th multiple of 1344 and $(-4) \times 3008$ is the (-4)-th multiple of 3008—or in more formal language, as an **integral linear combination** of 1344 and 3008. We have just given an illustration of a general procedure for expressing the gcd of two numbers as

35.2. Gcd as an Integral Linear Combination

an integral linear combination of the numbers themselves through repeated applications of division-with-remainder. This method came to us through Euclid's *Elements* (circa 300 BC), and is therefore referred to as the *Euclidean Algorithm*. The fact that the concept of gcd, which involves only *multiplicative* properties of the two given numbers, should be expressible as a *sum* of multiples of these numbers is not without an element of surprise.

Let us illustrate with another example this method of expressing the gcd of two numbers as an integral linear combination of the numbers themselves. Consider $\gcd(884, 374)$. As usual, we have the division-with-remainder:

$$(35.5) \qquad 884 \;=\; 2 \times \boxed{374} + \boxed{136}.$$

We know that

$$\gcd(884, 374) = \gcd(374, 136).$$

Next, we perform the division-with-remainder on 374 and 136:

$$(35.6) \qquad 374 \;=\; 2 \times \boxed{136} + \boxed{102}$$

and of course,

$$\gcd(374, 136) \;=\; \gcd(136, 102).$$

Now perform the division-with-remainder on 136 and 102:

$$(35.7) \qquad 136 \;=\; 1 \times \boxed{102} + \boxed{34}$$

and

$$\gcd(136, 102) \;=\; \gcd(102, 34).$$

But

$$102 \;=\; 3 \times \boxed{34} + \boxed{0},$$

and $\gcd(34, 0) = 34$, so $\gcd(102, 34) = 34$. Thus $\gcd(884, 374) = 34$. It remains to express 34 as an integral linear combination of 884 and 374 by retracing equations (35.7), (35.6), and (35.5) backwards while keeping the boxed numbers intact at each step. Thus,

$$\begin{aligned}
34 &= 136 - (1 \times 102) & \text{(by (35.7))} \\
&= 136 - \bigl(1 \times (374 - 2 \times 136)\bigr) & \text{(by (35.6))} \\
&= 136 - (1 \times 374) + (2 \times 136) \\
&= (3 \times 136) - (1 \times 374) \\
&= 3 \times \bigl(884 - (2 \times 374)\bigr) - (1 \times 374) & \text{(by (35.5))} \\
&= (3 \times 884) - (6 \times 374) - (1 \times 374) \\
&= (3 \times 884) - (7 \times 374).
\end{aligned}$$

We have therefore obtained

$$\gcd(884, 374) \;=\; 34 \;=\; (3 \times 884) + \bigl((-7) \times 374\bigr).$$

Could you have guessed what m and n should be if you were told that there are integers m and n so that $34 = m \times 884 + n \times 374$? Thus even when the gcd of two numbers is known, it does not follow that it would be easy to express this gcd as an integral linear combination of the two numbers themselves. To do the latter, one still has to perform the Euclidean Algorithm. For example, it is clear that, because 23 is a prime and $16 < 23$, the gcd of 23 and 16 is 1. Yet it is far from obvious how to express 1 as an integral linear combination of 23 and 16. Let us therefore do the Euclidean Algorithm quickly (again by paying special attention to the boxed numbers):

$$23 = 1 \times \boxed{16} + \boxed{7},$$
$$16 = 2 \times \boxed{7} + \boxed{2},$$
$$7 = 3 \times \boxed{2} + \boxed{1},$$
$$2 = 2 \times \boxed{1} + \boxed{0}.$$

We also have, in succession,

$$\gcd(23,16) = \gcd(16,7) = \gcd(7,2) = \gcd(2,1) = \gcd(1,0) = 1.$$

Therefore,

$$\begin{aligned}
1 &= 7 - (3 \times 2) \\
&= 7 - \big(3 \times (16 - 2 \times 7)\big) \\
&= 7 - (3 \times 16) + (6 \times 7) \\
&= (7 \times 7) - (3 \times 16) \\
&= \big(7 \times (23 - 1 \times 16)\big) - (3 \times 16) \\
&= (7 \times 23) - (7 \times 16) - (3 \times 16) \\
&= (7 \times 23) - (10 \times 16).
\end{aligned}$$

Thus,

$$1 = (7 \times 23) + \big((-10) \times 16\big).$$

Activity. Express the gcd of 14 and 82 as an integral linear combination of 14 and 82.

With all these computations behind us, the general case of the Euclidean Algorithm is now almost an afterthought. Given two whole numbers a and b, we want to express their gcd as an integral linear combination of a and b themselves. Let $a > b$. By division-with-remainder,

$$a = q\boxed{b} + \boxed{r},$$

for whole numbers q and r, with $0 \leq r < b$. We observe that *the set of common divisors of a and b is the same as the set of common divisors of b and r*. Indeed, if c divides a and b, then c divides a and qb, and therefore divides r, by Lemma 32.1. So c is a common divisor of b and r. Conversely, if c is a common divisor of b and r, then it divides the right side of $a = qb + r$ and therefore also the left side, i.e., a. So c is a common divisor of a and b

35.2. Gcd as an Integral Linear Combination

as well. This proves the observation, and consequently the fact that the gcd of a and b is the same as the gcd of b and r. Thus,

$$\gcd(a,b) = \gcd(b,r).$$

The main point of this equality is that the pair (a,b) is strictly bigger than the pair (b,r) in the sense that the first (respectively, second) member of (a,b) is bigger than the first (respectively, second) member of (b,r). Now perform the division-with-remainder on b and r,

$$b = q_1\boxed{r} + \boxed{r_1},$$

obtaining whole numbers q_1 and r_1, with $0 \le r_1 < r$, and

$$\gcd(a,b) = \gcd(b,r) = \gcd(r,r_1).$$

we then repeat the division algorithm on r and r_1, obtaining

$$r = q_2\boxed{r_1} + \boxed{r_2},$$

where q_2 and r_2 are whole numbers and $0 \le r_2 < r_1$, etc. As before, we have

$$\gcd(a,b) = \gcd(b,r) = \gcd(r,r_1) = \gcd(r_1,r_2).$$

In going from $\gcd(a,b)$ to $\gcd(b,r)$ to $\gcd(r,r_1)$ to $\gcd(r_1,r_2)$, notice that the second member of each pair—b, r, r_1, and r_2—is always a whole number and gets *strictly* smaller. If this procedure is repeated, it would become zero after a finite number of steps (in fact, at most b steps). For definiteness, let us say that in the next step, we have

$$r_1 = q_3 r_2 + 0,$$

so that $\gcd(r_1,r_2) = \gcd(r_2,0) = r_2$. Then

$$\gcd(a,b) = \gcd(b,r) = \gcd(r,r_1) = \gcd(r_1,r_2) = \gcd(r_2,0) = r_2.$$

We are going to express r_2 as an integral linear combination of a and b by writing the sequence of division algorithms backwards:

$$\begin{aligned}
\gcd(a,b) = r_2 &= r - q_2 r_1 \\
&= r - q_2(b - q_1 r) \\
&= (1 + q_1 q_2)r - q_2 b \\
&= (1 + q_1 q_2)(a - qb) - q_2 b \\
&= (1 + q_1 q_2)a + (-q - q_2 - q q_1 q_2)b,
\end{aligned}$$

where q, q_1, and q_2 are integers. We have therefore proved:

Theorem 35.1. (The Euclidean Algorithm) *If a and b are any two whole numbers and not both are equal to 0, then their* gcd *is an integral linear combination of a and b. Furthermore, this* gcd *can be obtained from a and b by an iteration of division-with-remainder.*

We can now give the **proof of Theorem 34.2** (see page 460). Recall that statement:

If p is a prime and $p|ab$ for whole numbers a and b, then $p|a$ or $p|b$.

Proof. Given that p is a prime and $p|ab$ for some whole numbers a and b, we must show $p|a$ or $p|b$. If $p|a$, then we are done. So suppose that $p \nmid a$, and we must show $p|b$. Because p is a prime, only 1 and p are the possible divisors of p, and a fortiori, 1 and p are the only possible common divisors of a and p. Because $p \nmid a$, the only common divisor of a and p is 1, i.e., $\gcd(a,p) = 1$. By the Euclidean Algorithm, $1 = mp + na$ for some integers m and n. Multiply both sides by b and we get

$$b = mbp + nab.$$

Now $p|(mbp)$; also $p|(nab)$ because $p|(ab)$. Therefore p divides the right side of the equation, and it must then also divide the left side, i.e., p divides b, as desired. □

A corollary of the Euclidean Algorithm (and Lemma 32.1) is that

If k is a common divisor of both a and b, then k divides $\gcd(a,b)$.

This is unexpected because the definition of $\gcd(a,b)$ only tells us that $k \leq \gcd(a,b)$, but says nothing about k being a divisor of $\gcd(a,b)$.

We can get more out of the preceding proof. Indeed, the only place we needed the assumption that p is a prime number in the proof of Theorem 34.2 was in obtaining the conclusion that $\gcd(a,p) = 1$. Go over the proof one more time, and you will see that once $\gcd(a,p) = 1$ is known, the proof as it stands would support the statement that "$p|ab$ implies $p|b$". We have therefore actually proved a slightly more general theorem, which will turn out to be very useful:

Theorem 35.2. *Suppose a, b and k are whole numbers such that $k|ab$ and $\gcd(a,k) = 1$. Then $k|b$.*

Let us re-examine the previous examples of $6|(3 \times 8)$ and $25|(10 \times 10)$ from the perspective of Theorem 35.2. Both are examples of a number dividing a product without dividing either of the individual factors. This phenomenon is explained by the fact that $\gcd(6,3) = 3 \neq 1$ and $\gcd(6,8) = 2 \neq 1$, and also $\gcd(25,10) = 5 \neq 1$.

We now come to the **proof of the uniqueness part of FTA.** Let us consider only the case of a whole number n with a prime decomposition consisting of the product of three primes, $n = pqs$, where p, q, and s are primes. (Of course, any two or all three of p, q, and s could be equal, e.g., $125 = 5 \times 5 \times 5$ or $845 = 5 \times 13 \times 13$.) The general case involves no new ideas and, for this reason, we avoid it because the notation gets unwieldy.

35.2. Gcd as an Integral Linear Combination

We want to show that no other prime decomposition of n is possible. We begin by showing that if p' is a prime that divides n, then $p' = p$ or $p' = q$ or $p' = s$. (Therefore p, q, and s are the only primes that may be present in any prime decomposition of n.) Indeed, because $n = pqs$, p' being a divisor of n means $p'|pqs$. If $p' = p$, we are done. Otherwise, $p' \nmid p$, and $\gcd(p, p') = 1$. But $p'|p(qs)$, so by the Theorem 35.2, $p'|(qs)$. If $p' = q$, we are done again. If not, because p' and q are distinct primes, $\gcd(p', q) = 1$ and by Theorem 35.2, $p'|s$. Since p' and s are primes, this is possible only if $p' = s$. Thus we have proved that any other prime decomposition of n must be a product of p's and q's and s's. We note that this does not preclude the possibility that one of p or q or s is absent from the product.

Now let us see why, of all possible products involving p, q, and s, only pqs can equal n.

First consider the case where p, q, s are not distinct primes. Suppose $p = q$.[3] Then $n = p^2 s$ and the preceding paragraph shows that only p and s can be present in any prime decomposition of n. We have to show why none of the products ps^2, ps^3, ps, $p^3 s$, $p^4 s$, etc., can equal $p^2 s$. Take, for instance, $p^3 s$. If $p^3 s = p^2 s$, then multiplying both sides by $1/(p^2 s)$ leads to $p = 1$, which contradicts the primality of p. A similar argument applies to all the other possibilities above. We can also make the same argument if $p = s$ or $q = s$. We have therefore proved the uniqueness of the prime decomposition of n in case two or more of p, q, s are equal.

So we may assume p, q, and s are distinct primes. We want to show that any product of p, q, and s other than pqs, cannot be equal to n. Consider $pq^2 s$, for instance. If $n = pq^2 s$, then $pqs = pq^2 s$. Multiplying both sides of the last equation by $\frac{1}{pqs}$ leads to $1 = q$, which contradicts the primality of q. Or we can consider ps^3. Then we have $ps^3 = pqs$. Multiplying both sides by $1/(ps)$ gives $s^2 = q$, and q is not a prime. Contradiction again. A similar argument applies to any other product so long as only one or more of p, q, and s appear in it. At the end, we see that $n = pqs$ is the only possibility. The proof is complete.

---

[3]Note that this includes the possibility that $p = q = s$. However, if we only need the assumption that p and q are equal (and it does not matter whether s is equal to p and q or not), then $p = q$ is all we are going to assume.

Exercises

1. (This problem brings closure to the discussion of ratios in Chapter 22; see footnote 3 on page 337.) Let M, N, m, n be whole numbers and let $\frac{M}{N} = \frac{m}{n}$. Also let $\frac{m}{n}$ be a reduced fraction. Then prove that $\frac{M}{m}$ and $\frac{N}{n}$ are both whole numbers.

2. Find the gcd of each of the following pairs of numbers by listing all the divisors of each number and comparing them: 12 and 42, 34 and 85, 24 and 69, 102 and 289, 195 and 442.

3. Find the gcd of each of the following pairs of numbers, and express it as an integral linear combination of the numbers in question: 314 and 159, 343 and 280, 924 and 105, 345 and 253, 578 and 442.

4. Write out a self-contained proof of Theorem 35.2 with all the details in place.

5. Given whole numbers a and b, suppose there are integers x and y so that $xa + yb = 1$. Prove that a and b are relatively prime.

6. Without using FTA, prove directly that the prime decomposition of 231 ($= 3 \times 7 \times 11$) is unique.

7. Write a self-contained proof of the uniqueness part of FTA for the case of a whole number which is a product of two primes.

Chapter 36

Applications

In this chapter, we give diverse applications of the various concepts and techniques we have learned up to this point.

The sections are as follows:

Gcd and Lcm

Fractions and Decimals

Irrational Numbers

Infinity of Primes

36.1. Gcd and Lcm

Let us revisit the concept of the divisor of a number. Consider 360. The methods described in Chapters 32 and 33 give a systematic procedure to get all its divisors. In view of FTA, Chapter 34, however, we can now try a different tack. We will write down the prime decomposition of 360 as a product of distinct primes by using the exponential notation (cf. Chapter 11):

$$360 = 2^3 \times 3^2 \times 5.$$

It is clear that a number which is the product of a subcollection of the primes

$$2, 2, 2, 3, 3, 5$$

is a divisor of 360, e.g., 24 ($= 2^3 \times 3$). However, it is usually taken for granted that the converse also holds: any divisor of 360 has to be equal to a product of a subcollection of $2, 2, 2, 3, 3, 5$. The simple *proof* of this fact actually requires the uniqueness part of FTA, and it goes as follows. Suppose b is a whole number and $b|360$, then $360 = bc$ for some whole number c. We now look at the prime decompositions of both 360 and b and c. By the uniqueness in FTA, the primes in the prime decomposition of 360 are the same collection (up to rearrangement) as the union of the primes in the prime decompositions of b and c. In particular, the prime decomposition of b has to be a subcollection of $2, 2, 2, 3, 3, 5$, as claimed.

The same reasoning proves the following lemma.

Lemma. *Let b and n be nonzero whole numbers. Then b being a divisor of n is equivalent to the fact that b is a product of primes that are a subcollection of the primes in the prime decomposition of n.*

On the basis of the lemma, we gain a new understanding of the gcd of two whole numbers m and n: $\gcd(m, n)$ consists of the product of *all* the primes common to both prime decompositions of m and n. Thus, knowing that

$$6104371 = 7^3 \times 13 \times 37^2,$$
$$9596209 = 7^2 \times 37 \times 67 \times 79,$$

we can easily read off

$$\gcd(6104371, 9596209) = 7^2 \times 37.$$

There is related concept. Define the **least common multiple (lcm)** of two whole numbers a and b to be the smallest whole number m so that m is a multiple of both a and b (i.e., $a|m$ and $b|m$). In symbols, we write $\text{lcm}(a, b)$ for this m. You may recall that the concept of lcm is of interest in the teaching of fractions because the addition of fractions can *sometimes* be simplified by taking the lcm of the denominators as the denominator of the sum;[1] see section 14.4.

Consider 24 and 108. Let us agree that their gcd is 12. We now want their lcm. A very practical way is to list the multiples of both numbers,

24, 48, 72, 96, 120, 144, 168, 192, 216, 240, ...,
108, 216, 270, 324, ...,

and look for the smallest number common to both lists. In this case, it is 216.

With FTA at our disposal, however, we can also approach the search for the lcm from a different direction. Let us list the prime decompositions of both numbers:

$24 = 2^3 \times 3$,
$108 = 2^2 \times 3^3$.

Clearly, if we take the product of *all* the primes in 24 and 108, we would get a common multiple, i.e.,

$$2^3 \times 3 \times 2^2 \times 3^3 \ (= 24 \times 108).$$

---

[1]It would be a simplification only if the lcm is painlessly obvious.

36.1. Gcd and Lcm

However, if we want the *least* common multiple, we should discard some of the primes in the preceding product. Precisely, any prime common to both may be safely discarded without affecting the resulting number being a common multiple of both 24 and 108. Thus two copies of 2 (the 2^2 in 108) and one copy of 3 (the 3 in 24) can be discarded and we get

$$\text{lcm}(24, 108) = 2^3 \times 3^3 = 216.$$

However, the previous discussion reveals that what we discarded is precisely the gcd of 24 and 108: $12 = 2^2 \times 3$. Putting these discarded primes back into $\text{lcm}(24, 108)$ therefore gives us back 24×108. Thus,

$$24 \times 108 = \gcd(24, 108) \times \text{lcm}(24, 108).$$

Look at one more example: 693 and 210. We know that $\gcd(693, 210) = 21$ (see Chapter 35, page 463). Their multiples are, respectively,

693, 1386, 2079, 2772, 3465, 4158, 4851, 5544, 6237, **6930**, ...,

210, 420, 630, 840, 1050, 1260, 1470, 1680, 1890, 2100, 2520, 2730, 2940, 3150, 3360, 3570, 3780, 3990, 4200, 4410, 4620, 4830, 5040, 5250, 5460, 5670, 5880, 6090, 6300, 6510, 6720, **6930**, 7140, ...

We notice of course that 6930 is the lcm of 210 and 693. This example illustrates the fact that it is too tedious in general to find the lcm of two numbers by inspecting two lists of multiples. We would have saved ourselves some time if we had looked at the (comparatively short) first list (the multiples of the larger 693) and checked the divisibility by 210 of every member on this list. (In practice, we need not even bother checking until we hit 6930, because only numbers ending in 0 in the ones digit can hope to be a multiple of 210.)

But what about the prime decompositions of 210 and 693?

$210 = 2 \times 3 \times 5 \times 7$.

$693 = 3^2 \times 7 \times 11$.

If we take away from the product of all these primes,

$$2 \times 3 \times 5 \times 7 \times 3^2 \times 7 \times 11 \; (= \; 210 \times 693),$$

the primes common to both, 3 and 7, then we get

$$\text{lcm}(210, 693) = 2 \times 5 \times 3^2 \times 7 \times 11 = 6930.$$

Of course 3×7 is exactly the gcd of 210 and 693. So once again,

$$210 \times 693 = \gcd(210, 693) \times \text{lcm}(210, 693).$$

The preceding use of prime decompositions to discuss the gcd and lcm of two numbers is in fact perfectly general, and we proceed to give a systematic discussion. First:

Proposition 1. *If $d = \gcd(a, b)$, then $\gcd(\frac{a}{d}, \frac{b}{d}) = 1$.*

We are going to give two proofs of this proposition; each is revealing in its own way. The **first proof** essentially repeats the reasoning above using the prime decompositions of the two numbers. To formulate this proof in complete generality, however, would create a veritable notational nightmare. We shall instead present the proof for a special case, which is in essence not different from the general case. So suppose the numbers a and b have prime decompositions

$$a = p^2 q s^3 t^4,$$
$$b = p s^5 t^2 u^3,$$

where p, q, s, t, and u are all primes. The number

$$d = p s^3 t^2$$

divides both, and its prime decomposition $p s^3 t^2$ cannot contain another prime without losing the property of being a common divisor of both a and b. Therefore d is the gcd of a and b. Because the prime decomposition of d contains all the primes common to both a and b, the divisions $\frac{a}{d}$ and $\frac{b}{d}$ would leave behind, after cancellations, two numbers which have no prime factors in common and must therefore be relatively prime. One can also verify this fact by a direct computation:

$$\frac{a}{d} = pqt^2 \quad \text{and} \quad \frac{b}{d} = s^2 u^3,$$

and these two numbers pqt^2 and $s^2 u^3$ have no prime factors in common and are therefore relatively prime. □

A **second proof** of Proposition 1 uses the Euclidean Algorithm and may seem at first to be nonintuitive. But once you have gone through this proof, you will realize that it is very natural. Moreover, this technique is important in elementary number theory and is definitely worth learning. Let $a = da'$ and $b = db'$ for some whole numbers a' and b'. Then $\frac{a}{d} = a'$ and $\frac{b}{d} = b'$, and we have to prove that $\gcd(a', b') = 1$. By the Euclidean Algorithm, we know that $d = pa + qb$ for some whole numbers p and q. Therefore $d = pda' + qdb'$. Dividing through by d, we obtain $1 = pa' + qb'$. If a' and b' have a common divisor c, then c divides each of the terms pa' and qb' on the right, hence divides the right side, and hence divides the left side which is 1. It follows that $c = 1$ and the proof is complete. □

We now formalize the relationship between the gcd and lcm of any two whole numbers a and b.

36.1. Gcd and Lcm

Proposition 2. *If a and b are whole numbers, then*
$$ab = \gcd(a,b) \cdot \operatorname{lcm}(a,b).$$

Before embarking on a proof of the proposition, it would be helpful to first do an activity.

Activity. Suppose we are given the following prime decompositions,
$$26460 = 2^2 \times 3^3 \times 5 \times 7^2,$$
$$15225 = 3 \times 5^2 \times 7 \times 29.$$
Find the gcd and lcm of these two numbers.

Again we give two proofs. The **first proof** uses the prime decompositions of a and b as in the first proof of Proposition 1. As before, let
$$a = p^2 q s^3 t^4,$$
$$b = p s^5 t^2 u^3,$$
where p, q, s, t, and u are all primes. The number
$$\ell = p^2 q s^5 t^4 u^3$$
is clearly a multiple of both a and b, and furthermore, it is the smallest number which has this property because the exponents 2, 1, 5, 4, 3 of the primes p, q, s, t, u, respectively, cannot be any smaller if ℓ is to be a multiple of both a and b. Thus ℓ is the lcm of a and b. We have seen above that the gcd d of a and b in this case is $d = p s^3 t^2$. Therefore,
$$d\ell = p s^3 t^2 \cdot p^2 q s^5 t^4 u^3 = p^2 q s^3 t^4 \cdot p s^5 t^2 u^3 = ab,$$
as desired. □

The **second proof** makes strong use of Theorem 35.2 on page 470. Let us first prove it for the familiar case of 24 and 108. We already know $12 = \gcd(24, 108)$. Therefore $24 = 12 \times 2$ and $108 = 12 \times 9$, so that
$$\frac{24 \times 108}{12} = 12 \times (2 \times 9).$$
To prove Proposition 2 in this special case, it suffices to prove that the right-hand side $12 \times (2 \times 9)$ is the lcm of 24 and 108. There is no doubt that $12 \times (2 \times 9)$ is a common multiple of 24 and 108. To prove that it is the *least* of the common multiples, we will prove that if n is another common multiple of 24 and 108, then $\bigl(12 \times (2 \times 9)\bigr) \mid n$. To this end, we let $n = 24A = 108B$ for some whole numbers A and B. Let us reflect on this for a minute: as it stands, $\bigl(12 \times (2 \times 9)\bigr)$ already *almost* divides n because if we look at n as $24A$, then we can write $\bigl(12 \times (2 \times 9)\bigr)$ as 24×9. Hence if we can show that 9 divides A, then for sure $(24 \times 9) \mid (24A)$ and we would be finished. But we are given $24A = 108B$, which implies
$$(12 \times 2)A = (12 \times 9)B,$$

and we have
$$2A = 9B.$$
The fact that 9 divides the right side implies 9 divides the left side, but 2 and 9 being relatively prime implies, by Theorem 35.2, that $9|A$. This proves our claim. □

We can now easily generalize this reasoning. Let $d = \gcd(a, b)$ and let $a = da'$, $b = db'$ for some whole numbers a' and b'. Then $\frac{ab}{d} = da'b'$. We will prove that $da'b'$ is the lcm of a and b. It is clear that $da'b'$ (which is equal to $ab' = a'b$) is a common multiple of a and b; what we have to show is that it is the smallest among these common multiples. We do so by proving that if n is *any* common multiple of a and b, then $da'b'$ divides n. Once this is done, then $da'b' \leq n$. So let $n = aA = bB$ for some whole numbers A and B. Now $a = da'$, $b = db'$, so $da'A = db'B$ and therefore $a'A = b'B$. Because b' obviously divides $b'B$, we now see that b' divides $a'A$. By Proposition 1, $\gcd(a', b') = 1$, so Theorem 35.2 implies that $b'|A$, let us say $A = b'k$ for some whole number k. So $n = aA = ab'k$. Recall that $a = da'$. Hence $n = ab'k = da'b'k = (da'b')k$ and therefore $da'b'$ divides n. This proves that $da'b'$ is smaller than or equal to n, and Proposition 2 is proved.

The second *proof* of Proposition 2 (i.e., not just the statement of Proposition 2 itself) contains a remarkable fact: if n is any common multiple of a and b, then $da'b'$ divides n. Therefore:

Proposition 3. *The* lcm *of a and b divides any common multiple of a and b.*

Proposition 3 would be unexpected if we only look at the definition of lcm because, by definition, all we know is that $\text{lcm}(a, b)$ is less than or equal to any common multiple of a and b, and there is no mention of any divisibility properties. However, our discussion of the lcm's of 24 and 108, and of 210 and 693, makes clear that in terms of the prime decompositions of a and b, their lcm is the number with the smallest number of the prime factors of a and b which is also a multiple of a and b. So the other common multiples of a and b can differ from the lcm only in having more prime factors. From this point of view, therefore, Proposition 3 is not surprising.

36.2. Fractions and Decimals

The purpose of this section is to prove two fundamental facts about fractions. The first (Theorem 36.1 below) says that every fraction is equal to a unique reduced fraction; this is usually assumed without proof. Recall that a fraction $\frac{a}{b}$ is said to be **reduced** or **in lowest terms** if $\gcd(a, b) = 1$. For example, the fraction $\frac{81}{54}$ can be reduced to $\frac{27}{18}$ (cancel 3 from both numerator and denominator), which can be further reduced to $\frac{9}{6}$ by doing the same, and finally to $\frac{3}{2}$. What was done was to remove all possible common factors

36.2. Fractions and Decimals

from both numerator and denominator, which is the same as dividing both numerator and denominator by their gcd. Intuitively, the process of reducing a fraction can go no further after dividing out by this gcd. This is the essential content of the next theorem.

Theorem 36.1. *Every fraction $\frac{m}{n}$ is equal to a fraction in lowest terms obtained by dividing m and n by their gcd. Furthermore, this fraction in lowest terms is unique in the sense that if $\frac{m}{n} = \frac{a}{b} = \frac{A}{B}$, and both $\frac{a}{b}$ and $\frac{A}{B}$ are fractions in lowest terms, then $a = A$ and $b = B$.*

Proof. Let $\gcd(m,n) = d$ and let $m = da$, $n = db$ for whole numbers a and b. Therefore, $\frac{m}{n} = \frac{da}{db} = \frac{a}{b}$ by the cancellation law in Chapter 13. Furthermore, Proposition 1 shows that $\frac{a}{b}$ is a fraction in lowest terms.

Now suppose $\frac{A}{B}$ is another fraction in lowest terms and is equal to $\frac{m}{n}$. We have to prove that $A = a$ and $B = b$. The proof is merely a repeated application of Theorem 35.2 on page 470, exploiting the hypothesis that $\gcd(c,d) = 1$ and $\gcd(A,B) = 1$. Thus, from $\frac{a}{b} = \frac{A}{B}$ (both equal to $\frac{m}{n}$), we have $aB = Ab$ by the cross-multiplication algorithm (Chapter 13). Now $a|Ab$ (because $a|aB$ and $aB = Ab$), and $\gcd(a,b) = 1$ by assumption, therefore $a|A$ by Theorem 35.2. Similarly, $A|aB$ and $\gcd(A,B) = 1$ by assumption, so the same Theorem 35.2 implies that $A|a$. Thus we have $A|a$ and $a|A$ simultaneously, and so $A \le a$ and $a \le A$, respectively. Hence $a = A$. Recall that we also have $aB = Ab$, so $b = B$, as desired. □

The second fact about fractions characterizes those that are equal to finite decimals.

Theorem 36.2. *A fraction $\frac{a}{b}$ in lowest terms is equal to a finite decimal exactly when the denominator b is equal to $2^m 5^n$, where m and n are whole numbers.*

Note that it is crucial that $\frac{a}{b}$ in Theorem 36.2 be in lowest terms, as otherwise $\frac{3}{6} = 0.5$ but the denominator 6 contains 3 in its prime decomposition.

Theorem 36.2 gives a convincing illustration of why knowing the *definition* that a finite decimal is a fraction whose denominator is a power of 10 clarifies one's thinking. What we are trying to prove is that, for a reduced fraction $\frac{a}{b}$,

$\frac{a}{b} = \frac{m}{10^k}$ *for some whole numbers m and k* $\iff$ *the prime decomposition of b involves only 2's and 5's.*

In this form, clearly the cross-multiplication algorithm (section 13.5) is involved and, knowing that $10^k = 2^k 5^k$ makes the conclusion plausible and much of the mystery surrounding the theorem is gone.

Proof. Given the fraction $\frac{a}{b}$ in lowest terms, we must prove two things: (i) if $\frac{a}{b}$ has a finite decimal expansion, then $b = 2^m 5^n$ for some whole numbers m and n, and (ii) if $b = 2^m 5^n$ for some whole numbers m and n, then $\frac{a}{b}$ has a finite decimal expansion.

Let us begin with the proof of part (i). As in the case of Theorem 36.1, instead of writing out a proof for a fraction with the most general decimal expansion, the proof would be more understandable if we look at a fraction $\frac{a}{b}$ equal to 0.5726, say. It will be obvious at the end of the proof that even if 0.5726 is replaced by an arbitrary finite decimal, the proof would remain essentially the same.

By the definition of a decimal, we have $\frac{a}{b} = \frac{5726}{10000}$. By the cross-multiplication algorithm (section 13.5), $10000\,a = 5726\,b$. So $b|(10000\,a)$. By the assumption of $\frac{a}{b}$ being in lowest terms, $\gcd(a,b) = 1$. Therefore Theorem 35.2 on page 470 implies that $b|10000$. Thus b is a divisor of $10^4 = 2^4 5^4$. By the lemma of the preceding section, b is a product of 2's and 5's (one of 2 and 5 could be absent of course).

We next prove part (ii). Let $b = 2^m 5^n$, where m and n are whole numbers, and we want to show that $\frac{a}{b}$ has a finite decimal expansion. To get a feel for what we are up against, suppose $b = 2^3 5^7$. The key point is that b is not that far from a power of 10 because, noting the difference in the powers of 2 and 5 in $2^3 5^7$, we multiply b by 2^4 to get $2^4 b = 2^4 2^3 5^7 = 2^7 5^7 = (2 \times 5)^7 = 10^7$, so that

$$\frac{a}{b} = \frac{2^4 a}{2^4 b} = \frac{2^4 a}{10^7},$$

and the last fraction is by definition a decimal fraction. We have therefore expressed $\frac{a}{b}$ as a finite decimal. The general case is no different. Suppose $b = 2^m 5^n$. For definiteness, we may assume $m \leq n$. Then $2^{n-m} b = 2^n 5^n = 10^n$, and

$$\frac{a}{b} = \frac{2^{n-m} a}{2^{n-m} b} = \frac{2^{n-m} a}{10^n}.$$

Because $2^{n-m} a$ is a whole number, $(2^{n-m} a)/10^n$ is a finite decimal with at most n decimal digits. Therefore so is $\frac{a}{b}$. □

36.3. Irrational Numbers

We are going to prove that certain numbers are irrational. First we need a lemma. A whole number n is said to be a **perfect square** if n is the square of another whole number. For example, 25 is a perfect square because $25 = 5^2$, and so are 1936 and 11025 because $1936 = 44^2$ and $11025 = 105^2$. Note the special feature of the primes in the prime decompositions of these numbers: $25 = 5 \times 5$, $1936 = 2 \times 2 \times 2 \times 2 \times 11 \times 11$, and $11025 = 3 \times 3 \times 5 \times 5 \times 7 \times 7$.

36.3. Irrational Numbers

The feature in question is that each prime appears an even number of times. This turns out to characterize perfect squares.

Lemma 36.3. *A whole number is a perfect square $\iff$ every prime in its prime decomposition appears an even number of times.*

Proof. If n is a perfect square, let $n = m^2$ for some whole number m. Let $m = p_1 p_2 \cdots p_k$ be the prime decomposition of n. Then $n = m^2 = p_1 p_2 \cdots p_k p_1 p_2 \cdots p_k = p_1^2 p_2^2 \cdots p_k^2$. Because all the p_i's are primes, the equality $n = p_1^2 p_2^2 \cdots p_k^2$ is the prime decomposition of n (because there is only one prime decomposition by FTA). We have therefore proved that all the primes in the prime decomposition of a perfect square come in an even number of times. It remains to show the converse, i.e., if the primes in the prime decomposition of a whole number n all come in evenly, then n is a perfect square. So suppose the prime decomposition of n is $p_1 p_1 p_2 p_2 \cdots p_\ell p_\ell$, where all the p_i's are primes. Let m be the whole number $p_1 p_2 \cdots p_\ell$, then $n = m^2$, and n is a perfect square. This proves Lemma 36.3. $\square$

If n is a perfect square, say $n = m^2$ for some whole number m, then clearly $\sqrt{n}$ is a fraction, in fact a whole number, namely, m. We will eventually show that this is the only time that the square root of a whole number is a fraction (Theorem 36.5 following). First, we show something simpler.

Theorem 36.4. *If p is a prime, then $\sqrt{p}$ is never a fraction, i.e., it is irrational.*

Before giving the proof, let us look at the special cases of the first two primes (i.e., 2 and 3) to see why their square roots are irrational. We are going to give proofs by contradiction. Suppose $\sqrt{2}$ is a fraction, so $\sqrt{2} = \frac{a}{b}$ for some whole numbers a and b. We will deduce a contradiction using only elementary facts without resorting to FTA, as follows. By using the cancellation law, we may keep reducing $\frac{a}{b}$ by dividing both numerator and denominator by 2 (see section 13.1) until only one of a and b is even and the other is odd. Because $\sqrt{2} = \frac{a}{b}$ implies $\sqrt{2}b = a$,[2] which in turn implies $2b^2 = a^2$, we see that a^2 is even. This means a is even (see Exercise 1(a) on page 444), and therefore b is odd. However, if a is even, $a = 2c$ for some whole number c and therefore $2b^2 = a^2$ implies that $2b^2 = 4c^2$. Multiplying both sides by $\frac{1}{2}$, we get $b^2 = 2c^2$. But this means b^2 is even, and therefore b is even. This contradicts the fact that b is odd. We conclude that $\sqrt{2}$ is irrational.

The preceding proof of the irrationality of $\sqrt{2}$ is designed to be as elementary as possible, using only equivalent fractions, the fact that every whole number must be either even or odd, and some simple facts about

---
[2] It is worth pointing out that here we are using FASM of Chapter 21, and are assuming that (c) on page 310 is valid for arbitrary real numbers.

evenness and oddness as encoded in Exercise 1 on page 444. For the irrationality of $\sqrt{3}$, we will have to make use of FTA. So again, suppose $\sqrt{3}$ is a fraction, and we shall deduce a contradiction. Then the principle of proof-by-contradiction would allow us to conclude that $\sqrt{3}$ is irrational. Let then $\sqrt{3} = \frac{a}{b}$ for some whole numbers a and b. Squaring both sides, we get $3 = \frac{a}{b}\frac{a}{b} = \frac{a^2}{b^2}$, so that $3b^2 = a^2$. Let $p_1 p_2 \cdots p_k$ be the prime decomposition of b^2, where each p_i is a prime. Since 3 is a prime, the expression of $3b^2$ as a product,

$$(36.1) \qquad 3b^2 = 3p_1 p_2 \cdots p_k,$$

is a prime decomposition of $3b^2$. We claim that the number of times 3 appears on the right side of (36.1) is odd. Indeed, if 3 is not equal to any of the $p_1, p_2, \ldots, p_k$, then 3 appears exactly once. If, however, 3 is equal to one of $p_1, p_2, \ldots, p_k$, then 3 appears in the prime decomposition $p_1 p_2 \cdots p_k$ of b^2 an even number of times according to Lemma 36.3. Therefore 3 will appear on the right side of (36.1) an odd number of times because of the extra 3 in $3p_1 p_2 \cdots p_k$. This proves the claim. But $3b^2 = a^2$, so by FTA, $3p_1 p_2 \cdots p_k$ is also the prime decomposition of a^2. However, Lemma 36.3 now shows that 3 must appear in the prime decomposition of a^2 an even number of times, which then implies that 3 appears in $3p_1 p_2 \cdots p_k$ an even number of times. This contradiction shows that $\sqrt{3}$ cannot be a fraction and is therefore irrational.

We now give the general **proof of Theorem 36.4**. It is essentially the same as the proof for $\sqrt{3}$. We will prove that if $\sqrt{p}$ is a fraction, there is a contradiction. Then the principle of proof-by-contradiction shows that $\sqrt{p}$ is irrational. So suppose $\sqrt{p}$ is a fraction $\frac{a}{b}$ for some whole numbers a and b. Then $p = \frac{a^2}{b^2}$, so that (by multiplying both sides by b^2) we get $pb^2 = a^2$. Let the prime decomposition of b^2 be $b = q_1 q_2 \cdots q_k$, where each q_i is a prime. Then

$$(36.2) \qquad pb^2 = p q_1^2 q_2^2 \cdots q_k^2$$

is a prime decomposition of pb^2. We claim that the number of times p appears on the right side of (36.2) is odd. Indeed, if p is not equal to any of the $q_1, q_2, \ldots, q_k$, then p appears exactly once on the right side of (36.2). If, however, p is equal to one of $q_1, q_2, \ldots, q_k$, then p appears in the prime decomposition $q_1 q_2 \cdots q_k$ of b^2 an even number of times according to Lemma 36.3. Therefore p will appear on the right side of (36.2) an odd number of times because of the extra p in $pq_1 q_2 \cdots q_k$. This proves the claim. But $3b^2 = a^2$, so by FTA, $pq_1 q_2 \cdots q_k$ is also the prime decomposition of a^2. However, Lemma 36.3 now shows that p must appear in the prime decomposition of a^2 an even number of times, which then implies that p appears in $pq_1 q_2 \cdots q_k$ an even number of times. This contradiction shows that $\sqrt{p}$ cannot be a fraction and is therefore irrational. Theorem 36.4 is proved.

We now generalize Theorem 36.4 slightly by proving in general:

Theorem 36.5. *If n is a whole number which is not a perfect square, then $\sqrt{n}$ is irrational.*

Proof. To avoid excessive notation, we shall prove that $\sqrt{175}$ is irrational; the general case does not require new ideas. As in the proof of Theorem 36.4, we only have to prove that if $\sqrt{175}$ were a fraction, there would be a contradiction. Suppose $\sqrt{175} = \frac{a}{b}$, where a and b are whole numbers, then by squaring both sides, we get

$$175 = \frac{a^2}{b^2} \quad \text{so that} \quad a^2 = 175 b^2.$$

Now $175 = 5^2 \times 7$, and the fact that the prime 7 enters the prime decomposition of 175 an odd number of times is no accident because 175 is not a perfect square (see Lemma 36.3). Thus we have

$$a^2 = 7 (5b)^2.$$

Now take the prime decompositions of both a^2 and $(5b)^2$, and all the primes will appear an even number of times. Putting these prime decompositions into $a^2 = 7(5b)^2$ and the primes on both sides of this equality must be the same because of the uniqueness in FTA. But since 7 appears on the right side an odd number of times, 7 appears also an odd number of times on the left, which implies that 7 appears an odd number of times in the prime decomposition of a^2, contradicting Lemma 36.3. The proof is complete. □

36.4. Infinity of Primes

We conclude these applications with one of the most famous proofs in the history of mathematics: Euclid's proof of 23 centuries ago that there is an infinite number of primes.

The proof is by contradiction. We assume that there is only a finite number of primes, say $p_1, p_1, \ldots, p_k$. We shall show that such an assumption leads to a contradiction. We proceed on the assumption that these k primes are all the primes among the whole numbers. Consider then the whole number $N = (p_1 p_2 \cdots p_k) + 1$. By FTA, this N has a prime decomposition. There are two possibilities: either N is itself a prime, or it is a product of two or more primes. The first possibility is impossible because N is bigger than each of $p_1, \ldots, p_k$ by definition and these p's are supposed to exhaust *all* the primes. What about the second possibility? If N is a product of two or more primes, then it is a product of two or more of these $p_1, \ldots, p_k$ because they are supposed to be *all* the primes in existence. But this too is impossible because, by the definition of N, N has remainder 1 when it is divided by each of these p's. Our conclusion: assuming there is only a finite number of primes is wrong. The number of primes is therefore infinite.

Of course everybody would prefer a direct proof, so why not just exhibit an infinite number of primes and be done with it? The sad fact is that people have tried, but that no one has succeeded in producing such a sequence thus far, and this is where we are.

Exercises

1. Find the lcm's of the pairs of whole numbers in Exercise 2 on page 472.
2. (i) Let a, b be two whole numbers with the prime decompositions $a = 3^5 \times 11^2 \times 17^4$ and $b = 3 \times 7^2 \times 17^8 \times 23^4$. What is their gcd and what is their lcm? (ii) The same question if $a = 17^3 \times 19 \times 67^4 \times 157^2$ and $b = 17 \times 19^5 \times 67^2 \times 97 \times 157^3$?
3. Let a, b be two whole numbers with the prime decompositions $a = p^2 q^7 r^3$ and $b = p^6 q s^4$, where p, q, r, s denote distinct primes. Write out the gcd and lcm of a and b in terms of p, q, r, and s.
4. Give a direct proof of Proposition 3 using FTA, without appealing to Proposition 2.
5. Find the fraction in lowest terms which is equal to each of the following:
$$\frac{132}{72}, \quad \frac{160}{256}, \quad \frac{273}{156}, \quad \frac{221}{323}, \quad \frac{144}{336}.$$
6. Give a direct proof, without using Theorem 36.5, that $\sqrt{720}$ is irrational.
7. Prove that $\sqrt[3]{2}$ is irrational, where $\sqrt[3]{2}$ (the so-called **cube root** of 2) denotes the positive number so that its cube is 2. Assuming that there is always a cube root of any positive number, state and prove the analogue of Theorem 36.5 for cube roots.

Chapter 37

Pythagorean Triples

If we discuss number theory in school mathematics at all, it would be almost immoral not to at least say a few words about **Pythagorean triples**, which are by definition a triple of nonzero *whole numbers* a, b, c so that $a^2+b^2=c^2$. For example, $3^2+4^2=5^2$. The fact that we require a, b, c to be whole numbers is because it takes no effort to get (noninteger) real numbers a, b, and c to satisfy $a^2+b^2=c^2$. For example, we may let a, b be arbitrary numbers, e.g., $a=21$ and $b=124$, then $c=\sqrt{15817}$ would satisfy $a^2+b^2=c^2$ because $21^2+124^2=15817$. Without entering into any discussion of the geometric background of the Pythagorean theorem, we may safely assume that Pythagorean triples can be enjoyed on their own terms without the geometric baggage. For example, who would not be impressed—after years of seeing nothing but $3^2+4^2=5^2$ or $5^2+12^2=13^2$—to find that in fact, $7809^2+7760^2=11009^2$? (If you really want to understand how remarkable this equality is, you should try to verify it without the use of a calculator, which the Babylonians did back in 1700 BC or thereabouts. Could you have gotten this triple by guess-and-check? Probably not.)

We are going to write down a precise set of formulas to produce *all* Pythagorean triples. But first, we want to make sense of what is meant by "all" such triples. We do not try, nor is there any reason for trying, to get all Pythagorean triples in a literal sense, because if 3, 4, 5 (for example) is a Pythagorean triple, then so would $3n$, $4n$, and $5n$ be one for any whole number n. We call the latter a **multiple** of the triple 3, 4, 5. Clearly, if you already have 3, 4, and 5, then you do not need to be told how to get all its multiples. In order to eliminate such redundancy, we define a Pythagorean triple a, b, c to be **primitive** if $\gcd(a,b,c)=1$. (We have already defined the gcd of two numbers, and the gcd of three or more numbers is defined in the expected manner: the **gcd** of k whole numbers is the largest whole number that divides all k of them.) Our formulas will concern only primitive Pythagorean triples.

Theorem 37.1. *The collection of all primitive Pythagorean triples coincides with the triples given by $\{m^2 - n^2, 2mn, m^2 + n^2\}$, where m, n are whole numbers so that $m > n$, $\gcd(m, n) = 1$, and exactly one of m and n is even.*

For example, the triple 7809, 7760, and 11009 given earlier corresponds to $m = 97$ and $n = 40$. Observe also that with $m = 2$ and $n = 1$, we get the ubiquitous 3, 4, 5. With $m = 3$ and $n = 2$, we get 5, 12, 13. With $m = 4$ and $n = 3$, we get 7, 24, 25. And so on. With a calculator at hand, you should have fun getting hold of some Pythagorean triples that you have never seen before (which you can then use to impress your students!) The validity of this theorem can be demonstrated in an elementary way using standard arguments in number theory; see [**Dud78**, section 16].

Activity. (a) Check that the triple 9, 12, 15 forms a Pythagorean triple. (b) Check that there are no whole numbers m and n so that 9, 12, 15 are given by $\{m^2 - n^2, 2mn, m^2 + n^2\}$. (c) Does the triple 9, 12, 15 contradict Theorem 37.1?

Repeat (a), (b), and (c) for the triple 15, 36, 39.

What is more interesting is how one could have discovered this theorem, i.e., how does one arrive at the representation of Pythagorean triples as $\{m^2 - n^2, 2mn, m^2 + n^2\}$? Of course no one knows for sure, but with hindsight, one can say that there are at least three distinct ways that these formulas could have been found. It will take too long to describe them, but we can refer you to the following three sources. One way makes use of standard number-theoretic arguments such as those in this chapter, and this can be found in almost any textbook on number theory. A particularly approachable one is the above-cited volume, [**Dud78**]. A second approach makes use of the method due to Diophantus (circa AD 250) which may be described as an unusual way to represent the unit circle in the plane; this can be found in [**Lan88**, section 4 of Chapter 8] or [**Rot98**, pages 56–71]. A third approach is to retrace the steps of the Babylonians circa 1700 BC, and get these formulas by solving simultaneous linear equations in two variables; see [**Wu10a**, pages 103–108].

Exercises

1. Show that for a Pythagorean triple a, b, c, each of the following assertions gives exactly the same information: (i) it is a primitive triple; (ii) $\gcd(a, b) = 1$; (iii) $\gcd(b, c) = 1$; and (iv) $\gcd(a, c) = 1$. (*This means that if any one of (i)–(iv) is true, then all the rest are true.*) Moreover, if a, b, c is a primitive Pythagorean triple, and c is the largest of the three numbers, then c is odd, and exactly one of a and b is odd.

2. Is $\{14, 48, 50\}$ a Pythagorean triple? Is it primitive? Can it be expressed in the form of Theorem 37.1?

3. Do the same for $\{25, 60, 65\}$.

4. Produce a *primitive* Pythagorean triple so that each of the three numbers has at least five digits.

Part 5

More on Decimals

Part Preview

Up to this point, we have considered only finite decimals and therefore, with no fear of confusion, we simply abbreviated "finite decimal" to "decimal". Now, the time has come for us to take a serious look at the concept of a *decimal* itself; for this reason, we will henceforth draw a clear distinction between a decimal and a finite decimal. The definition of a *decimal* will be given in Chapter 41 below; it will be a number ≥ 0.

In Part 4, Chapter 36, we characterized those fractions which are equal to finite decimals (see section 36.2). Otherwise, our treatment of finite decimals has been confined in Part 2 as part of the study of fractions. This is as it should be. Let us not forget for one moment that a finite decimal is a fraction. However, for reasons given in Chapter 38 below, finite decimals also occupy a position of singular importance in science and in everyday life, and it would be appropriate to have a coherent account of finite decimals without the distractions of other aspects of fractions. We give such a summary in Chapter 39.

The two main purposes of Part 5 are to describe a particular formalism for the use of finite decimals in science, the so-called *scientific notation*, and to give a general introduction to *infinite decimals*, with an emphasis on the relationship between fractions and finite and infinite decimals (see Chapter 42).

Chapter 38

Why Finite Decimals Are Important

Recall that a finite decimal (understood to be ≥ 0) is a fraction whose denominator is a power of 10 (i.e., equal to 10^n where n is a positive integer), and that the introduction of the *decimal point* notation is nothing more than an abbreviation (see section 12.3). Thus we choose to write, e.g.,

$$\frac{287}{10^2} \text{ as } 2.87 \quad \text{and} \quad \frac{65}{10^5} \text{ as } 0.00065.$$

Among fractions, finite decimals are distinguished for at least three reasons.[1] One is that insofar as we are in a *decimal* numeral system, anything related to a power of 10 will automatically stand out. See, for example, the expanded form of a whole number in Chapter 1.

A second reason has been brought up in Chapter 20 in the discussion of percent: finite decimals provide a transparent standard of comparison. We can illustrate with a fictitious example. Suppose all cookies are sold in one-pound packages. If one package says, "Fat content $\frac{2}{9}$ lb" and another says, "Fat content $\frac{3}{11}$ lb", consumers would have nightmares dealing with this kind of information. Which has more fat? If, however, the same information is expressed as finite decimals, then one package would read "Fat content 0.22 lb" and the other one, "Fat content 0.27 lb". Now everybody would know that the second package contains more fat because it is easy to see (and one can even learn how to do this by rote) that $0.27 > 0.22$. The ease with which one can make comparisons between decimals is one of the virtues of the decimal notation.

---

[1]This short discussion is not self-contained, but the remaining chapters do not depend on it.

A bit more can be said about the preceding example. Our rewriting of $\frac{2}{9}$ and $\frac{3}{11}$ as 0.22 and 0.27, respectively, is only an approximation, as is well known. We learn from school that, by long division, $\frac{2}{9} = 0.2222222\cdots$ and $\frac{3}{11} = 0.27272727\cdots$. This is a mysterious process, to be sure, and we will give an explanation in Chapter 42 below. Assuming this for the moment, what we did above was to approximate the *repeating decimals*[2] $0.22222222\cdots$ and $0.27272727\cdots$ by the finite decimals 0.22 and 0.27, and the rule of the approximation that is being applied here is easily described: *round off to two decimal digits* (see section 10.1). What this demonstrates is the ease with which we can specify the comparison of two decimals, finite or infinite, up to a given number of decimal digits.

When we push this advantage of finite decimals over fractions to its logical conclusion, we would arrive at the principle behind *scientific notation*; see Chapter 40.

The third reason for the importance of finite decimals is deeper. As is well known, certain numbers such as π, $\frac{1}{\pi}$, and $\sqrt{7}$ are not expressible as fractions (for the case of $\sqrt{7}$, see Theorem 36.5 on page 483). They are examples of "infinite decimals" ($\sqrt{7} = 2.64575\cdots$, $\pi = 3.14159\cdots$, $\frac{1}{\pi} = 0.318309\cdots$) which will be taken up in Chapter 41 below. It so happens that "most" positive real numbers[3] are not equal to any fractions, much less fractions whose denominators are powers of 10. It is therefore a striking fact that every real number can be approximated arbitrarily closely by finite decimals,[4] so that in practical terms, the small collection of numbers—the finite decimals—are all we ever need use to express magnitudes, no matter how big or how small.

---

[2]A precise definition of "repeating decimal" will be given in Chapter 41.

[3]The meaning of "most" can be made very precise using more advanced concepts.

[4]This fact is closely related to the fact, mentioned already in Chapter 21, that every positive number can be arbitrarily approximated by fractions. For an elementary discussion of this and related issues, see [**Wu**, Chapter 17].

Chapter 39

Review of Finite Decimals

We summarize what we learned about finite decimals in Part 2. We use the notation

$$7.20 \text{ and } 0.0050 \quad \text{to denote} \quad \frac{720}{10^2} \text{ and } \frac{50}{10^4}, \text{ respectively.}$$

In general, the notational convention is that

> *If the denominator of the decimal fraction is* 10^n, *then the decimal point is placed in front of the n-th digit of the numerator of the decimal fraction* counting from the last displayed digit on the right, *which could be zero.*

A first remark is that the "0" in front of 0.4126, 0.0050, etc. is optional and is there merely to call attention to the presence of the decimal point. To understand this remark, it may be recalled that every whole number, e.g., 8126, 7, or 5, can be written with any number of zeroes in front of it, e.g., 00008126, 0007, or 00005 (see Chapter 1), but that we normally suppress the writing of these zeroes. The "0" in front of, e.g., 0.4126, then serves the purpose of highlighting the presence of the decimal point. This line of thinking then brings us to the second remark, which is that if we regard 5 as 005, then the placement of the decimal point in front of the third digit counting from 5 makes the notation 0.0050 for $\frac{50}{10^4}$ entirely logical. A third remark is that adding zeros after the decimal point does not change the decimal at all, e.g., 2.005000 is the same as 2.005 because by definition,

$$2.0050000 = \frac{20050000}{10^7},$$

so that by use of equivalent fractions,
$$2.0050000 = \frac{2005 \times 10^4}{10^3 \times 10^4} = \frac{2005}{10^3} = 2.005.$$

A final remark is about the equality of finite decimals. Because they are just fractions, we say **two finite decimals A and B are equal, $A = B$ in symbols,** if they are the same point on the number line. Compare page 197 in Chapter 12 on the equality of fractions.

Pedagogical Comments. *It is impossible to over-emphasize the fact that, by definition, a finite decimal is a fraction with 10^n as denominator. The lack of a clear concept of what a finite decimal is seems to be one of the most common failings among school textbooks.* **End of Pedagogical Comments.**

For a finite decimal such as 41.15, we call 41 the **integer part** of the decimal, i.e., the whole number which is to the left of the decimal point. Thus the integer part of 657.0008 is 657, and the integer part of 0.265 is 0. Note that because the 0 in 0.265 is optional, the integer part of .265 is still 0; same for .028.

Two special features of the integer part of a finite decimal should be pointed out. One is that *every finite decimal with a nonzero decimal part can also be written as a mixed number with the integer part of the decimal appearing as the whole number in the mixed number*, e.g.,
$$41.15 = 41\frac{15}{10^2}, \qquad 6.028 = 6\frac{28}{10^3}.$$

To see why this is so, *we simply appeal to the definition of each of the concepts involved*:
$$41.15 = \frac{4115}{10^2} = \frac{4100 + 15}{10^2} = 41 + \frac{15}{10^2} = 41\frac{15}{10^2},$$
$$6.028 = \frac{6028}{10^3} = \frac{6000 + 28}{10^3} = 6 + \frac{28}{10^3} = 6\frac{28}{10^3}.$$

Please note that this conversion between finite decimals and mixed numbers has been achieved *strictly by applying mathematical reasoning to the relevant definitions* and is not a rote procedure sanctified by tradition.

A second special feature of the integer part of a finite decimal is that *every finite decimal can be written in exactly one way as the sum of a whole number and a decimal whose integer part is zero.* We can be more precise: the whole number in question is the integer part of the decimal. For example:
$$41.15 = 41\frac{15}{10^2} = 41 + \frac{15}{10^2} = 41 + 0.15,$$
$$6.028 = 6\frac{28}{10^3} = 6 + \frac{28}{10^3} = 6 + 0.028,$$
$$0.0752 = 0 + 0.0752.$$

39. Review of Finite Decimals

Please note again that the preceding equalities are not mindless symbolic manipulations; every step has been precisely explained. Be sure that your students understand this.

You may wonder why we are so insistent on *explaining* these notational matters. It is because many textbooks tend to give the impression that something like $4.15 = 4 + 0.15$ is just a matter of "taking the notation apart"; it is so tempting to regard it as "detaching the symbol 4.15 into two parts, 4 and 0.15". But mathematical symbols are supposed to convey a mathematical thought and cannot be put together or dismantled at will like a LEGO product. If students get the idea that that symbols can be randomly taken apart or put together, the next step for them would be to do $\frac{4}{5} + \frac{1}{5} = \frac{4+1}{5+5} = \frac{5}{10}$.

> Why do we say that the writing of a decimal as the sum of a whole number and a decimal with zero integer part can be done in only one way? A picture would make it clear. Take 41.15, for instance. It is bigger than 41 because $41.15 = 41 + 0.15$ so that (by the definition of the addition of fractions) the segment of length 41.15 is longer than the segment of length 41. On the other hand, 41.15 is less than 42 because $0.15 = \frac{15}{100} < 1$ so that $41.15 = 41 + 0.15 < 41 + 1 = 42$. Therefore the number 41.15 is between 41 and 42, as shown:

$$\begin{array}{ccc} 41 & 41.15 & 42 \\ \vdash & \vdash \kern-4pt \rule[0.5ex]{8cm}{0.4pt} \kern-4pt \dashv \end{array}$$

> It is now clear that if we want to write 41.15 as $n + d$, where n is a whole number and d is a decimal with zero integer part, then we are asking for a way to break up the segment $[0, 41.15]$ as the concatenation of two segments, one from 0 to n where n is a *whole number* and the other of length d which is less than 1. The picture tells us that $n = 41$ and $d = 0.15$.

Finite decimals enjoy a very pleasant property: they can be added, subtracted, and multiplied *essentially* the same way as whole numbers, and to a large extent they can also be divided as if they were whole numbers. Before going into the details, we need to recall yet another piece of notation (see section 31.2). Instead of writing

$$\frac{1}{10}, \quad \frac{1}{10^2}, \quad \frac{1}{10^3}, \quad \ldots \quad \frac{1}{10^n}, \quad \ldots,$$

we shall henceforth write

$$10^{-1}, \quad 10^{-2}, \quad 10^{-3}, \quad \ldots \quad 10^{-n}, \quad \ldots.$$

Using this notation, we may rewrite

$$41.15 = 4115 \times \frac{1}{10^2} = 4115 \times 10^{-2},$$

$$0.8126 = 8126 \times \frac{1}{10^4} = 8126 \times 10^{-4},$$

$$11.4608 = 114608 \times \frac{1}{10^4} = 114608 \times 10^{-4}, \text{ etc.},$$

and in general,

$$\frac{a}{10^n} = a \times 10^{-n}$$

for any whole number a and for any nonzero whole number n. The advantage of this notation is that one of the so-called **laws of exponents**, in the form of equation (1.4) on page 30, can now be extended to *integer* exponents. Thus we claim that, for any *integers m and n*, we have

(39.1) $$10^m \times 10^n = 10^{m+n}.$$

In proving equation (1.4), we have already shown (39.1) to be true when m and n are positive. So it remains to treat the remaining cases where (a) $m > 0$ but $n < 0$, (b) $m < 0$ but $n > 0$, or (c) both $m < 0$ and $n < 0$. The reasoning in any of these three cases is not particularly inspiring, so we will just check case (b) as the other two cases are similar. If $m = -8$ and $n = 3$, then by definition,

$$10^{-8} \times 10^3 = \frac{10^3}{10^8} = \frac{10^3}{10^5 \times 10^3} = \frac{1}{10^5} = 10^{-8+3}.$$

If $m = -5$ and $n = 9$, then

$$10^{-5} \times 10^9 = \frac{10^9}{10^5} = \frac{10^5 \times 10^4}{10^5} = 10^4 = 10^{-5+9}.$$

In general, if $m = -k$ for a whole number k, either $k > n$ or $k \leq n$. If $k > n$, we may write $k = n + p$ for some nonzero whole number p. Then

$$10^m \times 10^n = \frac{1}{10^k} \times 10^n = \frac{1}{10^{n+p}} \times 10^n$$

$$= \frac{10^n}{10^n \times 10^p} \qquad \text{(by (1.4) in section 1.6)}$$

$$= \frac{1}{10^p} = 10^{-p}$$

$$= 10^{n-k} = 10^{n+m}.$$

If $k \leq n$, let $n = k + q$ for some whole number q. Then

$$10^m \times 10^n = 10^{-k} \times 10^{k+q}$$

$$= \frac{10^k \times 10^q}{10^k} \qquad \text{(by (1.4) in section 1.6)}$$

$$= 10^q = 10^{n-k} = 10^{n+m}.$$

39. Review of Finite Decimals

So in both cases, (39.1) holds.

Now we return to arithmetic operations on finite decimals. Let us first take up addition. How do we add, for instance, $0.135 + 0.0486$ or $6.0053 + 20.411$? The rule given in most school texts follows.

> Align the finite decimals by the decimal points, add them as if they were whole numbers, and then pull down the decimal point to convert the sum back to a decimal.

To illustrate, we would get

(39.2) $$0.135 + 0.0486 = 0.1836$$

because according to the above rule:

(39.3)
$$\begin{array}{r} 0.\ 1\ 3\ 5 \\ +\ \ 0.\ 0\ 4\ 8\ 6 \\ \hline 0.\ 1\ 8\ 3\ 6 \end{array}$$

(As in all whole number algorithms, the blank space in the upper right corner is always understood to be a 0.)

This simple rule is easily explained *once we know the definition of a finite decimal* and the distributive law of fractions, as follows.

$$\begin{aligned} 0.135 + 0.0486 &= 0.1350 + 0.0486 \\ &= 1350 \times 10^{-4} + 486 \times 10^{-4} \\ &= (1350 + 486) \times 10^{-4} \qquad (*) \\ &= 1836 \times 10^{-4} = 0.1836 \end{aligned}$$

We therefore see that $(*)$ corresponds to (39.3) in regard to aligning the decimals by the decimal points and adding decimals as if they were whole numbers. The reasoning is of course entirely general and is applicable to the addition of any number of decimals.

The subtraction of finite decimals is handled in a similar manner: line up the decimals by the decimal point, perform the subtraction algorithm on the two numbers as if they were whole numbers, and pull down the decimal at the end. For example, $3.145 - 2.8675$ is computed as follows (remember that $3.145 = 3.1450$).

(39.4)
$$\begin{array}{r} 3.\ 1\ 4\ 5 \\ -\ \ 2.\ 8\ 6\ 7\ 5 \\ \hline 0.\ 2\ 7\ 7\ 5 \end{array}$$

(As before, the blank space in the upper right corner is understood to be a 0.)

The explanation is the usual one:
$$\begin{aligned}3.145 - 2.8675 &= 3.1450 - 2.8675 \\ &= 31450 \times 10^{-4} - 28675 \times 10^{-4} \\ &= (31450 - 28675) \times 10^{-4} \quad (\dagger) \\ &= 2775 \times 10^{-4} = 0.2775,\end{aligned}$$
and ($\dagger$) corresponds to (39.4).

Multiplication may be the simplest of the four arithmetic operations for finite decimals. The basic fact is that every finite decimal such as 11.4608 can be written as the product of a negative power of 10 and a whole number. Some common sense is involved in writing this. For example, there is no point in writing
$$11.4608 = 11460800 \times 10^{-6},$$
when it is simpler to just write
$$11.4608 = 114608 \times 10^{-4}.$$
But no matter. We use this fact to reduce decimal multiplication to one involving only whole numbers. For example,
$$\begin{aligned}11.4608 \times 0.397 &= 114608 \times 10^{-4} \times 397 \times 10^{-3} \\ &= 114608 \times 397 \times 10^{-(4+3)}.\end{aligned}$$
Now 114608×397 can be computed with the whole number multiplication algorithm, which equals 45499376, so that when multiplied by 10^{-7} ($= 10^{-(4+3)}$), it becomes:
$$11.4608 \times 0.397 = 4.5499376.$$
This is a typical illustration of the general rule:

> To multiply two finite decimals with n and m decimal digits,[1] respectively, simply multiply the two decimals as whole numbers by ignoring the decimal point, and then put a decimal point in front of the $(m+n)$-th digit from the right of the product.

The division of two finite decimals does not always lead to a finite decimal. Let us separate this statement into two parts. First,

> the division of two finite decimals is always equal to the division of two whole numbers.

This is clear because, for example,
$$\frac{0.0045}{0.14} = \frac{45 \times 10^{-4}}{1400 \times 10^{-4}} = \frac{45}{1400}.$$

---

[1] See section 12.2 for the terminology of *decimal digit*.

39. Review of Finite Decimals

Part two then says that *the division of two whole numbers is not always a finite decimal*. This is because of Theorem 36.2 on page 479. For example, the preceding division of 45 by 1400 cannot be a finite decimal because

$$\frac{45}{1400} = \frac{3^2 \times 5}{2^3 \times 5^2 \times 7} = \frac{3^2}{2^3 \times 5 \times 7}.$$

Now the last fraction is in reduced form and since the prime 7 is in the denominator, the above-mentioned theorem implies that $\frac{45}{1400}$ is not a finite decimal.

The discussion of the division of finite decimals in general will be completed at the end of Chapter 42 in the context of infinite decimals.

Exercises

1. Give reason for each step of the derivation that $41.15 = 41 + 0.15$ and $6.028 = 6 + 0.028$.

2. (You may use a four-function calculator for parts (f)–(i) below.)
   (a) $0.842 + 0.7 + 6.00093 = ?$ (b) $4.1 + 2.0806 + 1234 = ?$ (c) $5 - 3.7482 = ?$ (d) $76.0021 - 4.8 = ?$ (e) $3.52 \times 127 - 1.52 \times 127 = ?$ (f) $0.32 \times 16 + 64 \times 0.075 = ?$ (g) $12.3 \times 5.6 - 0.015 \times 8 = ?$ (h) $0.32 \times 15 + 64 \times 0.075 = ?$ (i) $4.21 \times (56.7 - 29.004) = ?$

3. Imagine that you are doing parts (f)–(i) of the preceding exercise for your students. How would you explain your calculations to them step-by-step?

4. Compute the following:
   (a) $\dfrac{52.2}{0.006}$. (b) $\dfrac{2.08}{0.64}$. (c) $\dfrac{40.5}{27}$.

5. Verify case (a) and case (c) of equation (39.1) on page 500.

Chapter 40

Scientific Notation

In science, the magnitude of a quantity is often of overriding importance. This chapter introduces the so-called scientific notation, which is a particular way of exhibiting the magnitude of a positive number that makes it easy to compare the magnitudes of any two positive numbers.[1]

The sections are as follows:

Comparing Finite Decimals

Scientific Notation

40.1. Comparing Finite Decimals

We begin with a discussion of the comparison of two finite decimals. To this end, we recall the concept of *decimal digit* introduced in Chapter 10 (page 142): the n-th digit after the decimal point of a finite decimal is called the *n-th decimal digit*. For example, the third decimal digit of 5.48706 is 7 and the fifth decimal digit is 6. Given two decimals A and B, by adding zeros after the decimal point to one of them if necessary, we may always assume that both decimals have the same number of decimal digits. For example, if we are to compare 0.4 with 0.38972, we compare instead 0.40000 with 0.38972 so that both decimals now have 5 decimal digits.

> From now on, we automatically assume that any two finite decimals under comparison have the same number of decimal digits. (See FFFP for decimals on page 214.)

---

[1] By a trivial modification, we can also write negative numbers in scientific notation. The added complication serves no purpose in the present context, however.

If finite decimals A and B are given, then for some whole number k, $A \times 10^k$ and $B \times 10^k$ are both whole numbers. The basic rule of the comparison of finite decimals is the following.

Let finite decimals A and B be given. If the whole numbers $A \times 10^k$ and $B \times 10^k$ satisfy

$$A \times 10^k < B \times 10^k,$$

then the decimals A and B satisfy

$$A < B.$$

The proof is a matter of knowing how to compare fractions with the same denominator:

$$A = \frac{A \times 10^k}{10^k} < \frac{B \times 10^k}{10^k} = B.$$

In specific situations, the basic rule can be made simpler. For example, if two finite decimals have unequal integer parts, then

the decimal with the larger integer part is the larger decimal.

We will not give a proof of this assertion in general because this is another one of those situations where the abstract symbolic notation is a hindrance rather than a help in human communication (cf. the similar situation in the discussion of the multiplication algorithm for whole numbers in Chapter 6). We will look at one or two concrete cases instead, and the implicit assumption is that you will see the general reasoning as a result. Let us go through, step-by-step, the reason why knowing that $120 < 121$ allows us to conclude $120.89 < 121.4$. We have $120.89 + 0.11 = 121$, so $120.89 < 121$. Thus the decimal 120.89 is less than the whole number 121. But $121 < 121.4$ because $121 + 0.4 = 121.4$. So putting all this together, we get

$$120.89 < 121 < 121.4.$$

Next, let us see why knowing $76 < 79$ allows us to conclude that $76.9 < 79.0001$. We know $76.9 < 77$ because $76.9 + 0.1 = 77$. Now $77 < 79$, so that $76.9 < 79$. Finally, $79 + 0.0001 = 79.0001$, so $79 < 79.0001$. Putting things together, we get

$$76.9 < 77 < 79 < 79.0001.$$

Next, consider the comparison of two finite decimals A and B with equal integer parts. We shall soon see that the most important case is the comparison of two finite decimals with 0 integer part, e.g., 0.2341927 and 0.2342. In this case, the rule is

Suppose two finite decimals have equal integer parts. If we compare the decimal digits of both decimals one-by-one, from left to right, then there will be a first time when the decimal digit of one

of them, say B, is bigger than the corresponding decimal digit of A. Then $A < B$.

For example, let us compare 0.2341927 and 0.2342. The first three decimal digits of both are equal, being 2, 3, and 4, respectively. However, the fourth decimal digit of 0.2342 is 2, whereas the fourth decimal digit of 0.2341927 is 1, so the rule says $0.2341927 < 0.2342$. The reason is the following:

$$0.2341927 = \frac{2341927}{10^7} < \frac{2342000}{10^7} = 0.2342,$$

where the middle inequality follows from a known fact about comparing whole numbers (item (ii) on page 27).

In general, let us say A and B agree on the first three decimal digits but that B has the larger fourth decimal digit. Then the two *whole numbers* $A \times 10^k$ and $B \times 10^k$ agree on the first three digits (counting from the left) but B has the larger fourth digit. By (ii) on page 27, $A \times 10^k < B \times 10^k$, and therefore

$$A = \frac{A \times 10^k}{10^k} < \frac{B \times 10^k}{10^k} = B.$$

We now round out the picture by explaining why *comparing two finite decimals with equal integer parts is reduced to the comparison of two decimals with zero integer parts* in general. Let A and B be two such decimals and we write $A = a + A'$ and $B = a + B'$, where a is a whole number and A' and B' are two decimals with zero integer parts. We may use the method of the preceding paragraph to determine which of the two, A' and B' is bigger. Let us say $A' < B'$. Then it follows that

$$A = a + A' < a + B' = B.$$

Thus knowing $A' < B'$ determines $A < B$, as claimed.

40.2. Scientific Notation

We now translate this information about comparing finite decimals into the language of scientific notation. A finite decimal is said to be written in **scientific notation** if it is expressed as a product $a \times 10^k$, where a is a finite decimal satisfying $1 \leq a < 10$, and k is an integer. The meaning of the condition $1 \leq a < 10$ on a is that the integer part of a is 1, 2, ..., or 9. Here are some examples of numbers in scientific notation:

$$751.2 = 7.512 \times 10^2,$$
$$5.7 = 5.7 \; (= 5.7 \times 10^0),$$
$$0.0000287 = 2.87 \times 10^{-5}.$$

The virtue of scientific notation is that the exponent of 10 clearly displays the rough magnitude of a number. For instance, if the exponent k in $a \times 10^k$

is positive, then the number is a $(k+1)$-digit number (because 10^k is the smallest whole number with $(k+1)$ digits). For example, 5.149×10^7 is roughly $50{,}000{,}000$. On the other hand, if k is a negative integer, then the first nonzero decimal digit of $a \times 10^k$ (going from left to right, as usual) is the $(-k)$-th decimal digit. For example, given 7.624×10^{-4}. It is equal to 0.000724 and therefore the first nonzero decimal digit is the fourth (and it is 7, of course).

Now we go further. We want to recognize at a glance which of two numbers in scientific notation is bigger. If two finite decimals in scientific notation, $a \times 10^k$ and $b \times 10^\ell$, are given, the following rule for comparing them is of interest:

(40.1)   If $k = \ell$, then $a < b$ implies $a \times 10^k < b \times 10^k$,

but if $k < \ell$, then $a \times 10^k < b \times 10^\ell$.

In connection with the assumption $k < \ell$, recall that the integers are ordered in the following way: $\cdots -6 < -5 < -4 < -3 < -2 < -1 < 0 < \cdots$.

First, consider the case of $k = \ell$. The rule says, for example, that because $3.9 < 4.002$, it must be true that $3.9 \times 10^{12} < 4.002 \times 10^{12}$ and $3.9 \times 10^{-8} < 4.002 \times 10^{-8}$. Clearly, we can obtain these conclusions by multiplying both sides of the inequality $3.9 < 4.002$ by 10^{12} and 10^{-8}, respectively (see inequality (D) on page 422). The reasoning shows that, in general, if two finite decimals in scientific notation, $a \times 10^k$ and $b \times 10^k$, are given, and if $a < b$, then $a \times 10^k < b \times 10^k$. This proves the first part of (40.1).

Next, consider the case $k < \ell$ in (40.1). It says that, because $34 < 35$, $8.5 \times 10^{34} < 2.3 \times 10^{35}$. Let us see why this is true. Although $8.5 > 2.3$, we know that $10^{35} = 10 \times 10^{34}$, so that $2.3 \times 10^{35} = 2.3 \times 10 \times 10^{34}$, which is 23×10^{34}. Therefore,

$$8.5 \times 10^{34} \;<\; 23 \times 10^{34} \;=\; 2.3 \times 10^{35}.$$

Another example along the same lines is this: $9.2 \times 10^{-8} < 1.1 \times 10^{-7}$ because $1.1 \times 10^{-7} = 1.1 \times 10 \times 10^{-8} = 11 \times 10^{-8}$, so that

$$9.2 \times 10^{-8} \;<\; 11 \times 10^{-8} \;=\; 1.1 \times 10^{-7}.$$

The reason in general is almost as simple. Let two finite decimals in scientific notation, $a \times 10^k$ and $b \times 10^\ell$, be given so that $k < \ell$. We will prove that $a \times 10^k < b \times 10^\ell$. First note that $a < b \times 10$, because $1 \leq b$ implies $10 \leq b \times 10$, so that combining this with $a < 10$, we obtain $a < b \times 10$. Multiply the last inequality by the fraction[2] 10^k to get $a \times 10^k < b \times 10^{k+1}$ (see inequality (D) on page 422). Now the assumption that $k < \ell$ means

---

[2]If k is a whole number, then of course 10^k is a whole number. But k could be negative, say -6, then 10^k would be the fraction $1/10^6$. Therefore we have to refer to 10^k in general as a "fraction" (also recall the fact that every whole number is a fraction).

40.2. Scientific Notation

that $k+1 \leq \ell$, so that $10^{k+1} \leq 10^\ell$. Therefore
$$a \times 10^k \ < \ b \times 10^{k+1} \ \leq \ b \times 10^\ell.$$
Thus $a \times 10^k < b \times 10^\ell$. This proves the second part of (40.1).

Activity. Which is bigger: 1.92×10^{-7} or 2.004×10^{-7}? 1.92×10^{-6} or 2.004×10^{-7}? 1.92×10^6 or 2.004×10^7?

Exercises

1. Compare which is bigger: 24.8799 or 25.000001. Explain each step.
2. Order the following numbers: $\frac{8}{7}$, 0.9998, $\frac{6}{5}$, 1.13.
3. Given a finite decimal A, show why there is a whole number n so that $10^{-n} < A$.
4. Write each of the following in scientific notation.
   (a) 23456
   (b) 0.000054
   (c) 386.2×10^{-7}
   (d) $46 + \frac{2}{1000} + \frac{7}{100000}$
5. Which is bigger?
   (a) 2.8×10^{-6} or 3.1×10^{-6}
   (b) 2.8×10^{-6} or 8.2×10^{-8}
   (c) 5.221×10^4 or 4.6×10^5
   (d) 1.0002×10^{-3} or 1.08×10^{-2}

Chapter 41

Decimals

A finite decimal has a complete expanded form. The general concept of a *decimal* is a natural extension of the complete expanded form. Among decimals, the *repeating decimals* are distinguished because they are the fractions. In this chapter, we prove that every repeating decimal is a fraction. The converse will be the subject of the next chapter.

The sections are as follows:

Review of Division-with-Remainder

Decimals and Infinite Decimals

Repeating Decimals

41.1. Review of Division-with-Remainder

We first recall how to write a finite decimal in its "complete expanded form". Given 35.2647, for example, we use the expanded form of a whole number (sections 1.2 and 1.6 of Chapter 1) to get

$$35.2647 = \frac{352647}{10^4}$$
$$= \frac{(3 \times 10^5) + (5 \times 10^4) + (2 \times 10^3) + (6 \times 10^2) + (4 \times 10^1) + (7 \times 10^0)}{10^4}.$$

By virtue of equation (39.1), we get

$$35.2647 = (3 \times 10^1) + (5 \times 10^0)$$
$$+ (2 \times 10^{-1}) + (6 \times 10^{-2}) + (4 \times 10^{-3}) + (7 \times 10^{-4})$$
$$= 35 + \frac{2}{10} + \frac{6}{10^2} + \frac{4}{10^3} + \frac{7}{10^4}.$$

This simple reasoning applies to any finite decimal, and we see that every finite decimal can be expressed as a sum of products of *single*-digit whole numbers and integer powers of 10. For example,

$$0.00004975 = (4 \times 10^{-5}) + (9 \times 10^{-6}) + (7 \times 10^{-7}) + (5 \times 10^{-8})$$
$$= \frac{0}{10} + \cdots + \frac{0}{10^4} + \frac{4}{10^5} + \frac{9}{10^6} + \frac{7}{10^7} + \frac{5}{10^8}.$$

Such an expression for a finite decimal is called its *complete expanded form* (section 14.2). It generalizes the expanded form of a whole number introduced in Chapter 1. The complete expanded form is a natural extension of the concept of place value for whole numbers. Notice that the digits of the integer part of such a finite decimal are associated with 10^m for a whole number m, whereas the decimal digits are associated with 10^k for a negative integer k. The exponent of 10 in each case is the **generalized place value** of the digit. With this extended concept of place value for finite decimals, the algorithms for the four arithmetic operations of finite decimals can now be understood in terms of the corresponding operations for whole numbers. For example, *the alignment of the finite decimals by their decimal point in the addition or subtraction algorithms for finite decimals is merely the addition of digits with the same generalized place value.*

We would like to draw attention to the careful logical sequencing of the concepts in the preceding paragraphs. In order to talk about the generalized place value of a finite decimal, we first defined the complete expanded form of the decimal. Since the latter is a sum of fractions, the concept of the addition of fractions had to be in place before we could make sense of the complete expanded form of a decimal. This then harks back to a recurrent theme in Part 2, to the effect that in order to make *mathematical* sense of decimals, one must begin with the study of fractions. Against this background, we can now reflect on the usual practice of discussing the "extended place value of finite decimals" without first defining the complete expanded form of a decimal, or sometimes even without defining what a fraction is; we realize that this way of dealing with finite decimals makes no sense at all and should be avoided at all costs. Of course this comment is about the logical development of fractions and decimals that we should be teaching students in grade 5 and beyond. Should we talk about the "extended place value of finite decimals" *informally* with students in the primary grades in the context of money (dollars, dimes, and pennies)? Of course, provided we can make sure they understand that we will bring mathematical closure to such a discussion later on.

The complete expanded form of a finite decimal brings conceptual closure to one aspect of the discussion of our decimal numeral system started in Chapter 1. We mentioned there that the use of place value allows for the easy writing of arbitrarily large numbers. We now see that it also allows for

41.2. Decimals and Infinite Decimals

the writing of arbitrarily small numbers, namely, 10^{-n} gets arbitrarily small as n gets arbitrarily large. For the justification of the latter assertion, see Exercise 3 on page 510.

41.2. Decimals and Infinite Decimals

There is a deeper reason for introducing the complete expanded form of a finite decimal: it leads to the general concept of a *decimal*. Let us consider a finite decimal from a different point of view. If we have a whole number w and a finite sequence of single-digit numbers, *not all of them equal to zero*,

$$a_1, a_2, a_3, \ldots, a_n$$

(thus each a_i is equal to 0, 1, 2, ..., or 9), then we get a finite decimal to be denoted by

$$w.a_1 a_2 \cdots a_n ,$$

which is by definition the number

$$w + \frac{a_1}{10} + \frac{a_2}{10^2} + \cdots + \frac{a_n}{10^n}.$$

If $w = 35$, $n = 4$ and $a_1 = 2$, $a_2 = 6$, $a_3 = 4$, and $a_4 = 7$, then we retrieve the finite decimal 35.2647 above. Now suppose we have a whole number w and an *infinite* sequence of single-digit numbers

$$a_1, a_2, a_3, \ldots, a_n, a_{n+1}, \ldots.$$

It can happen that beyond some positive integer m, all the a_{m+1}, a_{m+2}, $a_{m+3}, \ldots$ are equal to 0. In that event, we can again form the finite decimal

$$w + \frac{a_1}{10} + \frac{a_2}{10^2} + \cdots + \frac{a_m}{10^m}.$$

In general, though, an infinite number of the $a_1, a_2, a_3, \ldots$ will be nonzero, and it would make no sense to write down

$$w + \frac{a_1}{10} + \frac{a_2}{10^2} + \cdots + \frac{a_n}{10^n} + \frac{a_{n+1}}{10^{n+1}} + \cdots$$

because there is no such thing as "adding an infinite number of whole numbers". Addition is only meaningful for a finite collection of numbers. Nevertheless, we can form the corresponding *sequence* of finite decimals, $s_1, s_2, s_3, \ldots$ defined as follows.

(41.1)
$$\begin{cases} s_1 &= w.a_1 \\ s_2 &= w.a_1 a_2 \\ s_3 &= w.a_1 a_2 a_3 \\ &\vdots \\ s_n &= w.a_1 a_2 a_3 \cdots a_n \\ &\vdots \end{cases}$$

The standard notation for a sequence of numbers s_1, s_2, s_3, ... is $\{s_n\}$. The surprising thing is that we can still get a number out of this situation because there is the following general theorem, Theorem 41.1, in advanced mathematics. For its statement, let $\{s_n\}$ be an arbitrary sequence of numbers (i.e., not necessarily a sequence of finite decimals such as (41.1) above). We say a number s is the **limit of** $\{s_n\}$ if the distance (on the number line) between each s_n and s gets closer and closer to 0 as $n \to \infty$. In this case, we also say that the sequence $\{s_n\}$ **converges to s**.

At this point, we are making an excursion into higher mathematics, so we cannot be as precise as need be, e.g., we have offered the statement "if the distance (on the number line) between each s_n and s gets closer and closer to 0 as $n \to \infty$" to be taken in an intuitive sense without also giving a precise definition. In addition, we will ask you to accept on faith Theorems 41.1, 41.2, and 41.3 below without proof. The reason for such omissions is that these concepts and statements are highly believable while, at the same time, their full explanations require a level of technical sophistication that is not appropriate for K–12. With this understood, we have

Theorem 41.1. *Let w be a whole number, and let a_1, a_2, a_3, ... be an infinite collection of single-digit numbers. Let the sequence of finite decimals $\{s_n\}$ be defined as in (41.1). Then there is always a unique positive number s which is the limit of $\{s_n\}$.*

With w and a_1, a_2, a_3, ... as in Theorem 41.1, we now define the **decimal**

$$w.a_1 a_2 a_3 \cdots$$

to be the limit s of the sequence of finite decimals $\{s_n\}$ in (41.1). If an infinite number of the a_i's are nonzero, we will sometimes say $w.a_1 a_2 a_3 \cdots$ is an **infinite decimal** for emphasis. We call the sequence $\{s_n\}$ in (41.1) **the sequence of finite decimals associated with the decimal $w.a_1 a_2 a_3 \cdots$**.

We see that the symbol $w.a_1 a_2 a_3 \cdots$ stands for the limit of a sequence (the sequence in (41.1), to be precise) and is not "a number with an infinite number of decimal digits".

Following the case of finite decimals, we will continue to refer to a_n as the ***n*-th decimal digit** of the decimal $w.a_1 a_2 a_3 \cdots$.

As an example, consider the following familiar decimal expansion of π:

(41.2) $\qquad \pi \;=\; 3.14159\ 26535\ 89793\ 23846\ 26433\ 83279\ 50288 \cdots.$

41.2. Decimals and Infinite Decimals

We now know what this means. Namely, if we introduce a sequence of finite decimals $\{t_n\}$ so that

$$t_1 = 3.1$$
$$t_2 = 3.14$$
$$t_3 = 3.141$$
$$t_4 = 3.1415$$
$$t_5 = 3.14159$$
$$t_6 = 3.141592$$
$$\vdots \quad \vdots$$
$$t_{27} = 3.14159\ 26535\ 89793\ 23846\ 26433\ 83$$
$$\vdots \quad \vdots$$

and in general, for any positive integer n,

(41.3) $\begin{cases} t_n = \text{the finite decimal obtained from (41.2)} \\ \quad\quad \text{by discarding all decimal digits after the } n\text{-th,} \end{cases}$

then the number π is the limit of this sequence $\{t_n\}$.

Decimals, such as those associated with π above, appear in school textbooks around grade 7, usually without any instruction as to what they mean beyond the spurious statement that "you just keep adding more and more terms". This lack of information about a piece of seductive notation such as $3.14159\cdots$ naturally sparks heated debates that, as a rule, make no sense, e.g., why is $0.9999999\cdots$ equal to 1? The reason such debates make no sense is that it is impossible to discuss mathematics without *precise definitions*. If we want to say the two numbers $0.9999999\cdots$ and 1 are equal, then we first need a definition of a number, and in addition, *a definition of* $0.9999999\cdots$ *as a number*, and finally, the definition of when two numbers are equal. Because school textbooks do not discuss any of these definitions, such debates are bound to generate plenty of heat but no light.

We do have all the relevant definitions at this point. Let us see, for example, why the two numbers, $0.999999\cdots$ and 1, which "look" so different, could be the same number. This is because we have to look at the *limit* of the sequence of finite decimals associated with $0.999999\cdots$ and *not* the symbol $0.999999\cdots$ itself. It so happens that the limit of the sequence 0.9, 0.99, 0.999, 0.9999, ... is easily seen to be exactly 1 (do this as an exercise), and this is why $0.999999\cdots = 1$. We will have more to say about this equality at the end of this chapter.

Now that we have firmly established a decimal as a number (a point on the number line), we have to deal with the nitty gritty of computations with decimals. By this we mean that if we take the *formal* view that a decimal is just a number with an infinite number of digits, can we compute with these digits the same way we compute with whole numbers or finite decimals? In general, the answer is no, because the arithmetic of infinite decimals is complicated and cannot be encoded in a few simple rules. The simplest way to understand this difficulty is to consider the multiplication of π by a single-digit number. Let us say we want the first five decimal digits of 2π. We know $\pi = 3.14159265358979323\cdots$. If we take the value of π up to five decimal digits, 3.14159, for computation, we get $2 \times 3.14159 = 6.28318$ while $2\pi = 6.2831853071\cdots$. So we get the correct answer. But is this a correct algorithm in general? Let us see. If we try to get the first five decimal digits of 9π, for example, then $9 \times 3.14159 = 28.27431$, while the correct value is $9\pi = 28.274333882\cdots$. We get an error in the fifth decimal digit this time. Let us use six decimal digits of the value of π for computation to see if we do better: $9 \times 3.141592 = 28.274328$, still an error in the fifth decimal digit. It turns out we have to use seven decimal digits of the true value of π to get the correct value of 9π.

An even more revealing example is to compute π^2. Because the standard multiplication algorithm starts from the right, we are faced with the first obstacle of not having a "last" digit as the decimal digits of π are infinite in number. Let us try to circumvent it by computing the squares of successive finite decimal approximations:

$$3.14^2 = 9.8596,$$
$$3.141^2 = 9.865881,$$
$$3.1415^2 = 9.86902225,$$
$$3.14159^2 = 9.8695877281,$$
$$3.141592^2 = 9.86960029446,$$
$$3.1415926^2 = 9.86960406437476, \quad \text{etc.}$$

Suppose we accept that the correct value of π^2 is

$$\pi^2 = 9.869604401089358\cdots.$$

Comparing with this value, we see that if we want to get the first four digits of π^2 (9.869), we can just compute 3.1415^2, but if we want seven digits of π^2 (which is 9.869604), we would have to use at least the first eight digits of π and compute 3.1415926^2.

Now imagine that we have to compute $A^3(B+\frac{C}{D})$, where A, B, C, and D are infinite decimals. How one ensures an answer that is correct up to the first 100 decimal digits then becomes something of a nightmare. You can see that there would be *no simple and efficient digit-by-digit algorithm* for computing with infinite decimals as in the case of whole numbers.

Having described the difficulty of doing arithmetic with infinite decimals, we hope you will now appreciate *anything* that we can do with decimals.

41.2. Decimals and Infinite Decimals

For example, you may not consider the following to be very exciting, but you would be relieved to know that at least it is true:

$$w.a_1a_2a_3\cdots = w + 0.a_1a_2a_3\cdots.$$

Let us make sure that this equality makes sense: Theorem 41.1 guarantees that each of $w.a_1a_2a_3\cdots$ and $0.a_1a_2a_3\cdots$ is a number, so the above is a statement about the equality of numbers. Of course, the equality is trivial to verify if all the a_i are equal to 0 for i bigger than a certain integer m, because in that case this would be an assertion about finite decimals.

The following fact gives another operation on decimals that is universally valid. The proof is not suitable for school mathematics.

Theorem 41.2. *If we multiply the decimal $w.a_1a_2a_3\cdots$ by a power of 10, say 10^k for an integer k, we get the decimal which is obtained by shifting the decimal point of $w.a_1a_2a_3\cdots$ k places to the right or k places to the left, depending on whether k is positive or negative, respectively.*

An explanation of the last statement in Theorem 41.2 is in order. We first explain what it means by an example. Suppose we have an infinite decimal $x = 0.1246\cdots$. Then according to Theorem 41.2, $10^3 x = 124.6\cdots$ while $10^{-4}x = 0.00001246\cdots$, etc. Now, it would seem that the usual distributive law suffices to "distribute" 10^k across the "infinite sum" in (41.1) to each term. For example, if x is a finite decimal, $x = 0.1246$, then

$$\begin{aligned} 10^3 x &= 10^3 \times \left((1 \times 10^{-1}) + (2 \times 10^{-2}) + (4 \times 10^{-3}) + (6 \times 10^{-4})\right) \\ &= (1 \times 10^2) + (2 \times 10^1) + (4 \times 10^0) + (6 \times 10^{-1}) \\ &= 124.6. \end{aligned}$$

So $10^3 x$ simply shifts the decimal point of x three places to the right. In the same way, we see that if x continues to denote the finite decimal 0.1246, then $10^{-4}x = 0.00001246$, which then shifts the decimal point of x four places to the left. One would like to believe that, by analogy, this computation *should* also work for an "infinite sum". However, one must remember that these arguments are valid for the finite decimal 0.1246 only because we are dealing with a finite sum, whereas what is asserted in the last statement in Theorem 41.2 is for *the limit of a sequence*, and we have yet to learn how to deal with limits of sequences. The main thrust of the last statement in Theorem 41.2 is therefore that, this gap in our knowledge notwithstanding, *the formal operation of multiplying a decimal by a power of 10 still behaves as if the decimal were a finite decimal*.

We may ask why we should bother with infinite decimals at all if they are so troublesome? The fundamental reason is that we cannot avoid them, as the following theorem shows.

Theorem 41.3. *Every positive number is equal to a decimal in the sense of Theorem 41.1.*[1]

As an illustration of Theorem 41.3, recall that $\sqrt{2}$ is a number (see the discussion in Chapter 33), and therefore Theorem 41.3 guarantees that it must be equal to a decimal. This decimal is most likely already familiar to you, the so-called decimal expansion of $\sqrt{2}$:

$$\sqrt{2} = 1.41421\ 35623\ 73095\ 04880\ 16887\ 24209\ \cdots.$$

In this case, there is little mystery about how to get this decimal expansion: you simply approximate $\sqrt{2}$ decimal digit by decimal digit, in the sense of the following sequence of inequalities:

$$\begin{aligned} 1.4 &< \sqrt{2} < 1.5, \\ 1.41 &< \sqrt{2} < 1.42, \\ 1.414 &< \sqrt{2} < 1.415, \\ 1.4142 &< \sqrt{2} < 1.4143, \\ 1.41421 &< \sqrt{2} < 1.41422,\ \text{etc.} \end{aligned}$$

Each of these inequalities can be verified by using the Lemma 33.2 on page 448. We leave as an exercise the proof that, on the basis of these inequalities and Theorem 41.1, the preceding decimal expansion of $\sqrt{2}$ is correct.

By the way, as an illustration of Theorem 41.2, we have

$$10^5 \times \sqrt{2} = 141421.\ 35623\ 73095\ 04880\ 16887\ 24209\ \cdots.$$

41.3. Repeating Decimals

It can happen that, beyond a certain decimal digit, the decimal digits of a given decimal simply repeat a fixed block of digits indefinitely, e.g.,

$$16.419\ 76\ 76\ 76\ 76\ 76\ 76\ 76\ \cdots.$$

There is a standard shorthand notation for these repeating blocks of digits, and this notation can be simply explained through some typical examples such as the following.

$$\begin{aligned} 0.888888\cdots &\equiv 0.\overline{8}, \\ 0.15\,15\,15\,15\cdots &\equiv 0.\overline{15}, \\ 16.419\,76\,76\,76\,76\cdots &\equiv 16.419\overline{76}, \\ 4.00876\,523\,523\,523\,523\cdots &\equiv 4.00876\overline{523}, \\ 0.285714\,285714\,285714\cdots &\equiv 0.\overline{285714}. \end{aligned}$$

---

[1]This theorem would be true also for negative numbers if we had taken the trouble to define negative decimals. For a proof of Theorem 41.3 that is in the spirit of this book, see [**Wu**, Chapter 17].

41.3. Repeating Decimals

Here the symbol "≡" serves as a reminder that a notation is being defined, and the bar (e.g., $\overline{523}$) indicates that the digits under the bar are repeating. We adopt the CONVENTION that *at least one of the digits under the bar is nonzero*. Such infinite decimals are called **repeating decimals**. Observe that $0.\overline{15}$, $0.\overline{1515}$, and $0.1\overline{51}$ all denote the same repeating decimal. We will simply refer to any collection of consecutive repeating decimal digits as a **repeating block** of the repeating decimal, and call the number of digits in the block its **length**. It is quite easy to see that there is a minimum among the lengths of repeating blocks of a given repeating decimal, and we call this minimum length the **period** of the repeating decimal. For example, in the case of $0.\overline{15}$, both $0.\overline{15}$ and $0.1\overline{51}$ have length 2, and 2 is seen to be the period of $0.\overline{15}$.

A main goal of this chapter is to show

Theorem 41.4. *Every repeating decimal is equal to a fraction.*

To this end, we show how each of the preceding repeating decimals is equal to a fraction. Once that is done, it will be clear from the pattern of the computations how to proceed in the general case. The important point to note is that *the computations make strong use of Theorem 41.2 at almost every step*.[2]

We start with $0.\overline{8}$. A general comment about the computation that is to follow not only here but in all subsequent examples is that, by the theorem, $0.\overline{8}$ is a number, so that by FASM (Chapter 21), we can apply the distributive law to $0.\overline{8}$. With this understood, we use Theorem 41.2 to get

$$10 \times 0.\overline{8} = 8.8888\cdots = 8 + 0.8888\cdots = 8 + (1 \times 0.\overline{8}).$$

Thus $10 \times 0.\overline{8} - 1 \times 0.\overline{8} = 8$. By the distributive law (see (17.4) on page 270 and FASM) we have $(10 - 1) \times 0.\overline{8} = 8$ so that $9 \times 0.\overline{8} = 8$, and therefore

$$0.\overline{8} = \frac{8}{9}.$$

Similarly,

$$100 \times 0.\overline{15} = 15.151515\ldots = 15.\overline{15}$$
$$= 15 + (1 \times 0.\overline{15}),$$
$$(100 - 1) \times 0.\overline{15} = 15,$$
$$99 \times 0.\overline{15} = 15,$$

so that

$$0.\overline{15} = \frac{15}{99} = \frac{5}{33}.$$

---

[2] The fact that such a computation is not possible without the support of the nontrivial Theorem 41.2 is traditionally overlooked in school mathematics textbooks and professional development materials.

Lest you be misled into believing that every repeating decimal is equal to a fraction whose numerator is equal to the repeating decimal digits and whose denominator is equal to 9 or 99 or 999, etc.—as many school texts would have you believe—we now take a look at $16.419\overline{76}$.

Because
$$16.419\overline{76} = 16.419 + 0.000\overline{76},$$
it suffices to find the fraction equivalent of $0.000\overline{76}$. We have (by Theorem 41.2, of course)
$$0.000\overline{76} = 0.\overline{76} \times 10^{-3}.$$
Therefore it suffices to find the fraction equal to $0.\overline{76}$. We proceed as before:
$$100 \times 0.\overline{76} = 76.\overline{76} = 76 + (1 \times 0.\overline{76}),$$
$$99 \times 0.\overline{76} = 76,$$
$$0.\overline{76} = \frac{76}{99}.$$

Hence (again take note of how we use Theorem 41.2 and treat $0.\overline{76}$ as a number),
$$\begin{aligned}
16.419\overline{76} &= 16 + 0.419 + 0.000\overline{76} \\
&= 16 + (419 + 0.\overline{76}) \times 10^{-3} \\
&= 16 + \left(419 + \frac{76}{99}\right) \times 10^{-3} \\
&= 16 + \left(\frac{41557}{99} \times 10^{-3}\right) \\
&= 16 + \frac{41557}{99000} \\
&= 16 \frac{41557}{99000} = \frac{1625557}{99000}.
\end{aligned}$$

Similarly,
$$\begin{aligned}
4.00876\overline{523} &= 4.00876 + 0.00000\overline{523} \\
&= 4 + 0.00876 + 0.00000\overline{523} \\
&= 4 + (876 \times 10^{-5}) + (0.\overline{523} \times 10^{-5}) \\
&= 4 + (876 + 0.\overline{523}) \times 10^{-5}.
\end{aligned}$$

We can find the fraction equivalent of $0.\overline{523}$ as before:
$$10^3 \times 0.\overline{523} = 523.\overline{523} = 523 + 0.\overline{523},$$
$$999 \times 0.\overline{523} = 523,$$
$$0.\overline{523} = \frac{523}{999}.$$

41.3. Repeating Decimals

Hence,

$$4.00876\overline{523} = 4 + \left(876 + \frac{523}{999}\right) \times 10^{-5}$$
$$= 4 + \frac{875647}{99900000} = 4\frac{875647}{99900000}.$$

Finally, the case of $0.\overline{285714}$ is of particular interest:

$$10^6 \times 0.\overline{285714} = 285714.\overline{285714} = 285714 + 0.\overline{285714}.$$

So by the usual reasoning,

$$(10^6 - 1) \times 0.\overline{285714} = 285714,$$
$$999999 \times 0.\overline{285714} = 285714,$$
$$0.\overline{285714} = \frac{285714}{999999}.$$

It is easy to check that 142857 divides both the numerator and denominator of the last fraction. Therefore, we have the rather surprising result that

$$0.\overline{285714} = \frac{2}{7}.$$

Activity. (a) We have just come across the whole number 142857. Can you guess what fraction is equal to $0.\overline{142857}$? (b) Multiply 142857 successively by 2, 3, 4, 5, 6 and examine carefully the numbers you get. What do you notice about these numbers? (c) How are the numbers in (b) related to $\frac{2}{7}$, $\frac{3}{7}$, ..., $\frac{6}{7}$? (See [**Ros10**] for further information.)

It remains to point out *why we never see $\overline{9}$ in a repeating decimal*. This is because

(41.4) $$0.\overline{9} = 1.$$

In an exercise, you are asked to explain (41.4) by making use of Theorem 41.1. Here we follow the preceding procedure for converting a repeating decimal to a fraction to give a computational proof. Indeed,

$$10 \times 0.\overline{9} = 9.\overline{9} = 9 + 0.\overline{9},$$
$$(10 - 1) \times 0.\overline{9} = 9,$$
$$9 \times 0.\overline{9} = 9,$$
$$0.\overline{9} = 1.$$

It follows from (41.4) that, for example,

$$2.045\overline{9} = 2.045 + (0.\overline{9} \times 10^{-3}) = 2.045 + 10^{-3} = 2.046.$$

This example illustrates the general fact that if a decimal ends in $4\overline{9}$, then it in fact ends in 5, or if it ends in $7\overline{9}$, then it in fact ends in 8, etc. For example, $8.745\overline{9} = 8.746$. In view of this fact, *we shall henceforth agree never to write a decimal which ends in $\overline{9}$.*

Exercises

1. Write down the complete expanded form of (a) 425.00410, (b) 0.00008, (c) 400.00001.

2. Explain to a 7th grader, without computations but assuming Theorem 41.1 on page 514, why $0.\overline{9} = 1$.

3. Write each of the following decimals as a fraction: (a) $0.\overline{3}$, (b) $0.\overline{6}$, (c) $2.11\overline{4}$, (d) $0.4\overline{25}$, (e) $5.00\overline{123}$.

4. Explain to fifth graders why $\frac{1}{3}$ is equal to the repeating decimal $0.\overline{3}$, why $0.\overline{5}$ is equal to $\frac{5}{9}$, and why $0.2\overline{1}$ is equal to $\frac{19}{90}$. (In other words, you have to master the art of saying enough, but not saying too much. And of course, you cannot say anything that is obviously wrong even if you know you are lying by omission.)

5. Write each of the following decimals as a fraction in lowest terms: (a) $1.\overline{6}$, (b) $0.58\overline{3}$, (c) $1.1\overline{6}$, (d) $1.\overline{285714}$, (e) $0.\overline{142857}$, (f) $0.\overline{846153}$.

6. Show directly by computation why the decimal $0.5412\overline{9}$ is equal to 0.5413.

7. Write down the fraction equal to $0.\overline{124}$, and then write down the fraction equal to $0.12\overline{412}$. Compare your answers and explain your observation.

8. Assuming Theorem 41.1, write a complete and self-contained proof of the fact that $\sqrt{2} = 1.41421\ 35623 \cdots$.

Chapter 42

Decimal Expansions of Fractions

The goal of this chapter is to prove the theorem that every fraction is equal to a finite or a repeating decimal (recall that our repeating decimals are infinite, by definition). The proof of the finite case is simple, but the proof of the repeating case is predictably more difficult. As in section 41.2 of the preceding chapter, certain aspects of repeating decimals (all the concepts related to limits) have to be treated on an intuitive level.

The sections are as follows:

> The Theorem
> Proof of the Finite Case
> Proof of the Repeating Case

42.1. The Theorem

We have seen that a finite or repeating decimal is equal to a fraction (Theorem 41.4 on page 519). It is natural to ask whether the converse is also true. The following theorem answers this question in the affirmative.

Theorem 42.1. *Every fraction is equal to a finite or repeating decimal. Moreover, if the fraction is $\frac{m}{n}$, then the decimal digits of this decimal are given by the quotient of the long division of $m \times 10^k$ by n, where k is any whole number $\geq n$.*

The (finite or repeating) decimal in Theorem 42.1 is called the **decimal expansion** of the given fraction $\frac{m}{n}$. There is some ambiguity in the part of the theorem that says, "the decimal digits of this decimal are given by the quotient of the long division of $m \times 10^k$ by n". What this means is that if the decimal is a finite decimal, then, ignoring the decimal point, it is equal to the quotient of the long division of $m \times 10^k$ by n for some well-determined positive integer k (see the proof below). If, however, the decimal D is a repeating decimal, then the whole number that comprises all the decimal digits of D up to and including a first complete repeating block (e.g., the number 26 in $0.\overline{26}$ or 246187 in $0.246\,\overline{187}$), will be the quotient of the long division of $m \times 10^k$ by n for any $k \geq n$. All this will be clear from the proof below.[1]

With a little more care, one can actually extract from the proof of the theorem the fact that, in the case of a repeating decimal, a repeating block must appear within the first n decimal digits and the period of the repeating decimal is $\leq (n-1)$ (see Exercise 2 on page 539). Thus one knows ahead of time, for example, that the decimal expansion of $\frac{2}{17}$ would have a period of ≤ 16.

The following proof of the theorem relies on the long division algorithm and, in fact, furnishes a main justification for studying the latter.

Activity. Make use of Theorems 41.1 and 42.1 to show that there are irrational numbers. (This argument complements Theorem 36.5 in section 36.3.)

Now Theorem 42.1 prompts a further question: how can we tell whether a fraction has a finite or a repeating decimal expansion? The answer has been given already in Theorem 36.2 on page 479. We recall that theorem:

Theorem 36.2. *A fraction $\frac{m}{n}$ in lowest terms is equal to a finite decimal exactly when the denominator n is equal to $2^k 5^\ell$, where k and ℓ are whole numbers.*

We preface the proof of Theorem 42.1 with an informal discussion. First, there is a similar discussion of part of this theorem in Chapter 18. However, we will not quote directly from that discussion in order to keep the exposition here as self-contained as possible.

An integral part of this theorem is the explanation of the common algorithm taught in school for converting a fraction to a decimal by using long division to divide the numerator (with an arbitrary number of zeros added to its right) by the denominator. We wish to call attention to the need for giving *some* explanation of this algorithm in the school classroom, because students should not be left wondering why the decimal produced by this quaint procedure is actually equal to the fraction itself. For example,

---

[1]But see Exercise 2 on page 539.

42.1. The Theorem

consider the long division[2] of the whole number 2 by 11 (students are told to add as many zeros to the right of 2 as necessary to continue the long division):

$$
\begin{array}{r}
1\ 8\ 1\ 8 \\
11\ )\ \overline{2\ 0\ 0\ 0\ 0} \\
\underline{1\ 1} \\
9\ 0 \\
\underline{8\ 8} \\
2\ 0 \\
\underline{1\ 1} \\
9\ 0 \\
\underline{8\ 8} \\
2
\end{array}
$$

This algorithm would have you conclude that $\frac{2}{11} = 0.\overline{18}$. Aside from the seeming arbitrariness of the placement of the decimal point, the most critical question here is, in what sense is the fraction $\frac{2}{11}$ actually equal to the infinite decimal $0.\overline{18}$? *Why should students believe that $0.\overline{18}$ has anything to do with $\frac{2}{11}$?* To partially answer this question, one can offer the following heuristic argument which is an oversimplification of the proof to follow.

We have, by the product formula ((17.2) on page 267),

$$\frac{2}{11} = \frac{2 \times 10000}{11} \times \frac{1}{10000} = \frac{20000}{11} \times \frac{1}{10000}.$$

By the long division algorithm carried out above, we have $20000 = (1818 \times 11) + 2$. Therefore,

$$\frac{2}{11} = \frac{(1818 \times 11) + 2}{11} \times \frac{1}{10000} = \left(1818 + \frac{2}{11}\right) \times \frac{1}{10000}.$$

By the distributive law,

$$\frac{2}{11} = \frac{1818}{10000} + \left(\frac{2}{11} \times \frac{1}{10000}\right),$$

so that

$$\frac{2}{11} - \frac{1818}{10000} = \left(\frac{2}{11} \times \frac{1}{10000}\right).$$

By the definition of finite decimals, $\frac{1818}{1000} = 0.1818$. Furthermore, $\frac{2}{11}$ is smaller than 1 so that $\frac{2}{11} \times \frac{1}{10000}$ is smaller than $\frac{1}{10000} = 0.0001$. Therefore,

$$\frac{2}{11} - 0.1818 < 0.0001.$$

---

[2]The remainder of this chapter will draw heavily on the understanding of the relationship between the long division algorithm and the concept of division-with-remainder. Please review sections 7.2 and 7.3 if necessary.

So the fraction $\frac{2}{11}$ differs from 0.1818 by very little. If we use instead the division of 2×10^8 by 11, then we would get $2 \times 10^8 = (18181818 \times 11) + 2$, and would obtain $\frac{2}{11} - 0.18181818 < 0.00000001$. So $\frac{2}{11}$ differs from 0.18181818 by even less, and the repeating phenomenon would be even more obvious. So it is believable that $\frac{2}{11} = 0.\overline{18}$.

42.2. Proof of the Finite Case

We now get down to the business of giving a proof of the theorem. A main feature of the proof is that it explicitly exhibits the equality between the fraction and the decimal, and explains why the decimal point of the latter is where it is supposed to be.

Let a fraction $\frac{m}{n}$ be given. There are some obvious simplifications to be made.

First of all, it suffices to get the decimal expansion of a fraction $\frac{m}{n}$ so that the denominator n is not a multiple of 10. Indeed, suppose $n = n' \times 10^k$ for some positive integer k and some whole number n' which is not a multiple of 10. Then,

$$\frac{m}{n} = \frac{m}{n' \times 10^k} = \frac{m}{n'} \times \frac{1}{10^k} = \frac{m}{n'} \times 10^{-k}.$$

If we know that $\frac{m}{n'}$ is equal to a finite or a repeating decimal, by Theorem 41.2, we see that $\frac{m}{n}$ is equal to the decimal of $\frac{m}{n'}$ with the decimal point shifted k places to the left. So we are done.

Next, we may assume that the given fraction $\frac{m}{n}$ is a reduced proper fraction, i.e., $m < n$ and m and n have no common divisor other than 1. In fact, if $m \geq n$, we can use division-with-remainder of m by n, $m = qn + r$, to write $\frac{m}{n}$ as $q + \frac{r}{n}$, where q is a whole number and $\frac{r}{n}$ is a proper fraction. Then we need only show how to write the proper fraction $\frac{r}{n}$ as a decimal. If $\frac{r}{n}$ is not reduced, it is equal to a reduced fraction, by Theorem 36.1 on page 479. So we may as well assume $\frac{r}{n}$ is reduced to begin with.

From now on, we will assume that $\frac{m}{n}$ is reduced, $m < n$, and n is not a multiple of 10.

Proof of Theorem 42.1: Finite Case. We first tackle the case in which $n = 2^k 5^\ell$, where k and ℓ are whole numbers. In this case, we want to show that $\frac{m}{n}$ is a finite decimal, i.e.,

$$\frac{m}{n} = \frac{q}{10^p}$$

for some whole numbers p and q. Because n is not a multiple of 10, and because $2 \times 5 = 10$, we see that not both k and ℓ are positive. For definiteness, let us say $k > 0$ and $\ell = 0$. Therefore it suffices to show that

42.3. Proof of the Repeating Case

a fraction of the form $\frac{m}{2^k}$ is equal to $\frac{q}{10^k}$, where q is the quotient of the long division of $m \times 10^k$ by 2^k.

We first verify that $m \times 10^k$ is in fact a multiple of 2^k. This is because $m \times 10^k = m \times 2^k \times 5^k$. Therefore the number q above can be written as

$$q = \frac{m \times 10^k}{2^k}.$$

Now we have, by the product formula ((17.2) on page 267),

$$\frac{m}{2^k} = \left(\frac{m \times 10^k}{2^k}\right) \times \frac{1}{10^k} = q \times \frac{1}{10^k} = \frac{q}{10^k}.$$

Our assertion is now proved. □

Remarks. (1) Would it have mattered if we had chosen a higher power of 10, e.g., 10^{k+4} instead of 10^k, in the preceding computation? Not at all, because in that case we would get

$$\frac{m}{2^k} = \frac{m \times 10^{k+4}}{2^k} \times \frac{1}{10^{k+4}} = \frac{m \times 10^k}{2^k} \times \frac{10^4}{10^{k+4}} = \frac{q}{10^k}.$$

(2) Notice once again the critical role played by the product formula in this proof.

(3) Notice also the fact that $\frac{m}{2^k}$ is a decimal with k decimal digits because it is equal to $\frac{q}{10^k}$, which is *by definition* a decimal with k decimal digits.

Example. Let us find the decimal expansion of $\frac{21}{8}$. Because $21 = (2 \times 8) + 5$, we see that $\frac{21}{8} = 2 + \frac{5}{8}$. Knowing $8 = 2^3$, we have $k = 3$, so that

$$\frac{5}{8} = \left(\frac{5 \times 10^3}{8}\right) \times \frac{1}{10^3} = \frac{625}{10^3} = 0.625$$

```
          6 2 5
      8 ) 5 0 0 0
          4 8
            2 0
            1 6
              4 0
              4 0
                0
```

Therefore, we get the desired decimal expansion:

$$\frac{21}{8} = 2 + \frac{5}{8} = 2 + 0.625 = 2.625.$$

42.3. Proof of the Repeating Case

We continue to assume that the given fraction $\frac{m}{n}$ is reduced, $m < n$, and n is not a multiple of 10.

Next we consider the decimal expansion of a reduced proper fraction whose denominator is divisible by a prime other than 2 and 5. We expect to get an infinite decimal on account of Theorem 36.2. In this situation a symbolic proof is less transparent than an explanation in a concrete case (cf. the explanations of the standard algorithms for whole numbers in Part 1). So we will explain how to find the decimal expansion of two specific fractions.

Proof of Theorem 42.1: Repeating Case. First, consider $\frac{3}{7}$. We introduce a decimal
$$s = 0.a_1 a_2 a_3 \cdots$$
that will turn out to be equal to $\frac{3}{7}$ in the sense of Theorem 41.1. To define s, it suffices to define for each integer n the finite decimal $s_n = 0.a_1 a_2 \cdots a_n$ so that

(42.1) $$s_n = \frac{q_n}{10^n},$$

where q_n is the quotient of the long division of 3×10^n by 7. (A word about the notation: 3 and 7 are the numerator and denominator of $\frac{3}{7}$, and the subscript n in q_n serves to indicate that the quotient of the division-with-remainder of 3×10^n by 7 depends on the exponent n of 10^n.) Then s is by definition the limit of the sequence $\{s_n\}$.

Remark. The number s_n so defined is automatically a decimal with n decimal digits, but to say that $s_n = 0.a_1 a_2 \cdots a_n$ is to imply that s_n has integer part equal to 0, i.e., $s_n < 1$. Let us prove this. By the definition of s_n, this is equivalent to proving $q_n < 10^n$. The definition of q_n is that
$$3 \times 10^n = (q_n \times 7) + r_n, \quad \text{where } 0 \leq r_n < 7.$$
Thus $7q_n = (3 \times 10^n) - r_n < (3 \times 10^n)$, so that $q_n < \frac{3}{7} \times 10^n < 1 \times 10^n = 10^n$, as desired.

Let us look at $n = 4$ and $n = 5$. Then we have $q_4 = 4285$ and $q_5 = 42857$, respectively, as shown.

```
              4 2 8 5                          4 2 8 5 7
      7 ) 3 0 0 0 0                    7 ) 3 0 0 0 0 0
          2 8                                  2 8
          ─────                                ─────
            2 0                                  2 0
            1 4                                  1 4
            ─────                                ─────
              6 0                                  6 0
              5 6                                  5 6
              ─────                                ─────
                4 0                                  4 0
                3 5                                  3 5
                ─────                                ─────
                  5                                    5 0
                                                       4 9
                                                       ─────
                                                         1
```

42.3. Proof of the Repeating Case

Note that the 5-digit number q_5 is just the 4-digit number q_4 with an additional digit (7, to be exact) attached to the right end of q_4; this is by the very nature of the long division algorithm (see both long divisions above).[3] For the same reason, the $(n+1)$-digit number q_{n+1} is just the n-digit number q_n with an additional digit attached to the end of q_n so that the first n decimal digits (from the left) of s_n and s_{n+1} coincide. Without this property, the above definition of the decimal s would not make any sense.

We claim that $s = \frac{3}{7}$.

To prove the claim, we note that s is the limit of the s_n's. If we can show that $\frac{3}{7}$ is also the limit of the s_n's, in the sense that the distance on the number line between s_n and $\frac{3}{7}$ goes to 0 as $n \to \infty$, then by the uniqueness statement in Theorem 41.1, we would have $s = \frac{3}{7}$.

We now prove that $\frac{3}{7}$ is the limit of the s_n's. By the product formula, we have

$$\frac{3}{7} = \frac{3 \times 10^n}{7} \times \frac{1}{10^n}.$$

If we do the long division of 3×10^n by 7 to get the quotient q_n, then we have

$$3 \times 10^n = (q_n \times 7) + r_n, \quad \text{where } 0 \leq r_n < 7.$$

Therefore

$$\frac{3}{7} = \frac{(q_n \times 7) + r_n}{7} \times \frac{1}{10^n}$$
$$= \left(q_n + \frac{r_n}{7}\right) \times \frac{1}{10^n}$$
$$= \frac{q_n}{10^n} + \left(\frac{r_n}{7} \times \frac{1}{10^n}\right)$$
$$= s_n + \left(\frac{r_n}{7} \times \frac{1}{10^n}\right),$$

where the last step is by the definition of s_n in (42.1). We then have

$$\frac{3}{7} - s_n = \left(\frac{r_n}{7} \times \frac{1}{10^n}\right).$$

But $0 \leq r_n$, so

$$\frac{r_n}{7} \times \frac{1}{10^n} \geq 0,$$

and therefore

$$\frac{3}{7} - s_n \geq 0.$$

Thus $\frac{3}{7} \geq s_n$. On the number line, s_n is therefore to the left of $\frac{3}{7}$, as shown.

---

[3]This is true in a more general setting; see Exercise 3 on page 539.

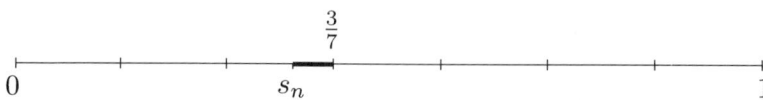

We want to show that the length of the thickened segment (= the distance between $\frac{3}{7}$ and s_n) shrinks down to 0 as $n \to \infty$. For this purpose, we are going to make use of the fact that $r_n < 7$. It implies that

$$\frac{r_n}{7} \times \frac{1}{10^n} < 1 \times \frac{1}{10^n} = \frac{1}{10^n},$$

so that

$$\frac{3}{7} < s_n + \frac{1}{10^n},$$

which implies that

$$\frac{3}{7} - s_n < \frac{1}{10^n}.$$

Together with $\frac{3}{7} \geq s_n$, we see that

$$0 \leq \frac{3}{7} - s_n < \frac{1}{10^n}.$$

This means that the length of the thickened segment is smaller than $1/10^n$. Intuitively, $1/10^n$ goes to 0 as $n \to \infty$, so the distance between $\frac{3}{7}$ and s_n goes to 0 as $n \to \infty$. So the limit of the sequence of finite decimals $\{s_n\}$ is $\frac{3}{7}$. The uniqueness part of Theorem 41.1 therefore affirms that $s = \frac{3}{7}$, as desired.

It remains to prove that s is a repeating decimal. More precisely, we prove

$$s = 0.\overline{428571}.$$

Let us consider the long division of 3×10^{10} by 7:

42.3. Proof of the Repeating Case

```
              4 2 8 5 7 1 4 2 8 5
       7 ) 3  0 0 0 0 0 0 0 0 0 0
           2  8
              2 0
              1 4
                6 0
                5 6
                  4 0
                  3 5
                    5 0
                    4 9
                      1 0
                        7
                        3 0
                        2 8
                          2 0
                          1 4
                            6 0
                            5 6
                              4 0
                              3 5
                                5
```

You may notice that there are two 2-by-2 blocks, consisting of 3 0 in one row and 2 8 in the next, that are in boldface italic fonts. We will return to these blocks below. In any case, we have

$$3 \times 10^{10} = (4285714285 \times 7) + 5.$$

Since q_{10} is the quotient of the division-with-remainder of 3×10^{10} by 7, we have $q_{10} = 4285714285$.

The quotient clearly shows the (beginning of the) repetition of the digits 428571. We want to explain why the repetition is inevitable. For clarity's sake, we now give the procedural description of the preceding schematic presentation of the long division of 30,000,000,000 by 7 (cf., e.g., (7.12) on page 121):

(42.2)
$$\begin{cases} 3 \times 10 = \boxed{4} \times 7 + \boxed{2} \\ 2 \times 10 = \boxed{2} \times 7 + \boxed{6} \\ 6 \times 10 = \boxed{8} \times 7 + \boxed{4} \\ 4 \times 10 = \boxed{5} \times 7 + \boxed{5} \\ 5 \times 10 = \boxed{7} \times 7 + \boxed{1} \\ 1 \times 10 = \boxed{1} \times 7 + \boxed{3} \\ 3 \times 10 = \boxed{4} \times 7 + \boxed{2} \\ 2 \times 10 = \boxed{2} \times 7 + \boxed{6} \\ 6 \times 10 = \boxed{8} \times 7 + \boxed{4} \\ 4 \times 10 = \boxed{5} \times 7 + \boxed{5} \end{cases}$$

Notice as usual that the quotient 4,285,714,285 appears (vertically) in the first column of the right-hand side, and that the *ten* digits of the quotient correspond to the quotients of *ten* divisions-with-remainder all with the same divisor 7 in (42.2). Now concentrate on the last column of remainders in the first eight divisions-with-remainder. Because these divisions all have the same divisor 7, the only possible remainders are 0, 1, 2, 3, 4, 5, 6. However, we have to use the seven numbers 0, 1, 2, 3, 4, 5, 6 for each of these eight remainders, so at least one of these seven numbers will have to be used twice. In other words, *among the remainders of the first eight divisions-with-remainder in (42.2), at least two of the remainders must be the same*. In this case, 2 appears in the remainder of both the first and seventh divisions. Because the long division algorithm takes the remainder r of a division-with-remainder and uses $10r$ as the dividend of the next division-with-remainder (see Chapter 7), the second and eighth divisions-with-remainder in (42.2) must coincide. Indeed, we observe that they do:

$$2 \times 10 = \boxed{2} \times 7 + \boxed{6}.$$

This means all the divisions after the seventh must repeat those that follow the first, and therefore the digits 285714 in the quotient of the long division of 3×10^k by 7 must repeat for any k provided $k \geq 8$. Recalling that the decimal digits of the infinite decimal s are those of these quotients (see (42.2)), we see that the decimal expansion of $\frac{3}{7}$ will have a repeating block consisting of 285714. (We have more to say about the repeating block, but we will only address that later so as not to lose the main thread of our argument.)

Now it is clear why we used 3×10^{10} (and not 3×10^5) as the dividend for the long division to begin with, because the exponent 10 guarantees that we will get more than eight divisions-with-remainder in the long division in (42.2) so that we can witness the repetition of one of 0, 1, ..., 6 in

42.3. Proof of the Repeating Case

these eight divisions. Equally clearly, this phenomenon of repetition does not depend on this particular fraction $\frac{3}{7}$. If we are given $\frac{2}{107}$, then we would consider the long division of 3×10^{108} (or 3×10^k for any $k > 107$) in order to get at least 108 divisions-with-remainder. Then because we have to use the 107 numbers 0, 1, 2, ..., 105, 106 for the 108 remainders, at least one of the remainders will have to be used twice and then the same repeating phenomenon will appear again.

Finally we tie up a loose end about the repeating block in the quotient of the division-with-remainder of 3×10^k by 7. Look closely at (42.2) again and you will observe that it is not just the second division-with-remainder that is identical to the eighth, but that, *by accident*, the first division-with-remainder
$$3 \times 10 = \boxed{4} \times 7 + \boxed{2}$$
is already identical to the seventh. See the bold-faced, 2-by-2 blocks in the schematic presentation of the long division of 30,000,000,000 by 7 above (42.2).[4] So in this case, the repeating block in the quotient of this division-with-remainder is actually 428571, so that s is simply $0.\overline{428571}$ rather than $0.4\overline{285714}$.

We have completely proved the theorem in the case of $\frac{3}{7}$.

Next we turn to the decimal expansion of $\frac{1}{28}$. This time, let the decimal
$$s = 0.a_1 a_2 a_3 \cdots$$
be defined by the requirement that, for each integer n, the finite decimal $s_n = 0.a_1 a_2 \cdots a_n$ is given by

(42.3) $$s_n = \frac{q_n}{10^n},$$

where each q_n is the quotient of the long division of 1×10^n by 28 (the 1 here is the numerator of $\frac{1}{28}$). In other words,

(42.4) $$1 \times 10^n = (q_n \times 28) + r_n \quad \text{where } 0 \leq r_n < 28.$$

Before proceeding further, we should clarify the sequence $\{s_n\}$. First, we must verify that, while s_n is a decimal with n decimal digits by its definition, its integer part is equal to zero (so that we can write $s_n = 0.a_1 a_2 \cdots a_n$). This is equivalent to showing that $s_n < 1$, or what is the same thing, $q_n < 10^n$ by (42.3). But by (42.4), $28 q_n = 10^n - r_n < 10^n$, so that (using inequality (D) on page 422),
$$q_n < \frac{1}{28} \times 10^n < 1 \times 10^n < 10^n,$$

---

[4]To get an idea that this doesn't happen often, see the next example or do Exercise 1 on page 539.

as desired. Second, we note as before that, because the q_n's are the quotients of the long division of 10^n by 28, by the very nature of long division, the first n digits (from the left) of the $(n + 1)$ digits of q_{n+1} are exactly the n digits of q_n. It follows that the first n decimal digits of s_{n+1} are exactly the decimal digits of s_n. This is the reason that the infinite decimal s is well defined.

Now back to the proof of Theorem 42.1 for the fraction $\frac{1}{28}$, we claim that

(42.5) $$s = 0.03\overline{571428}.$$

For the proof of (42.5), let us compute s_3 and s_9 to get a feel for (42.5):

```
            0 3 5
      28 ) 1 0 0 0
            0
            1 0 0
              8 4
              1 6 0
              1 4 0
                2 0
```

$1 \times 10^3 = (35 \times 28) + 2$

$q_3 = 35$

$s_3 = \dfrac{35}{10^3} = 0.035$

```
                0 3 5 7 1 4 2 8 5
      28 ) 1 0 0 0 0 0 0 0 0 0
            0
            1 0 0
              8 4
              1 6 0
              1 4 0
                2 0 0
                1 9 6
                    4 0
                    2 8
                    1 2 0
                    1 1 2
                        8 0
                        5 6
                        2 4 0
                        2 2 4
                            1 6 0
                            1 4 0
                                2 0
```

$1 \times 10^9 = (35714285 \times 28) + 20$

$q_9 = 35714285$

$s_9 = \dfrac{35714285}{10^9}$

$= 0.035714285$

(Observe that s_3 is just the first three decimal digits of s_9.)

42.3. Proof of the Repeating Case

All the steps in the preceding long divisions are summarized in the following sequence of divisions-with-remainder:

(42.6)
$$\begin{cases} 1 \times 10 &= \boxed{0} \times 28 + \boxed{10} \\ 10 \times 10 &= \boxed{3} \times 28 + \boxed{16} \\ 16 \times 10 &= \boxed{5} \times 28 + \boxed{20} \\ 20 \times 10 &= \boxed{7} \times 28 + \boxed{4} \\ 4 \times 10 &= \boxed{1} \times 28 + \boxed{12} \\ 12 \times 10 &= \boxed{4} \times 28 + \boxed{8} \\ 8 \times 10 &= \boxed{2} \times 28 + \boxed{24} \\ 24 \times 10 &= \boxed{8} \times 28 + \boxed{16} \\ 16 \times 10 &= \boxed{5} \times 28 + \boxed{20} \end{cases}$$

With the decimal expansion of $\frac{3}{7}$ as a template, what we should do is perform the long division of 1×10^{29} by 28 and express it as 29 divisions-with-remainder which will then include (42.6). Note trivially that these 29 divisions-with-remainder all have the same divisor 28. Then with only 28 possible numbers 0, 1, ..., 26, 27 to be used as remainders for these 29 divisions-with-remainder, at least one of these 28 numbers must be repeated and therefore a division-with-remainder will also be repeated. Consequently, a repeating block will appear in the quotient. In this case, however, we use only 1×10^9 and not 1×10^{29}, because the remainder 16 of the second division-with-remainder in (42.6),

$$10 \times 10 = \boxed{3} \times 28 + \boxed{16},$$

already reappears in the remainder of the eighth division-with-remainder,

$$24 \times 10 = \boxed{8} \times 28 + \boxed{16}.$$

The long division algorithm then dictates that the third and ninth divisions-with-remainder must coincide, and they do:

$$16 \times 10 = \boxed{5} \times 28 + \boxed{20}.$$

This implies that the corresponding digits in the quotient, 571428, are a repeating block of the decimal s, which is exactly the claim in (42.5).

It is worth noting that, in this case, although the second and the eighth divisions-with-remainder have the same remainder, these two divisions-with-remainder are themselves not identical; see above. Therefore, unlike the case of $\frac{3}{7}$, the digit 3 in the second division-with-remainder of (42.6) is not part of a repeating block.

Finally, we prove that

$$\frac{1}{28} = 0.03\overline{571428}.$$

We begin the proof the usual way:

$$\frac{1}{28} = \frac{1 \times 10^n}{28} \times \frac{1}{10^n}$$

$$= \frac{(q_n \times 28) + r_n}{28} \times \frac{1}{10^n} \quad \text{(see (42.4))}$$

$$= \left(q_n + \frac{r_n}{28}\right) \times \frac{1}{10^n}$$

$$= \frac{q_n}{10^n} + \left(\frac{r_n}{28} \times \frac{1}{10^n}\right)$$

$$= s_n + \left(\frac{r_n}{28} \times \frac{1}{10^n}\right) \quad \text{(see (42.1))}.$$

Let the last quantity in parentheses be denoted by R_n. Then we have

$$\frac{1}{28} = s_n + R_n.$$

The fact that $0 \leq r_n < 28$ leads to the fact that $0 \leq R_n < (1/10^n)$, so we have

$$0 \leq \frac{1}{28} - s_n < \frac{1}{10^n}.$$

Thus the distance between $\frac{1}{28}$ and s_n is at most $1/10^n$. As $n \to \infty$, it is intuitively clear that this distance gets closer and closer to 0. Thus $\frac{1}{28}$ is the limit of the sequence $\{s_n\}$. Because $0.03\overline{571428}$ is also the limit of $\{s_n\}$ (by (42.5)), the uniqueness statement in Theorem 41.1 says $0.03\overline{571428} = \frac{1}{28}$.

The proof of Theorem 42.1 is complete. □

We conclude with some simple remarks about the period (i.e., the minimum length of a repeating block) of a repeating decimal (see page 518). It is known that the period of the decimal expansion of a fraction $\frac{m}{p}$, where p is a prime, is always a divisor of $p-1$. We have observed that the period of the decimal expansion of $\frac{3}{7}$ is exactly 6, as $\frac{3}{7} = 0.\overline{428571}$. In fact it follows that the decimal expansion of any $\frac{k}{7}$ has an identical repeating block and therefore has the same period 6, e.g., $\frac{4}{7} = 0.571\overline{428571}$. This phenomenon of "maximum period" does not happen often, however. Take the prime 13, for example. One can check that $\frac{1}{13}$ has the decimal expansion $0.\overline{076923}$ so that its period is 6, and 6 is a divisor of 12.

As in the case of divisibility rules, these facts about the periods of decimal expansions have to do with the behavior of powers of 10 when they are divided by a prime p (see section 32.3, page 439). For further information, see [**Kal96**].

Exercises

1. Get the first eight digits of the decimal expansion of $\frac{22}{35}$. Notice that in this case, a repeating block shows up within the first eight decimal digits rather than the first 36 decimal digits.

2. Go through the proof of Theorem 42.1 carefully and show why a repeating block of the decimal expansion of a reduced, proper fraction $\frac{m}{n}$ (in the case the decimal is infinite) must show up in the first n decimal digits (rather than the first $n+1$ decimal digits) of the decimal expansion, and that the period of the (repeated) decimal expansion of $\frac{m}{n}$ is at most $n-1$ (rather than n). Can it equal $n-1$?

3. Let a, b be positive integers, and let q, r be the quotient and remainder of the division-with-remainder of a by b, respectively, i.e.,
$$a = qb + r, \quad \text{where } 0 \leq r < b.$$
Suppose q' and r' are the quotient and remainder of the division-with-remainder of $a \times 10$ by b, respectively, i.e.,
$$a \times 10 = q'b + r', \quad \text{where } 0 \leq r' < b.$$
Prove that there is a *single-digit number* k so that
$$q' = (q \times 10) + k.$$
(*Hint*: Use the uniqueness part of Theorem 7.1 on page 105 wisely.)

4. Explain to a sixth grader why the decimal expansion of $\frac{5}{187}$ must be repeating.

5. Determine if each of the following fractions has a finite decimal expansion:
$$\frac{142}{60}, \quad \frac{37}{60}, \quad \frac{81}{625}, \quad \frac{91}{224}, \quad \frac{684}{1125}.$$

6. Find the decimal expansions of the following fractions: (a) $\frac{5}{2}$, (b) $\frac{2}{3}$, (c) $\frac{13}{5}$, (d) $\frac{2}{9}$, (e) $\frac{5}{16}$, (f) $\frac{2}{13}$, (g) $\frac{5}{7}$, (h) $\frac{19}{7}$, (i) $\frac{7}{12}$, (j) $\frac{13}{15}$, (k) $\frac{5}{11}$, (l) $\frac{3}{1250}$, (m) $\frac{1}{1280}$. What did you notice about the answers to (a) and (e), or (b) and (d), or (c) and (j)?

7. If you are a patient person, try to find the repeating block in the decimal expansion of $\frac{2}{17}$. You may in fact replace the numerator 2 by any whole number.

8. Give a direct argument without using Theorem 36.2 on page 479 to convince a sixth grader that $\frac{1}{35}$ cannot have a finite decimal expansion.

9. Compute the decimals equal to the following:
$$\frac{2.2}{0.009}, \quad \frac{1.3}{6}, \quad \frac{2.0002}{0.4}, \quad \frac{2.4}{75}.$$

Bibliography

[Ash02]   Mark H. Ashcraft, *Math Anxiety: Personal, Educational, and Cognitive Consequences*, Current Directions in Psychological Science **11** (2002), no. 5, 181–185.

[BC89]   Nadine Bezuk and Kathleen Cramer, *Teaching about Fractions: What, When, and How?*, National Council of Teachers of Mathematics 1989 Yearbook: New Directions For Elementary School Mathematics (Paul R. Trafton, ed.), NCTM, Reston, VA, 1989, pp. 156–167.

[BGJ94]   Carne Barnett-Clarke, Donna Goldenstein, and Babette Jackson (eds.), *Mathematics Teaching Cases: Fractions, Decimals, Ratios, & Percents; Hard to Teach and Hard to Learn?*, Heinemann, Portsmouth, NH, 1994.

[Bru02]   John T. Bruer, *Avoiding the pediatrician's error: how neuroscientists can help educators (and themselves)*, Nature Neuroscience **5** (2002), 1031–1033.

[Bur07]   David M. Burton, *The History of Mathematics*, 6th ed., McGraw Hill, New York, 2007.

[Bus59]   Douglas Bush, *Literature*, The Case for Basic Education (James D. Koerner, ed.), Little, Brown, Boston, 1959, pp. 106–120.

[CAF99]   *Mathematics Framework for California Public Schools*, California Department of Education, Sacramento, 1999.

[Cle95]   Herb Clemens, *Can university math people contribute significantly to precollege mathematics education (beyond giving future teachers a few preservice math courses)?*, Changing the Culture: Mathematics Education in the Research Community (Naomi D. Fisher, Harvey B. Keynes, and Philip D. Wagreich, eds.), AMS and CBMS, Providence, RI, 1995, pp. 55–59.

[CRE96]   *Creative Math Teaching in Grades 7–12*, Newsletter of the Mathematics Department, University of Rhode Island **2** (1996), no. 1, 1–5.

[Dud78]   Underwood Dudley, *Elementary Number Theory*, 2nd ed., W.H. Freeman, San Francisco, 1978.

[Euc56]   Euclid, *The Thirteen Books of the Elements*, vol. I–III, Dover, New York, 1956, translated and edited by Thomas L. Heath.

[Fre83]   Hans Freudenthal, *Didactical Phenomenology of Mathematical Structures*, D. Reidel, Boston, 1983.

[Gea06]   David C. Geary, *Development of Mathematical Understanding*, Handbook of Child Psychology (William Damon, Richard M. Lerner, Deanna Kuhn, and Robert S. Siegler, eds.), vol. 2, John Wiley & Sons, New York, 6th ed., 2006, pp. 777–810.

[Gin28]   Jukuthiel Ginsburg, *On the Early History of the Decimal Point*, American Mathematical Monthly **35** (1928), no. 7, 347–349.

[GW00]   Rochel Gelman and Earl M. Williams, *Enabling constraints for cognitive development and learning: Domain-specificity and epigenesis*, Handbook of Child Psychology (William Damon, Deanna Kuhn, and Robert S. Siegler, eds.), vol. 2, John Wiley & Sons, New York, 5th ed., 2000, pp. 575–630.

[GW08]   Julie Greenberg and Kate Walsh, *No Common Denominator: The Preparation of Elementary Teachers in Mathematics by America's Education Schools*, National Council on Teacher Quality, Washington, DC, 2008, available at http://www.nctq.org/p/publications/docs/nctq_ttmath_fullreport_20080626115953.pdf.

[Har00]   Kathleen Hart, *Mathematics Content and Learning Issues in the Middle Grades*, Mathematics Education in the Middle Grades, The National Academies Press, Washington, DC, 2000, pp. 50–57.

[Hui98]   DeAnn Huinker, *Letting Fraction Algorithms Emerge Through Problem Solving*, National Council of Teachers of Mathematics 1998 Yearbook: The Teaching and Learning of Algorithms in School Mathematics (Lorna J. Morrow and Margaret J. Kenney, eds.), NCTM, Reston, VA, 1998, pp. 170–182.

[Kal96]   Dan Kalman, *Fractions with Cycling Digit Patterns*, The College Mathematics Journal **27** (1996), no. 2, 109–115.

[KCL99]   Shen Kangshen, John N. Crossley, and Anthony W.-C. Lun (eds.), *The Nine Chapters on the Mathematical Art: Companion & Commentary*, Oxford University Press, Oxford, 1999.

[LA92]   Lay Yong Lam and Tian Se Ang, *Fleeting Footsteps: Tracing the Conception of Arithmetic and Algebra in Ancient China*, World Scientific, Singapore, 1992.

[Lam99]   Susan J. Lamon, *Teaching Fractions and Ratios for Understanding*, Lawrence Erlbaum Associates, Mahwah, NJ, 1999.

[Lan88]   Serge Lang, *Basic Mathematics*, 1st ed., Springer, New York, Heidelberg, Berlin, 1988.

[LB98]   Glenda Lappan and Mary K. Bouck, *Developing Algorithms for Adding and Subtracting Fractions*, National Council of Teachers of Mathematics

	1998 Yearbook: The Teaching and Learning of Algorithms in School Mathematics (Lorna J. Morrow and Margaret J. Kenney, eds.), NCTM, Reston, VA, 1998, pp. 183–197.
[Mat82]	Helena Matheopoulos, *Maestro, Encounters with Conductors of Today*, Harper & Row, New York, 1982.
[Moy96]	Joanne Moynahan, *Of-ing Fractions*, What's Happening in Math Class? (Deborah Schifter, ed.), Teachers College Press, New York, 1996, pp. 24–36.
[NMP08a]	*Foundations for Success: The Final Report of the National Mathematics Advisory Panel*, U.S. Department of Education, Washington, DC, 2008, available at http://www2.ed.gov/about/bdscomm/list/mathpanel/report/final-report.pdf.
[NMP08b]	*Chapter 3: Report of the Task Group on Conceptual Knowledge and Skills*, Reports of the Task Groups and Subcommittees of the National Mathematics Advisory Panel, U.S. Department of Education, Washington, DC, 2008, available at http://www2.ed.gov/about/bdscomm/list/mathpanel/report/conceptual-knowledge.pdf.
[NMP08c]	*Chapter 4: Report of the Task Group on Learning Processes*, 2008, available at http://www.ed.gov/about/bdscomm/list/mathpanel/report/learning-processes.pdf.
[PSS00]	*Principles and Standards for School Mathematics*, NCTM, Reston, VA, 2000.
[Ros10]	Kenneth A. Ross, *Repeating decimals: A period piece*, Mathematics Magazine **83** (2010), no. 1, 33–45.
[Rot98]	Joseph Rotman, *Journey into Mathematics*, Prentice Hall, Englewood Cliffs, NJ, 1998.
[RSLS98]	Robert E. Reys, Marilyn N. Suydam, Mary M. Lindquist, and Nancy L. Smith, *Helping Children Learn Mathematics*, 5th ed., Allyn and Bacon, Boston, 1998.
[Sha98]	Janet Sharp, *A Constructed Algorithm for the Division of Fractions*, National Council of Teachers of Mathematics 1998 Yearbook: The Teaching and Learning of Algorithms in School Mathematics (Lorna J. Morrow and Margaret J. Kenney, eds.), NCTM, Reston, VA, 1998, pp. 204–207.
[Shu86]	Lee S. Shulman, *Those who understand: Knowledge growth in teaching*, Educational Researcher **15** (1986), no. 2, 4–14.
[Wil02]	Daniel T. Willingham, *Inflexible Knowledge: The First Step to Expertise*, American Educator **26** (2002), no. 4, 31–33, 48–49.
[Wu]	Hung-Hsi Wu, *Mathematics of the Secondary School Curriculum*, vol. III, to appear.
[Wu98]	_____, *Teaching fractions in elementary school: A manual for teachers*, available at http://math.berkeley.edu/~wu/fractions1998.pdf, April 1998.

[Wu99a] _____, *Basic Skills Versus Conceptual Understanding: A Bogus Dichotomy in Mathematics Education*, American Educator **23** (1999), no. 3, 14–19, 50–52, available at http://www.aft.org/pdfs/americaneducator/fall1999/wu.pdf.

[Wu99b] _____, *Preservice professional development of mathematics teachers*, available at http://math.berkeley.edu/~wu/pspd2.pdf, March 1999.

[Wu01] _____, *How to Prepare Students for Algebra*, American Educator **25** (2001), no. 2, 10–17, available at http://www.aft.org/pdfs/americaneducator/summer2001/algebra.pdf.

[Wu02] _____, *Chapter 2: Fractions (Draft)*, available at http://math.berkeley.edu/~wu/EMI2a.pdf, September 2002.

[Wu05] _____, *Must Content Dictate Pedagogy in Mathematics Education?*, available at http://math.berkeley.edu/~wu/Northridge2004a2.pdf, May 2005.

[Wu06] _____, *How mathematicians can contribute to K–12 mathematics education*, Proceedings of the International Congress of Mathematicians (Madrid), vol. III, International Congress of Mathematicians, European Mathematical Society, 2006, available at http://www.mathunion.org/ICM/ICM2006.3/Main/icm2006.3.1673.1696.ocr.pdf, pp. 1676–1688.

[Wu08] _____, *Fractions, Decimals, and Rational Numbers*, available at http://math.berkeley.edu/~wu/NMPfractions4.pdf, February 2008.

[Wu09a] _____, *From arithmetic to algebra*, available at http://math.berkeley.edu/~wu/C57Eugene_3.pdf, February 2009.

[Wu09b] _____, *What's Sophisticated about Elementary Mathematics?*, American Educator **33** (2009), no. 3, 4–14, available at http://www.aft.org/pdfs/americaneducator/fall2009/wu.pdf.

[Wu10a] _____, *Introduction to school algebra*, available at http://math.berkeley.edu/~wu/Algebrasummary.pdf, July 2010.

[Wu10b] _____, *Teaching Fractions: Is it Poetry or Mathematics?*, available at http://math.berkeley.edu/~wu/NCTM2010.pdf, April 2010.

[Wu11] _____, *The Mis-Education of Mathematics Teachers*, Notices of the American Mathematical Society **58** (2011), no. 3, 372–384.

Index

A page number in **boldface type** indicates where a formal definition of the concept may be found.

+, 17
−, 72
×, 27
÷, 98
=, 18, 197, 498
<, 25
≤, 54, 72
>, 25
≥, 72
$\stackrel{\text{def}}{=}$, 28
≡, 518
4 R 1, 106
$\frac{m}{n}$, 184
m/n, 185
⊂, 376
⟹, 421
⟺, 76
p^*, 375
$\vec{x}$, 382
B^{-1}, 289
$0.\overline{8}$, 518
$y|x$, 436
$y \nmid x$, 436
x^n, 425
x^{-n}, 425
$|x|$, 426
$n!$, 230
$(-b, b)$, 426
$[a, b]$, 23

absolute error, **144**, 326
absolute value, **426**
addition
    fractions, **222**, 230–231
    rational numbers, **386**
    whole numbers, **17**, 24
addition algorithm for finite
    decimals, **224**, 501, 512
    explanation of, 224
addition algorithm for whole
    numbers, 60, **62–63**, 67
    explanation of, 63–65
addition formula for fractions, **222**,
    227–229
addition formulas for rational
    numbers, 388, 389, 399–400
addition table in base 7, 161
algorithm, **57**, 464, 516
    addition, 60, **62–63**, 67
    Euclidean, 464–469, **469**, 476
    long division, 3, **108–109**, 526,
        529, 531–534, 536–537
    multiplication, **84–87**, 87–91
    subtraction, **71–74**, 75–79, 81
alternating sum of the digits of a
    whole number, 442
area
    model, 45, 190–195
    of rectangle, 45

545

properties of, 191
arithmetic progression, 450
associative law
   of addition, **42**, 51, 68, 223, 386, 387, 392
   of multiplication, 43, 269, 404
at bat, **251**
average rate in a given time interval
   of lawn mowing, 348
   of water flow, 345
   unit of, 345, 348–349
average speed in a given time interval, **292**, 293

Babylonian numeral system, 33
Barnett, C., 201
base b
   expansion, **156**
   representation, **156**
base (of numeral system), 4, **156**
base 12 expansion, **156**
base 7 expansion, **156**
Basic Facts, **397**–**399**, 400, **409**, 414, 466
batting average, **251**
Berkeley, population of, 143
Bezuk, N., 367, 370
bigger than, 25
binary
   expansion, **157**
   numbers, **157**
   representation, **157**
Bouck, M. K., 367, 368, 370
Bruer, J. T., 22
Burton, D. M., 33

calculator, use of, 55, 58, 92, 150, 156, 229–230, 433, 447, 453, 455–456
cancellation law, **204**
cancellation phenomenon, **271**, **311**, 417
carrying (for addition), 64
Cauchy–Schwarz inequality, 428
Chinese
   counting board numerals, **33**
   rod numerals, **33**
Clavius, C., 187

Clemens, H., 367
coefficient, **31**, 155, **156**, 225
Collins, D., 306
common divisor, **450**, 463, 468
commutative law
   of addition, **42**, 51, 68, 223, 387, 392
   of multiplication, **43**, 269, 404
   of vector addition, **385**
compare
   finite decimals, 242, 505–507
   fractions, **197**, 239–245
   whole numbers, **25**, 48
complete expanded form of decimal, **225**, 512
complex fractions, **309**
   importance of, 315–316
composite, **446**
concatenate, 24, 131
constant speed, **293**, 296
   for whole numbers, **100**
converge (sequence), 514
conversion of fractions to decimals, 207
cooperative work, 349–354
   assumptions on, 349
copies (of a fraction), **186**, **274**, 285
counting, 3, 5, **7**, 17
   consecutive, **17**
   continued, **17**
   step of, **7**, 13, 24
counting board numerals (Chinese), **33**
Cramer, K., 367, 370
cross-multiplication algorithm, **214**, 214–216, **240**, 240–242
cube root, 485
cubed, **425**

decimal, 174, **187**, **514**
   digit, **142**, **187**, **514**
   equal, 498
   finite, **187**, 493, 495–503
   fraction, **187**
   infinite, **514**
   point, **187**
   repeating, **519**

terminating, **187**
decimal expansion of fraction, **156**, 302, **525**
decimal numeral system, **6**
denominator, **185**, **310**, **417**
difference
    of fractions, **253**
    of whole numbers, **72**
different types, **344**
digit, **6**, **13**
Diophantus, 488
direction (of vector), **382**
Dirichlet, P. G. L., 450
distance, **426**
distributive law, **44**, 47, **270**, 404
    for division, 125, 281
    for subtraction, **73**, 270, 411
divide
    one whole number by another, 97
dividend, **105**
divisibility rules, 435–443
division
    analogy with subtraction, 97
    finite decimal, 296–302, 503
    fractions, **289**, 291, 303–305
    integers, **436**
    measurement interpretation, **98**
    partitive interpretation, **99**
    rational numbers, **415**
    relationship with multiplication, 97–100, 286, 290
    whole numbers, **97**, 132–133, **239**, 284, 286
division of finite decimals, 296–302, 503
division-with-remainder, 3, **105**, 195, 435, 464–469, 511
    incorrect notation for, 106
    theorem on, **105**
divisor
    common, **450**, 463, 468
    greatest common, *see* gcd
    of a division, **97**, 436
    of a division-with-remainder, **105**
    proper, **445**
double inequality, **104**, 426
Dudley, U., 488

Einstein, 151
endpoint, **23**, 183, **382**
equal
    decimals, 498
    fractions, **197**
    parts, 180
    sign, **18**, 39–40
    whole numbers, **40**
equi-spaced points, **23**
equivalent (statements), 49
equivalent fractions, **197**
    theorem on, **203**, 209–212, 247
Eratosthenes, 449
    Sieve of, **449**
estimate, 3, 139–150
    reasons for, 147–150
Euclid, 134, 450, 460, 483
Euclidean Algorithm, 464–469, **469**, 476
even (number), **69**
exactly when, **49**
expanded form (of number), 4, **20**, 30, 155
exponent, **30**, **156**, **425**
exponential notation, **30**, 69

factor, **28**, **436**
factorial, **230**, 452
factorization, **446**
FASM, **332**, 336, 447, 519
FFFP, 212, **213**, 216–218, 240
    for decimals, **214**, 296
finite decimal, 174, **187**, 493, 495–503
    comparing, 242, 505–507
fraction, 173, 177, **183**, 187, 236–238
    complex, **309**
    decimal, **187**
    decimal expansion, **525**
    improper, **186**
    proper, **186**
    reduced, **205**, 478, 479
fraction as division, 239
fraction of fraction, *see* of
fraction symbol, 185
fractional multiple, **288**

fractions equal to finite decimal, 479, 528–529
fractions equal to repeating decimal, 529–538
fractions in lowest terms, **206**, 478, 479
Freudenthal, H., 369
FTA, **457**, 457–459, 470–471
Fundamental Assumption of School Mathematics, *see* FASM
fundamental assumptions
  on addition of rational numbers, 396–397
  on multiplication of rational numbers, 404
Fundamental Fact of Fraction Pairs, *see* FFFP
fundamental hypothesis
  about speed of light, 148
Fundamental Theorem of Arithmetic, *see* FTA

gcd, **463**, 464–470, 473–478, **487**
Geary, D. C., 22
Gelman, R., 22
generalized place value, **512**
Ginsburg, J., 187
Goldbach Conjecture, **451**
Goldbach, C., 451
Goldenstein, D., 201
greater than, **25**, **197**
greatest common divisor, *see* gcd
Green, B., 452

Hamlet, 52
harmonic mean, **307**, 358, 359
have the same sign, **421**
Hindu-Arabic numeral system, 3, **6**, 33–34, 185
Homo sapiens, survival of, 137
Huinker, D., 368

if and only if, **49**
in lowest terms, **206**, 478, 479
inequality, **27**, 48, 50, 256, 278, 302, 421–426
  Cauchy–Schwarz, 428
  double, **104**, 426

  triangle, 428
infinite ruler, 24
infinity of primes, 483–484
integer, **376**
integer part of decimal, **498**
integral linear combination, **466**
inverse, **289**
invert-and-multiply, 286, **289**, 305, **418**
  explanation of, 271–272, 286–288
irrational number, **331**, 480–483

Jackson, B., 201

Kalman, D., 538

Lam, L. Y., 33
Lamon, S., 367–370
Lang, S., 488
Lappan, G., 367, 368, 370
laws of exponents, **156**, **500**
lcm, **474**, 476–478
leading digit (of whole number), **145**
leading term, **32**
least common multiple, *see* lcm
leitmotif, **59**, 65, 66, 74, 78, 79, 86, 107, 117
length, **24**, **186**, **382**, **519**
less than, **25**, **197**
limit (of sequence), **514**
long division algorithm, 3, **108–109**, 526, 529, 531–534, 536–537
  explanation of, 110–122
  relationship with division-with-remainder, 107, 117–119

Martian rovers, 153
Mathematics Framework for California Public Schools, 371
Mayan numeral system, 33
measurement interpretation of whole number division, **98**
meter, history of, 134, 150
Michelson–Morley experiment, 150
minus sign, **391**
mirror reflection, **375**
mixed number, 225

Index

motion, 100
   constant speed, **101**, **293**
   speed of, **293**
   uniform, 100
Moynahan, J., 279, 368
multiple
   fractional multiple, **288**
   integer, **436**
   of a point, **99**, 142, 288, 415
   of whole number, **97**
   proper, **445**
   Pythagorean triple, **487**
   rational, **415**
multiplication
   fractions, **263**, 263–268, 272–275
   history of, 134
   importance of multiplication table, 85
   relationship with division, 97–100, 286, 290
   whole numbers, **28**, 97, 131
multiplication algorithm for finite decimals, **269**, 502
   explanation of, 269
multiplication algorithm for whole numbers, **85–86**
   explanation of, 87–91
multiplication of fractions as fraction of fractions, 272–275
multiplication table
   importance of knowing, 85
   in base 2, 165
   in base 7, 163
multiplicative inverse, **289**, **414**

Napoleon, 150
national debt, 6, 27, 146, 148
National Mathematics Advisory Panel, 22, 175, 372
necessary and sufficient, **49**
negation (of a statement), **439**
negative number, **376**
negative sign, **376**, **390**
negative times negative, 403, 405–410
number, **128**, 135–137, 375
   irrational, **331**, 480–483

negative, **376**
positive, **376**
rational, **376**
real, **128**, 332
number line, **22**, 128, 375
   for addition, 24, 129
   for division, 286–288
   for multiplication, 272–275
   for subtraction, 71–73, 79–81, 129
   mirror reflection on, **375–376**
numbers
   one-digit, **7**
   two-digit, **8**, **13**
   three-digit, **10**, **14**
   up to two digits, **8**, **13**
   up to three digits, **10**, **14**
   up to four digits, **11**, **14**
   in base 7, 161
numerator, **185**, **310**, **417**

odd (number), **69**
of (a fraction), **245**, 248, 272–275, 278–279
Olympic sprinters, 137
on equal footing, **212**
order, **25**, **197**, **376**, 421–424, *see also* compare
order of magnitude, **146**
order of operations, 45

parentheses, **30**, 391
partitive interpretation of whole number division, **99**
pattern blocks, 189, 201
pave (a region), 45
percent, **319**, 321–323
perfect square, **480**
perimeter, **307**
period of repeating decimal, **519**, 538
   maximum, 538
pizza parlors, 306
place, **6**
   ones, **7**
   tens, **8**
   hundreds, **10**
   thousands, **11**
place value, 5, **19**, 66

generalized, **512**
position, **6**
positive (number), **25**, **376**
positive square root, **446**
power, **30**, **156**, **425**
power of 10, **439**
  multiplied by, 32, 44
prime, **446**
  as building blocks of whole
    numbers, 457
  infinity of primes, 483–484
prime decomposition, **455**, 482–483
  existence, 458–459
  uniqueness, 459
prime divisor, **458**
prime number, **446**
Principles and Standards for School
  Mathematics, 370
product, **28**, 44
product formula for fractions,
  263–267, **267**, 269, 529, 531
proof by contradiction, **439**
proper divisor, **445**
proper multiple, **445**
public key cryptosystems, 456
Pythagorean triple, **487**
  multiple of, **487**
  primitive, **487**

quantity, **336**
quotient
  division-with-remainder, **105**, 435
  fraction division, 289
  integer division, 416
  rational number division, **415**
  whole number division, **97**

rate, **344**
ratio, **336**, **342**
  interpretations of, 338
  need of simple definition, 339
rational number, **375**
rational quotient, **416**
real number, **128**, **332**
reciprocal, **243**
reducing fractions, **205**, 479
relative error, **144**, 325–327
  importance of knowing, 146

relatively prime, **450**, 464
remainder, **105**, 436, 464–469
Rembrandt, 372
removing parentheses, 392, 398
repeating block, **519**
  length of, **519**
repeating decimal, **519**
  ending in $\overline{9}$, 521
  period of, **519**
  repeating block of, **519**
Representation Theorem
  base b representation, 158
Reys, B., 368
rod numerals (Chinese), **33**
Roman numerals, 20
Ross, K., 57, 521
Rotman, J., 488
rounding decimals, 142
  to the nearest hundredth, **142**
rounding whole numbers, 140–144
  to the nearest ten, **140**
  to the nearest hundred, 141
  to the nearest thousand, **141**
  algorithm for, 141
  error of, 144
  need of, 143–144, 148

scientific notation, **507**
  comparing numbers in, 508–509
sequence of finite decimals
    associated with a decimal, **514**
setting up a proportion, 201, 296,
    346, 361
sexagesimal
  expansion, **157**
  numbers, **157**
Sharp, J., 368
Sieve of Eratosthenes, 448, **449**
skip-count, 12, 16–17
smaller than, 25
solution (to an equation), **379**, 420
special vector, **383**
  addition of, **383**
speed of light, 148, 150, 427
  fundamental hypothesis on, 148
square (of a number), **51**, **425**
square root, **446**, 481–483

standard algorithm, 3, 57–60
standard representation of a
    fraction, **181**
starting point (of vector), **382**
step (of counting), **7**, 13, 24
subtraction
    algorithm, **71–74**, 75–79, 81
    analogy with division, 97, 284
    fractions, **253**, 284
    rational numbers, 381, **390**, 392, 401
    whole numbers, **72**, 82, 98, 402
subtraction algorithm for finite
    decimals, **256**, 501, 512
    explanation of, 256
subtraction algorithm for whole
    numbers, 71–74, 81
    explanation of, 75–79, 402
sum
    of fractions, **222**
    of rational numbers, **387**
    of special vectors, **383**
    of whole numbers, **18**
supermarket, 137, 150, 321
symbolic notation, 21–22

Tao, T., 452
theorem
    on division-with-remainder, **105**, 103–105
    on equivalent fractions, **204**, 209–212, 247
tile (a region), 45
trading (for subtraction), **74**, 254
transitivity, **25**, 239
triangle inequality, 428
trichotomy law, **26**, 240
triplet primes, **451**
twin primes, **451**

unique, **104**, 272, 413, 435, 457, 459, 470, 479, 483, 514, 531, 532
unique up to rearrangement, **457**
unit, **23**
    importance of, 129–131, 188–190, 200
    need of precision in definition of, 189

unit segment, **23**, 183
unit square, **45**

vector, **382**
    addition of, 383
    direction of, **382**
    endpoint of, **382**
    left-pointing, **382**
    length of, **382**
    right-pointing, **382**
    special, **383**
    starting point of, **382**

weak inequality, **54**, 240
when and only when, **49**
whole (the unit in fractions), 180
whole numbers, **3**, 128
whole-number multiple, **142**
Williams, E. M., 22
work, *see* cooperative work
WYSIWYG, 107